**PROGRAMMED
GRAPHICS**

PROGRAMMED GRAPHICS

William F. Schneerer

Associate Professor of Engineering Graphics
Director of the Department of Instructional Support
Case Western Reserve University

McGraw-Hill Book Company

New York   St. Louis   San Francisco   Toronto
London   Sydney

PROGRAMMED GRAPHICS

Copyright © 1967 by McGraw-Hill, Inc. All Rights Reserved. Printed in the United States of America. No part of this publication may be reproduced, stored in a retrieval system, or transmitted, in any form or by any means, electronic, mechanical, photocopying, recording, or otherwise, without the prior written permission of the publisher.

55402

1 2 3 4 5 6 7 8 9 0   (BABA)   7 4 3 2 1 0 6 9 8 7

PREFACE

Programmed Graphics is a book on how to draw. It is not an artistic treatment of the subject of drawing. An art student would work through the pages of this book only if he or she were interested in learning about the geometry of drawing, for this is the subject of the book—graphics, the geometry of technical drawing.

In another sense, this is a book on how to observe. To be able to draw any detail of the physical world, the student must first be able to visualize—to picture mentally the exact nature of the detail around him. In his imagination he relates his subject to a standard spatial framework and then translates this three-dimensional image to the two dimensions of the drawing paper.

The painter and illustrator have this ability to observe and to visualize. It is a motivating force leading them into careers in art. The student entering a career in the technical world does not necessarily have a keen sense of observation. Unfortunately, preparatory education treats this human ability only cursorily. The young person who develops a sharp ability to observe the world around him does so through personal interest and motivation.

Drawing in science and technology is a means of communication. A creative technical person must communicate first with himself and then with a host of other people. The languages of communication are:

The spoken word
The written word
Mathematics
Drawings

"A picture is worth 1000 words." This old saw is often repeated but seldom questioned. The radio stations of the American Broadcasting Company, in a recent trade journal advertisement said,

> "You give me 1000 words and I'll take The Lord's Prayer, the twenty-third psalm, the Hippocratic oath, a sonnet by Shakespeare, the Preamble to the Constitution, Lincoln's Gettysburg Address, and I'd still have enough words left over for just about all of the Boy Scout oath. And I wouldn't trade you for any picture on earth."

This is a worthy and laudable statement, especially for a communications medium that relies exclusively on sound transmission. But suppose your problem is to explain the propulsion and control systems of a Saturn missile to an audience ranging from the machinist who must build the missile to interested taxpayers who wish to learn more about this complex and expensive device. Certainly no group of 1000 words would be adequate. For such a problem, the drawing and the symbolism of mathematics are worth many thousands of words, and this book is addressed to the subject of drawing as a powerful language for communicating technical information.

The programmed nature of the text represents a unique effort to use this interesting development in the teaching-learning process to help the student learn to draw. *Programmed Graphics* is designed specifically as a course-connected learning aid. It permits the student to study fundamentals of technical drawing at his own time, place, and pace and to verify and practice these fundamentals as he studies. It allows the teacher to devote his time to the problems of the individual student in the scheduled laboratory or work session and relieves him of the necessity of lecturing extensively on drawing fundamentals.

The psychological studies that have led to the development of programmed instruction show that learning is most efficient:

(1) When we study the subject matter in a logical order.
(2) When we become actively involved in the learning process.

Historical evolution of the subject matter, graphics, has divided it into definite units. The logical order of presentation of these units is, has been, and will always be a subject of controversy among authorities. *Programmed Graphics* presents the subject matter units in the following order:

Orthographic Projection
Pictorial Drawing
Drafting Standards
Descriptive Geometry
Graphic Mathematics

This order proceeds from the simple to the complicated. The emphasis is on developing the student's ability to translate readily and confidently between the orthographic and pictorial forms both in reading and in making a drawing. Dealing first with relatively simple physical objects, he proceeds to more abstract and geometric space problems.

The book is divided into two parts, the graphic description of objects and the graphic solution of problems. Part One (Chapters 1 through 14) carries the student through orthographic projection, pictorial drawing, and drafting standards. This is the material normally covered in one or two semesters of engineering graphics. Part Two presents descriptive geometry (Chapters 15 through 28) and graphic mathematics (Chapters 29 through 33). The material coverage of this problem-solving aspect of graphics represents more advanced courses.

Chapters 15 through 18 present the basic operations of descriptive geometry. They show the use of the orthographic projection system for operating on points, lines, and plane surfaces in space. These four chapters could be added to the fourteen chapters of Part One for a course which normally would not include any problem-solving uses of drawing. From this introduction to descriptive geometry, the student gains precision in orthographic projection, especially in the exact use of auxiliary views.

The logical order of presentation of the ideas and concepts within each unit of the subject matter is more important than the order of the units themselves. It is here that good teaching and efficient learning is indispensable. The programmed format in this text strives to lead the student carefully through the logic behind each idea much in the way a master teacher at the chalkboard leads his students step-by-step through a graphic solution. The student is not left merely to "understand" each new idea. He is actively involved in its development. He contributes both words and pictures.

The text presents the subject matter broken up into many short steps called frames, which lead logically through the material. Sections of straight text are interspersed throughout, along with many illustrations. Seventy-eight sketch exercises are included on loose-leaf sheets. These exercises permit the student to verify each drawing principle as it is presented and to gain some practice in the application of these principles. The sketches are not equivalent to conventional drawing homework or laboratory problems. They are doodling sheets provided to let the student try for himself an idea on drawing theory before proceeding to the next idea.

Key words are missing from most of the frames. The student is asked to supply these words. The background information required to do this without error comes from an analysis of a specified illustration or from information presented in previous frames or text. Other frames give step-by-step instructions on drawing specific lines to achieve the logical solution of a drawing problem. Answers for both the word and line contributions are immediately available so that the student is never building an idea or a sketch on false information.

Thus the student participates both in the written discussion of the subject matter and in the development of the supporting illustrations. At the end of each area of subject matter, the student has an opportunity to make a complete drawing solution of a typical problem. Again, a correct solution is available.

When completed, the sketch exercises will furnish a good review, covering the entire subject of the book in its own medium—drawing. Since they were developed by the student himself, recall should be high upon review.

Technical drawing is about one-third theory and two-thirds detail or standard practice. The many excellent conventional drawing texts cover the entire field, but with heavy emphasis on the details and standard practices of the industrial design and drafting scene. They are ideal reference books. *Programmed Graphics* concentrates on the theory—how to draw. This ability is the first need of anyone entering the technical professions, whether as draftsman, technician, machinist, engineer, or scientist. Knowledge of details and standards must come through concurrent formal course work offering wider problem experience and closer student-teacher relations.

For this text to be successful as a self-learning aid, student motivation must be high. The individual would have to be resourceful in developing his own exercises to extend experience and practice beyond the range of the simple drawing exercises presented.

Learning to draw involves:

1. Understanding the relatively simple concepts of spatial geometry.
2. Verification that these concepts yield the desired results (a recognizable drawing).
3. Visualization—the instant recognition of these concepts for any drawing requirement.
4. Practice! One learns to draw by drawing.

Programmed Graphics develops competence in the first three of these requirements. The fourth, practice, whether guided or not, is the realm of the school, the course, the teacher, and the student himself.

Computer graphics holds the promise of forcing revolutionary changes in the use of graphics in engineering design and allied fields. When large time-sharing computers with interactive graphic terminals are universally available, the engineer and the designer will have the most powerful tool ever available to their professions. *Programmed Graphics* is designed to elicit the type of thinking needed to use computer graphics effectively. The computer is capable of converting three-dimensional (*xyz*) data into a two-dimensional (*xy*) display in any drawing form and from any viewing position. The user's job is to define the points that make up the object and the sequence in which the points are to be connected by lines. This requires a careful analysis of the drawing and structuring of the large amount of data required to define a complicated object. Notation and symbolism will assume a new role in graphics. To prepare himself to use this powerful tool, the student must gain an appreciation of the role of points in describing an object and the sequencing of lines between points to create the plane and curved surfaces which form the solid.

Many people have assisted me in this project. Much credit goes to my wife, Shirley, who not only provided moral support during the four years this book was in preparation, but also accurately typed and proofread three versions of the manuscript. To William Martin go my thanks for drawing most of the 600 illustrations in the text and in the sketch pad.

Feedback from student users is a most important ingredient in the writing of programmed material. One thousand members of the classes of 1968 and 1969 of Case Institute of Technology used *Programmed Graphics* in its preliminary edition and provided the feedback needed for the preparation of this final version. Finally, five young people deserve special thanks for their reading of and comment on rough drafts of the manuscript. I hope that Carol Schneerer, Mark Barden, Alan Kepner, Sanford Bucklan, and Elliot Soloway learned something about graphics, because I learned much about the teaching of graphics from them.

My interest is in helping people learn to draw, and continued feedback on the use and effectiveness of this book is still important to me. I invite all users—students and teachers—to send comments and criticisms on both the general aspects and the details of *Programmed Graphics* to me at Case Western Reserve University, School of Engineering, University Circle, Cleveland, Ohio 44106.

William F. Schneerer

HOW TO USE THIS BOOK

A programmed textbook actively involves the reader in the development of the subject matter. *Programmed Graphics* has been designed to help you learn graphics efficiently by participating directly in the development of the text and its supporting illustrations.

The portfolio contains a book and a pad of sketch exercises. A quick inspection of the book will show that the pages contain conventional text material, numbered frames, and illustrations. The sketch pad contains 78 exercises for you to complete.

FRAMES

Read each numbered frame completely and carefully and then decide which word, phrase, or number should be inserted in the blank to make the frame statement correct. Information for correct responses should have been gained from prior frames or text, or from an analysis of referenced figures.

BLANKS

Write your answer down. Write in the blanks themselves or on a separate sheet of paper. Experience shows that positive action (writing an answer rather than merely thinking the answer) promotes learning.

The number of lines in the blanks are a clue to the number of words or numbers in the expected answer. In most cases, just one word or number is required. Some frames ask you to state an answer in your own words. These are marked (your words) at the end of a long blank. In some frames, a selection of words is offered and you must discriminate between them. These are marked (*Choose one:* larger/smaller; left/right).

ANSWERS

Answers for frame blanks are printed in a column at the left side of each page. Before starting with Chapter 1, cut or fold a strip of paper about 1½-in. wide to cover the answer column. The answer for each frame is aligned with the last type line of the frame. After exposing one answer, hold the cover strip in place until you are ready to check the answer to the next frame. Do this by sliding the cover strip downward until the desired answer is exposed.

ANSWER FORMAT

Frame answers are printed according to the following plan:
   (1) Single words, phrases, or numbers: The only (or most) acceptable answer.
   (2) Words, phrases, or numbers in parentheses: Alternate acceptable answers.
   (3) Semicolon: Two or more words, phrases, or numbers separated by a semicolon are answers to multiblank frames where the order of the answer words corresponds directly to the order of the blanks.
   (4) Comma: Two or more words separated by commas are answers where the order is not important.
   (5) Other answer formats: A few other answer arrangements will be found. The order as related to blank order should be evident.

CORRECT ANSWERS

Answers are correct according to the terminology used in this book. You must use your own judgment in deciding whether your answer is equivalent if yours disagrees with the printed answer. If you feel that your answer supplies the same meaning to the frame as the given answer does, then proceed to the next frame.

INCORRECT ANSWERS

If your answer is not equivalent to the given answer, pause to consider why it is not. Do not move on until you understand the statement of the frame. It is often useful to back up and review preceding frame or text material. In working a program, it is useless (and quite discouraging) to proceed if you are making many errors. You are not learning and should stop. If you do make an error, correct it in the written or sketched form before moving on. This helps you to consolidate the right answer rather than the wrong one.

Approach this work thoughtfully and seriously. Your goal should be to learn the subject matter, not to guess correct answers.

SCALE

All figures are printed full scale where scale is important to the discussion. Some have an inch scale printed with the figures. On all others, the exact scale is not important to the understanding of the figure or the accompanying text.

Sketch exercises are full scale on a ¼-in.-grid background. Sketch solutions in Part Three are printed ¾-size or 75 percent of the size of the exercises. Chapter 6, a discussion of proportion and scale, should help you if you have difficulty relating two drawings of different size.

SKETCH EXERCISES

Seventy-eight sketch exercises are included in a separate pad. Instructions for each exercise will be found in frames marked by a dark square preceding the frame number (example: ■ 18).

The only tools you will need are a soft, sharp pencil and an eraser. An HB drawing pencil or a No. 2 or 2½ office-grade pencil is best. Chapter 2 will tell you more about sketching materials and techniques.

SKETCH SOLUTIONS

Solutions to sketch problems are located in Part Three of the textbook. A letter "S" enclosed in a square on a sketch form signals that a solution to this particular sketch is available in Part Three. If this sign is missing, there is no printed solution, and the correct solution should be evident. Both the sketches and the sketch solutions carry references by chapter and frame number to the instruction frame in the text.

As with word answers, you must use your own judgment of the correctness of your sketches. If you prove yourself wrong, erase and correct your drawing before proceeding. The ideal programmed book on the subject of graphics would include only sketch responses and no word responses. It is through the sketch exercises that you have the opportunity to verify drawing theory and prove to yourself that you can draw.

CONTENTS

PREFACE v
HOW TO USE THIS BOOK ix

PART 1/ Graphic description of objects 3

Drawing—A Language for Communication 3
The Graphic Language 5

Unit 1 / Seeing and Drawing 7

1 How we see and draw objects 8
Introduction 8
Graphics—The Geometry of Technical Drawing 10
Visualization 11
Reference Surface—The Picture-plane Concept 15
   Pictorial Viewing 16
   Orthographic Viewing 18
Technical Drawing 22
Summary 23

2 Freehand sketching 24
Tools of Sketching 24
   Pencils 24
   Paper 25
Developing Skills in Sketching 26
   Sketching Straight Lines 26:
     Arm and Hand Control 28, Doodling 29, Straight-line Combinations 29, Parallel Lines 30, Perpendicular Lines 32, Grids 33, Angular Lines 33
   Sketching Curved Lines 35:
     The Circle 36, The Ellipse 37
   Instrument-aided Sketching 38:
     Compass and Template 39, Straightedge 39

Irregular Curves 40
Estimating Distances 41
Grid Paper 43
Developing Style in Sketching 44
Accuracy 44
Shading 45
Lettering 46:
  Letter Forms 46, Word Composition 48, Style in Lettering 49

Unit 2 / *Orthographic Projection* 51

3 Principal orthographic views 52
The Enclosing Box 53
Position of Orthographic Views 54
Coordinates 56
Frame of Reference 59
Adding a Third View 63
The Six Principal Orthographic Views 68
Summary 72

4 Straight lines 74
Visible Lines 75
Hidden Lines 77
Centerlines 79

5 Object orientation, auxiliary views, and sectional views 81
Object Position and Orientation 81
Auxiliary Views 85
Sectional Views 90
Summary—Orthographic Projection 93

6 Proportion and scale 94
Where Are We Going? 94
Introduction 94
Definitions 95
Proportion 95
Scale 95
Proportion versus Scale 95
Two-dimensional Proportions 96
Three-dimensional Proportions 98
Scale, Scales, and Scaling 102
Measuring with a Scale 102
Proportioning Distances with a Scale 107

Unit 3 / *Pictorial Drawing* 109

A Backward and Then Forward Look 109

7 The theory of perspective 111
Introduction 111
Sketching the Basic Box 111
Sketching in Perspective 115
Criterion Test 117
The Horizon Line 117
The Vanishing Points 118

First Rule of Perspective 121
　　Two-point Perspective 122
　　One-point Perspective 123
　　Three-point Perspective 123
　　Setting the Horizontal Angle θ 123
Foreshortening 130
　　Sketching the Box 130
　　An Intuitive Method of Foreshortening 134
　　Basic-square Method of Foreshortening 135
　　Perspective Sketching without Vanishing Points 141:
　　　　Sketching Converging Lines 142
Summary 142

8 Perspective sketching 143
Adding Details 143
　　Units of Measurement 143
　　Shading 145:
　　　　Useful Shading Techniques 148
　　Multiple Boxes 149
　　Sloping Lines and Surfaces 151
　　Circles 154
　　　　Circular Arcs (Partial Circles) 158
　　Irregular Curves 160
Summary 165

9 Perspective sketching—size and orientation 166
Object Size 166
Object Orientation 171

10 Projected perspective 175
Basic Construction 175
　　Measuring Lines 179
　　Curved Lines 181
A Note on Size 182

11 Other pictorial drawing forms 183
Isometric Drawing 183
　　Non-Isometric Lines 188
　　Isometric Circles 189
　　Instrument Isometric Circles 190
　　Arcs or Partial Circles 193
　　Irregular Curves 196
Summary of Isometric Considerations 197
Dimetric and Trimetric Drawing 198
　　Dimetric Forms 202:
　　　　Summary of Dimetric Considerations 203
　　Trimetric Forms 204:
　　　　Evaluation of Trimetric Considerations 205
Oblique Drawing 205
　　Cavalier versus Cabinet Form 207
Advantages of Perspective Sketching 209

Unit 4 / Drafting Standards 211
The Professional Uses of Graphics 211
   Design Process 211
   Manufacturing Process 213
Rules and Conventions—Drafting Standards 213

12 Working drawings—dimensioning 214
Introduction 214
Working Drawings 215
Detail Drawings 219
  Design of the Drawing 220
  Line Specification 222
  Placement of Lines 225:
    Thumbnail Sketch 226, The Five Basic 3-Dimensional Shapes 227, Dimensioning Holes 227, Dimensioning Rounded Ends and Corners 228, Miscellaneous Placement Rules 230, Placement of Notes and Leaders 233
  Selection of Letters and Numbers 234:
    Freehand Lettering 234
  Placement of Letters and Numbers 236

13 Working drawings—conventional practices 238
Sections 238
Rotation 240
Precision 242
Terminology 244
  Clearance and Interference Fits 245
Screw Threads 246

14 Drawing instruments and their use 250
Introduction 250
Instrument Drawing 250
Drawing Instruments 251
  Part 1 251:
    Class 1 Instruments 251, Class 2 Instruments 253, Class 3 Instruments 256, Class 4 Instruments 257, Class 5 Instruments 259
  Part 2 260:
    The Compass 263, French Curve 263, Circle Template 265, Scales 265

PART 2/ Graphic solution of problems 269

Introduction 269
Descriptive Geometry and Graphic Mathematics 269
Drawings and the New Design 270
Evaluating and Testing a New Design 270

Unit 5 / Descriptive Geometry 271

15 Graphic operations on points and lines 272
Freehand Solutions for Sketch Exercises 272
Points 273
  Directions in Space 275
  Reference Line Nomenclature 276
Lines and Planes 281
Analysis of Lines 283
  True Length of a Line 283
  Point View of a Line 286

16   Graphic operations on plane surfaces   290
Analysis of Planes   290
　Edge View of a Plane   290
　True Shape of a Plane   294
Summary   298
　Points   298
　Lines and Planes   298

17   Visibility of lines and planes   299
Visibility in Orthographic Views   299
Perspective Sketches to Aid Visibility   302

18   Angular relations and intersections   305
Bearing and Slope of a Line   305
True Angle between a Line and a Plane   308
　Specific Cases   308
　General Case   310
Perpendicularity   311
Angular Relation between Planes   313
Intersections between Lines and Planes and between Planes   315
Planes as Cutting Devices   316
Cutting-plane Method   317
　Point of Intersection   317
　Line of Intersection   319
Auxiliary Views of Physical Objects   320

19   Generating lines and surfaces   322
Concept of Line and Surface Generation   322
　Line Generation   322
　Surface Generation   323
　Nomenclature   324

20   Points—projecting shade and shadow   326
Points in Space   326
Projected Shade and Shadow   326
　Standard Direction of Light   329

21   Straight lines   338
True Length of a Line by Rotation   338

22   Vector geometry   343
Concurrent Coplanar Vectors   345
　Components   348
Nonconcurrent Coplanar Vectors   349
　The String Polygon   350
　Parallel Forces   351
Concurrent Noncoplanar Vectors   352

23   Curved lines   357
Single Curved Lines   357
Double Curved Lines   363

24 Plane surfaces 365
Planes as Boundaries of Solids 365
Polyhedrons 366
Surface Development 367
Intersection of Polyhedrons 370

25 Curved surface 373
Single Curved Surfaces 373
   Summary and Pretest 374
Cones and Cylinders 375
   Cones 376
   Cylinders 379
Intersection of a Line with a Cone or a Cylinder 384
   Summary and Pretest 384
   Intersection of Line and Cylinder 385
   Intersection of Line and Cone 387
Surface Description 388
   Summary and Pretest 388

26 The conic sections 393
Intersection of a Plane and a Surface of Revolution 393
   Summary and Pretest 393
   Circle 394
   Ellipse 395
   Parabola 396
   Hyperbola 397
True Shape of the Line of Intersection 399
   Summary and Pretest 399

27 Development of curved surfaces 404
Cylinders 405
Cones 406
Truncated Cylinders and Cones 408
Transition Pieces 412

28 Intersection of curved surfaces 415
Two Cylinders 417
Cone and a Cylinder 419
Concept of the Cutting Sphere 423
Surfaces of Revolution (Axes Intersecting) 424
Summary of Descriptive Geometry 426

Unit 6 / Graphic Mathematics 427

29 Graphs and charts 429
Introduction 429
Graphing Standards 431
Graph Paper 434
Trend Graphs 436
Charts 436
Summary 438

30 Scales 439
Scale Nomenclature 439
The Scale Equation 440
   Scale Modulus 440
   Logarithmic Scales 443
Graphic Graduation 444
Conversion Scales 445

31 Nomography 447
Concurrency Graphs and Nomographs 447
Basis for Constructing a Nomograph 450
   Nomograph for $u + v = w$ 451
   Nomograph for $uv = w$ 453
   Four or More Variables 456
   Others Forms for Nomographs 457

32 Empirical equations 458
Equation of a Straight Line: $y = mx + b$ 458
Best Curve and Scale Moduli 460
Power Equation: $y = bx^m$ 462
Exponential Equation: $y = bm^x$ 464

33 Graphic calculus 466
Integration and Differentiation 466
Graphic Integration 467
   Pole-and-Ray Method 468:
     Selecting a Pole Distance 469, Integrating a Closed Curve 471
Graphic Differentiation 472
   Pole-and-Ray Method 472:
     Drawing Tangents 474, Graphic Differentiation of a Given Curve 475, Slope of a Curve 476

GRAPHICS AND THE COMPUTER 479

PART 3/ Sketch solutions 483
   DECIMAL EQUIVALENTS OF INCH FRACTIONS (inside back cover)

PROGRAMMED GRAPHICS

A ARTISTIC – Emotional

B TECHNICAL – Unemotional

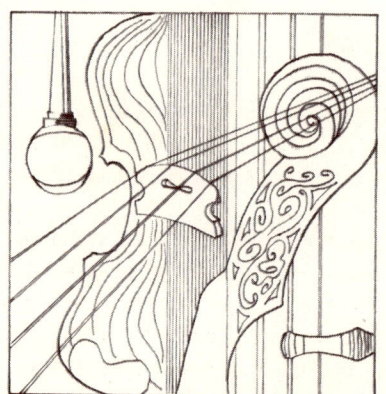
C ARTISTIC – Interpretive

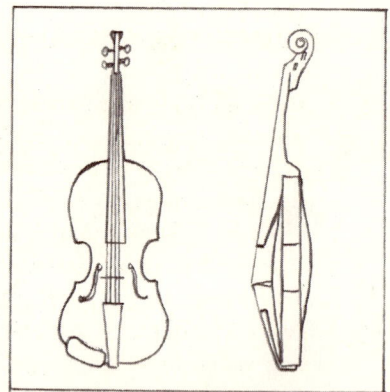
D TECHNICAL – Unambiguous

PART ONE *graphic description of objects*

DRAWING—A LANGUAGE FOR COMMUNICATION

There are two major divisions of the subject of drawing, artistic drawing and technical drawing. Our subject is technical drawing, but it is interesting to pause and explore the differences between the two areas. In the accompanying figures, the principal object is a violin.

Figure A is an artistic treatment of the object. It is a still-life picture. The artist wished to create an *emotion* in the viewer. He used emotion in drawing the picture, and objects other than the violin were included to help create a mood. The purpose of the drawing was not to show a realistic, detailed picture of a violin. In general, then, we can say that the artistic approach to drawing is emotional rather than realistic.

Figure B, on the other hand, is a realistic picture of a violin—a pictorial drawing. The intent here is to show the true details of the instrument. This can be called a technical drawing since it is realistic, factual, and drawn to communicate information rather than a mood or an emotion. In general, a technical approach to drawing is unemotional.

Now look at Figures C and D. These figures show another difference between artistic and technical drawing. Figure C could be an abstractionist's *interpretation* of a violin. He has taken the realistic shapes of the instrument and combined them into a pleasing design. His intent is still emotional. A viewer may or may not notice that the basic shapes are those of a violin; his judgment of the picture will be based upon a personal like or dislike of the design. The artist has interpreted the violin shape as the mood urged him when he drew it. Artistic drawing is interpretive drawing.

Figure D is an example of the most precise and *unambiguous* form of technical drawing. It is a two-view, orthographic projection—front and side. If a statement of the scale of the drawing were included, the violin could be

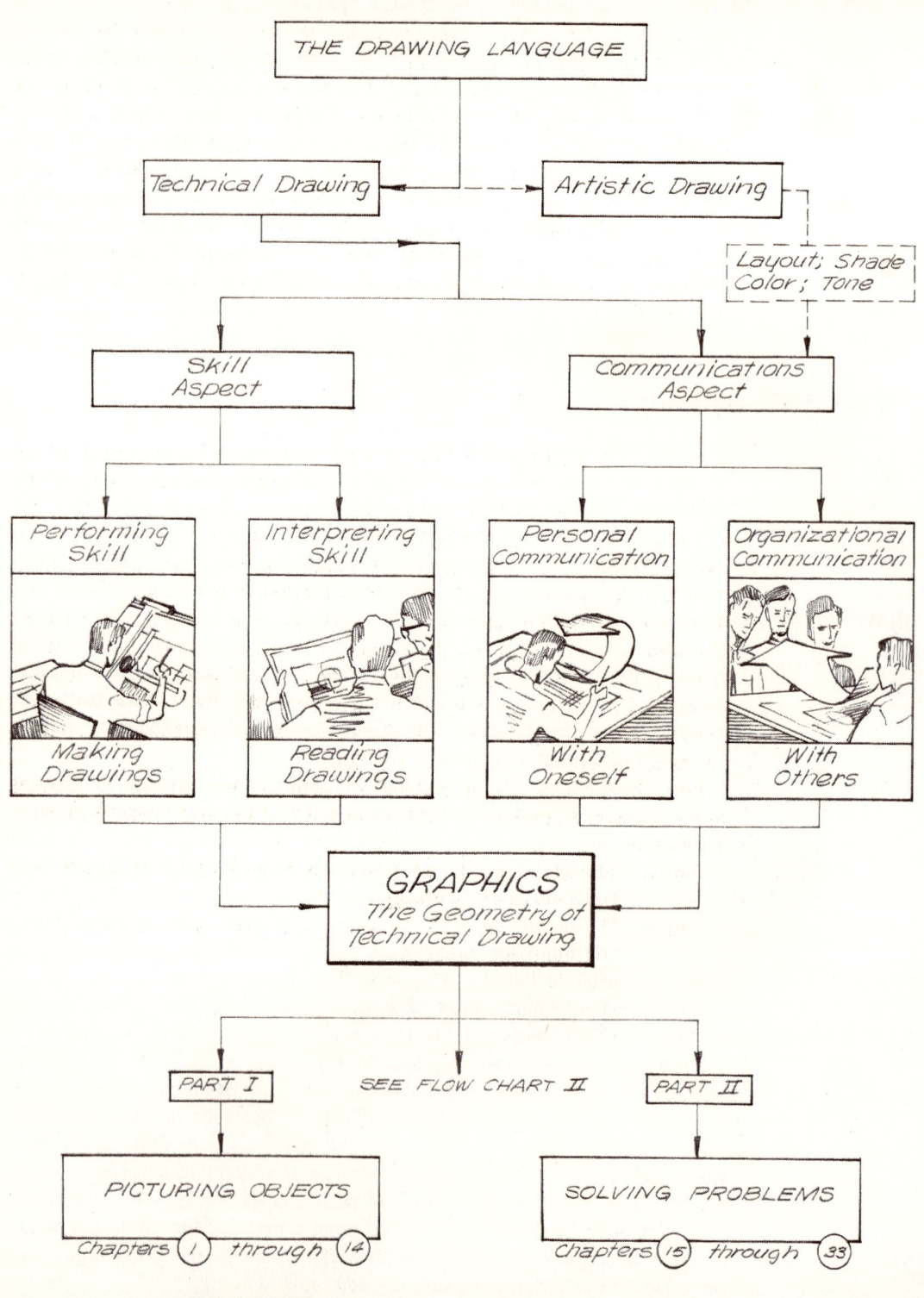

FLOW CHART I

duplicated by taking measurements from the drawing. Because the drawing is exact, it cannot be interpreted in any but the one way intended by the originator. From this we may say that technical drawings are unambiguous.

Technical drawing is an important language of communication in the industrial and scientific world. The kinds of drawing that make up this language may go by many names: idea sketches, technical drawings, design layouts, mechanical drawings, shop drawings, working drawings, engineering drawings, etc. But whatever they may be called, the knowledge required to make and read technical drawings comes from an understanding of graphics—the geometry of technical drawing—the science and art of drawing according to mathematical rule.

THE GRAPHIC LANGUAGE

Graphics is a language and thus—as with any type of language—a means of communication. We think of the act of communication mainly as the passing of information from one person to another, from a person to a machine, or, perhaps, from a machine to a person. This text is dedicated to the use of graphics as a language for communicating technical ideas and information. It is directed to a wide variety of persons in the technical world who have need to read or make technical drawings—the scientist and engineer, the designer and draftsman, the supervisor and foreman, the machinist, inspector, and purchaser to name only a few. *Programmed Graphics* is especially directed to the creative people who design and develop new things and whose need to communicate with themselves is as great as their need to communicate with other people or machines.

Flow Chart I is a graphic outline which relates the language of drawing to the subject of graphics. It then points out that this text presents graphics in two parts:

    Part I.   The use of graphics for picturing three-dimensional objects on a two-dimensional surface.

    Part II. The use of graphics for solving problems—both space problems and numerical problems.

UNIT 1   *seeing and drawing*

Chapters 1 and 2 are an introduction to graphics—the geometry of technical drawing. The two forms of drawing that must be mastered by the student of graphics are orthographic projection and pictorial drawing. Any three-dimensional object can be considered to be composed of a number of points in space. To draw an object, we must know how to plot these points on a two-dimensional surface. Chapter 1 relates orthographic projection and pictorial drawing by showing how we view an object and how we can then translate what we see into a series of points on the drawing paper. Connecting these points with straight or curved lines produces a drawing in the desired form.

Programmed Graphics emphasizes freehand sketching. Chapter 2 shows some techniques of sketching with emphasis on sketching straight and curved lines between points. It ends by showing refinements which can be added to a technical drawing through hand-lettered notes and simple shading techniques.

chapter 1 *how we see and draw objects*

■ 1 Refer to Sketch 1. This is a dot-to-dot drawing similar to those found in children's books. There is a series of dots starting with No. 1 and ending with No. 55.

Starting at point (dot) No. 1, sketch straight lines from point to point. Use a soft pencil to obtain a dark line.

NOTE: Numbers 8 and 15 have circles around them and represent the ends of line sequences. Pick up the next sequences at 9 and 16 respectively.

automobile (car)

2 You have just made a freehand sketch of a(n) _____.

INTRODUCTION

Admittedly, a dot-to-dot drawing is kid stuff. But Sketch 1 presents an important introductory point to the subject of graphics.

You could not have made this exact sketch without the dots. Sketching the lines was easy. Most of you have done it before—years ago.

The main purpose of this volume is to help you to learn where to place the dots to draw any object or concept you wish.

two (2)

3 How many points were used to define each straight line on the automobile? _____.

shortest

4 In geometry, we recognize that two points define a straight line by observing that a straight line is the _____ distance between two points in space.

8

curved

5 Any object can be represented by a series of straight lines. Figure 1.1 supports this statement. It is a sketch of a man's hand. We must consider this an abstract (or unreal) representation because straight lines replace the actual _____ lines of a human hand.

6 Nevertheless, Figure 1.1 is unmistakably a picture, or at least a graphic interpretation, of a human hand. Let's consider curved lines as they relate to our statement that all objects can be drawn by a series of straight lines.

wheels

In Sketch 1, the straight lines that you sketched betweens points 1 through 8 and 9 through 15 represent the _____ of the automobile.

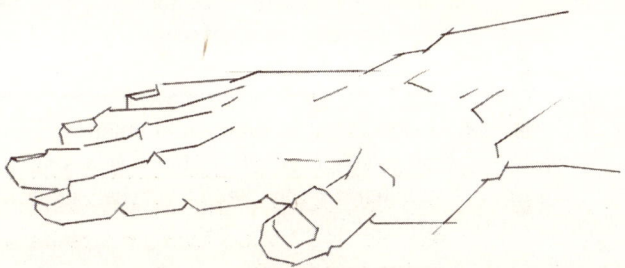

Figure 1.1 Any object can be represented by a series of straight lines.

curved

7 We know intuitively that wheels are round and, therefore, should be drawn with _____ lines.

8 A circle is the most common curved line. It is defined as a series of points lying in a plane and equidistant from a single point called the center.

Figure 1.2 shows a circle drawn using only straight lines. At 1.2A, four points equidistant from a center were used to define four straight lines. The resultant figure is not a circle but a(n) _____.

square

9 In Figure 1.2B, eight points were used yielding an octagon which is beginning to suggest a circle.

circle

Finally, at 1.2C, 16 points were used and the resulting polygon is truly beginning to *look like* a(n) _____.

NOTE: A polygon is a many-sided plane figure.

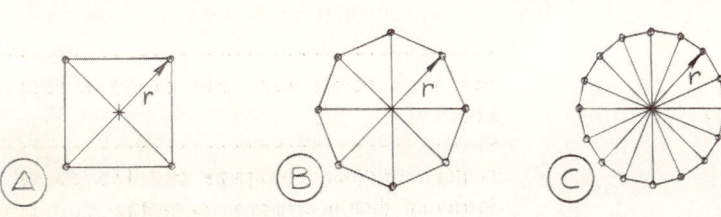

Figure 1.2 A straight-line circle.

10　SEEING AND DRAWING/UNIT 1

decreasing

10 We have approached a circle by _____ the lengths of the sides of a regular polygon. (*Choose one:* increasing/decreasing.)

zero
(nothing)

11 A true circle would be reached by continuing to shorten the line segments (by adding more points each the same distance from the center) until the length of each straight line segment approached _____ .

12 Thus we can draw an object by laying out points and joining them in a prescribed sequence with straight lines. The distances between points determine whether the line is straight or curved.

straight;
curved

If points in a sequence are far apart, they define _____ lines. If they are close together, they define _____ lines. (*Choose one:* straight/curved.)

■ **13** Sketch 2 is an exercise in creating a drawing of an object from a series of given points. The object is a simple mechanical block which is unfamiliar to you. See if you can draw it by connecting the points with straight lines.

SUGGESTION: Connect all the outside points first. This defines the outline of the sketch and should give you some insight as to the sequence of joining the rest of the points.

Check your sketch with the answer sketch.

GRAPHICS—THE GEOMETRY OF TECHNICAL DRAWING

Graphics is the science and art of drawing according to mathematical rule. It is an important *language* for technical persons and those in industry and research who support the technical work force. To use this language effectively, these people must be able to make drawings and to read drawings. This text is concerned with the theory of graphics—fundamental material for all who use drawings to communicate technical information.

We have begun by presenting the concept that any real or imagined object, no matter how complex, can be drawn with a series of straight lines. In the remainder of this introductory chapter we shall consider two more concepts that are fundamental to success in understanding the theory of graphics:

(1) The power of *visualization* is a necessary preliminary to interpretation of the three-dimensional world on the two-dimensional drawing surface (a sheet of paper).

(2) A *reference surface* must be established for each of the two forms of technical drawing, *orthographic projection* and *pictorial drawing.*

HOW WE SEE AND DRAW OBJECTS 11

VISUALIZATION

The ability to form mental interpretations of what we see or what we imagine is called *visualization*. When we read a book, our imagination gives us a personal visual interpretation of the characters and settings. Our minds transform words into mental images. Movies and television are visual means of communication. In addition to explaining or telling a story, they provide a visual interpretation of the spoken words. Our imagination of the story is assisted by the visual presentation. Since we began with a discussion of straight lines, let us continue by considering what we visualize when confronted with a straight line drawn on a piece of paper.

47

14 Look back at the automobile of Sketch 1 and, in particular, the line joining point 47 and point 48. Which of these two points is nearer to you as the observer? _____.

---

don't know

15 Line 47/48 is the left-hand roofline of the automobile. Usually, it is receding from us in space even though we know that it is just a line of a definite length drawn on a piece of paper. We are *visualizing* it as a spatial line because it represents part of a familiar object.

Consider line AB in Figure 1.3. Which end of the line is nearer? _____ (*Choose one:* A/B/don't know.)

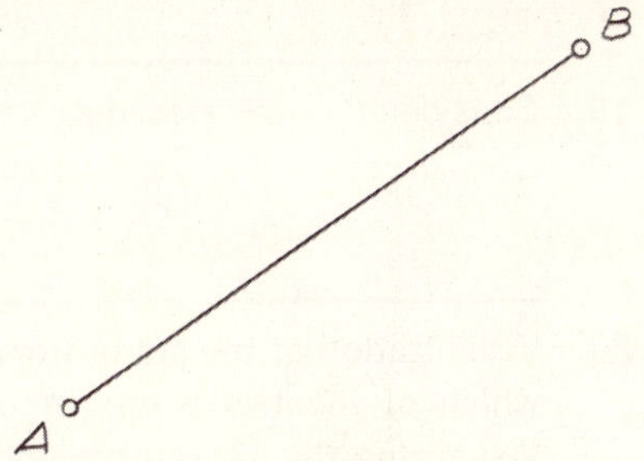

Figure 1.3 Two points define a straight line in space.

---

visualize

16 We don't know which point is nearer since AB is just a 2-in. line drawn on the paper. Since it is not part of a familiar object, we cannot _____ it as a particular spatial line.

---

right;
left

17 There are three possible interpretations of line AB in Figure 1.3.
    (1) A 2-in. line in the plane of the paper.
    (2) A line AB receding away from us (beyond the plane of the paper) from point A to point B to the _____. (*Choose one:* left/right.)
    (3) A line BA receding beyond the plane of the paper from point B to point A to the _____. (*Choose one:* left/right.)

12 SEEING AND DRAWING/UNIT 1

18 In Figure 1.4, line AB has been included with a group of other lines. If we visualize AB as a 2-in. line on the paper, then Figure 1.4 is just a geometric design.

Here, if we visualize AB as a line receding to the right, we see a flight of stairs as viewed from _____. (Choose one: above/underneath.)

above

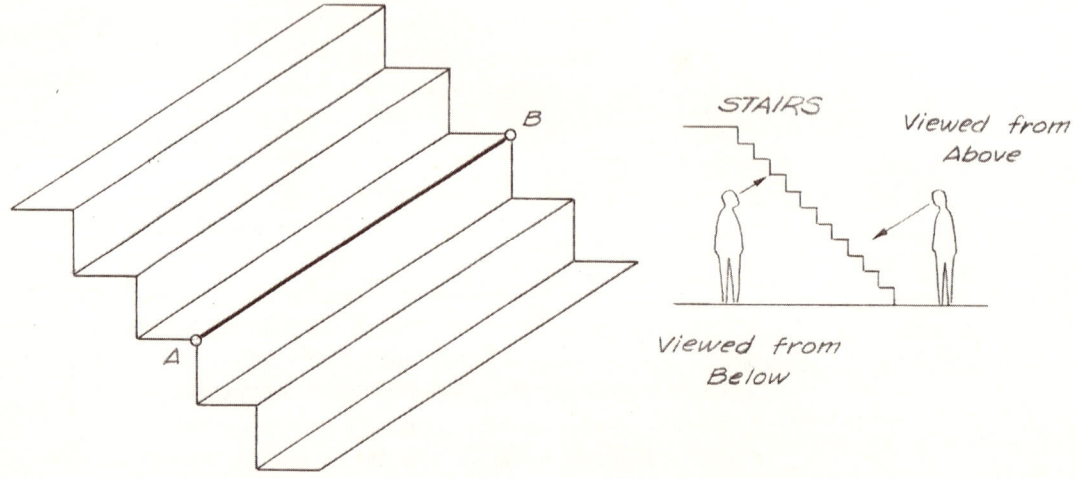

Figure 1.4 An optical illusion—geometric design or picture of an object?

19 Consider line BA (receding to the left). What do you see? _____ _____ (your words).

flight of stairs as viewed from underneath.

20 Visualization of the stairs from above or from underneath depends upon which of the two points, A or B, we imagine as being closer to us. Visualizing the stairs from underneath puts point _____ closer to us.

B

21 Stare at the lines of Figure 1.4. You should be able to make them change through the three interpretations by forcing points A and B to lie on the surface of the paper or alternately to come forward and recede.

This ability to see spatial relationships between elements of a flat, two-dimensional drawing is the ability to _____.

visualize

Figure 1.4 is an optical illusion. The auto roofline, 47/48, of Sketch 1 does not carry different interpretations because it is part of a familiar object. Visualizing it as line 48/47 (with 48 closer) would distort the object.

Let us now form a conclusion about the visualization of straight lines: We can visualize a straight line drawn on paper in three possible ways. Its meaning is *ambiguous* (capable of being interpreted or understood in two or more senses) unless the line is used in some familiar or logical way.

HOW WE SEE AND DRAW OBJECTS 13

22 In Figure 1.5, line CD can be visualized in one way and is therefore _____. (*Choose one:* ambiguous/unambiguous.)

Line XY can be visualized in three ways and is _____. (*Choose one:* ambiguous/unambiguous.)

unambiguous;
ambiguous

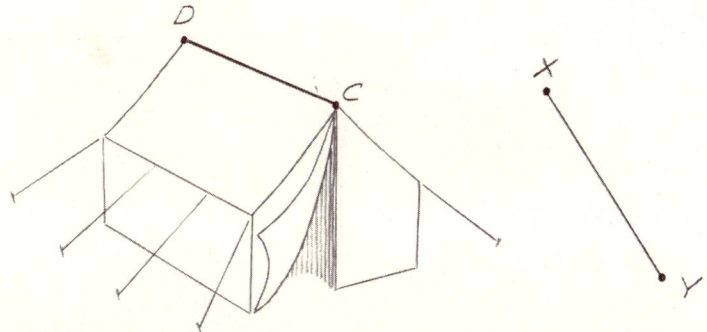

Figure 1.5 When a line is part of a familiar object, it is easily visualized as a line in space.

Performance in the field of technical drawing (graphics) requires this ability to visualize. Visualization in graphics involves picturing points, lines, and surfaces in space. The points are the means of arriving at the finished picture, which consists of a series of lines. Sketches 1 and 2 presented only arrays of points. It wasn't until you had supplied the lines that you could see the object represented by the points. If your power of visualization is sharp, you may have been able to see the objects before you added the lines.

23 Figure 1.6A is a reproduction of the automobile you drew in Sketch 1. Three basic dimensions have been added. They are:

(1) _____, (2) _____, (3) _____.

height;
width (or breadth);
depth (or length)

24 The three dimensions of height, width, and depth are basic to every physical object. Figure 1.6B shows a front view and a side view of the automobile. Three dimension lines with arrowheads are shown. Label each dimension line, indicating it as height H, width W, or depth D. Check your solution in the answer section.

NOTE: Throughout the rest of this text, we will use the capital letters H, W, and D to designate height, width, and depth respectively.

14 SEEING AND DRAWING/UNIT 1

25   Figure 1.6 illustrates the two important forms of technical drawing, *pictorial drawing* (1.6A) and *orthographic projection* (1.6B).

Examine these two drawings and then tell how many separate views are needed to show the principal dimensions of height H, width W, and depth D in:

(1) An orthographic projection; _____ view(s);
(2) A pictorial drawing; _____ view(s).

(1) two (2);
(2) one (1)

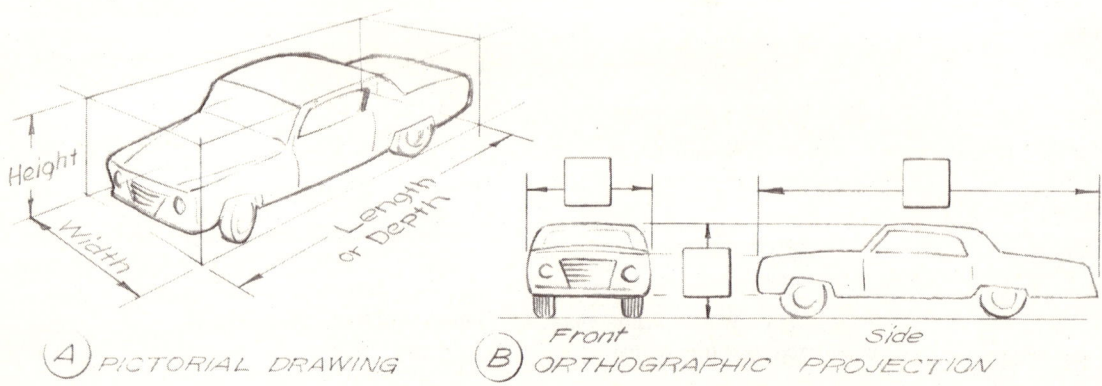

Figure 1.6 Pictorial drawing and orthographic projection.

26   A pictorial drawing shows H, W, and D in one view. It shows an object as we actually see it. On the other hand, an orthographic projection requires at least two views to describe fully the H, W, and D dimensions of a single object.

In Figure 1.7, which sketch is an orthographic view and which is a pictorial view?

(A): pictorial
(B): orthographic

(A) _____   (B) _____

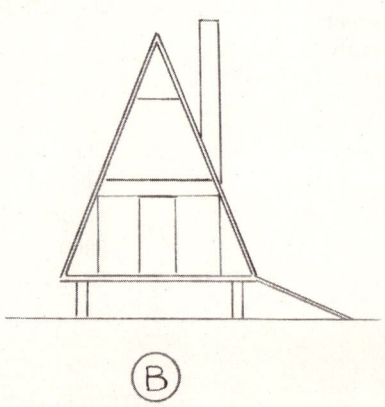

Figure 1.7 An A-frame house.

HOW WE SEE AND DRAW OBJECTS   15

H; W

27 Check which of the three principal dimensions of the A-frame house are shown in the orthographic projection at 1.7B.
   H _____   W _____   D _____

add a second view (draw another view, draw a side view, etc.)

28 To show the depth D of the house of 1.7B you would have to _____ _____ (your words).

orthographic projection, pictorial drawing

29 The two important forms of technical drawing (graphics) are _____ _____ _____ and _____ _____.

REFERENCE SURFACE—THE PICTURE-PLANE CONCEPT

Visualize drawing as a process of viewing an object through a transparent plane and then recording what you see on the plane. Figure 1.8 shows one of the easiest ways to make an accurate drawing. A flagpole is seen from the window of a house. If you stood in front of the window and, without moving your head, traced the details of the flagpole on the glass, you would create an accurate, but reduced-size, image of the object.

The window glass is a *picture plane*.

In technical drawing, however, we wish to draw on paper, not on a window pane. We can use the concept of the picture plane to serve as a *reference surface* as we attempt to translate a visual image or a mental image into a drawing, either orthographic or pictorial. Later, in Chapter 3, you will see how two or more reference surfaces are combined to form a frame of reference around which we build a theory of orthographic projection.

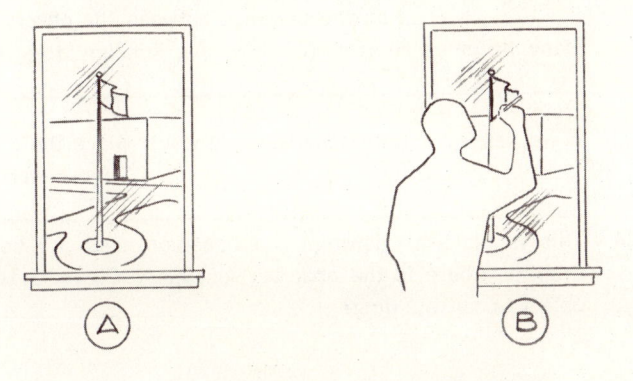

Figure 1.8   The window pane as a picture plane.

16 SEEING AND DRAWING/UNIT 1

Pictorial Viewing

projection

30 Consider Figure 1.9A. An observer stands and views a flagpole from a distance. Straight lines are projected from his eyes to the top and bottom of the flagpole. These lines are labeled _____ lines.

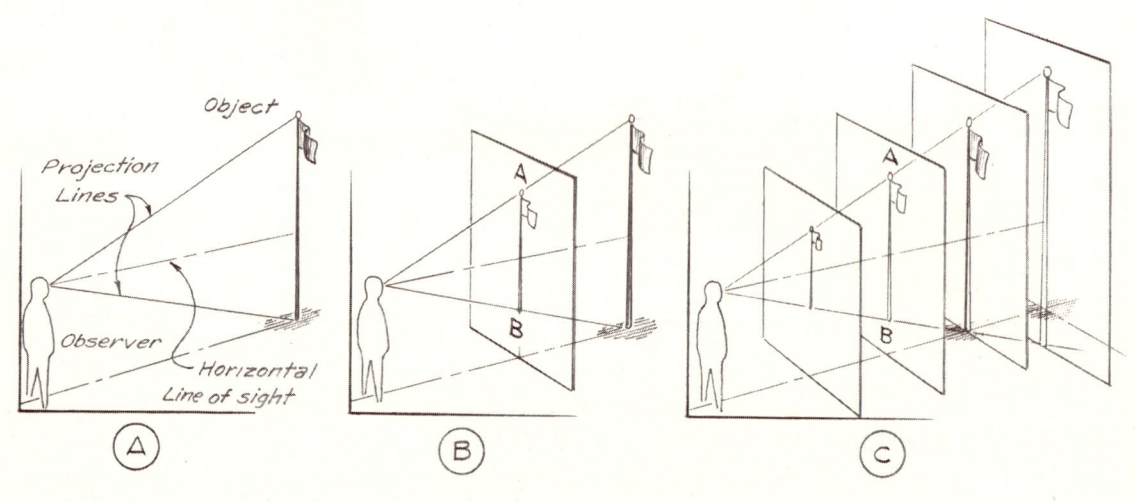

Figure 1.9 The picture-plane concept.

image (picture, projection)

31 In Figure 1.9B a picture plane has been placed about midway between the observer and the flagpole. The picture plane is perpendicular to the observer's horizontal line of sight. Points A and B mark the intersection of the two projection lines with the picture plane.

Points A and B define a reduced-size _____ of the flagpole.

smaller

32 Four picture planes are shown in Figure 1.9C, yielding four different-sized images. From this figure, we can deduce that the image becomes _____ as the distance between the observer and the picture plane becomes shorter. (*Choose one:* smaller/larger.)

at the flagpole

33 If we desired a full-size image, we would place the picture plane _____ _____ (your words).

in the air
(in space; atop a taller building)

34 This is pictorial viewing. The observer need not be standing on the ground. Where is the observer located in Figure 1.10 as he views the object, a tall building? _____ (your words).

HOW WE SEE AND DRAW OBJECTS 17

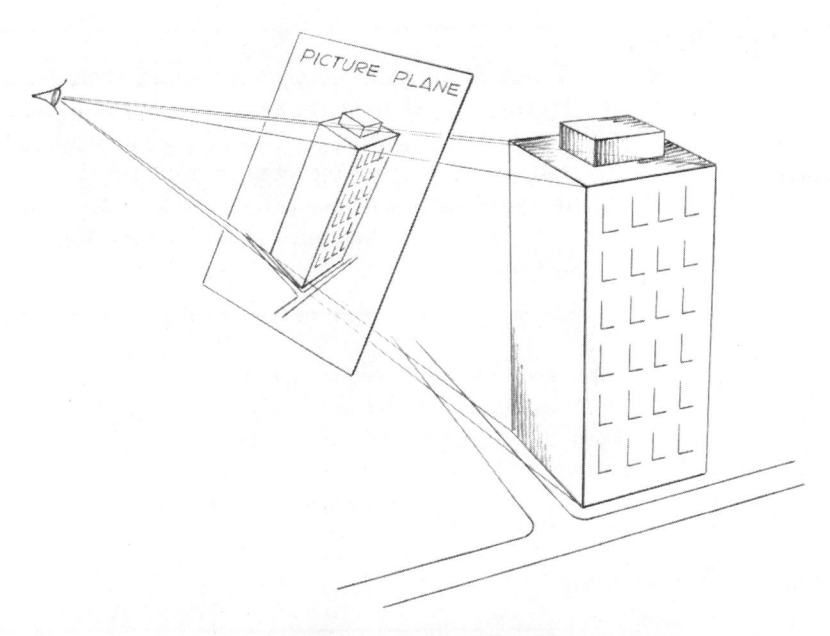

Figure 1.10   An observer can be positioned anywhere in space.

pictorial drawing	35 The observer is imagining that the picture plane with its image of the building exists. He uses the concept of the picture plane as a reference surface to relate the actual three-dimensional scene to his two-dimensional drawing paper. Figure 1.11 is the result. It is a(n) _____ _____ of the building.

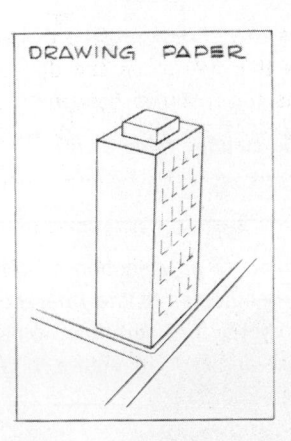

Figure 1.11   Pictorial drawing of a building.

18 SEEING AND DRAWING/UNIT 1

■ **36** Sketch 3 shows an observer viewing a cube through a picture plane. Projection lines from the eye to seven visible corners of the cube are drawn. The intersections of these lines with the picture plane are given and are numbered. The observer's sketch pad is shown. Point No. 1 is located. Do the following:
  (1) Complete the sketch of the cube on the sketch pad.
  (2) Complete the title of the drawing. Check your solution.

We have just developed a picture-plane concept which supplies us with a reference surface to aid in making a pictorial drawing. We might ask now, "Does this concept relate to orthographic projection as well as to pictorial drawing?" The answer is yes—but with one important difference. This difference is *the manner in which we view the object through the picture plane.*

Orthographic Viewing

37 Figure 1.12 shows the flagpole being viewed from three different vertical positions (*a*, *b*, and *c*). A picture plane has been placed between the object and the observer positions. The angle between the picture plane and each of the three projection lines is _____ degrees.

90

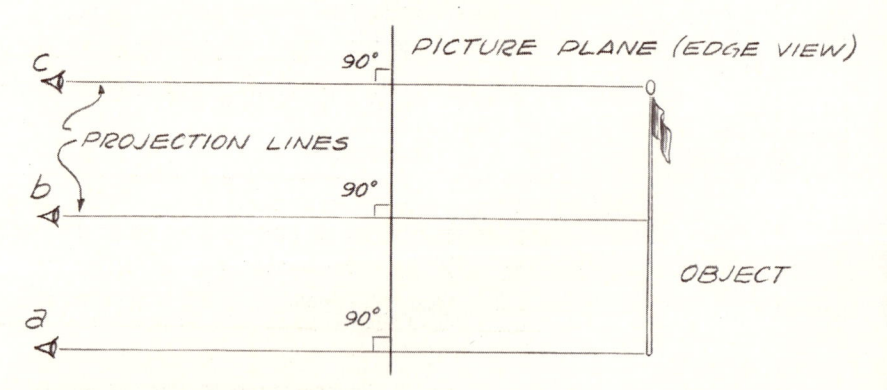

Figure 1.12 Orthographic viewing (side view).

38 *Orthographic projection is a projection system in which the projection lines are always perpendicular to the reference surface (picture plane).*
   By definition, the word "orthographic" means "pertaining to perpendicular lines or right angles." In Figure 1.12 the projection lines are perpendicular to the _____ _____.

picture plane
(reference
surface)

HOW WE SEE AND DRAW OBJECTS   19

39 Figure 1.13 is a pictorial sketch of the scene of 1.12. Superimposed on the figure, in dashed lines, is the geometric arrangement of lines we saw in Figure 1.9.

What are the size relationships between the two images, AB and CD, on the picture plane and the actual flagpole?

AB is _____ the object.

CD is _____ the object (*Choose one:* smaller than/same size as/larger than.*)

smaller than;
same size as

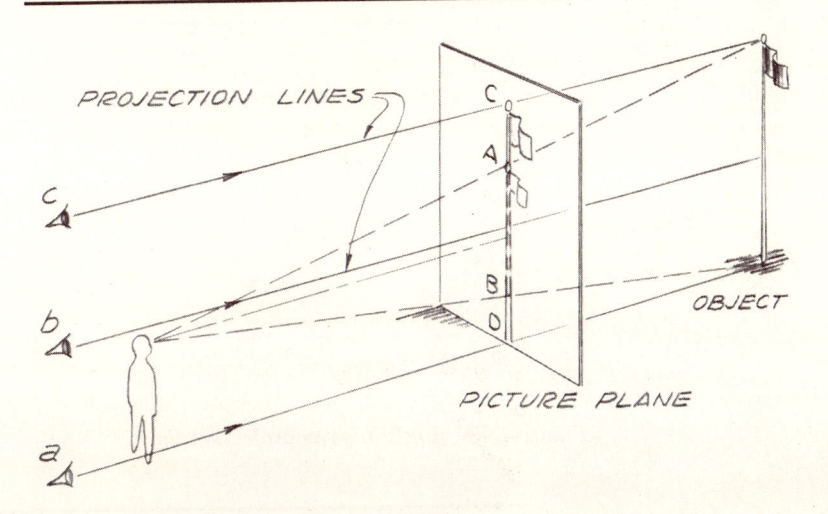

Figure 1.13 Orthographic viewing (pictorial).

40 The projection lines emanating from the three observer positions at *a*, *b*, and *c* are _____ to the picture plane.

perpendicular
(at right angles)

41 If the three lines are each perpendicular to the same plane surface, they are considered to be parallel to each other. *This is another criterion of orthographic projection*—that the projection lines form a system of _____ lines.

parallel

Figures 1.12 and 1.13 show three observer positions. This is an unrealistic way of looking at a flagpole. Normally, we view the flagpole from a single position and record what we see. We may change our position with respect to the object, but if we do, we will obtain a different picture. Let's see then if the three observer positions at *a*, *b*, and *c* can be considered as one single position rather than as three separate positions.

In the world of science, we consider that straight lines emanating from a point an infinite distance away are parallel. Rays of light from the sun are parallel to each other. Although the distance between the earth and the sun is finite, it is so great compared to the size of the earth (93,000,000 miles vs 8,000 miles in diameter) that it can be considered infinite.

20 SEEING AND DRAWING/UNIT 1

	42	The sun and stars are considered to be an infinite distance from the earth. Rays of light coming from the sun and the stars are considered to be _____.
parallel		

	43	Figure 1.14 illustrates the point using light rays as projection lines. In A a candle shines light on a small globe. A picture plane is placed between the two. Are any of the rays from the candle perpendicular to the picture plane? _____. If you answered "yes," how many rays are perpendicular? _____.
yes; one		

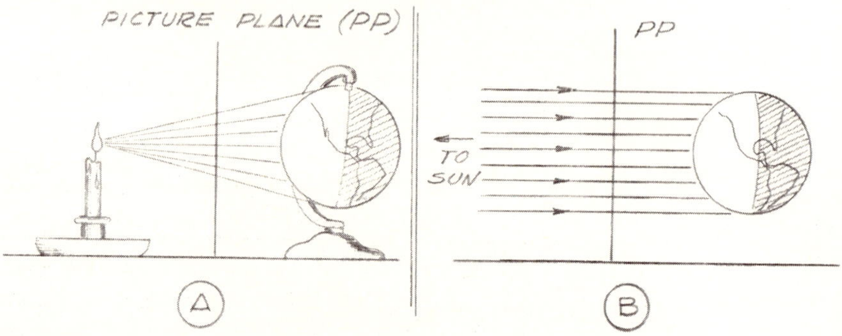

Figure 1.14 Light rays as projection lines.

	44	Only one ray can be perpendicular to the picture plane since the candle is but a short (finite) distance away. Now look at 1.14B. The sun's rays are shining on the earth. A picture plane is introduced. Are any rays perpendicular to the picture plane? _____. If "yes," how many? _____.
yes; all (every)		

	45	Look at Figure 1.15A. A cube is positioned in space and you, as the observer, are stationed an infinite distance away in a direction such that you are looking directly at (perpendicular to) one face of the cube labeled F. Since your station point is at infinity, projection lines (rays) from your eye to the cube will be _____.
parallel		

	46	Three picture planes have been placed in front of the cube and are parallel to face F. The four projection lines shown are _____ to the three picture planes.
perpendicular		

	47	Since the object is a cube, the shape of face F is _____.
square		

	48	The four projection lines are directed to the four corners of the square face of the cube. The intersections of these lines with the three picture planes give us images of three identical _____.
squares (pictures, objects)		

HOW WE SEE AND DRAW OBJECTS 21

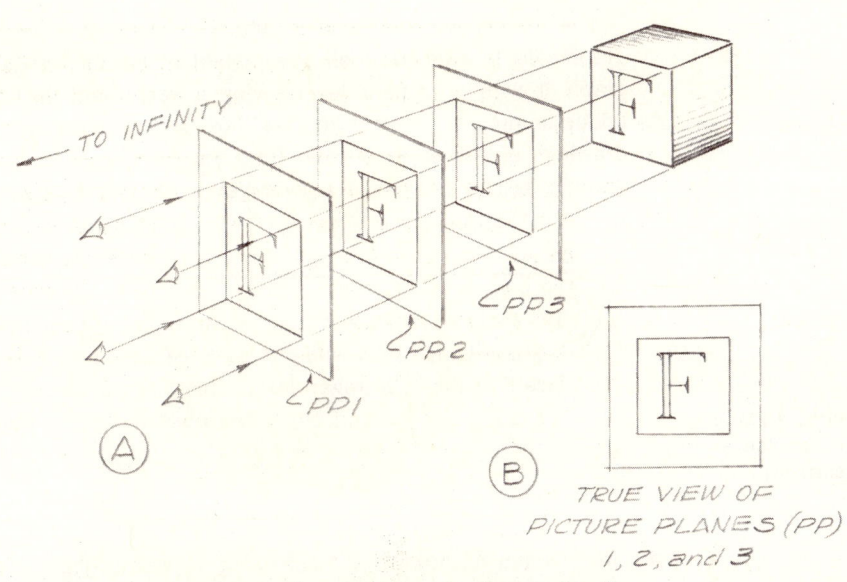

Figure 1.15 *One face of a cube viewed orthographically.*

reduction, enlargement	**49** The size of each image is identical to the size of the original square face of the cube. Thus this system of viewing an object gives no _____ or _____ in size of image regardless of where the picture planes are positioned with respect to either the object or the observer. (Refer to Figure 1.9C)
H, W	**50** The true view (Figure 1.15*B*) of the information recorded on picture planes 1, 2, and 3 shows a square which represents one face *(F)* of the cube. If we call this the front of the cube, what dimensions of the cube does the picture show? *(Check one or more:* H ____; W ____; D ____.)
does not	**51** Since a cube is a three-dimensional object, it has three principal dimensions of height, width, and depth. Thus, a single orthographic view _____ completely describe a three-dimensional object. *(Choose one: does/does not.)*
height, width, depth *(H, W, D)*	**52** As will be shown in Chapter 3, at least two orthographic views are required to describe a three-dimensional object completely. For a complete graphic description of an object, it is necessary to show the object's basic _____, _____, and _____ dimensions.
infinite	**53** The orthographic system of drawing is not a realistic system. We never see an object as it is shown orthographically since we cannot position ourselves at a(n) _____ distance from the object. And if we could, we would be too far away to recognize it.

22 SEEING AND DRAWING/UNIT 1

The closest we can come to actually seeing or picturing an object orthographically is when we look at a distant object through a telescope or photograph it with a camera lens having a very long focal length. Through an understanding of a few principles, however, we can easily make orthographic drawings of any object viewed from any position. For this reason, the orthographic system has many advantages in technical drawing.

---

picture plane
(reference
surface)

54 The main advantage is that the drawings are accurate and true to scale (size). Figure 1.15 showed that we obtained a true full-size view of face F of the cube regardless of the distance between the _____ _____ and either the observer or the object.

■ **55** Sketch 4 shows an object being viewed orthographically through a picture plane. Eight parallel projection lines are directed to eight numbered points on one face of the object. The intersections of these lines with the picture plane are given and numbered.

The observer's sketch pad is shown with a cross hatch of ¼-in. spaces. Point No. 1 is located. Do the following:

(1) Complete the sketch of the object as seen orthographically through the picture plane.

(2) Complete the title of the drawing.

You have now seen how we alter the manner in which we view an object through a picture plane to obtain either a pictorial drawing or an orthographic projection. As used in technical drawing, the picture plane is not real but imagined. We use it as an aid to visualization—an intermediate step between three-dimensional reality and a two-dimensional drawing.

TECHNICAL DRAWING

There are two sources of information for technical drawing:
(1) What we see
(2) What we imagine

If we see an object and draw what we see, we are imitating reality. If we imagine an object and make a drawing of it, we are creating an image of the object or, perhaps, recreating the image if our memories are strong and we can remember details of the object viewed in the past.

Both sources of information are vital to the needs of the student of graphics. If asked to make a drawing of a pencil sharpener, either you would find a pencil sharpener and draw it as you see it, or you would try to remember what a pencil sharpener looks like and then draw it from memory.

HOW WE SEE AND DRAW OBJECTS   23

SUMMARY

This chapter has been a presentation of some of the mental gymnastics required to understand fully the theory of graphics. The next few chapters present the detailed theory of orthographic projection and pictorial drawing using the concepts developed here.

There follows a programmed summary of this chapter. Use it as you wish. It is really a test on the main points presented.

straight lines

56 Any object or concept, no matter how complicated, can be drawn by a series of _____ _____.

short

57 Curved lines can be pictured as being built up from many _____ straight lines.

surface;
surface (plane)

58 A straight line drawn on a piece of paper can be visualized in two ways:
   (1) As a line lying on the _____ of the paper.
   (2) As a line receding beyond (in back of) the _____ of the paper.

height (H),
width (W),
depth (D)

59 All physical objects have three basic dimensions of _____, _____, and _____.

orthographic
projection,
pictorial drawing

60 The two forms of technical drawing are _____ _____ and _____ _____.

picture plane

61 To relate from three-dimensional reality to the two-dimensional drawing surface, we imagine a reference surface called the _____ _____, placed between the object and ourselves as observers.

one

62 In *pictorial viewing*, projection lines radiate from the observer's eye to all points on the object. How many of these projection lines are perpendicular to the picture plane? _____

every (all)

63 In *orthographic viewing*, the observer considers himself to be an infinite distance from the object. How many projection lines from observer to object are perpendicular to the picture plane? _____

imagine (create)

64 Two types of technical drawing, creative and imitative, differ in the source of the information used in making the drawing. In creative drawing we draw what we _____.

chapter 2   freehand sketching

Freehand sketching is a powerful tool in all technical work. Skill in freehand sketching ranks with skill in using the written and spoken language. Drawing is a language—a language employed both for self-communication and to communicate with other people.

All who set out to study graphics are communicating with themselves in the process. In this programmed text, principles of drawing are presented, examples are shown, and then simple drawing exercises are given. In doing these simple exercises you are testing yourself—finding out if you understand what you have read and seen. The program shows you the correct answer (or one of a number of correct answers) so that you are not left in the dark as to whether your solution is right or wrong.

The exercise problems should be solved by freehand techniques for the following reasons:

(1) You will save time since freehand drawing is faster than instrument drawing.

(2) Extreme accuracy will not be important.

(3) The practice will help you learn to sketch quickly and accurately and perhaps give you increased confidence in your own sketching abilities.

TOOLS OF SKETCHING

When setting out to learn any art, you must familiarize yourself with the tools. The tools of sketching are, simply, paper, pencil, and eraser. If you learn the art well and understand the principles of drawing, the eraser may fall into relative disuse.

Pencils

Drawing pencils come in 18 grades of hardness as shown in Figure 2.1A. The HB grade is a good sketching pencil. A No. 2 or No. 2½ office-grade pencil is also good. Do not use fountain or ball-point pens until you gain confidence in your ability to sketch. Pens give lines of characteristic quality, but the lines cannot be erased. If you prefer a mechanical pencil, use the thin-lead variety and choose HB grade or equivalent. The sketches in this book were done with an HB pencil. Lettering was with both H and F pencils. Instrument drawings were drawn with H, 2H, and 3H pencils.

Paper

Almost any paper will do for sketching. Many great ideas were developed from a quick sketch on the back of an envelope or a restaurant menu. Paper surface must have a "tooth" to wear away the pencil "lead" and must be hard enough to erase easily without digging up the surface.

---

paper, pencil, eraser

1 The tool requirements for freehand sketching are simple. They are _____, _____, and _____.

7B; 9H; HB

2 Drawing pencils come in 18 grades of hardness (Figure 2.1A) ranging from very hard, grade _____, to very soft, grade _____. The _____ grade is recommended as a good sketching pencil.

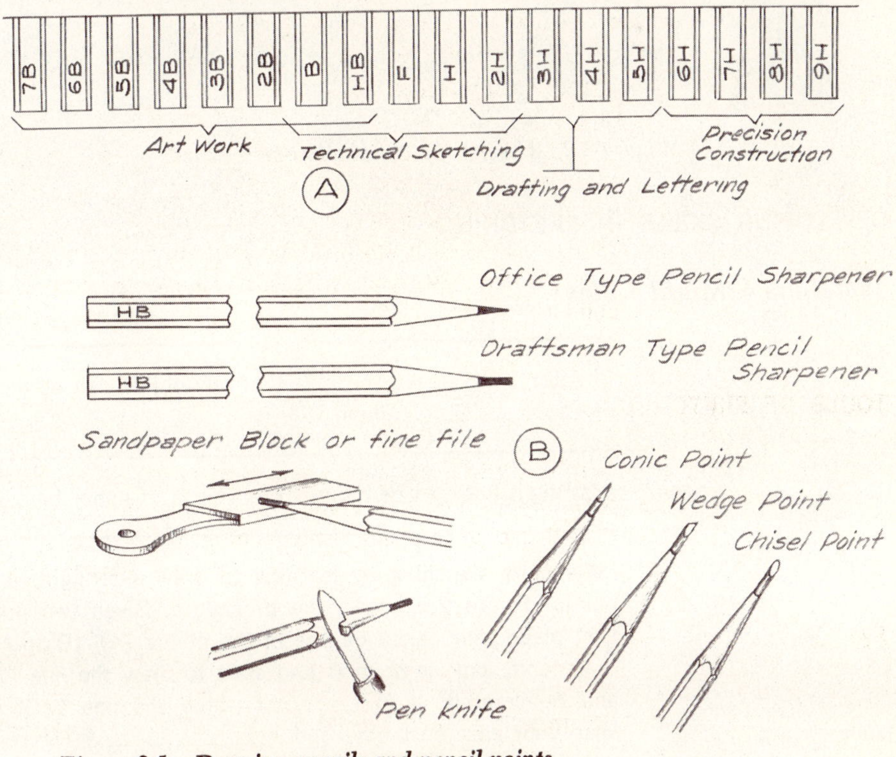

Figure 2.1 Drawing pencils and pencil points.

26 SEEING AND DRAWING/UNIT 1

light;
dark (heavy)

3. A soft pencil is best because you can sketch both light and dark lines with one instrument by varying the pressure. Light pressure yields a(n) _____ line but when you bear down heavily with a soft pencil you get a(n) _____ line.

sandpaper block (fine file)

4. Figure 2.1B shows how to sharpen a wooden drawing pencil. Most engineers and draftsmen use a draftsman's pencil sharpener which cuts the wood away from the graphite stick (lead) but does not shape the point. They then make any point they wish using a(n) _____.

thin (fine); broad (thick)

5. Three point styles used in graphics are the conic, the wedge, and the chisel. The chisel point (Figure 2.2) is most useful in freehand work because you can draw very _____ lines and very _____ lines with the same point.

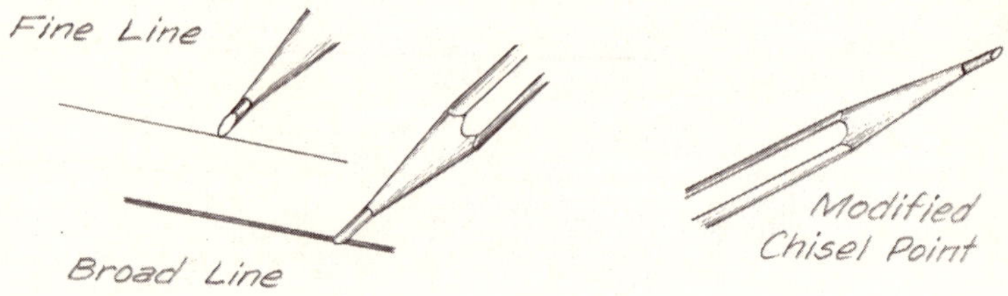

Figure 2.2 Uses of the chisel point on a pencil.

DEVELOPING SKILLS IN SKETCHING

Sketching Straight Lines

straight

6. Chapter 1 developed the idea that any object can be drawn by a series of _____ lines.

two (2)

7. A straight line can be defined on a piece of paper by _____ points.

point

8. Let's start sketching by learning to draw a straight line between two points. Figure 2.3 shows the procedure. Given two points, A and B, first place your pencil on one of the points (A). (Figure 2.3A.)

Second, *look at point B* and start to draw the line AB. Draw slowly and deliberately at first. Do not watch the line as it is formed, but keep your gaze on the second _____. (Figure 2.3B.)

FREEHAND SKETCHING 27

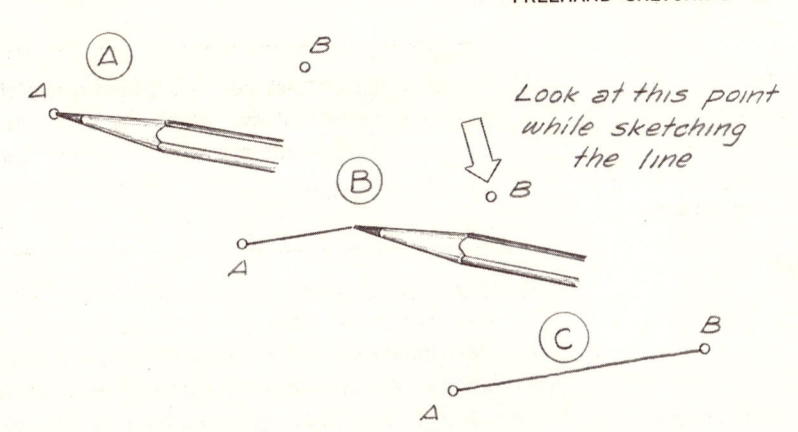

Figure 2.3 Sketching a straight line between two known points.

accuracy (precision)	9 Figure 2.3C shows the finished line AB. It is not instrument-perfect, but can be considered a straight line. In sketching, extreme _____ _____ is relatively unimportant.
No. 3	10 In sketching long straight lines, a few trial starts are often helpful to establish the correct direction. These are short lines sketched quickly and lightly. (See Figure 2.4.) Three such starts are shown. One appears headed in the right direction (No. _____) and two are false starts as far as sketching a straight line between A and B is concerned.

Figure 2.4 Quick, light trial starts aid in sketching long straight lines.

B	11 Finish line AB of Figure 2.4, remembering to keep your gaze on point _____ as you sketch slowly, lightly, and deliberately.
soft	12 Don't sketch straight lines in short, jerky steps. (See Figure 2.5A.) Use a(n) _____ pencil so that you do not have to exert much pressure to get a reasonably dark line. Draw the line in one easy motion.
pressure	13 Heavy pencil pressure causes the straight line to wander (Figure 2.5A). Always sketch lightly at first as at Figure 2.5B. (Go past the points if you want to.) If you achieve the desired results, then you can darken the correct line segment (between points). A soft pencil gives both light and dark lines by varying _____.

28 SEEING AND DRAWING/UNIT 1

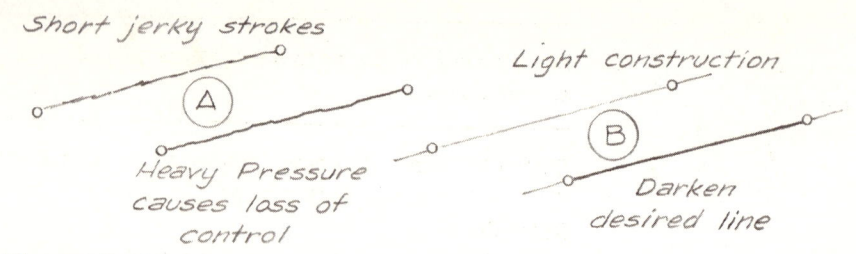

Figure 2.5 Line-sketching techniques.

This last point is an important one for technical sketching. Always sketch lightly at first. As your drawing develops, you will begin to see whether it is correct or not. If correct, you can then darken all the good lines to make the object stand out. If your light sketch proves incorrect, the light lines are easy to erase and correct. Don't be concerned about the existence of light construction lines. Keep them light and leave them. You are not preparing an illustration for a magazine advertisement.

Arm and hand control

Don't use stiff, finger motions only. Stay loose. Learn to employ wrist motion and arm motion when sketching. Control of the pencil in sketching comes from the combination of three motions, finger, wrist, and arm. Always position the paper so that you can use the most natural motion to draw a given line, turning the paper as required. *Never tack or tape the paper to a drawing board when sketching.*

---

short;
long

14 Figure 2.6 shows the motions used in sketching. Wrist and finger motions are good for _____ lines and arm motion for _____ lines. (*Choose one: long/short.*)

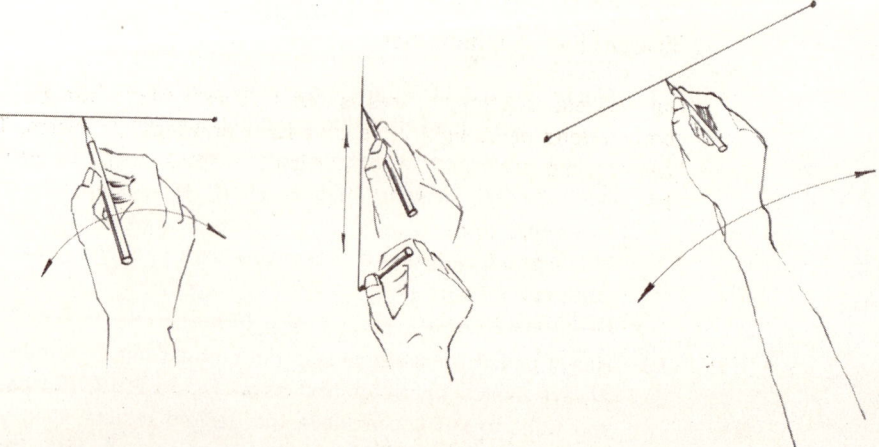

Figure 2.6 Finger, wrist, and arm motions combine to give pencil control.

FREEHAND SKETCHING 29

motion

15 Keep turning the paper to adapt to the natural ─────── of your arm and wrist.

■ **16** Sketch 5A shows 26 points (A through Z). Sketch straight lines between these points in sequence starting with point A.
REMEMBER:
   (1) Always direct your gaze to the next point.
   (2) Draw slowly, lightly, and deliberately.
   (3) Try some test starts on the longer lines.
   (4) Keep turning the paper to utilize the natural motion of your arm and wrist.

Use your own judgment to grade your efforts. Instrument accuracy is not important in freehand sketching. It isn't even desired because it would indicate that too much time was taken to make the sketch.

■ **17** Sketch 5B shows a series of points connected by straight lines in a random pattern. Another set of points is given. Sketch straight lines between these points in any order you wish.

Doodling

Many people doodle while talking on the telephone or idling away an hour or two watching TV. Try directing your doodling toward some of the exercises in this chapter. Place points on a piece of paper and practice sketching straight lines between them in the manner of Sketches 5A and 5B. *You learn to draw through practice.*

Straight-line combinations

Our interest in straight lines is not just with single straight lines, but with combinations of straight lines that form drawings of objects. There are four types of line combinations of particular interest to us in technical drawing. They are:
   (1) Parallel lines
   (2) Perpendicular lines
   (3) Angular lines
   (4) Curved lines

short

18 Remember that we consider curved lines to be made up of a series of ─────── straight lines.

Parallel lines

The first of the line combinations, parallel lines, are lines that do not converge (come together) or diverge (move apart) but are the same distance apart at all points along the lines.

Figure 2.7 shows four lines that have been sketched parallel to line AB. If they do not appear exactly parallel (instrument-perfect) to you, remember that this is a freehand sketch and that instrument accuracy is not expected from sketches.

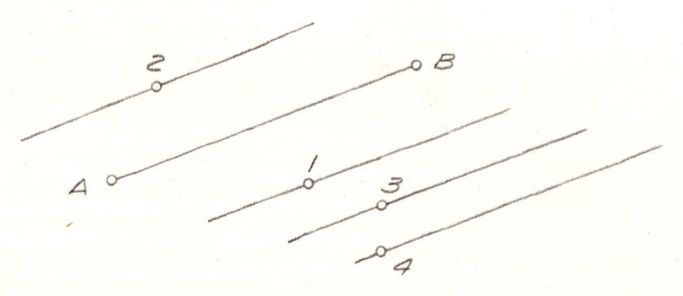

Figure 2.7   Parallel lines.

same

19   To sketch a line parallel to a given line you must keep the new line the _____ distance from the given line at all points along both lines.

The trick in sketching parallel lines is to direct your gaze to a point beyond (or behind) the surface of the paper as in a vacant stare. You will see both the original line and the new line as you draw and will find it easy to keep the new line parallel to the original.

one; two

20   Hold a pencil vertically in front of your eyes about 18 in. away and in line with a distant object such as a lamp, picture on the wall, etc. (See Figure 2.8.) When you focus your gaze on the pencil, you see _____ image(s) [pencil(s)]. However, when you transfer your attention to the distant object, you still see the pencil, but there is (are) _____ image(s).

parallel;
surface

21   The two images of the vertical pencil appear to be _____.
This experiment illustrates what we mean when we say to direct your gaze beyond the _____ of the paper when sketching parallel lines.

Actually, you do not transfer your gaze to another object beyond the paper surface, you merely take your focus off the detail of the drawing that you are making and gain an all-inclusive view.

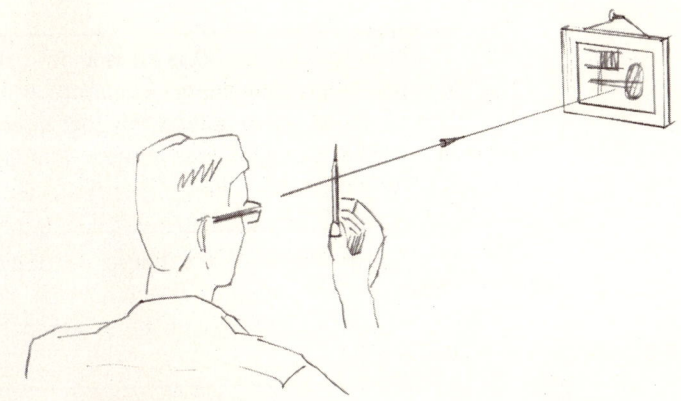

Figure 2.8 An experiment in vision.

■ 22 Sketch 5C shows two lines, *AB* and *CD*, and four points near each line. Sketch straight lines parallel to the given lines through the points using the techniques of Figures 2.7 and 2.8.

Sketch each line slowly and deliberately, with a single stroke. Take in an all-inclusive view by gazing beyond the lines and direct your attention to the end (terminal point) of the new line. Position the paper to take advantage of your natural arm motion.

■ 23 One useful application of this technique is in sketching rectangular blocks on a sheet of paper with the sides of the blocks parallel to the sides of the paper. (See Figure 2.9.) On a sheet of $8\frac{1}{2} \times 11$ unlined paper, sketch some rectangles similar to Figure 2.9. Choose any size and position you wish.

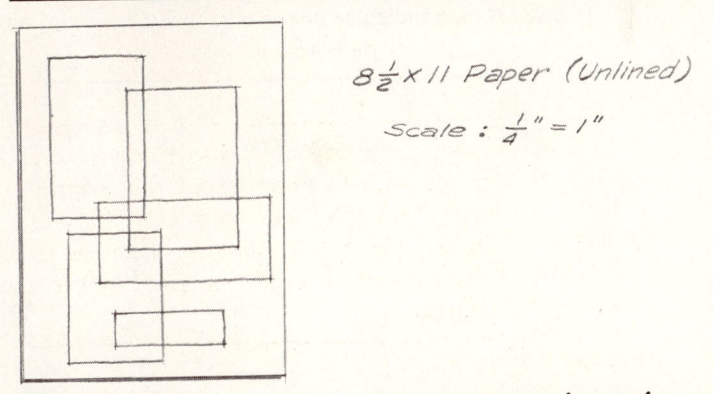

Figure 2.9 Using the edge of the drawing paper as one of a set of parallel lines.

■ 24 Sketch 5D is another exercise in sketching parallel lines. Three line segments (*A*, *B*, and *C*) are shown with a number of points (labeled *a*, *b*, or *c*). Sketch straight lines about two in. long starting at each of the points. Sketch the "*a*" lines parallel to line *A*, the "*b*" lines parallel to *B*, and the "*c*" lines parallel to *C*. Make light lines, that is, use light pressure on a soft pencil. Check your solution.

32 SEEING AND DRAWING/UNIT 1

house
(building,
shack)

25 The object of Sketch 5D is an isometric sketch of a(n) _____.
The meaning of the word "isometric" will be discussed in Chapter 11. It is of interest here to note only that a recognizable object was sketched by a series of parallel lines in three different directions.

dark

26 Finish the sketch of the house by darkening the important lines as shown in the answer sketch. When sketching with a soft pencil, _____ lines can be achieved by exerting more pressure on the pencil.

Perpendicular lines

If two lines are perpendicular, the angle between them is a right angle, or 90 degrees.

parallel,
angular,
perpendicular

27 Figure 2.10A shows the relationships that one line can have to a given line AB. The three possible relations are _____, _____, and _____.

180

28 A single straight line can be considered as two lines at an angle of _____ degrees to each other (Figure 2.10B). Lines XY and YZ represent a single line XZ.

90

29 A perpendicular line is a special case of an angular line where the angle is _____ degrees.

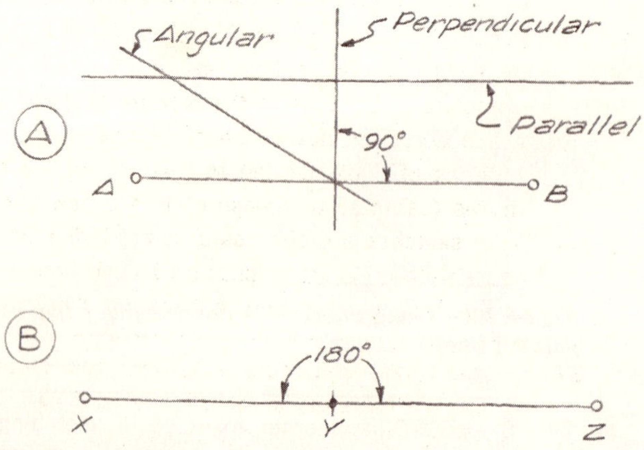

Figure 2.10 Angular relations between lines.

To sketch a line that is perpendicular to a given line, we must make sure that the new line bisects (divides equally) the *area* represented by an angle of 180 degrees. Figure 2.11A shows this idea.

In sketching, we must estimate distances, angles, and areas *visually*. In Figure 2.11A we visually estimate two equal areas in order to make line CD perpendicular to line AB. Often a short, quickly sketched, trial line will aid in this estimation as shown in Figure 2.11B.

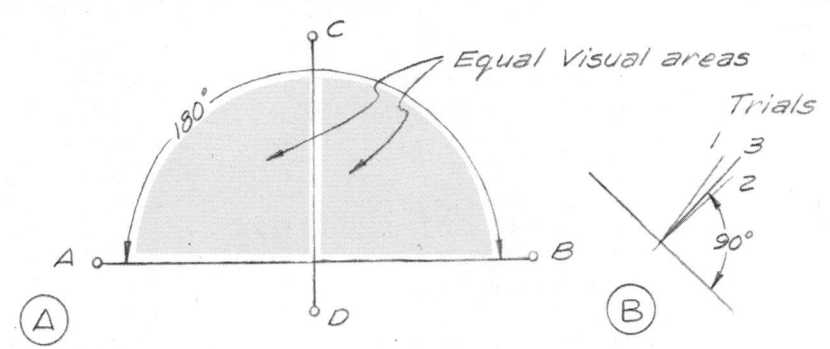

Figure 2.11 The visual concept of a "perpendicular."

■ 30 Sketch 6A shows five line segments (a, b, c, d, and e). There is a point on each line. Sketch a straight line perpendicular to each given line at the points marked.

Again, an aid to visualizing a 90-degree relationship in sketching is to gaze beyond the paper surface as in a vacant stare so that the two lines and their relationship appear as a whole. Check your own construction by using the corner of a sheet of paper for a 90-degree guide.

Grids

A grid is a most useful device in technical sketching. A grid is composed of two sets of parallel lines that are perpendicular to each other.

■ 31 Sketch 6B shows a portion of a ⅛-in. grid (⅛-in. squares). Also shown in Sketch 6B are the beginnings of a ¼-in., a ½-in., and a 1-cm grid. Complete these grids in the manner of the ⅛-in. grid. Start by sketching a line perpendicular to the given line at one of the points. Then sketch a series of lines parallel to this line at the other marks. Transfer distances (estimate or measure) to the new lines and complete the grids. Your sketches are successful if the individual blocks of the grid appear to be square (¼ in., ½ in., and 1 cm). Remember that extreme accuracy is relatively unimportant in freehand sketching.

32 Of the three relationships between two straight lines shown in Figure 2.10A, we have discussed parallel and perpendicular lines. This leaves the third, _____ lines.

angular

Angular lines

In graphics it is often necessary to sketch a line at some angle other than 90 degrees to a given line. Actually, the angle could be of any value between

0 and 360 degrees. (See Figure 2.12A.) We can view the geometry of each situation, however, in such a way that we can always think in terms of an angle between 0 and 90 degrees. (See Figure 2.12B.)

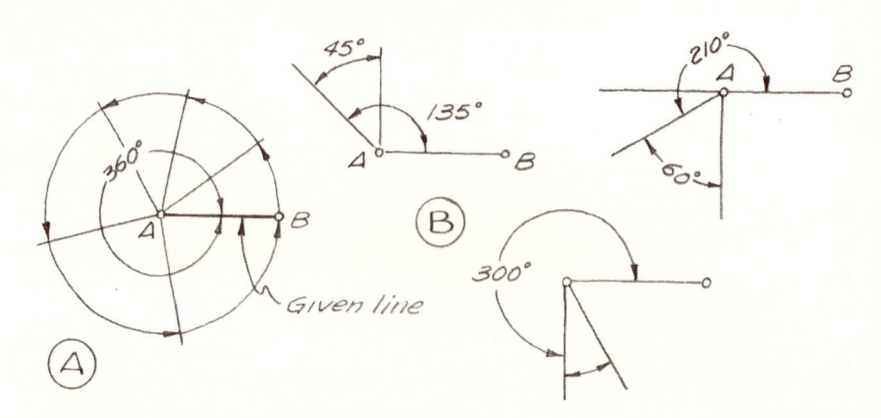

Figure 2.12   Viewing any angle as a subdivision of 90°.

---

33   Thus the problem of sketching angles becomes one of subdividing a 90-degree angle. A 90-degree angle has been subdivided in five ways in Figure 2.13. Dividing 90 degrees into:
   (1) Two equal parts yields _____° angles.
   (2) Three equal parts yields _____° angles.
   (3) Four equal parts yields _____° angles.
   (4) Six equal parts yields _____° angles.
   (5) Nine equal parts yields _____° angles.

45°; 30°;
22½°;
15°; 10°

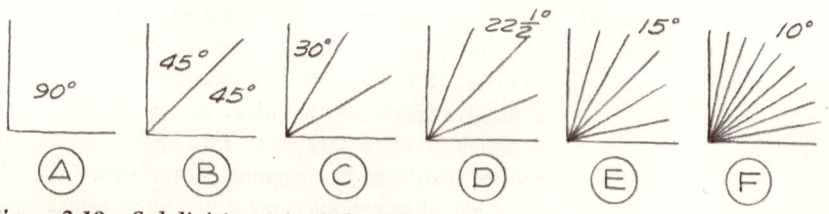

Figure 2.13   Subdivisions of a 90° angle.

---

34   From these figures we see that the first two are the important ones, since the subsequent divisions are merely further subdivisions of both _____ and _____ degrees.

45; 30

---

35   Figure 2.14 shows what to look for when dividing a 90-degree angle into both 45- and 30-degree segments by freehand techniques. The important consideration is to draw straight lines that divide the total area into either two or three equal _____ areas.

visual

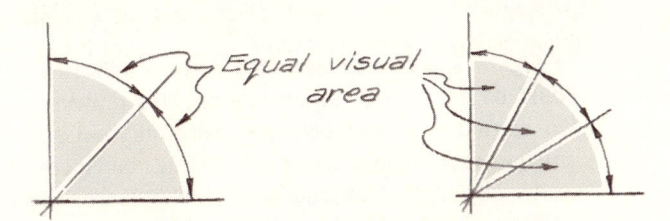

Figure 2.14   A visual concept of the subdivision of a 90° angle.

We did this in Figure 2.11 and Sketch 6A when we divided 180 degrees into two equal parts to find a perpendicular. The shaded areas of Figure 2.14 show the equal areas to look for.

As with all sketching problems, a few very light, quick, trial strokes are useful to establish a basic direction for a new line. If the trial line appears correct, it can then be darkened as a finished line.

■ 36   Complete the drawings of Sketch 6C by making the subdivisions of 90 degrees called for. Evaluate your own solutions.

Dividing a 45° angle into two equal parts yields two 22½° angles. Bisecting a 30° angle yields two 15° angles.

Further subdivision of the 30° angle gives us 10° and 5° angles. We know that freehand sketching is not a precisely accurate technique. In general, being able to sketch a line at a specified angle to a given line to an accuracy of ±5 degrees is considered sufficient for freehand work.

■ 37   Sketch 6D gives five exercises in sketching specific angles. Sketch the line AC that gives the required angles with line AB by the method of subdividing the appropriate 90-degree quadrant. Measure all angles in a clockwise direction from the given line AB. An example has been worked out to show the procedure for sketching a 75-degree angle. The meaning of the word "quadrant" is shown in the inset sketch. Check your solution.

Practice at sketching angles will make the subdivision of 90 degrees come instinctively.

Sketching Curved Lines

We pointed out earlier that curved lines are merely lines made up of a series of short, straight lines. The shorter we make the straight lines, the more accurate the curved line.

The circle

In Figure 1.2 we demonstrated how a circle (the most common curved line) could be drawn with straight lines. This drawing is repeated at Figure 2.15. We started with four points equidistant from a center and found we had a square rather than a circle. By adding four more points we drew an octagon which begins to look somewhat like a circle. Finally, we added eight more points and got a 16-sided polygon which closely resembles a circle.

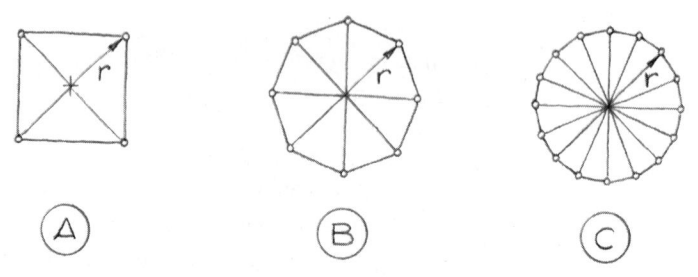

Figure 2.15 Creating a circle by reducing the lengths of the sides of a regular polygon.

38 Such a construction would, however, be too tedious and time-consuming to make by freehand techniques. We will sketch a circle by going back to four known points.

Consider Figure 2.16A. It shows a(n) _____ inscribed inside a(n) _____.

circle;
square

39 R, S, T, and U are the points where the circle is *tangent* to the sides of the square. Because of the symmetry of the two figures, these points occur at the midpoints of the _____ of the _____.

sides;
square

40 A point of tangency is the single point that is common to both a straight and a curved line when these lines are said to be tangent. Points R, S, T, and U of Figure 2.16A are _____ _____ _____.

points of
tangency

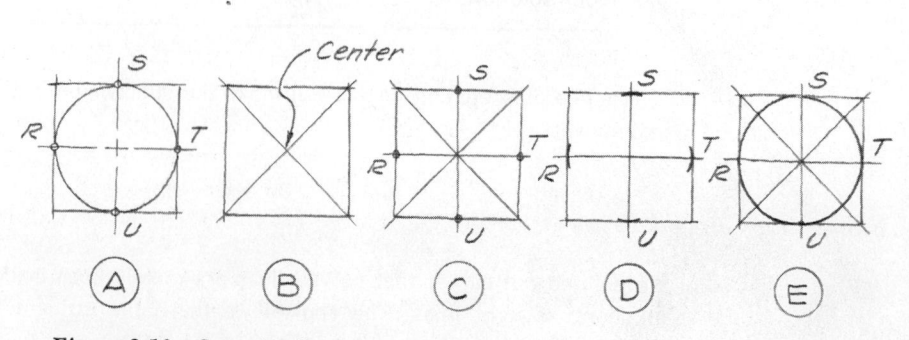

Figure 2.16 Steps in sketching a circle.

FREEHAND SKETCHING 37

equal
(the same)

41 A square is an easy figure to sketch since it is made up of two sets of parallel lines that are perpendicular to each other. The lengths of the sides are _____.

mid (center);
tangent

42 We can thus use the square to form the basis for sketching a circle. We need only to find the _____ points of the four sides and then sketch a smooth curve that is _____ at each of these four points.

diagonals

43 The center of a square can be found easily by sketching two diagonals. See Figure 2.16B. The intersection of the _____ is the center of the square.

Having the center, we may find the midpoints of the four sides by sketching two lines through the center parallel to the sides as shown in 2.16C.

Start sketching the circle by drawing four short tangents (arcs) at points R, S, T, and U. (See Figure 2.16D.) This sets the stage for sketching the final curve which yields a circle as shown in 2.16E.

A circle is a smooth curve—there are no sharp discontinuities or changes in direction. A common mistake in sketching circles is illustrated by Figure 2.17A. Here the tangents have been sketched with too small a radius, yielding a final curve that is not smooth. Figure 2.17B shows an enlarged view of correct and incorrect tangents for this use.

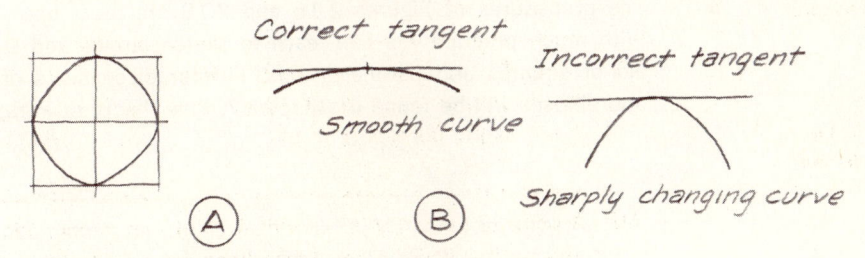

Figure 2.17 A common error in sketching circles.

The ellipse

An ellipse may be visualized as a circle viewed at an oblique angle, that is, at an angle other than 90°.

perpendicular;
circle

44 See Figure 2.18A. The observer's line of sight is _____ to the plane of the circle and the observed figure is a true _____.

ellipse

45 In Figure 2.18B, however, the line of sight is not perpendicular to the plane of the circle. The observed figure is a(n) _____.

38 SEEING AND DRAWING/UNIT 1

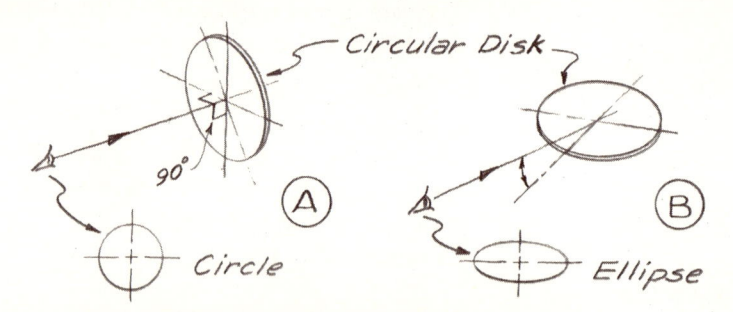

Figure 2.18 Seeing a circle as an ellipse.

parallelogram
or rhombus

46 An ellipse can be sketched in the same manner developed for a circle in Figure 2.16. Figure 2.19 shows the steps in sketching an ellipse. The only difference is that the enclosing square now becomes a "squashed" square or a(n) _____.

■ **47** Sketches 7A and 7B present some exercises in sketching circles and ellipses. Sketch the figures called for. Check your own results by comparison with Figures 2.16 and 2.19. You should have smooth curves with no sharp discontinuities.

Circles and ellipses are not easy to sketch. Squares and parallelograms are. The procedures of Figures 2.16 and 2.19 are most useful to the beginner. With much practice you can learn to sketch circles and ellipses without the aid of squares and parallelograms. Further discussions of sketching circles and ellipses will be found in Chapter 7, The Theory of Perspective.

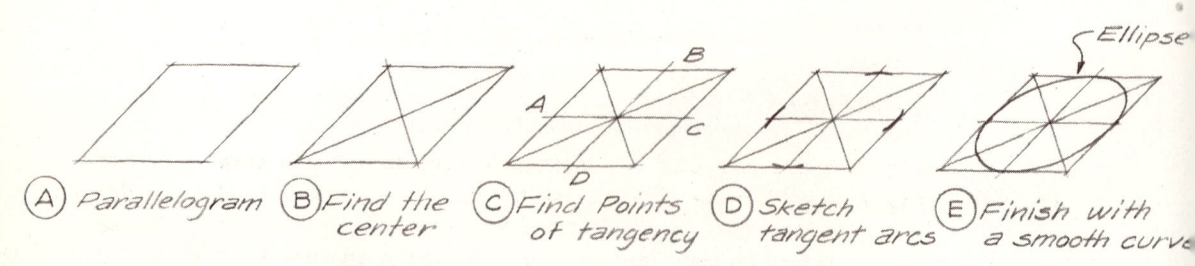

(A) Parallelogram (B) Find the center (C) Find Points of tangency (D) Sketch tangent arcs (E) Finish with a smooth curve

Figure 2.19 Steps in sketching an ellipse.

Instrument-aided Sketching

The tools of freehand sketching were listed as paper, pencil, and eraser. Let's add a few instruments to this list, the compass and the plastic circle template and the straightedge. (See Figure 2.20.) Since it is not easy to sketch circles and since it is so very easy to draw a circle with a compass or a circle template, it seems worthwhile to use these devices if only to save time.

FREEHAND SKETCHING 39

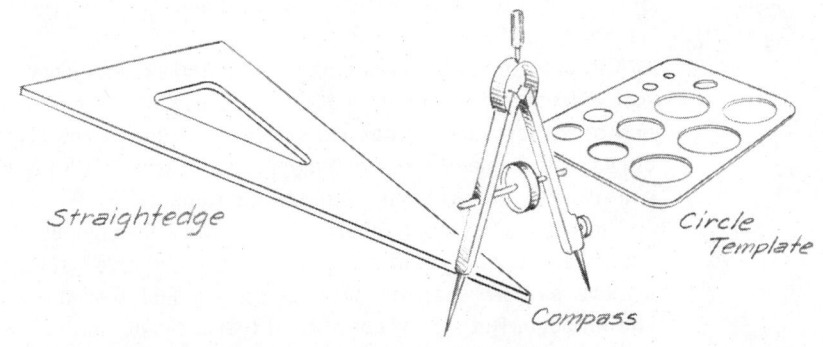

Figure 2.20 Tools for instrument-aided sketching.

Compass and template

Often the compass or template can be used to put in the correct construction and then the instrument-drawn lines can be traced freehand to make the line weight and character consistent with the rest of the sketch. Figure 2.21 illustrates this.

Ellipse templates are not as useful in sketching as circle templates since there are an infinite number of variations on the size and shape of the ellipse. However, an ellipse is easier to sketch than a circle. More about this in Chapter 7.

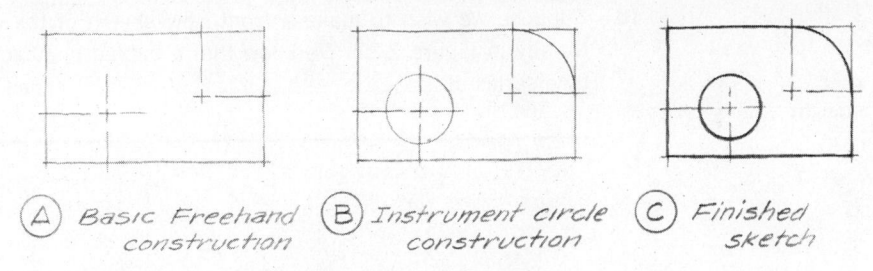

Figure 2.21 An instrument-aided freehand sketch.

Straightedge

A straightedge is a precision instrument used to assist in drawing perfect straight lines. A truly "freehand" sketch would preclude the use of a straightedge. However, there are many instances in technical sketching where the use of this instrument helps in both the construction and the finishing of the sketch.

The figures in this text are a good example. Every figure except for a few of the more complicated ones in later chapters was first sketched in true freehand form. They were sketched on grid paper to take advantage of the precision of horizontal and vertical lines and of measurements in these two directions. When the construction was complete, a sheet of tracing paper was placed on the sketch and the finished version was traced using a straightedge, compass, and plastic templates. This procedure was followed since the drawings were to be photographed and printed. If any one of them had been

40 SEEING AND DRAWING/UNIT 1

drawn for just a casual study of a problem, the first freehand sketch would have sufficed.

The procedures for sketching parallel and perpendicular lines that we have just discussed are greatly aided by the use of a straightedge. The new line can be visually aligned parallel or perpendicular to the given line before it is drawn. Clear plastic straightedges are especially useful since all parts of your drawing are visible through the instrument.

This text stresses freehand construction throughout. As you start each sketch exercise, keep in mind that instrument-aided sketching can help you achieve a more precise construction—so feel free to use a compass and straightedge. But also keep in mind that a facility in pure freehand sketching is an important asset to those engaged in technical work.

We are presenting freehand sketching as a fast and effective means of communicating technical information. Any device or technique that saves drawing time is worthwhile.

Irregular Curves

Curves which vary in their deviation from a straight line and which contain discontinuities and changes in radius and direction are called irregular. No template can be constructed to offer the infinite variations of form that irregular curves may assume. In Chapter 8 we adopt a simplified definition that describes irregular curves as curves which are not circles or circular arcs.

short straight

48 Suppose we wish to make a front-view sketch of the object shown pictorially in Figure 2.22. We know that a curved line can be approximated by a series of _____ _____ lines.

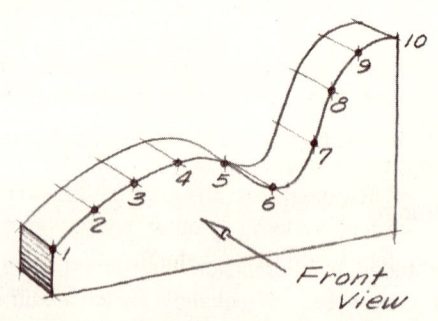

Figure 2.22 An irregularly curved surface (pictorial view).

curve

49 Accordingly, we have placed 10 points on the curved portion of the object. Figure 2.23A shows what happens to the desired drawing if we consider points 1, 3, 8, and 10 only. Whether we join these four points with straight lines or a smooth curve, the view does not look like the desired shape.

However, at 2.23B we use all 10 points and even the sketch showing these points connected by straight lines begins to look like the desired _____.

FREEHAND SKETCHING 41

Since we are dealing with a curve, we know that the straight-line treatment is inaccurate. We can modify this treatment by curving the straight lines between adjacent points, giving us an accurate picture as a finished sketch.

■ 50 We still haven't discussed where and how to place the points. This subject will come up in subsequent chapters. Sketch 7C presents four exercises in sketching curved lines between points. Sketch the best smooth curve connecting the points. (No solution is given.)

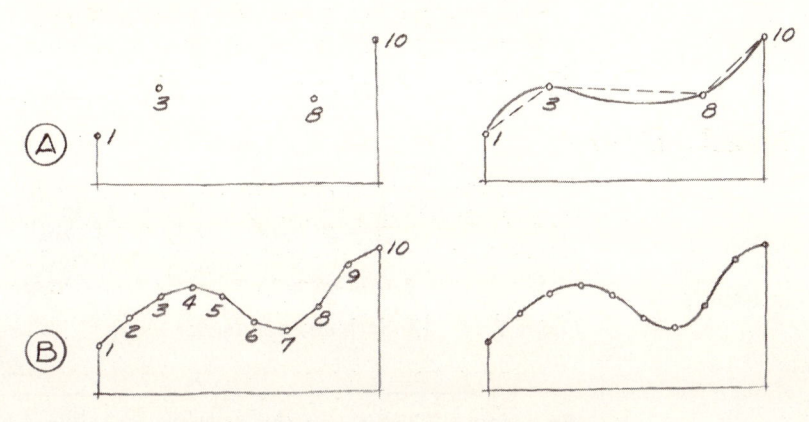

Figure 2.23 Using points to sketch an irregular curved line.

Estimating Distances

A part of learning to sketch is developing a facility for estimating distances along straight lines. The tools of sketching listed did not include a ruler or scale.

Estimation of distances along straight lines should be based on standard units of measurement. In the United States and Great Britain, this standard is the inch-foot system. In other countries of the world the standard is the metric system (centimeter-meter).

51 Sketching is an approximation to precise and accurate drawing so that making precisely true measurements is not required. However, you should learn to _____ distances with a fair degree of accuracy.

estimate

52 In this book measurements are given in the United States standard, the _____ _____ system.

inch-foot

42 SEEING AND DRAWING/UNIT 1

2.5+ cm (2.54);
⅜+ in. (0.394)

53 A conversion scale showing the relationship between inches and centimeters is shown in Figure 2.24. From this you can read the conversion unit that
  1 in. = _____ cm or, conversely,
  1 cm = _____ in. (approximate)

four (4)

54 In estimating distances for sketching purposes, learn to think in terms of the basic units of inches or centimeters. Any distance you wish to show is then just a multiple or a division of the basic unit. The inch scale of Figure 2.24 is _____ basic units long.

```
INCHES
0        1        2        3        4
|--|--|--|--|--|--|--|--|--|--|--|--|
0  1  2  3  4  5  6  7  8  9  10
CENTIMETERS
```

1 in. = 2.54 cm.   1 ft. = 12 in. = 30.48 cm.
1 cm. = 0.394 in. = ⅜ + in.   1 meter = 100 cm. = 39.37 in.

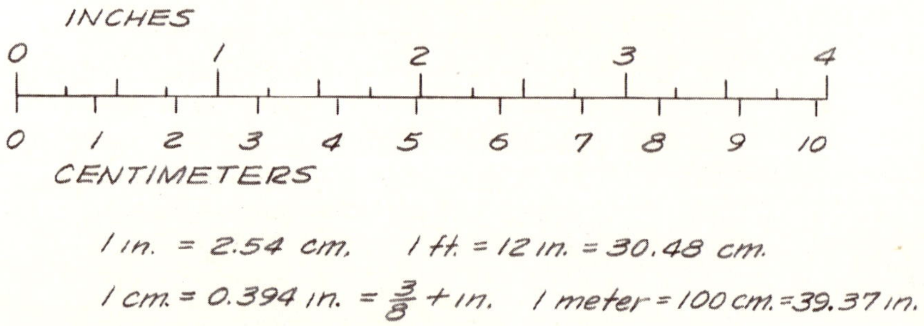

Figure 2.24 *The inch and centimeter scales.*

½, ¼, ⅛,
1/16, 1/32, 1/64

55 Figure 2.25A shows the divisions of one inch that are standard in the industrial world along with their decimal equivalents. The standard fractional divisions of the inch are _____, _____, _____, _____, _____, and _____.

½; 1/10;
millimeter (mm)

56 Figure 2.25B shows the divisions of the centimeter. The fractional divisions of the centimeter are _____, and _____. This latter is called the _____.

A:
1 in.
½ in.
¼ in.
⅛ in.
1/16 in.
1/32 in.
1/64 in.

B:
1 cm.
centimeter
½ centimeter
1/10 centimeter
millimeter

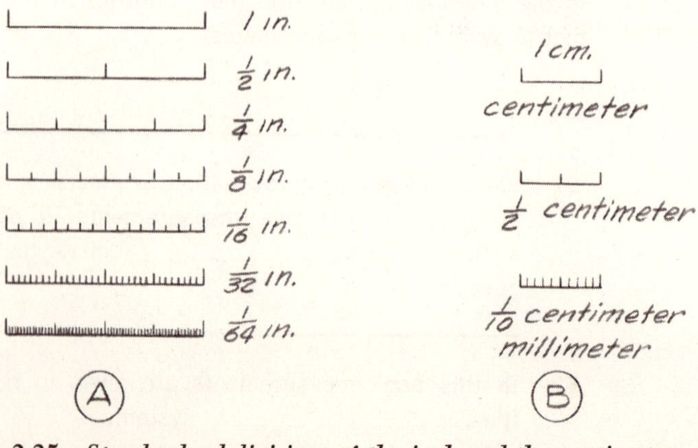

Figure 2.25 *Standard subdivisions of the inch and the centimeter.*

FREEHAND SKETCHING 43

Accuracy in freehand sketching is usually not much greater than ± (plus or minus) 1/32 in. per inch. If the drawing requires greater accuracy, it should be done carefully with instruments. The main purpose of freehand sketching is to get ideas down on paper quickly. Accuracy becomes relatively unimportant.

An accuracy of 1/32 in. per inch means that for each inch of length, you are allowed 1/32 in. longer or shorter than one inch (31/32 in. to 1 1/32 in.).

±4/32" (1/8");
±10/32" (5/16")

57 Thus the accuracy of a 2-in. line would be ±2 times 1/32 = ±1/16 in. The accuracy of a 4-in. line would be ± _____ in., for a 10-in. line ± _____ in., etc.

■ **58** Sketch 8 is an exercise in estimating dimensions. Line A calls for a measurement of 4 3/8 in. Procedure is as follows:
(1) Get in mind the length of one inch (look at a scale or refer to the inset figure).
(2) Mark off five inches (estimated) on line A.
(3) Divide the last inch into halves (by eye).
(4) Divide the first 1/2-in. segment into halves (1/4-in.).
(5) Divide the second 1/4-in. segment into halves (1/8-in.).
(6) This last mark should represent 4 3/8 in. from A.

Line A has been estimated according to the above procedure. Try the rest of the lines (B through G) yourself. For line B, first estimate three inches along the line, then divide the last inch into halves.

Check your solutions to the above exercise by measuring the line you estimated. (Compare with the inch scale of Figure 2.24 if you do not have a ruler.) If you can expect an accuracy of ±1/32 in. per inch, then your answer for B should be 2 1/2 in. ±3/32 in. (Consider the three-inch original estimate.)

To learn to sketch with a reasonable degree of dimensional accuracy, you must have skill at estimating distances in the basic unit of measurement—the inch. Further discussions of measurement systems will be found in Chapter 6, Proportion and Scale.

Grid Paper

One of the most useful aids to freehand sketching is grid paper. Sometimes called graph paper, this is paper on which is printed a grid of squares, each square being an accurate subdivision of the inch or the centimeter. Common grids useful for sketching purposes are 4 squares per inch, 10 squares per inch, 8 squares per inch, and the centimeter grid. Grids are usually printed in light colors (blue, orange, green) so that they can be seen but do not interfere with black pencil lines.

Grid lines are horizontal and vertical and serve as a guide for making parallel and perpendicular lines in these directions. Accurate measurements

44 SEEING AND DRAWING/UNIT 1

can, of course, be made since the grid represents accurate subdivision of the inch. Although it is useful to sketch on plain paper without the aid of a grid, you should use a grid whenever available because of the advantages of speed and accuracy.

Printed grids are provided for most of the exercises in this volume to give you the advantages of speed and accuracy.

DEVELOPING STYLE IN SKETCHING

Anyone who must make drawings for other people to see and use should work toward developing a personal style in his work. We all seek to demonstrate our own personal style or taste in the clothes we wear, the words we use (both spoken and written), and in many other ways. Your drawings, too, can reflect your personal style. A professional drawing is, first, neat and accurate. The purpose of a technical drawing is to convey information. Neatness implies clear, distinct lines and general cleanliness.

■ 59 A pencil is a versatile tool especially when used for freehand drawing. In a sketch there should be no attempt to imitate the precision of an instrument drawing. Strive for an expressive line quality. Sketch 9A illustrates expressive line sketching. Note that the three expressive lines shown give the impression that they have been sketched quickly and with varying pressure on the pencil. These lines have an expressive quality. There are three light guide lines at A', B', and C'. Sketch lines that duplicate the qualities of lines A, B, and C. Use a soft pencil (HB or equivalent) and vary the pressure as you trace across the lines.

■ 60 Sketch 9B is a cube sketched with expressive lines. Note that the lines are darkest around point A—the nearest point to the observer—and taper off as they recede in space. Another cube and a block are sketched with light construction lines. Go over these lines using an expressive line treatment. Compare results with the first cube.

It is often hard to keep a sketch clean. Erasures and movement of the hand over the lines tend to smudge the drawing. In sketching quickly you often draw many extraneous lines as you "feel" for the shape you want. All of these add confusion to a sketch. When a sketch gets to this stage and yet carries the information that you desire, place a sheet of translucent paper over the sketch and trace just the lines you want. Use quick, expressive lines and the resulting sketch will be a finished piece of work. Figure 2.26 illustrates this technique.

Accuracy

A sketch that is accurate in dimension, proportion, and drawing form will be appreciated as a piece of professional work. This chapter is concerned with

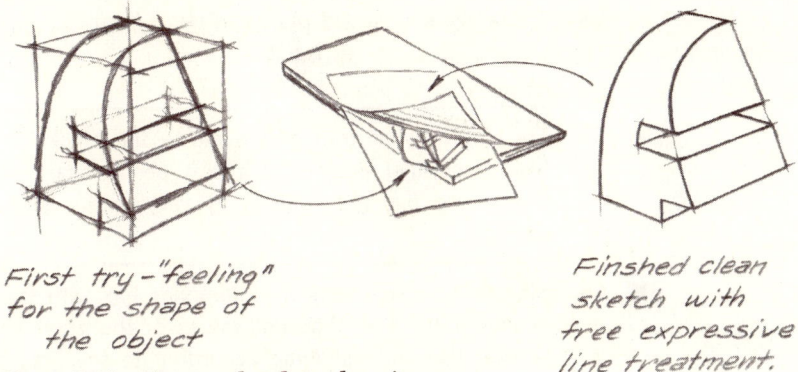

Figure 2.26 Trace a sketch to clean it up.

sketching techniques. Subsequent chapters will present the principles and rules of drawing that lead to accuracy.

Shading

Many sketches (especially pictorials) can be improved by adding some simple shading—variations of light and dark. Consider Figure 2.27. At 2.27A you see a line pictorial sketch of a vertical cylinder. At 2.27B the same cylinder is shown, but with some line shading to emphasize the curved nature of the surface.

curved

61 The shading of the _____ surface was achieved by sketching a few parallel straight lines.

close together; far apart

62 To show dark areas, the shade lines are _____ _____ and heavy (dark). To show lighter areas, the lines are _____ _____ and light.

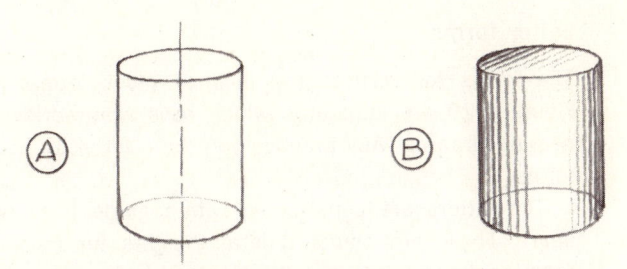

Figure 2.27 Simple line shading adds form to a pictorial sketch.

A soft pencil is versatile in that you can sketch both light and dark lines with the same instrument. Artists, printers, and photographers use a gray scale when creating or reproducing black and white drawings or photographs. Figure 2.28 is such a gray scale sketched with an HB pencil.

46 SEEING AND DRAWING/UNIT 1

white (light);
black (dark)

63 There are six shaded blocks ranging from _____ at No. 1 to _____ at No. 6.

■ 64 Try making your own gray scale in the blocks provided in Sketch 9C. Sketch parallel lines with a soft pencil sharpened to a chisel point. Obtain the five gray values by varying line spacing as shown in the example.

■ 65 Sketch 9C also shows a rectangular box with the three exposed faces shaded with three different values of the gray scale. An unshaded box is sketched. Add shading according to the gray scale number shown. Note that the parallel shade lines follow the principal (long) dimension of the rectangular faces.

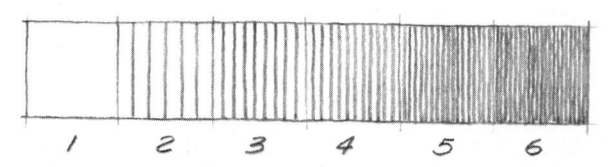

Figure 2.28 A white-to-black gray scale.

Lettering

No technical drawing stands by itself. Words or numbers must accompany it whether as a figure number, title, set of dimensions, explanatory note, or other descriptive phrases. The fastest way to add these words is by freehand lettering.

lettering

66 Communication through words and numbers is important to graphics as a means of making a drawing complete and self-explanatory. We will concentrate on freehand _____ as a means of adding words and numbers to drawings.

Letter forms

The forms for vertical and inclined block letters and numerals shown in Figure 2.29 are standards which have won worldwide acceptance for engineering drawing. Any practice you do in lettering should be directed to these forms.

The letters are constructed within guide lines which are ³⁄₁₆ and ⅛ in. apart. These are standard letter heights for most needs. Titles and other prominent notes may be taller (³⁄₁₆ to ⅜ in.).

vertical,
inclined

67 The two accepted standards for block letters and numerals in engineering drawing are the _____ style and the _____ style.

Figure 2.29 Standard inclined and vertical engineering block letters and numerals.

guide | **68** Freehand lettering is done between _____ lines.

We know that handwriting can vary from the beautiful and completely legible to the sloppy and completely illegible. Freehand lettering should always tend toward the former. Notes are put on drawings to communicate information not contained in the drawing itself. If the notes are illegible, little information is communicated.

information | **69** The fact that you can read your own handwriting is not necessarily proof that other people can. The same is true with engineering lettering, since the drawing and the notes are being made specifically to communicate _____ to others.

height | **70** The important consideration in learning to letter is to observe letter form as shown in Figure 2.29. Copy the letters exactly and train your hand to form them that way. The guide lines control the _____ of the letters.

48 SEEING AND DRAWING/UNIT 1

71 Forming the correct *letter width* is up to you. Note that most of the letters are approximately as wide as they are high. This is an important consideration in lettering. Figure 2.30 shows a common error in letter form that most beginners make. The letters are too _____.

narrow

72 Four letters, I, J, W, and M, are exceptions to the rule but, in general, the _____ of engineering block letters is approximately equal to the _____ as controlled by the guide lines.

width;
height

73 Whether you use vertical or inclined letters is a matter of personal preference. Some industries, however, standardize on one or the other for all drawings. Lettering of either style must be _____ so that the intended information is communicated.

legible
(readable)

■ **74** Sketch 10 is a lettering exercise. The standard forms of both inclined and vertical letters are shown along with blank guide lines for you to use. The sketching of letters, like the sketching of lines, should be done quickly but deliberately with single strokes. Pay attention to the form and proportion of each letter. Memorize them so future lettering will come instinctively.

ABCDEFGH - LETTERS TOO NARROW - KEEP LETTERS WIDE
BETTER TO EXAGGERATE WIDTH RATHER
THAN MINIMIZE IT.

Figure 2.30   A common error in forming letters.

Word composition

The second step in learning to letter is to form the letters into words. Again, the purpose is communication so do not do anything that will spoil the legibility of your lettering. Figure 2.31 shows some examples of word composition.

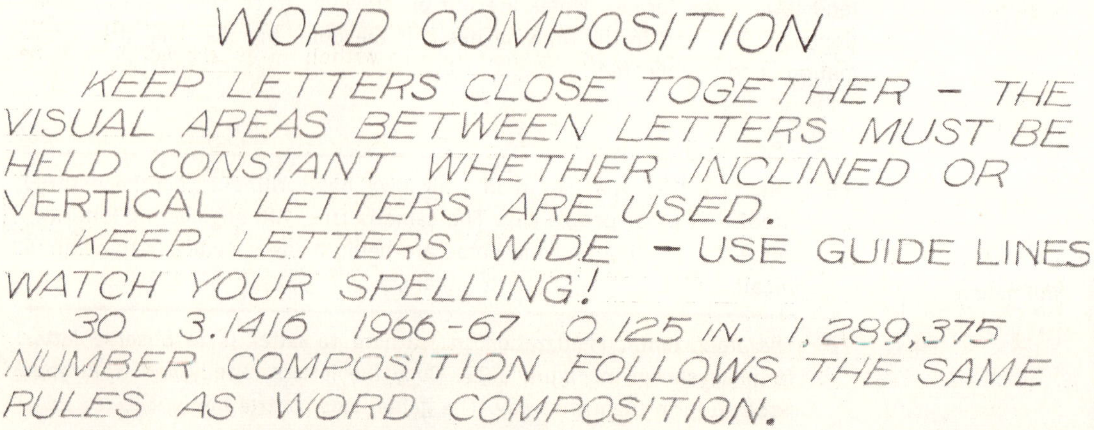

WORD COMPOSITION
   KEEP LETTERS CLOSE TOGETHER - THE
VISUAL AREAS BETWEEN LETTERS MUST BE
HELD CONSTANT WHETHER INCLINED OR
VERTICAL LETTERS ARE USED.
   KEEP LETTERS WIDE - USE GUIDE LINES
WATCH YOUR SPELLING!
   30   3.1416   1966-67   0.125 IN.   1,289,375
NUMBER COMPOSITION FOLLOWS THE SAME
RULES AS WORD COMPOSITION.

Figure 2.31   Word composition.

Note that individual letters in a word are kept close together. Separating letters destroys word recognition. The reader's eye has to work to decide whether the image is one word or perhaps two or three. Figure 2.32 shows some "Don'ts" in word composition.

Also note in Figure 2.32 that some letter combinations require different spacing because of the form of the letters involved. A, I, J, L, T, V, and Y are the letters to watch in this respect. A good rule of thumb in word composition is that *the space between letters must appear to be equal.*

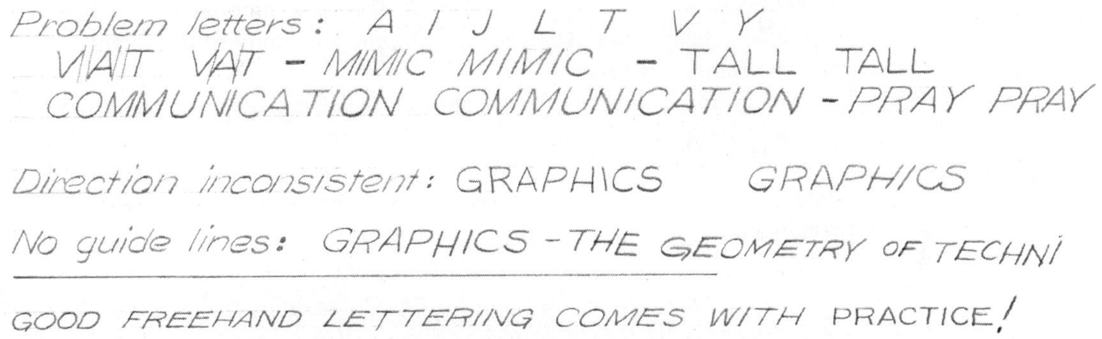

Problem letters: A I J L T V Y
WAIT WAT — MIMIC MIMIC — TALL TALL
COMMUNICATION COMMUNICATION — PRAY PRAY

Direction inconsistent: GRAPHICS GRAPHICS

No guide lines: GRAPHICS — THE GEOMETRY OF TECHNI

GOOD FREEHAND LETTERING COMES WITH PRACTICE!

Figure 2.32 Some common lettering errors.

Style in lettering

Even though letter form and use is standardized, you can and should strive for a personal style in hand lettering *as long as you do not affect legibility.* Put a bit of your own personality into your work, but keep in mind that other people must be able to read and understand your drawings.

■ 75 On the blank guide lines of Sketch 11 copy the phrases given. Pay close attention to letter form and proportion and to the spacing of letters and words.

76 To summarize, lettering is added to a drawing to _____ information.

communicate
(convey)

77 To communicate information through written words, the words must be _____.

legible
(readable)

78 The engineering profession has accepted standard letter forms. Memorize these forms and use them consistently. Guide lines control the _____ of the letters. The proportion of height to width for most letters is approximately _____ to _____.

height;
one to one
(1:1)

79 A group of letters forming a(n) _____ must be kept close together. The visual _____ between letters must appear equal or constant.

word; area (space)

50   SEEING AND DRAWING/UNIT 1

legibility

80   There is room for displaying a personal style in your lettering as long as you do nothing to decrease _____.

Confidence in your own ability to make good freehand sketches leads to a better reception of your ideas by other people and a greater satisfaction in your own work. Graphics is a language and freehand sketching is an important form of the language. When we talk of a "good" freehand sketch we do not necessarily mean "accurate." A good sketch is informative. The freedom of freehand techniques lets you concentrate on information content rather than on drafting precision.

Although this completes the chapter on freehand sketching, it by no means completes the discussion and practice of sketching in this volume. In subsequent chapters you will be asked to make many sketches. Refer to this chapter and especially to the seven sketch exercises you did to refresh your memory on the techniques of freehand sketching.

UNIT 2 *orthographic projection*

The fundamental concepts of orthographic projection are presented in Chapters 3, 4, and 5. Chapter 3 develops the theory of orthographic projection through the six principal orthographic views. A short examination of the state of lines (visible or hidden) in a drawing is given in Chapter 4. Chapter 5 expands on the flexibility of the orthographic system with a presentation of how any desired view of an object may be obtained.

Chapter 6 deals with proportion and scale, two concepts important to both orthographic projection and pictorial drawing.

chapter 3   *principal orthographic views*

■ 1 Sketch 12 shows a grid of ¼-in. squares. The horizontal lines are labeled *1* through *15*. The vertical lines are labeled *A* through *O*. Do the following:

(1) Plot (mark) points at intersections *2B, 2F, 6F,* and *6B*. (Use the numbers and letters in the same way you locate a city on a road map.)

(2) Connect these four points with straight lines in the order named, starting and ending with point *2B*.

(3) Plot points at *10B, 10F, 14F,* and *14B* and connect with straight lines in that order, ending with *10B*.

(4) Plot points at *10J, 10N, 14N,* and *14J* and connect.

Check your solution.

2 If you sketched the points and lines of Sketch 12 correctly, you have made a *three-view orthographic projection* of an object. Can you name the object?

It is a(n) _____.

cube
(more accurately, a 1-in. cube)

I don't know what it is _____.

3 Figure 3.1 is a pictorial sketch of the cube showing the three viewing directions that created Sketch 12. Can you identify the top, front, and right profile views in Sketch 12? Label them *T* for top, *F* for front and P_R for right profile.

|T|

|F| |P_R|

52

PRINCIPAL ORTHOGRAPHIC VIEWS 53

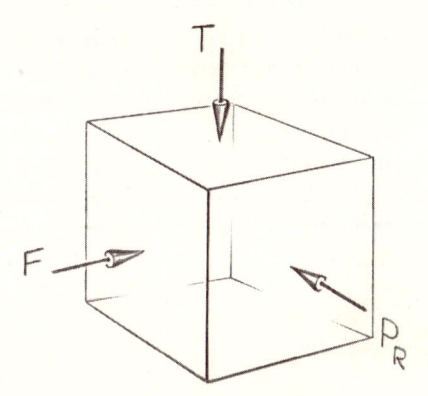

Figure 3.1 Pictorial view of a 1-in. cube.

Orthographic projection is the standard drawing form of the industrial world. It is a precise and unambiguous means for describing a physical object. The form is unreal in that we do not see an object as it is drawn orthographically. Chapter 1 showed that orthographic *viewing* is unreal. An orthographic form is abstract. We do not see a single object as multiple views. Pictorial drawing, on the other hand, has photographic realism.

The precision of the orthographic form serves the need for communicating detailed technical information accurately to the variety of people who design, develop, produce, and sell the products of industry.

THE ENCLOSING BOX

H, W, D	4 In Chapter 1 we discussed the subject of visualization and mentioned that any object has three principal dimensions of height, width, and depth. Other words may be used such as breadth for width and length for depth, but we will standardize and abbreviate the three principal dimensions with capitalized first letters as ____, ____, and ____.
D (and) H; W	5 H and W are the two dimensions that we see when we stand looking at an object from the front. D is the dimension that is receding from us. If we look at an object from the side (profile), then ____ and ____ are the two obvious dimensions and ____ is the receding dimension.
W (and) D; H	6 Viewing the object from the top, we see the ____ and ____ dimensions. The receding dimension is ____.
(1) one; (2) two	7 How many views are required to show H, W, and D in: (1) A pictorial drawing? ____ (2) An orthographic projection? ____

box

8 In Figure 3.2 an automobile is enclosed in a transparent box. The dimensions of the box are H, W, and D.

If we ignore details for the moment by saying that any object has three principal dimensions, H, W, and D (viz. the automobile of 3.2), then we can also say that any object can be enclosed in a rectangular _____ of dimensions H, W, and D. (See Figure 3.3.)

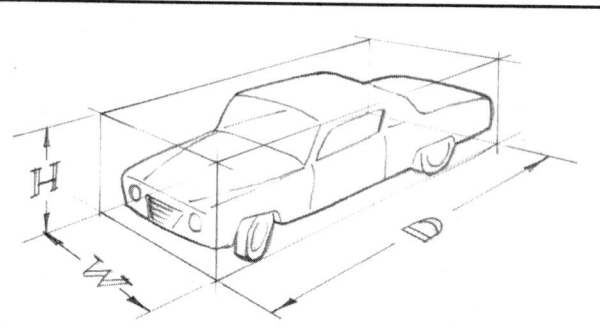

Figure 3.2 The basic H, W, and D dimensions of an automobile.

Figure 3.3 shows some complex objects sketched inside transparent boxes. This is an important concept in technical drawing. First the enclosing box must be drawn to scale and to the correct proportions. Then the details can be added inside the box to give an accurate drawing. In fact, the total object can be subdivided into smaller parts, which, themselves, can be enclosed in a box. An object such as the human figure can be considered as being composed of many boxes. Thus our study of orthographic projection resolves itself into two areas:

(1) How to draw the enclosing boxes.
(2) How to draw detail inside the boxes.

9 Sketch 13 is a three-view orthographic projection of an airplane. Sketch the following:

(1) The orthographic views of a box that *completely* encloses the airplane.
(2) A box that encloses just one wing.
Check your solution.

The answer to Sketch 13 shows that in either the orthographic or pictorial form, the drawing of the airplane could be started by first drawing a series of rectangular boxes and then adding details inside the boxes.

POSITION OF ORTHOGRAPHIC VIEWS

The two most commonly drawn orthographic views, top and front, ALWAYS have the same relation—top view above the front view, as in Figure 3.4. The top and front views and the front and profile views are connected by thin lines. These are called *projection lines*. We will learn more about them later.

PRINCIPAL ORTHOGRAPHIC VIEWS 55

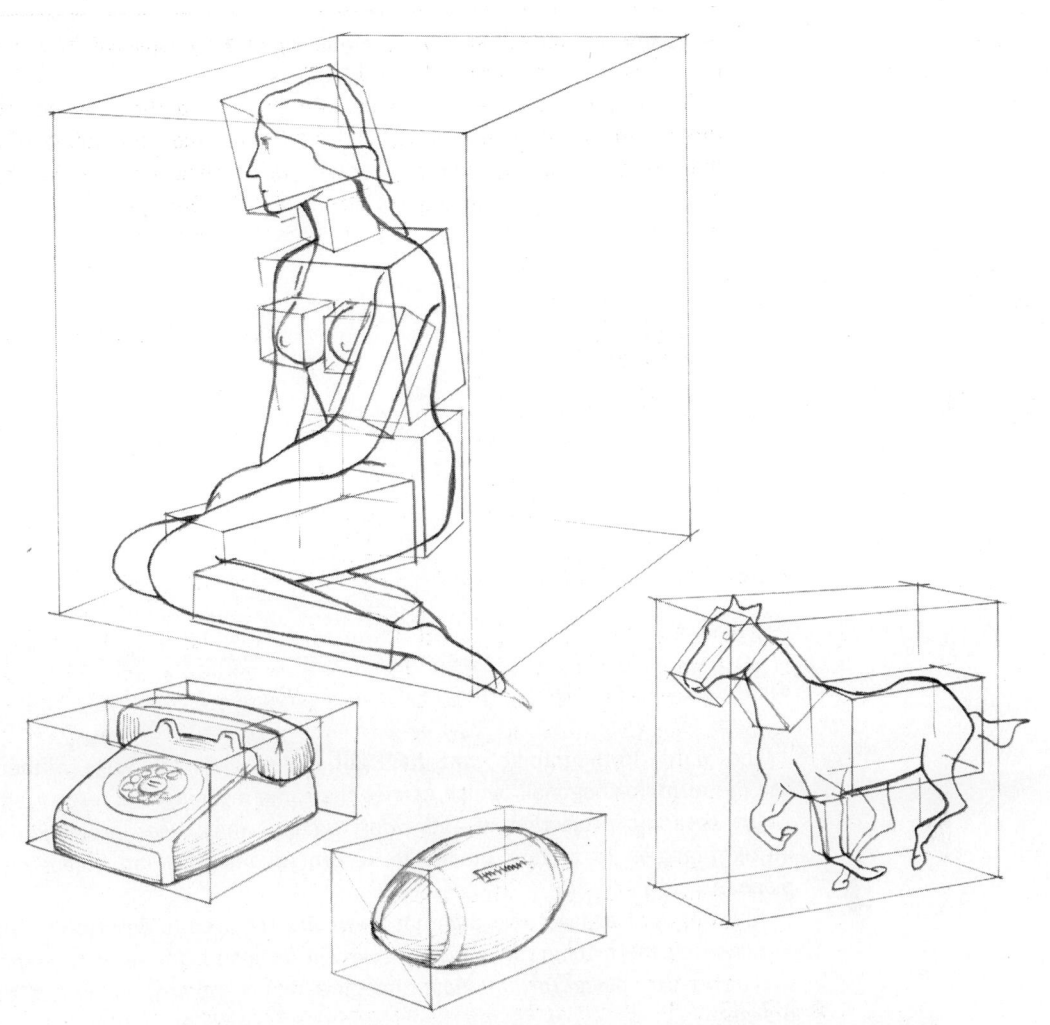

Figure 3.3 To illustrate that any object can be enclosed in a rectangular box or boxes which describe the basic H, W, and D dimensions of the whole or the parts.

above

10 Examine Figure 3.4. It is the same cube you constructed in Sketch 12. What are the *relative positions* of the top and front views?
   The top view is _____ the front view.

front *(F)*

11 The right profile view is *projected* to the right of the _____ view.

picture plane

12 The word "projection" implies the extending of lines or rays from the eye to various points on an object. In Chapter 1 we learned that the intersection of these lines with a(n) _____ _____ placed between the observer and an object defined the drawing of the object.

56 ORTHOGRAPHIC PROJECTION/UNIT 2

perpendicular
(90°)

13 We also learned that, in orthographic viewing, the projection lines are _____ to the picture plane.

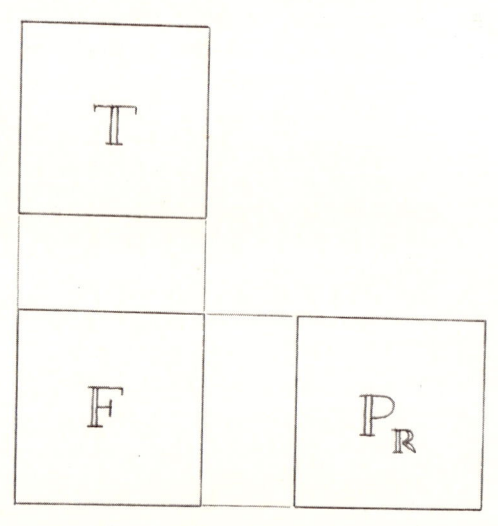

Figure 3.4   A three-view, orthographic projection of a 1-in. cube.

The word "orthographic" means right-angled drawing. Some authorities use the term "orthogonal" which carries the same meaning.

Orthographic projection is a drawing system which uses projection lines at right angles to picture planes to record an image of an object on the planes.

Knowing that at least two different views are required to describe an object in space by the orthographic system, we will develop a *frame of reference* in which we can place (or, at least, imagine that we place) an object to be drawn.

Our frame of reference will be a set of two or more reference surfaces (picture planes) that can be placed around an object in a *definite and standard manner*. Viewing the object orthographically (at right angles) through each picture plane will yield an orthographic projection.

The frame of reference is built from the system that scientists, engineers, and mathematicians use to describe a point in space. This is called the *coordinate system*.

Coordinates

A coordinate is defined as "any of two or more magnitudes which establish the position in space of such elements as points, lines, planes, etc."

x; y

14 Figure 3.5A shows a set of two-dimensional coordinate axes. From the origin O measurements can be made along the axes in both the plus (+) and minus (−) directions of the coordinates, _____ and _____.

PRINCIPAL ORTHOGRAPHIC VIEWS 57

coordinates

15 In Figure 3.5B, the magnitudes 5 and 3 represent measurements along the x axis and y axis respectively of the distances from the origin O to point A. The values 5 and 3 are the _____ of point A.

origin

16 The horizontal x axis and the vertical y axis form the frame of reference for describing point A on a planar (two-dimensional) coordinate system. Measurements are made from the common meeting point of these two axes which is called the _____. (See Figure 3.5.)

negative (−)
below (under)

17 The origin divides the two axes into positive and negative numbers. Points on the x axis to the left of the origin O are _____. Points on the y axis which are _____ the origin O are negative.

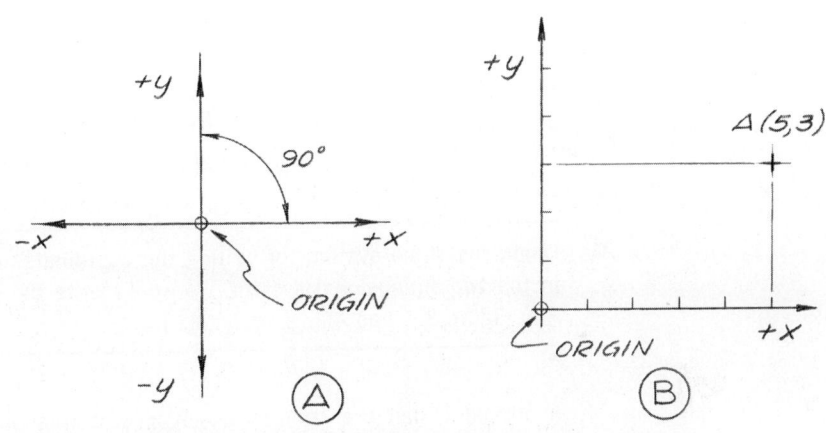

Figure 3.5 Locating a point in a 2-dimensional coordinate system.

18 Coordinates are written in parentheses as two numbers separated by a comma. They are written in the alphabetical order of the axes, x first and y second. In Figure 3.6, the coordinates of point A are written (5,3). Write the coordinates of points B, C, and D.
B: _____
C: _____
D: _____

B: (−2,4)
C: (−5,−4)
D: (0,−3)

x, y, and z

19 We are concerned, however, with points in *space*, not points on a plane surface. Let's add a third axis to the system, perpendicular to the x and y axes. See Figure 3.7A. We now can show the (+) and (−) directions of the space coordinates _____, _____, and _____.

x = 5
y = 3
z = 2

20 In Figure 3.7B, point A is described in space by the three coordinate magnitudes x = _____, y = _____, and z = _____.

58 ORTHOGRAPHIC PROJECTION/UNIT 2

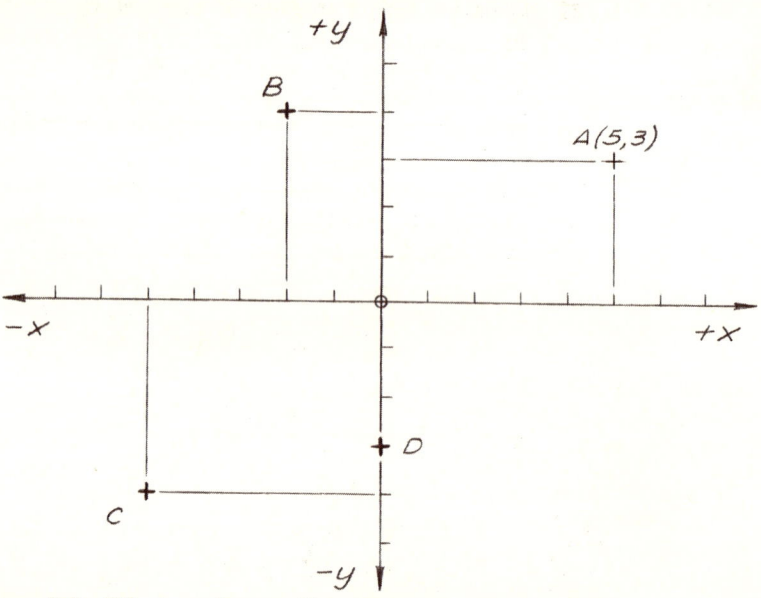

Figure 3.6 What are the coordinates of points B, C, and D?

---

21 Using the same system for writing the coordinates of a point as we did in the two-dimensional system, we would write the coordinates of point A in Figure 3.7B as A(_____).

A(5,3,2)

It is normal to list the coordinates in x, y, z (alphabetical) order. Note, however, that the position of A is described no matter what order is used in tracing a coordinate path from the origin to point A. A(z,x,y) and A(y,z,x) describe the same position in space as A(x,y,z).

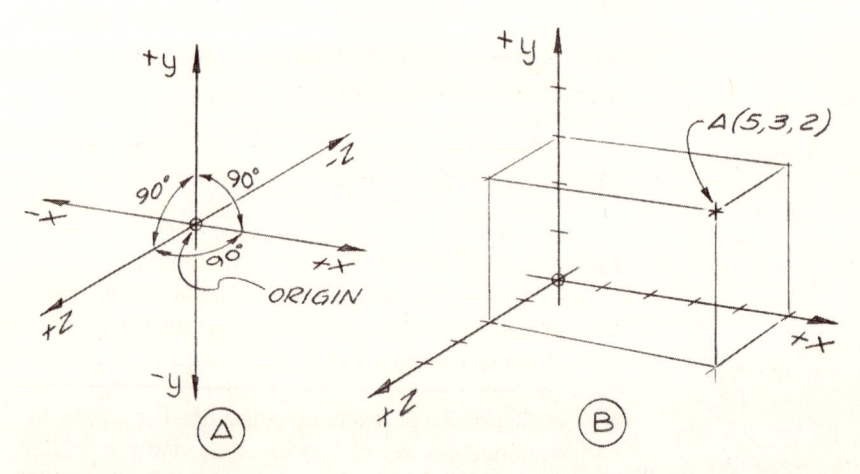

Figure 3.7 Locating a point in space using the 3-dimensional coordinate system.

PRINCIPAL ORTHOGRAPHIC VIEWS

22 What are the coordinates of points B, C, and D in Figure 3.8?
B(____,____,____); C(____,____,____); D(____,____,____).

B(3,5,−2)
C(−2,4,5)
D(−5,−3,−2)

But we are looking for a frame of reference around which we can establish a system for orthographic viewing. How can the x,y,z space coordinates be used to establish such a frame of reference?

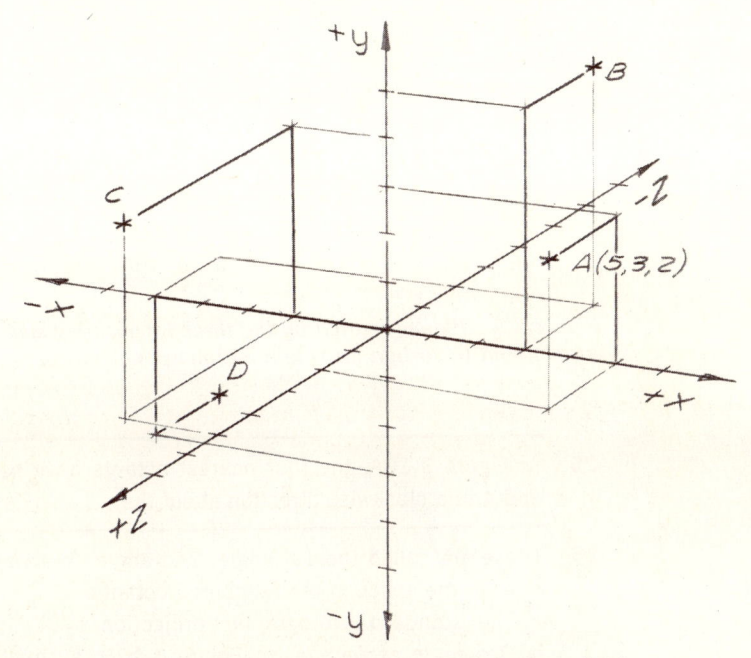

Figure 3.8 What are the coordinates of points B, C, and D?

Frame of Reference

Plane geometry shows us that two intersecting straight lines define a plane. See Figure 3.9A for confirmation of this.

23 Since the three axes x, y, and z all intersect at the origin, they define three plane surfaces, the _____ plane, the _____ plane, and the _____ plane. (See Figure 3.9B.)

xy; yz; xz
(any order)

24 These are three mutually perpendicular planes that divide the *space* around the origin into eight equal zones or octants. These zones are the spaces in which we can position an object for orthographic viewing. Each octant is formed by three planes that are _____ to each other.

perpendicular
(right angles,
90°)

60 ORTHOGRAPHIC PROJECTION/UNIT 2

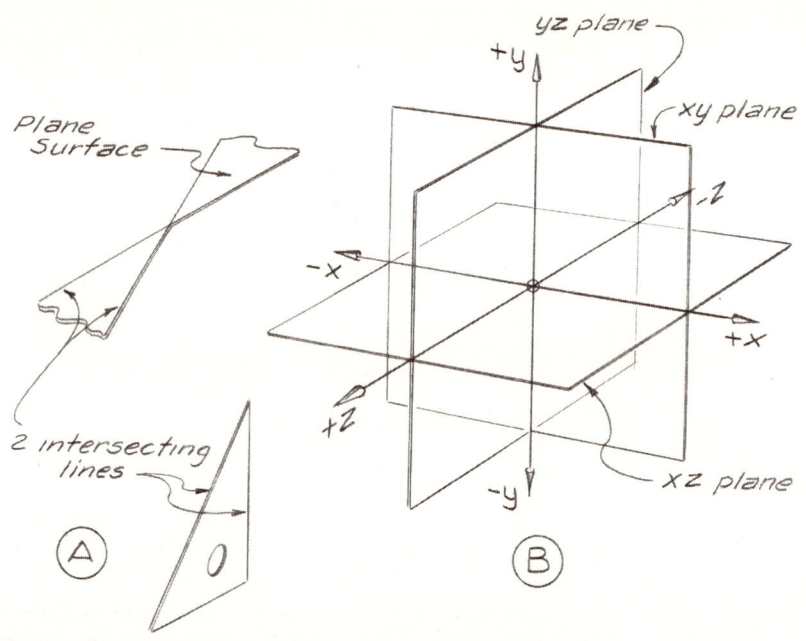

Figure 3.9 Planes created by the three intersecting space axes divide the space about the origin into eight 3-dimensional zones.

x	25	In Figure 3.10A, the four nearest octants have been numbered 1, 2, 3, and 4 in a clockwise direction about the _____ axis.
(5,−5,−5)	26	These are called the 1st angle, 2nd angle, 3rd angle, and 4th angle and refer to the *space* in the numbered octants. The standard orthographic projection system in the United States is *3rd-angle projection*. In Figure 3.10B, a small sphere (or ball) has been positioned in space 3 at x,y,z coordinates (_____, _____, _____).
perpendicular to the xy plane	27	Let's view the sphere orthographically from the *front*. This will make our viewing projection lines parallel to the z axis. What is the angular relation between these projection lines and the xy plane? _____ (your words).
intersect	28	The points where projection lines _____ the xy plane are used to define an image of the sphere.
front *(F)*	29	A circle on the xy plane is shown in the pictorial sketch (3.10B). This is the _____ orthographic view of the sphere.
parallel; xz	30	Similarly, let's view the sphere from the top. Here our projection lines are _____ to the y axis and perpendicular to the _____ plane.

PRINCIPAL ORTHOGRAPHIC VIEWS 61

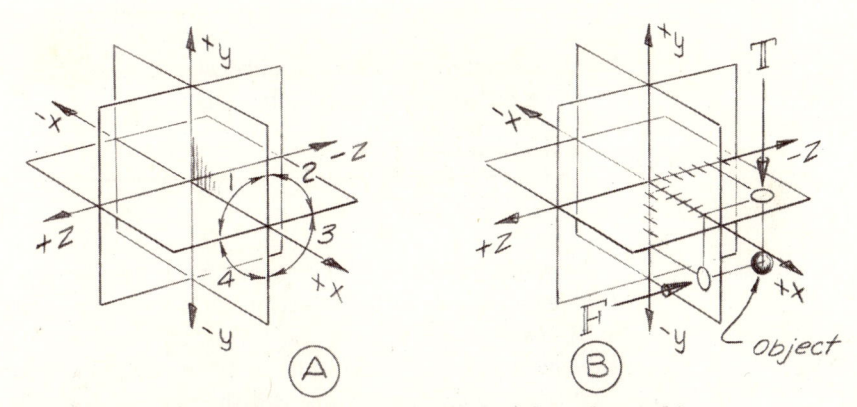

Figure 3.10 The physical description of 3rd-angle orthographic projection.

top (T)

90

top (T) and front (F)

31 The image we get on the *xz* plane is another circle which is the _____ view of the sphere.

32 Figure 3.11A shows the 3rd angle (*xy* and *xz* planes) isolated from the rest of the frame of reference. The dashed lines indicate that the *xz* or top plane is rotated _____ degrees to form a single plane surface.

33 Figure 3.11B is a true view (not pictorial) of the new plane. It is a correct, 3rd-angle, orthographic projection of the sphere.
Two views exist (_____ and _____) which, together, show the dimensions *H*, *W*, and *D* of the sphere.

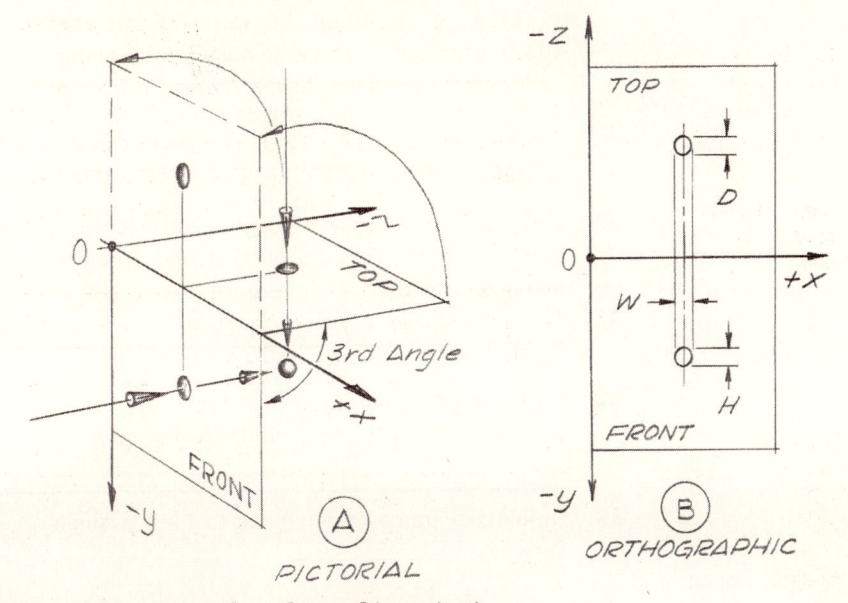

Figure 3.11 3rd-angle orthographic projection.

62 ORTHOGRAPHIC PROJECTION/UNIT 2

H and W;
W and D

34 The front view shows the dimensions _____.
The top view shows the dimensions _____.

The foregoing analysis of orthographic projection can be used for any one of the eight spaces (octants) surrounding the origin in the frame of reference. No. 3 space or 3rd angle is the United States standard and involves *looking through* transparent picture planes at an object placed in the 3rd angle.

1st-angle projection is the British and European standard. The viewing method differs in that you look at the object and project its image (or shadow) onto the picture plane. Many countries formerly using 1st angle are converting to 3rd-angle projection. The other angles are not useful in describing orthographic projection since, in unfolding the two principal planes, the images would coincide.

We will standardize on 3rd-angle projection throughout the rest of this book. Let's summarize:

perpendicular

35 Imagine a frame of reference consisting of two intersecting planes which are _____ to each other, one horizontal and one vertical.

(1) behind;
(2) below

36 An object (any object) is positioned within this frame of reference:
(1) _____ the vertical *(xy)* plane;
(2) _____ the horizontal *(xz)* plane.
(*Choose one:* above/behind/below/in front of.)

picture plane

37 View orthographically through each picture plane separately. Orthographic viewing involves parallel rays (projection lines) which are perpendicular to the _____ _____.

intersects

38 Each set of parallel projection lines _____ the picture plane and describes accurately the particular view.

plane
(2-dimensional)

39 Unfold the frame of reference to form a single _____ surface.

PRINCIPAL ORTHOGRAPHIC VIEWS 63

T (top), F (front)	40 The results are _____ and _____ orthographic views of the object.
dimensions (H,W,D)	41 Together the top and front views completely describe a three-dimensional object because they show the three principal _____ of the object.
	■ 42 Sketch 14 shows: (1) In 14A, a pictorial sketch of a rectangular object positioned in the 3rd-angle frame of reference. (2) In 14B, the xz and xy planes unfolded to form a single plane. Sketch the top and front views of the object on the diagram at B. Check your solution.

Adding a Third View

F; T (xy; xz)	43 Figure 3.12A shows a cube positioned in the frame of reference. The front (F) of the cube is parallel to the _____ (vertical) picture plane and the top (T) is parallel to the _____ (horizontal) picture plane.
orthographic projection (orthographic drawing)	44 In 3.12B, the picture planes are unfolded revealing a(n) _____ _____ of the cube.
yes	45 H, W, and D are all shown, which satisfies our requirement for picturing an object in space. But, do the two squares of 3.12B completely describe a cube? Can you imagine any other objects that would have the same top and front views? Yes _____; No _____.

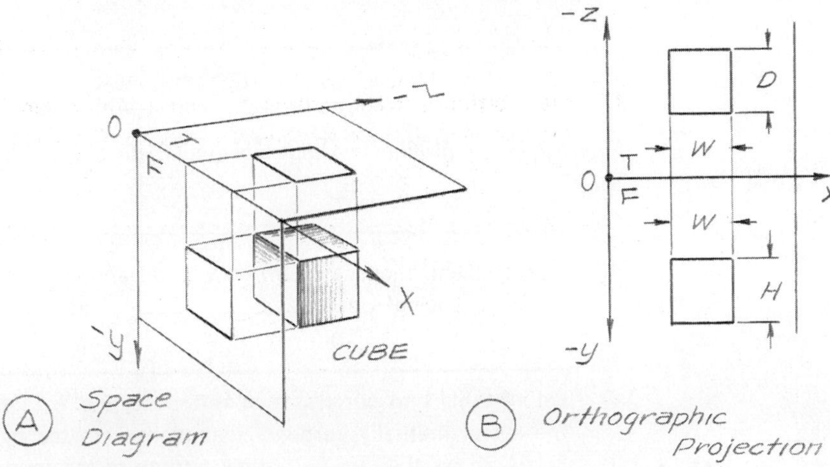

Figure 3.12   Top and front orthographic views of a cube.

64 ORTHOGRAPHIC PROJECTION/UNIT 2

Your response to the last frame should have been yes—there are other objects that have the same top and front views as a cube.

Figures 3.13A and B show two such objects. Top and front views taken in the direction of the arrows yield two-view orthographic projections identical to Figure 3.12B. Thus the two-view orthographic projection of the cube is not accurate because it can be interpreted in more than one way. Figure 3.12B is ambiguous. A study of Figures 3.12 and 3.13 should show you that a profile view of each of the objects would clear up the ambiguity since, especially in the latter two cases, the characteristic shape of the object is displayed in the profile (or side) view. See Figure 3.14.

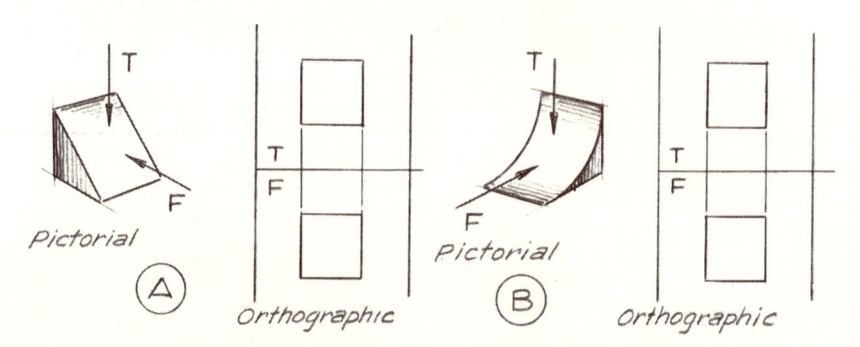

Figure 3.13 Two objects which can have top and front orthographic views identical to those of a cube.

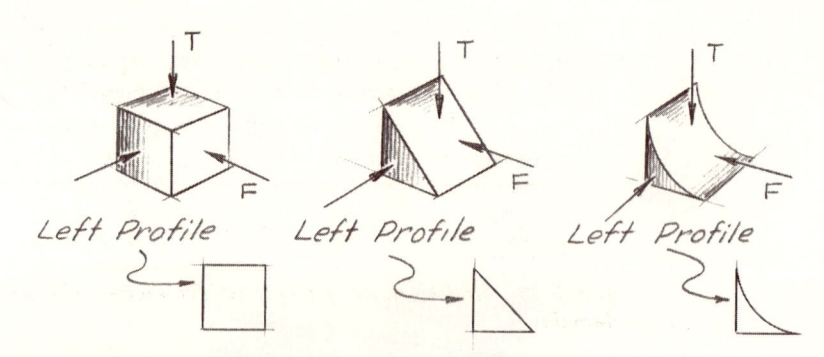

Figure 3.14 A profile view shows characteristic shape.

---

top,
front

46 Our problem, then, is "How do we add another view (a profile view) to the two principal views, _____ and _____?"

3rd

47 Let's go back to our frame of reference, the x,y,z coordinate axes and the set of mutually perpendicular planes created by these axes. Figure 3.15A shows the _____-angle portion of the frame of reference.

PRINCIPAL ORTHOGRAPHIC VIEWS 65

yz

48 In addition to the familiar top *(xz)* plane and the front *(xy)* plane, Figure 3.15A shows a third plane, the _____ plane.

parallel

49 Figure 3.15B shows a cube positioned in the 3rd angle. It is placed as before with the top and front of the cube parallel to the xz and xy planes respectively. The left side of the cube is _____ to the yz plane.

yz
(profile)

50 If we view the left side orthographically, we will get a correct orthographic profile (side) view of the cube by noting where the projection lines intersect the _____ plane.

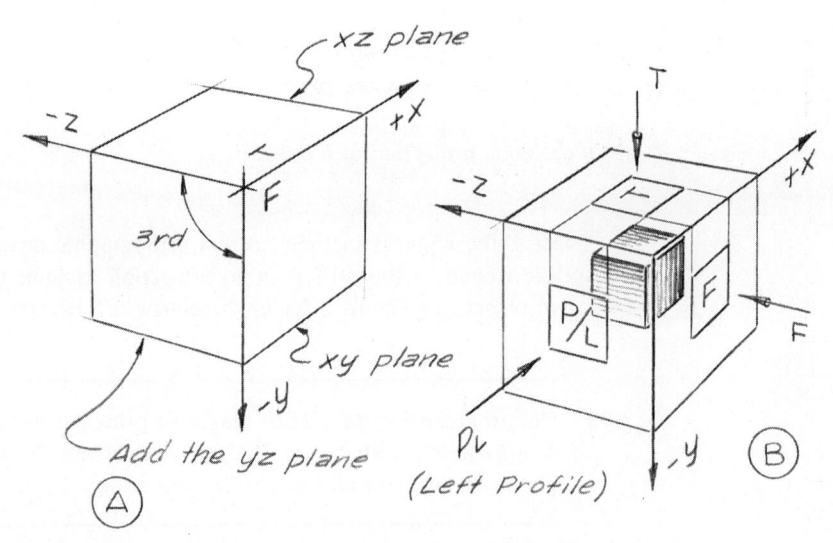

Figure 3.15 Adding the yz plane to obtain a left-profile view in the 3rd-angle.

x; y

51 Figure 3.16A shows the 3rd-angle frame of reference being unfolded to form a plane surface. The top plane *(xz)* is hinged about the _____ axis and the profile plane is hinged about the _____ axis.

squares

52 The completed orthographic views of the cube are shown in Figure 3.16B. The views are actually three _____ positioned orthographically.

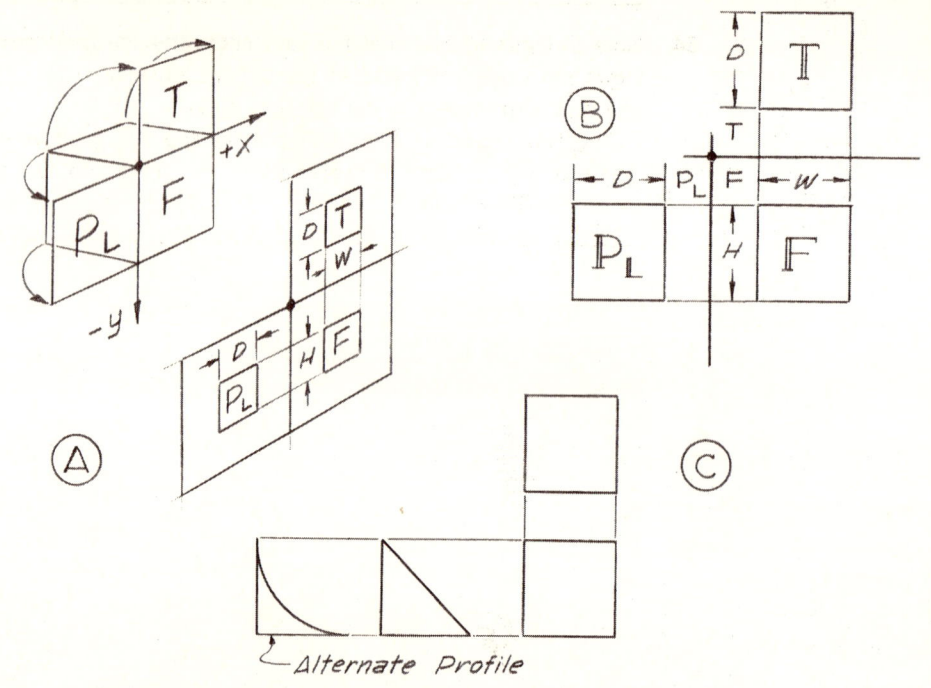

Figure 3.16 The addition of a third (profile) view clarifies the orthographic projection of a cube.

These three squares in their correct orthographic positions unambiguously describe a cube in the orthographic projection system. (Figure 3.16C shows the two objects of Figure 3.13 in three-view orthographic projection.)

H; D

53 Note that, in Figure 3.16B, the three principal dimensions of H, W, and D are each shown twice. The top view shows W and D; the front view shows W and H; and the profile view shows _____ and _____.

This introduces *redundancy* (repetition) into the drawing as far as communicating basic H, W, and D information. However, it is interesting to note that this redundancy is essential to communicate the information that the object is a cube.

As we will see later, controlled redundancy is useful in showing complicated objects clearly and unambiguously. The orthographic system of drawing lends itself to controlled redundancy because we may project and draw as many views of the object as we wish.

A three-view orthographic projection of a cube is shown in Figure 3.17A. The outlines of the picture planes have been omitted. They are not necessary to the drawing and have been used up to this point only to show the method of obtaining orthographic views.

54 Note in Figure 3.17A that the segments of the projection lines between views have been included as part of the drawing. These lines position one view with respect to the adjacent views.

The two projection lines between the top and front views define the dimension _____ and those between the front and left profile views define the dimension _____.

W; H

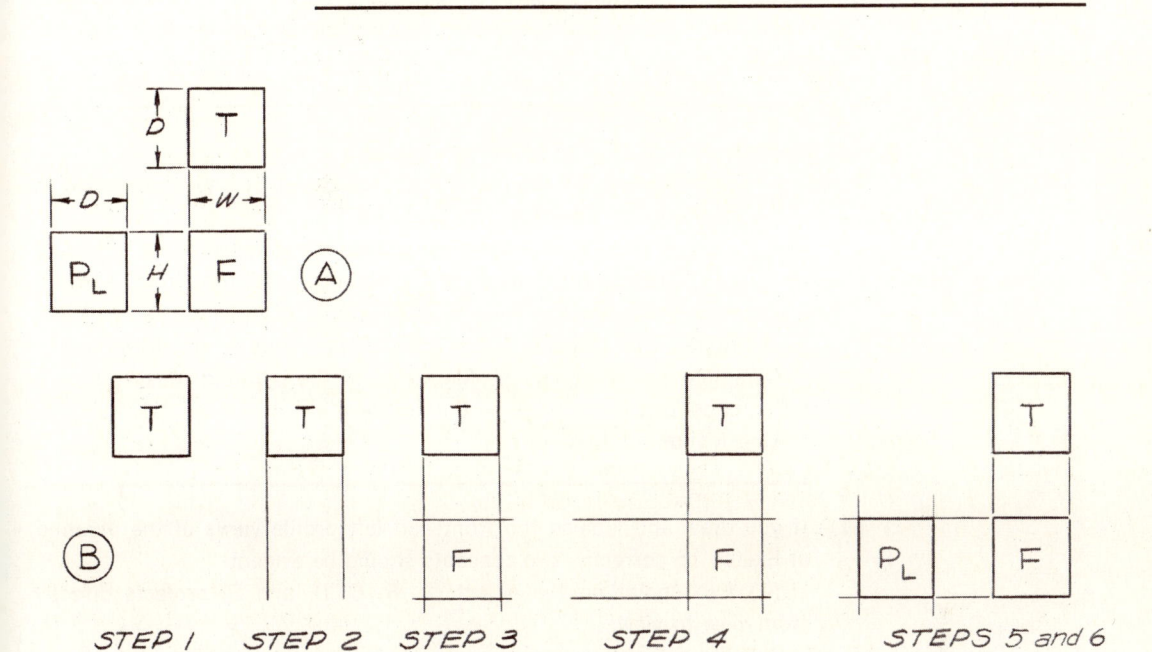

Figure 3.17 Steps in constructing three-view orthographic projection of a cube.

55 The procedure for making a three-view orthographic sketch of a cube is shown in Figure 3.17B. It is:
STEP 1: Establish the top view by drawing a square which shows dimensions _____ and _____.

D; W

56 **STEP 2:** Extend the two lines that represent the two sides of the cube downward. These lines project the W dimension into the _____ view.

F (front)

57 **STEP 3:** Sketch two horizontal parallel lines which cross the previous projection lines at a convenient distance below the top view. The distance between these parallel lines is the dimension _____. The _____ view is now defined by a square of dimensions H × W.

H; F

68 ORTHOGRAPHIC PROJECTION/UNIT 2

D

58 **STEP 4:** Extend the H dimension lines a convenient distance to the left of the front view.
STEP 5: Across these lines, sketch two vertical parallel lines that *repeat* the dimension _____.

profile (P_L)

59 **STEP 6:** This creates another square, the _____ view. Darken the three squares and the orthographic projection of the cube is complete.

■ 60 Refer to Sketch 15. The following are given:
 (1) A pictorial sketch of a square-based pyramid positioned in the frame of reference.
 (2) A pictorial sketch of the frame of reference being unfolded.
 (3) Orthographic views of the three planes, top, front, and left profile, with the top view of the pyramid complete. The front and profile views of point A, the apex of the pyramid, are drawn.
 Complete the front and profile views of the pyramid making $H = 5$ units. Label points B, C, D, and E in all views.
 Check your solution.

61 If you drew and labeled the front and left profile views of the pyramid of Sketch 15 correctly, two concepts should be evident:
 (1) Each point on the object (A, B, C, D, and E) *projects* directly from view to view.
 (2) The projection lines are _____ to the coordinate axes or folding lines as they are more familiarly called in technical drawing.

perpendicular (90°)

These two concepts are fundamentally important to orthographic projection. Any point projects directly from any one view to any adjacent view, and the projectors are always perpendicular to the coordinate axis (folding line) that separates the two views. As long as these two rules are observed, a true orthographic projection will result no matter how many views are drawn.

THE SIX PRINCIPAL ORTHOGRAPHIC VIEWS

There is no limit to the number of orthographic views that can be drawn of any physical object. In other words, an infinite number of views are possible. So far we have shown how three orthographic views are developed from a standard frame of reference.

PRINCIPAL ORTHOGRAPHIC VIEWS 69

62 In the orthographic system, two views are the minimum number since it takes two views to describe an object completely by _____ _____ (your words).

showing H, W, and D

63 A third view is redundant as far as showing H, W, and D of an object but is often needed to eliminate ambiguity. This is true in the case of the _____.

cube

64 The frame of reference has served well to give us a device for making an accurate drawing of an object in space. It consists of three mutually perpendicular _____ _____, xy, yz, and xz.

picture planes

65 In Figure 3.18A, the framework of the three regular picture planes has been modified to form a transparent box by adding three more planes; bottom, rear, and right profile.
The bottom plane is parallel to the _____ plane; the rear is parallel to the _____ plane; and the right profile is parallel to the _____ plane.

T (xz);
F (xy);
P_L (yz)

66 Let's now consider this closed transparent box as our frame of reference. In Figure 3.18B, a cube has been positioned within this frame of reference and _____ orthographic views have been projected onto the _____ sides of the box.

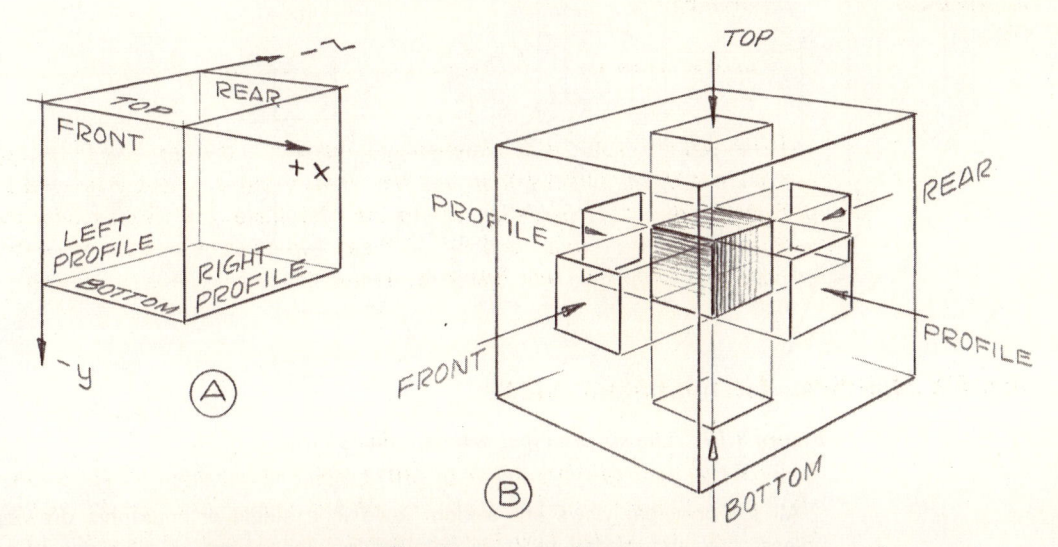

six; six

Figure 3.18 Adding three more planes to create a transparent box out of the frame of reference.

70 ORTHOGRAPHIC PROJECTION/UNIT 2

F

67 Figure 3.19A shows the box being unfolded. The _____ view is kept in its original (vertical) position.

68 When the six sides have been unfolded into a single plane surface, we get a six-view orthographic drawing as shown in Figure 3.19B. The six orthographic views are:

top (T); front (F);
left profile (P_L);
right profile (P_R);
bottom (B);
rear (R)

two

69 These six views are called the *principal orthographic views*. Any _____ adjacent views will show the basic dimensions of H, W, and D.

redundant
(repetitive)

70 The use of more than two views adds _____ information to the drawing.

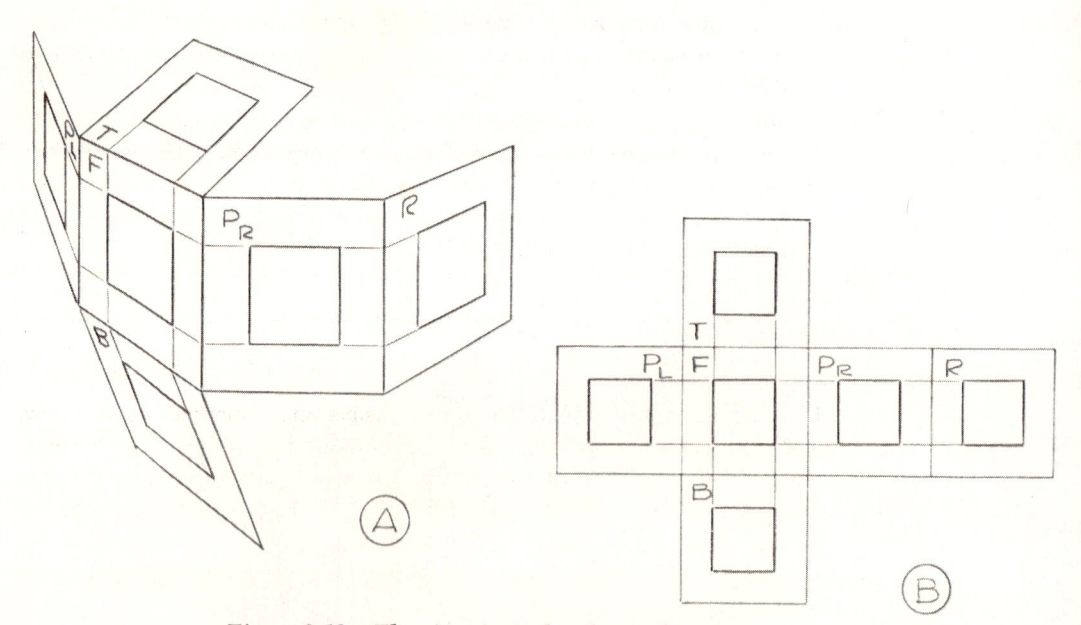

Figure 3.19 The six principal orthographic views.

All six principal views are seldom used in a single orthographic drawing unless they are needed to show detailed information on all six sides of an

PRINCIPAL ORTHOGRAPHIC VIEWS   71

object. Figure 3.20 is a simple example of such a case. The detail on the six sides of a child's block cannot be shown orthographically without drawing all six principal views.

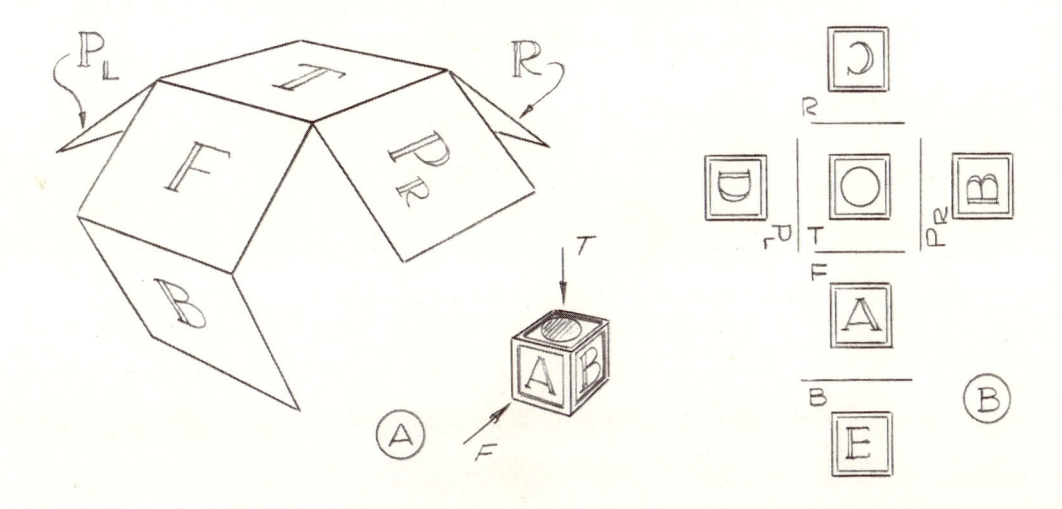

Figure 3.20   One alternate method of unfolding the frame of reference.

---

71   The transparent box can be unfolded in a variety of ways. Figure 3.20 is an example. Comparing this example with Figure 3.19, we see that the two profile planes are now projected from the _____ view rather than the _____. Also the rear view is projected from the _____ rather than the _____.

T; F;
T; P_R

This change in view orientation is permissible and sometimes useful. However, it does have one drawback when picturing a familiar object in the orthographic system. If the profile views are projected from the top view rather than the front view, the bottom of the object in these views is oriented unnaturally on the paper. Figure 3.21 illustrates this. It is more natural to see the bottom (water line) of the cabin cruiser as a horizontal line rather than a vertical line.

Bottom and rear views can be placed wherever convenient as long as they represent direct projections from adjacent views. For example, the rear view cannot be projected directly from the front view, but it can from the top, bottom, or profile views.

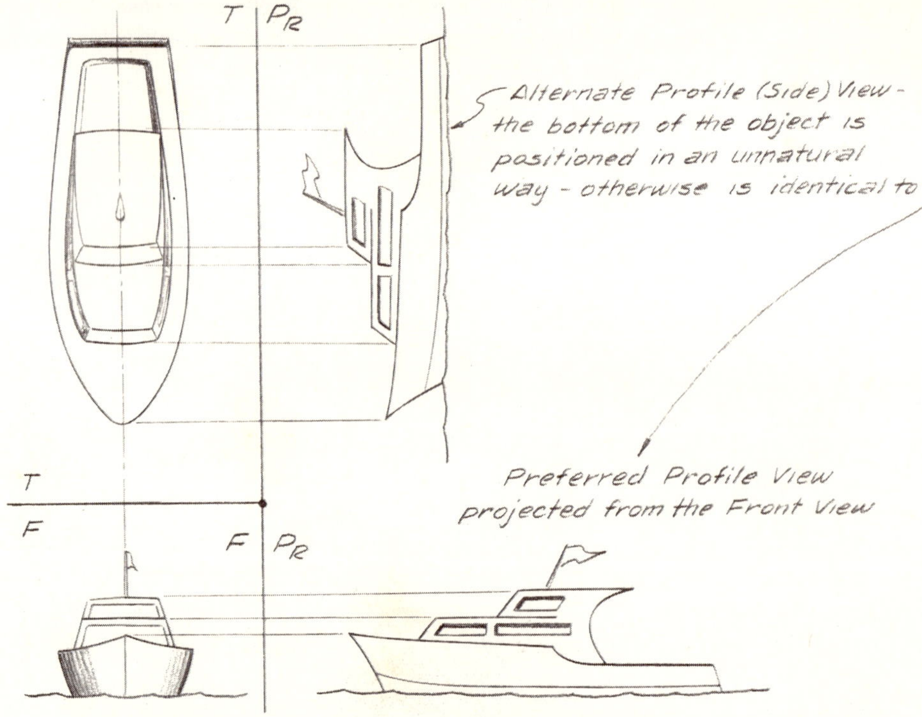

72 ORTHOGRAPHIC PROJECTION/UNIT 2

Figure 3.21 Alternate positions of the profile view.

■ **72** Sketch 16 is an exercise covering all the material of this chapter.
   Make a six-view orthographic projection of the object shown pictorially. Note that the front and top views are marked. Point A is at the center of the top sloping face. Find point A in all six views by projection.
   Check your solution.

SUMMARY

This chapter has been a study of the basis for making an orthographic projection of a physical object. Figure 3.22 is a graphic summary of the chapter. The process of making an orthographic drawing involves:

(1) Visualizing a standard frame of reference with the object placed inside (Figure 3.22A).

(2) Visualizing orthographic (right-angled) viewing of the object and subsequent recording of images on the picture planes which comprise the frame of reference (Figure 3.22B).

(3) Visualizing the frame of reference as being unfolded to bring all picture planes into a single plane surface. This surface is the drawing surface and the combined images are an orthographic projection (Figure 3.22C).

This system assures us that every point is projected directly from view to view and that the projection lines are perpendicular to the folding lines (coordinate axes) between adjacent planes.

PRINCIPAL ORTHOGRAPHIC VIEWS 73

SIX-VIEW ORTHOGRAPHIC PROJECTION

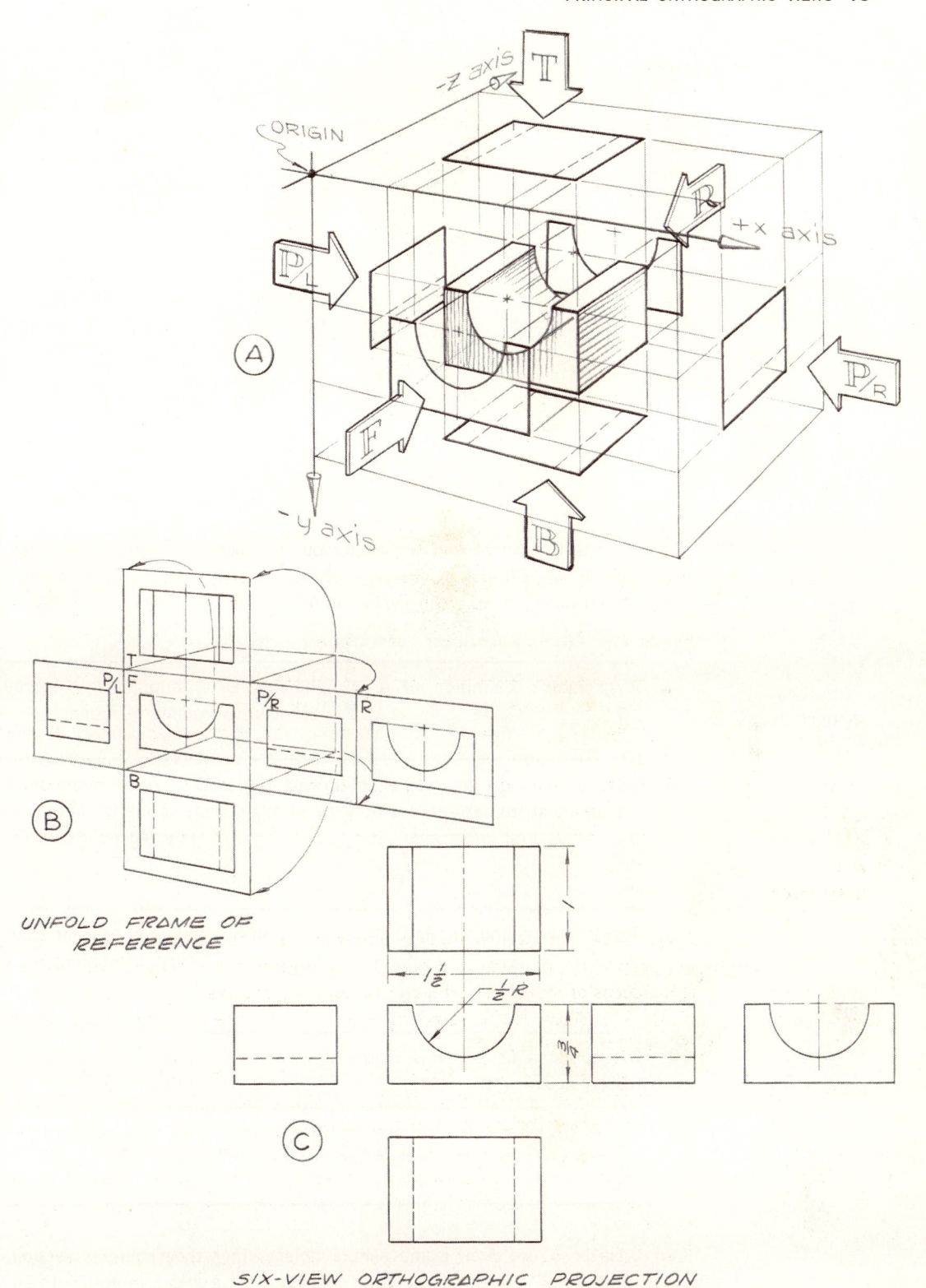

Figure 3.22 Summary: The basis for orthographic projection.

chapter 4   *straight lines*

Let us pause a moment and reconsider our concept of straight lines. Then, in Chapter 5, we will look at auxiliary orthographic views and sectioned views to round out our study of orthographic projection.

shortest distance

1. A geometric definition of a straight line is: A straight line is the _____ _____ between two points in space.

plot (place)

2. After introducing the dot-to-dot drawing (Figure 1.1) of an automobile, we said that we can make a drawing of any object, no matter how complex, if we first know where to _____ selected points on the object.

straight lines

3. The theory behind this is that a succession of points can be connected with _____ _____ which define the object on a two-dimensional surface.

short

4. Curved lines fit into this theory if we consider that a curved line is merely a series of _____ successive straight lines.

accurate (precise)

5. As we decrease the distance between points (make the straight-line segments shorter) the curved line becomes more _____.

In Chapter 3, we drew some simple objects in orthographic projection. Recall Sketch 12 in which you plotted 12 points on a grid to obtain a three-view orthographic projection of a cube. Theoretically, the orthographic

STRAIGHT LINES 75

sketches were obtained by plotting points and then connecting the points with straight lines. Actually, however, it is easier and more useful to consider that an object is made up of *lines*. We might unconsciously work with points, but the lines are of the most visual interest to us since, combined, they constitute the drawing.

VISIBLE LINES

We might ask here, "When should a straight line be included in an orthographic view?" There are three conditions under which a straight line appears (is visible) in an orthographic view:
    (1) The edge view of a plane surface appears as a straight line.
    (2) The intersection of two plane surfaces is a straight line.
    (3) The profile, or contour, of certain curved surfaces yields straight lines.

---

6 Study Figure 4.1. It shows the top and front views of a composite shape (a circular cylinder and a trapezoidal block). A pictorial sketch is also given. Four lines have been labeled A, B, C, and D.

On the table below, check one or more of the conditions that cause lines A, B, C, and D to be seen as straight lines in the orthographic views:

	A	B	C	D
(1) Edge view of a plane	___	___	___	___
(2) Intersection of two planes	___	___	___	___
(3) Profile of a curved surface	___	___	___	___

(1) B and C
(2) B and D
(3) A

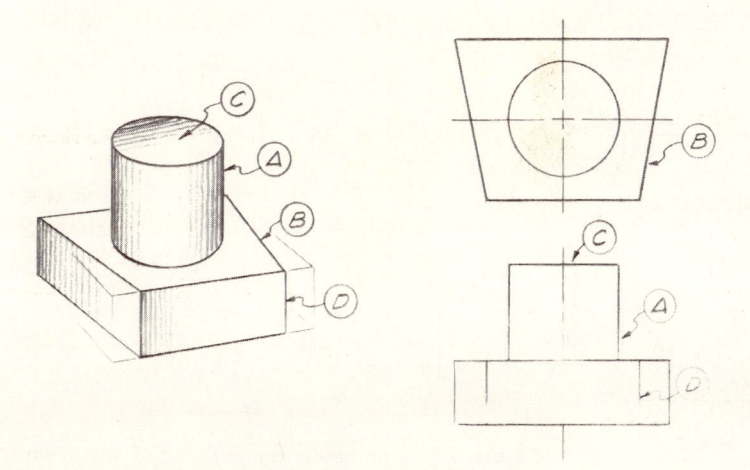

Figure 4.1   *Visible lines in orthographic views.*

Any object can be reduced to one, or a combination, of five basic shapes. These are shown in Figure 4.2. The five shapes are:
- (1) Rectangular polygon (or prism)
- (2) Cylinder
- (3) Cone
- (4) Sphere
- (5) Triangular polygon (or prism)—a pyramid

profile (contour)

7 Some of the lines on the objects of Figure 4.2 have been lettered. Lines labeled A show that the _____ of curved surfaces such as the cone and the cylinder yield straight lines in an orthographic view.

edge view

8 Condition B (lines labeled B) shows that a plane surface appears as a straight line if we see a(n) _____ _____ of the plane.

intersection; plane

9 In Condition C, the _____ of two _____ surfaces creates a straight line.

no

10 Cones and cylinders are curved surfaces that have straight-line profiles. Figure 4.2 shows another curved surface, the sphere. It has _____ straight lines no matter how it is viewed.

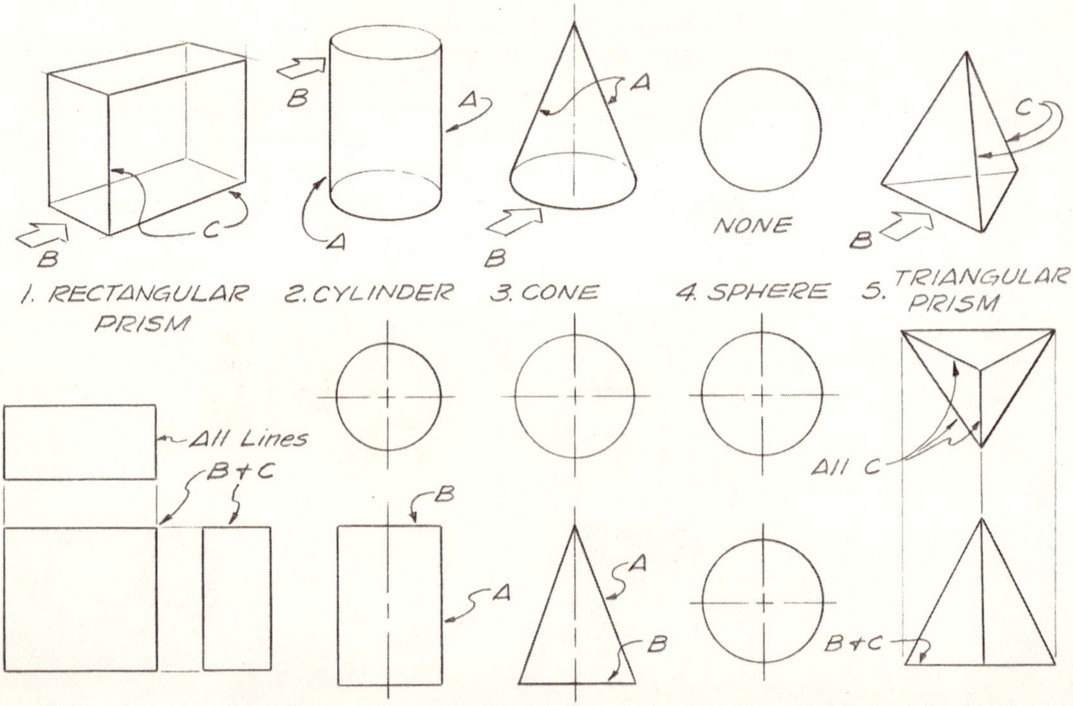

Figure 4.2 *Any object is composed of one or more of these five basic shapes.*

STRAIGHT LINES 77

	11	Cones and cylinders are single-ruled surfaces (surfaces formed by a moving straight line) and have _____-line profiles. A sphere is a double-ruled surface (a surface formed by a moving curved line) and thus has no straight lines.
straight		

HIDDEN LINES

We have been considering visible lines. Consider Figure 4.3. The triangular pyramid of Figure 4.2 has been reoriented and drawn orthographically. The object consists of six straight lines, AB, AC, AD, BC, CD, and DB. Some of these lines are not visible in the reoriented views.

	12	All the straight lines are formed by one or the other of the two conditions, _____ view of a plane or, _____ of two planes.
edge; intersection		
edge; ABC, ADC	13	In Figure 4.3A, one of the sloping sides (ABD) now shows as a(n) _____ in the front view. The sloping line AC in the front view represents the intersection of planes _____ and _____.
no	14	Is line AD visible in the front view of 4.3A? Yes_____ No_____
behind	15	Line AD is not visible because it coincides with line AB. Line AD is _____ line AB. (Choose one: in front of/behind.)

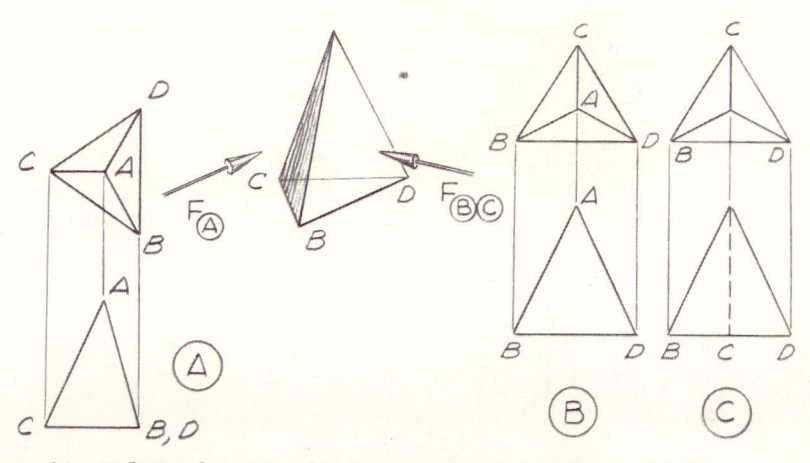

Figure 4.3 Relation between object orientation and visible and hidden lines.

78 ORTHOGRAPHIC PROJECTION/UNIT 2

visible

16 Figures 4.3B and C show yet another orientation of the prism. In 4.3B one of the sloping lines (AC) appears only in the top view. It does not appear in the front view because it lies on the rear side of the pyramid and therefore is not _____ from the front.

In 4.3C, line AC is included as a *dashed line*. This is called a *hidden line*. Theoretically, it should not be included in the drawing. If we look at a solid pyramid as in Figure 4.3C, we would not see line AC in the front view since it lies on the far side of the object.

However, it is standard procedure in orthographic drawing to show hidden lines in any view where, by doing so, we make the drawing clear and unambiguous. In Figure 4.3C, this line is included to clarify the drawing. It is a hidden line and therefore must be drawn as a dashed, rather than a solid, line.

dashed
(hidden)

17 Figure 4.4 shows some examples of hidden lines. In Figure 4.4A a pictorial sketch of a cube is shown. The three lines on the far side of the cube are shown as _____ lines to indicate that they are not normally seen.

visible
(seen)

18 In 4.4B a circular vertical hole has been drilled through the cube. In the front view, the sides of the hole are hidden lines since they exist within the center of the cube and will not be _____ from the outside.

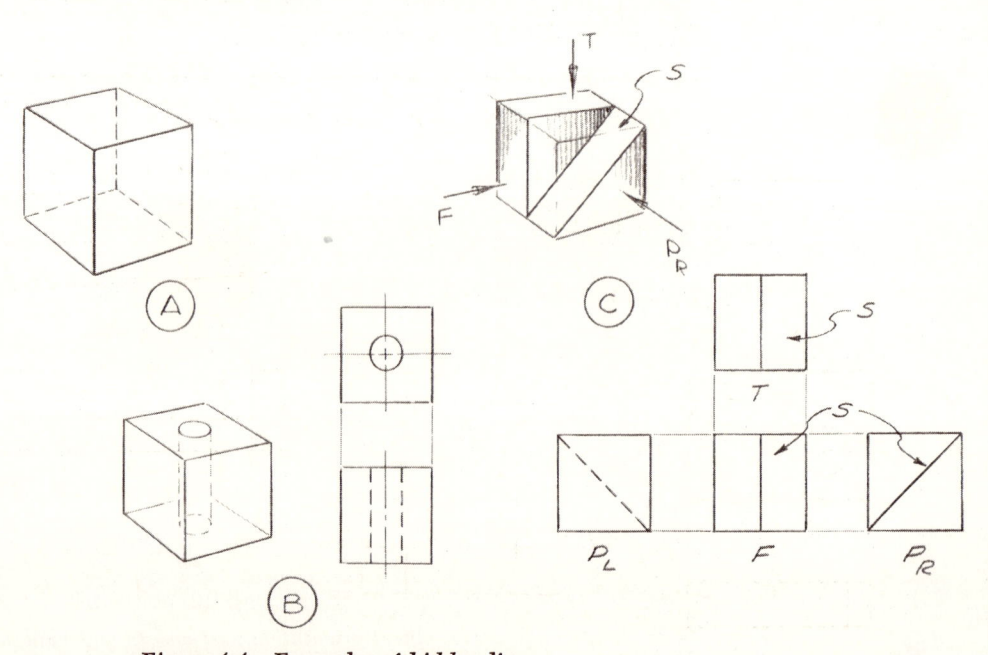

Figure 4.4 Examples of hidden lines.

STRAIGHT LINES

profiles
(contours)

19 The dashed lines in the front view of 4.4B are straight lines under the condition that a cylinder has straight-line _____. The cylinder in this case is a negative cylinder, that is, it is not a physical object, but a negative volume shaped like a cylinder and contained within the cube.

T (top);
F (front);
P_R (right profile)

20 The object of Figure 4.4C must have three orthographic views to describe it unambiguously. The three best views to choose are _____, _____, and _____.

edge

21 The right profile is chosen over the left to avoid a hidden line. If the left profile is drawn, the sloping dashed line represents a(n) _____ view of the sloping plane S.

CENTERLINES

Another standard broken line found on most orthographic drawings is the centerline. Figure 4.5 shows four examples of the use of centerlines on orthographic views.

long;
short

22 A centerline should always be included when showing circular elements, whether the part is positive (a cylinder) or negative (a hole). The centerline is drawn by making alternate _____ and _____ dashes. This is standardized and has specific meaning to the makers and users of orthographic drawings.

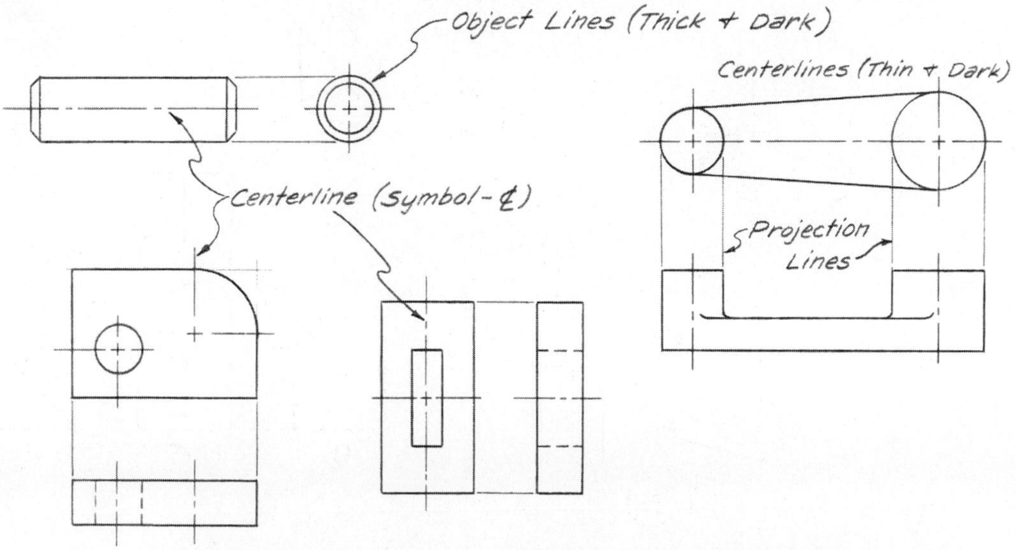

Figure 4.5   Centerlines—a standard line form in orthographic projection.

80 ORTHOGRAPHIC PROJECTION/UNIT 2

short

23 Note in Figure 4.5 that in views that show circles as true circles, the center is marked by a cross made up of two _____ dashes. This is standard procedure in the use of centerlines.

■ **24** Sketch 17 gives you an opportunity to select and draw visible lines, hidden (invisible) lines, and centerlines. A completed example showing accepted procedures and techniques is given. A display of accepted junctures for hidden lines and a table of line precedences are shown. Study this example and then complete the orthographic sketch of the stop block including all missing visible and hidden lines and centerlines. Check your solution.

chapter 5   object orientation, auxiliary views, and sectional views

To this point in our study of orthographic projection we have:
    (1) Developed a means for projecting the six principal orthographic views (Chapter 3)
    (2) Reconfirmed the visibility of lines in orthographic views and studied the use of standard line convention (Chapter 4)

Let us now look at the flexibility we can gain by
    (1) Varying the orientation of the object within the standard frame of reference
    (2) Projecting auxiliary views where needed
    (3) Taking sectional views to show internal details

Object orientation and auxiliary views will be discussed more thoroughly in Chapters 15 and 16, which take up the basic operations of descriptive geometry. The discussion in this chapter will be confined to the need for picturing physical objects in the orthographic form.

OBJECT POSITION AND ORIENTATION

1   Figure 5.1 shows a pictorial (A) and an orthographic (B) sketch of a cube positioned in the standard frame of reference.
    The front face of the cube is _____ to the front reference plane.

parallel

82 ORTHOGRAPHIC PROJECTION/UNIT 2

2 One corner of the cube has been designated Point A. The top, front, and left profile views of Point A are shown in both sketches. Dimensions *m*, *n*, and *p* relate point A to the three picture planes. Complete the following:

Point A is:
(1) _____ distance behind the front plane
(2) _____ distance below the top plane
(3) _____ distance to the right of the left profile plane

(1) *m;*
(2) *n;*
(3) *p*

3 These three dimensions are quite evident in the pictorial (realistic) sketch but somewhat less evident in the orthographic (abstract) drawing.

In the pictorial form, distances *m*, *n*, and *p* can be visualized as the _____ distances from point A on the object to the three reference planes.

perpendicular

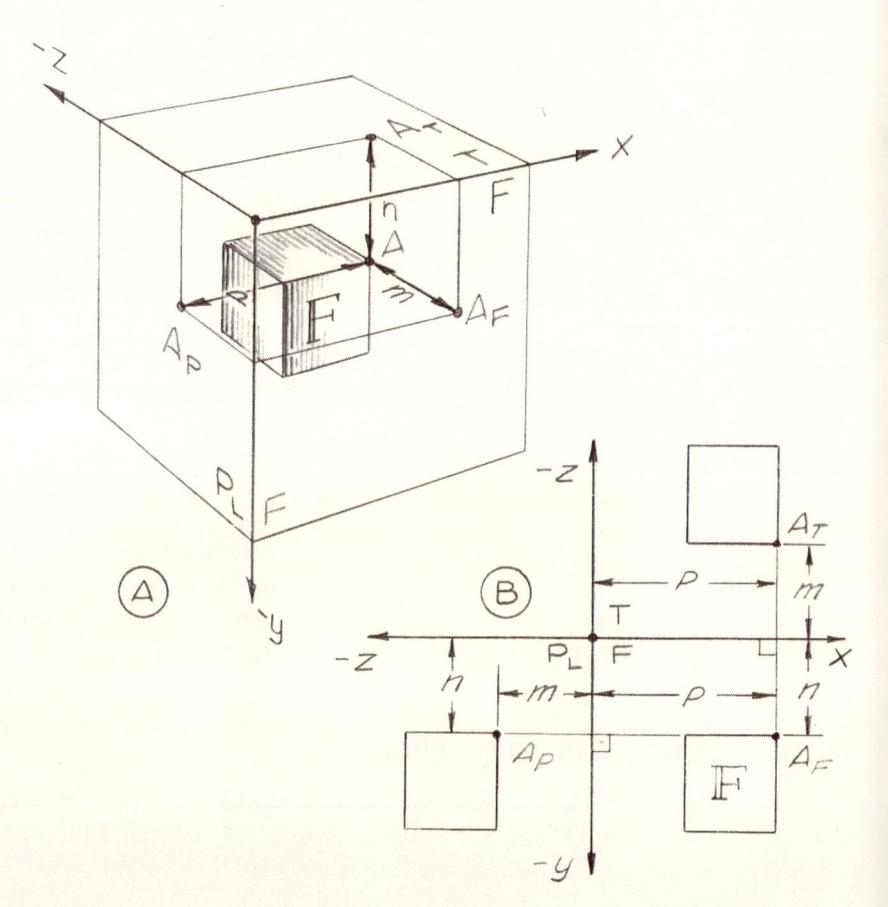

Figure 5.1 Relation between an object and the T, F, and P reference planes.

OBJECT ORIENTATION, AUXILIARY VIEWS, AND SECTIONAL VIEWS 83

axes
(folding lines)

4 However, in the orthographic drawing (5.1B), these planes are not evident because they have been folded out to form a single plane—the drawing surface. The only parts remaining of the three-dimensional frame of reference are the three space _____, x, y, and z.

edge
(straight-line)

5 When we look at the top view of 5.1B, the x axis (folding line) can be visualized as an *edge view* (straight-line view) of the front *(xy)* reference plane in the same manner that face F of the cube shows as a(n) _____ in the top view.

profile *(yz)*

6 Looking at just the top view, the y axis is visualized as an edge view of the _____ reference plane.

7 Complete the following table by checking the appropriate blanks:
(1) When looking at just the *front* view, which axis (or folding line) represents an edge view of the

	Top?	Front?	Profile?
x axis			
y axis			
z axis			

(2) When looking at just the *profile* view, which axis represents an edge view of the

	Top?	Front?	Profile?
x axis			
y axis			
z axis			

(1) x axis: top *(T)*;
y axis: profile *(P)*
(2) y axis: front *(F)*;
z axis: top *(T)*

Thus distances m, n, and p represent the shortest (perpendicular) distances from point A to the front, top, and left profile planes respectively. These three distances also control the distance between views in the orthographic sketch of Figure 5.1B.

The distance between views is not of great concern when we use the orthographic form for picturing objects. Later, however, in Chapter 15 when we begin the study of descriptive geometry, we will need to know and use these dimensions.

For the present, distance between views is a matter of convenience and usually the folding lines or axes are not even drawn. You should always be aware of them, however, and realize that the projection lines between views are always perpendicular to the folding lines, imaginary or real.

Figures 5.2A and B show two different orthographic sketches of a cylinder. The distances between views are different but both A and B are correct and unambiguous drawings of a cylinder.

84 ORTHOGRAPHIC PROJECTION/UNIT 2

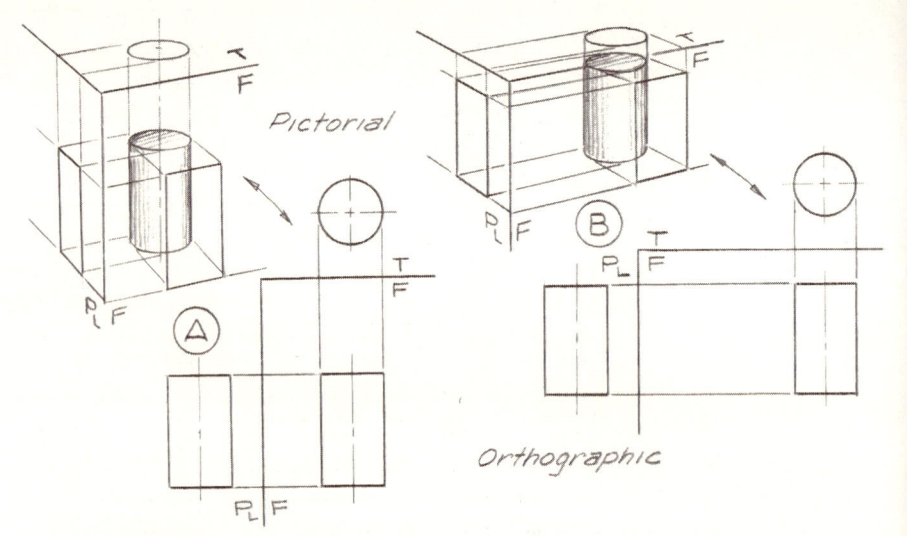

Figure 5.2 The effect of changing the position of the object within the orthographic frame of reference.

position
(location)

8 The distances between views varies in Figures 5.2A and 5.2B because the _____ of the cylinder within the frame of reference is varied.

9 We may also change the *orientation* of the object within the frame of reference. Study Figure 5.3. It shows two orthographic sketches of the same cylinder.
In 5.3A, the front view of the cylinder appears as a(n) _____.
In 5.3B, the front view appears as a(n) _____.

(5.3A) rectangle;
(5.3B) circle

10 It is still the same cylinder. We have merely re-_____ it with respect to the standard top, front, and profile views.

(re-)oriented

Note that changing the orientation of the cylinder does not change the position of the top and front orthographic views. We are merely redefining the top and front of the cylinder. *The frame of reference never changes.* Only the manner in which we position and orient the object within the frame of reference changes.

■ **11** Sketch 18 is an exercise in positioning an object in the frame of reference. The object is a triangular prism with H:W:D = 1″:1″:1″. Four arrows show the viewing directions for four separate *top* views. In the spaces provided, sketch the following:

(1) Four 2-view orthographic drawings of the object according to the specified top-view orientations. Include hidden lines.

(2) A profile view wherever needed to provide an unambiguous drawing (one showing the triangular shape characteristic of the object). Check your solution.

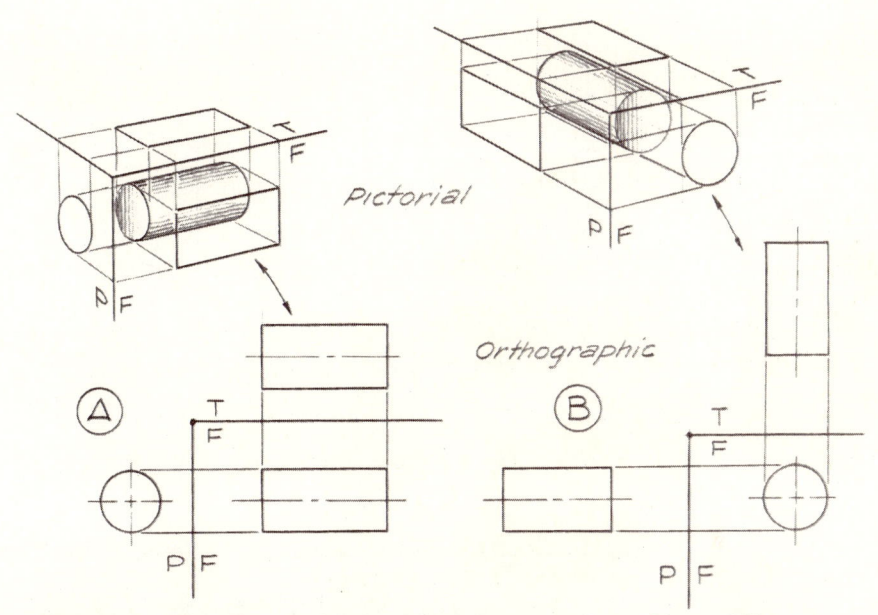

OBJECT ORIENTATION, AUXILIARY VIEWS, AND SECTIONAL VIEWS 85

Figure 5.3 The effect of changing the orientation of the object within the frame of reference.

AUXILIARY VIEWS

In Chapter 3 we developed a theory for drawing six principal orthographic views. We stated that these were the most common views but that an infinite number of orthographic views may be drawn of any object. Any view other than one of the six principal views is called an auxiliary view.

principal	12	Figure 5.4A shows a golf ball surrounded by six arrows that represent the six _____ orthographic views.
reference planes (picture planes)	13	In Figure 5.4B, the ball has been positioned in the frame of reference and the six views recorded on the _____ _____.
2; 3	14	Figure 5.4C shows the frame of reference unfolded to yield an orthographic projection of the golf ball. Actually, there is no need to have six views to describe the ball; _____ views would be enough for an intuitively correct drawing and _____ views would make it technically correct. (Remember the cube? A sphere presents the same problems.)
sphere	15	How many orthographic views of the golf ball can we take? Figure 5.5A shows the ball enclosed in a transparent _____. The six principal views are shown by arrows.
greater (larger)	16	Considering the space around the ball, we could take an infinite number of views. Seven of these views taken at random are shown by numbered arrows. If we filled the surface of the sphere with images, we could obtain still more by using a sphere of _____ diameter.

86 ORTHOGRAPHIC PROJECTION/UNIT 2

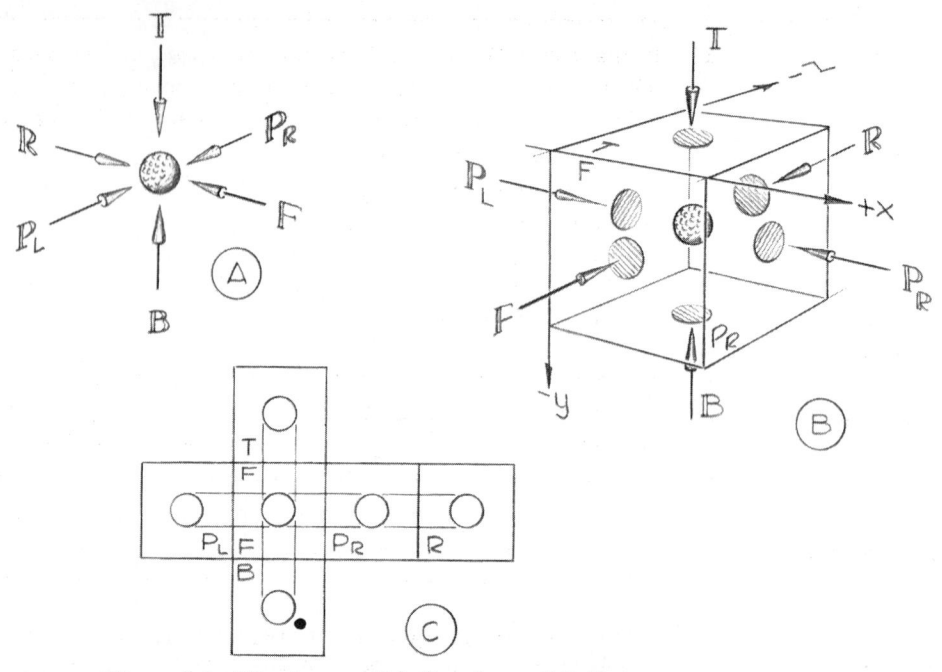

Figure 5.4 Viewing a golf ball orthographically.

---

17 Views 1 through 7 are auxiliary views. Each one represents a different aspect of the ball as viewed from a particular point with respect to the orthographic frame of reference. Any orthographic view taken in a direction other than the six principal orthographic directions is called a(n) _____ view.

auxiliary

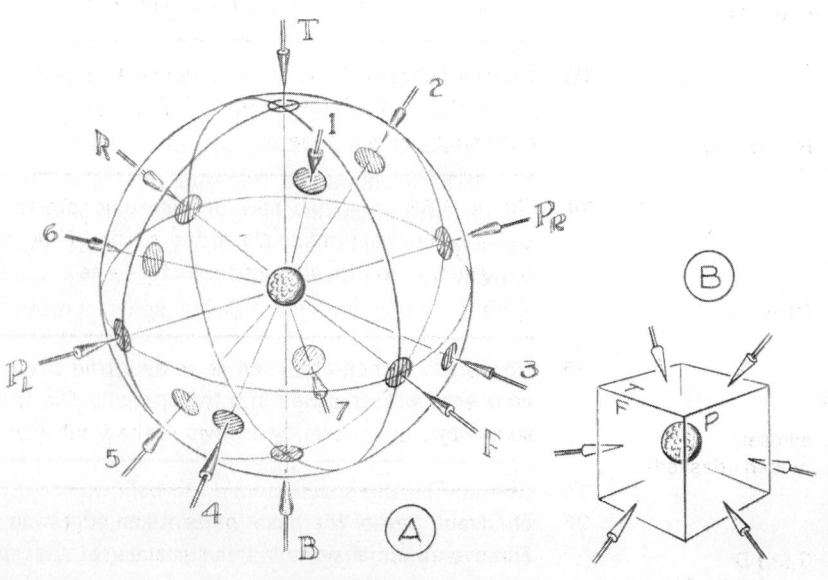

Figure 5.5 One concept of projecting auxiliary views.

OBJECT ORIENTATION, AUXILIARY VIEWS, AND SECTIONAL VIEWS 87

center	18 If the sphere were the frame of reference, we could establish any auxiliary view by defining geometrically a point on the surface of the sphere and then directing a projection line through the point to the _____ of the sphere.
picture (reference)	19 However, the orthographic frame of reference is not a sphere, but a cube (or rectangular prism) as in Figure 5.5B. If we attempt to take views in directions other than the principal directions, we find that we are looking obliquely at the _____ planes.
right(-angle)	20 Orthographic viewing is, by definition _____-angle viewing.
perpendicular	21 Therefore, we must modify our concept of the frame of reference to admit additional reference planes that are _____ to any desired direction of view.

These will be auxiliary views. The manner in which they are constructed is the subject of this chapter. We have stated that three orthographic views would be enough for a technically correct drawing of the golf ball. Each of the many auxiliary views of Figure 5.5A is essentially the same. None of them adds any new *information* about the golf ball that is not contained in any three of the principal views.

information	22 In a negative sense, this fact answers an appropriate question, "Why do we need auxiliary views?" Stated positively, we construct selected auxiliary views to add _____ to a drawing that is not shown or is not clear in any of the six principal views.
$H = W = D$	23 Consider Figure 5.6A. It is a perspective sketch of a block with an inclined face. A shallow hole is drilled perpendicular to the face. The enclosing box is a cube so _____ = _____ = _____.
front	24 Figure 5.6B is a two-view orthographic projection of the block. Two views completely define the basic shape of the block as long as one of the views shows its sloping character—the _____ view in this instance.
ellipse; hidden (dashed)	25 Note that the hole does not show as a true circle in either view. The top view shows it as a(n) _____. The true depth of the hole is shown by _____ lines in the front view.
S ($\times$) D (read S by D)	26 The front view of the block presents an edge view of the sloping surface. The overall dimensions of this surface are _____ $\times$ _____.

88 ORTHOGRAPHIC PROJECTION/UNIT 2

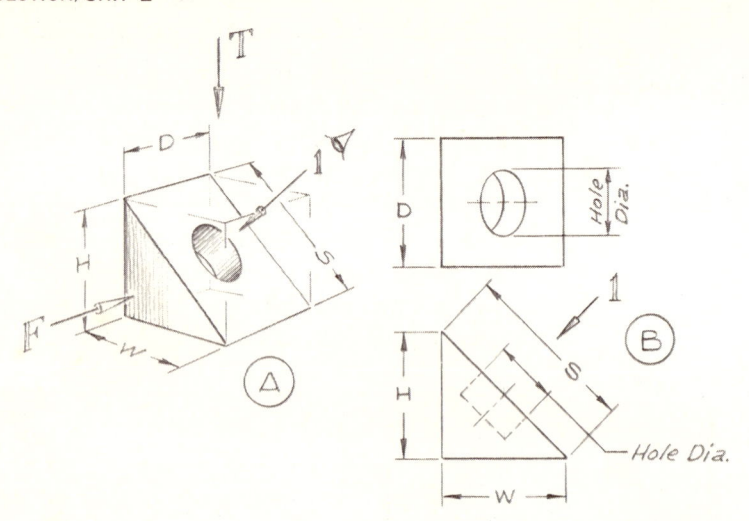

Figure 5.6 None of the six principal views will show the true shape of the hole.

perpendicular

27 We can imagine that a view in direction 1, _____ to the sloping face, should give us a picture of the true shape of the face and thus a picture of the hole as a true circle.

circle

28 Figure 5.7A shows this auxiliary view No. 1 added to the orthographic drawing. This view has added information to the drawing—information that the hole is a true _____.

Strictly speaking, this is not new information. By using information from both the top and front views, we can interpret the hole as being circular. Repeating this information by the addition of auxiliary view 1 is a form of controlled redundancy that is most useful in the communication of technical information. The auxiliary view makes the drawing clearer.

parallel

29 Figure 5.7B shows the modifications made in the orthographic frame of reference to obtain the auxiliary view. The upper right edge of the box has been removed and a new plane inserted _____ to the sloping face of the object.

perpendicular

30 Thus projection lines perpendicular to the new reference plane will be _____ to the sloping face of the object.

clearer

31 When the frame of reference is unfolded, the orthographic projection of Figure 5.7C results. View 1 repeats both dimensions S and D. The purpose of view 1 is to make the drawing _____.

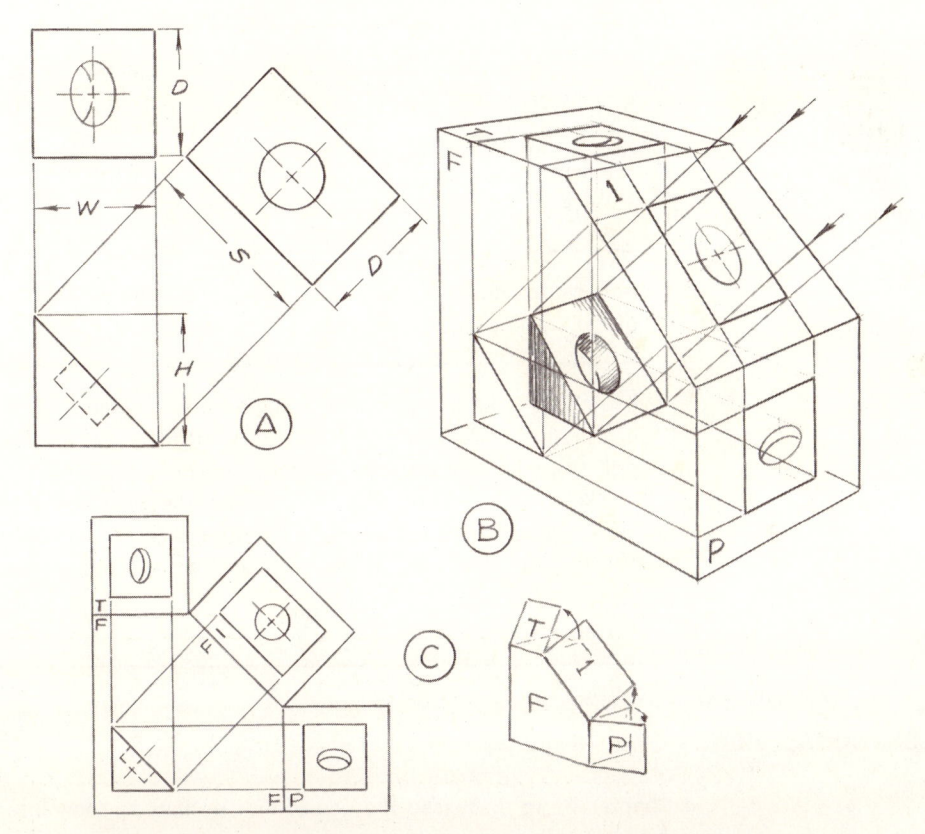

OBJECT ORIENTATION, AUXILIARY VIEWS, AND SECTIONAL VIEWS 89

Figure 5.7 Modifying the frame of reference to obtain auxiliary view No. 1.

---

32 The examples shown so far have attempted to show, in a simple manner, what an auxiliary view is. We have not shown how we project auxiliary views or how we select an appropriate view. We know that the reason for taking these views is to add or to clarify _____ in an orthographic drawing.

information

A precise theory of the projection of auxiliary views will be reserved for the study of points, lines, and planes in Chapters 15 and 16. There, in descriptive geometry, we need a precise theory since we will be using the orthographic projection system to *solve problems*. For the moment, we are only concerned with *picturing objects*.

A study of Figure 5.7 will show that the method for constructing an auxiliary view of the sloping face of the block (and consequently a true-circle view of the shallow hole) is as follows:

---

33 Construct two projection lines from the ends of the straight line that represents the edge view of the sloping face. These must be perpendicular to the edge view. They transfer the dimension _____ into the auxiliary view.

S

90 ORTHOGRAPHIC PROJECTION/UNIT 2

rectangle

34 At any convenient distance along these projection lines, measure the dimension D and complete a(n) _____ that has the dimension S × D.

circle

35 Find the center of the rectangle and draw a(n) _____.

auxiliary

36 Add centerlines and the _____ view showing the true shape of the sloping face is complete.

■ **37** Sketch 19 is an exercise in drawing an auxiliary view. An office intercom is shown pictorially and in a three-view orthographic sketch. The speaker grill and push buttons are on a sloping face.

Sketch an auxiliary view that shows the true shape of the sloping face. Sketch in the details by estimating size and position.
NOTE: The edge view of the sloping face appears in the right profile view.
Check your solution.

SECTIONAL VIEWS

In Chapter 4, we discussed the use of hidden lines to show interior detail on an orthographic drawing. Interior details are those details that are not visible as solid lines on any of the principal faces of an object.

If an object has complex interior detail, the use of hidden lines often results in a very complicated drawing. To simplify the drawing (make it easier to read) one of the orthographic views can be *sectioned*. Sectioning is a deliberate cutting away of a part of an object in at least one orthographic view so that interior detail may be shown as solid lines rather than hidden lines. Figure 5.8 is a simple example of the technique of sectioning.

hidden lines
(dashed lines)

38 Figure 5.8A shows a cube with a vertical hole drilled through it. The orthographic front view shows the hole outline with _____ _____.

front;
entire (complete)

39 Figure 5.8B shows both pictorial and orthographic views of the same object in *full section*. The front view is drawn as if the _____ half of the object has been removed. The top view shows the _____ object.

full section

40 The top view is marked with a *cutting-plane line* to show where the object is cut and in what direction the cut portion is to be viewed. This type of section is called a(n) _____ _____.

OBJECT ORIENTATION, AUXILIARY VIEWS, AND SECTIONAL VIEWS 91

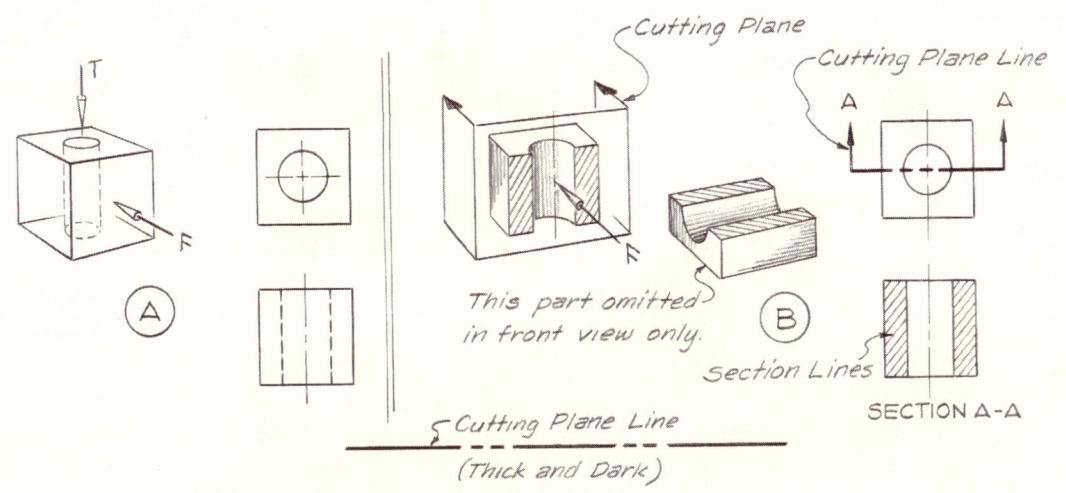

Figure 5.8 Sectioning—the full section.

long; short; arrowheads	41. The cutting-plane line is a broken line made up of alternate one _____ and two _____ dashes. It is a heavy, thick line so that it stands out against other lines. The ends are turned and are finished with _____ to show direction of view.
section	42. In the front view, the part of the object that has been cut is "shaded" with a series of closely but evenly spaced, inclined, parallel lines. These are called *section lines*. Section lines add a gray tone to the cut and serve as a visual aid in reading the drawing. They are always drawn at an angle to the horizontal and vertical object lines as a means of distinguishing between the different line uses. In the sectioned view, the part of the object which is imagined to be cut is given a gray tone by adding _____ lines.
SECTION A-A	43. Standard procedures call for labeling each arrowhead with the same capital letter, "A" in this case, and then labeling the sectioned view _____ _____. A complicated drawing may have many sections, so careful labeling is important.
offset (turned)	44. Sometimes a single cutting plane is not adequate to show the interior detail of a part. In this case, an *offset section* is used (see Figure 5.9). The cutting plane is passed through the left-hand hole and then _____ to pass through the right-hand hole.
cutting-plane	45. The top view contains the _____ _____ line that shows where the section is taken (how the object is cut apart).
arrowheads; B and B	46. The direction of view is indicated with _____ at the ends of the cutting-plane line and the section is labeled ____ and ____.

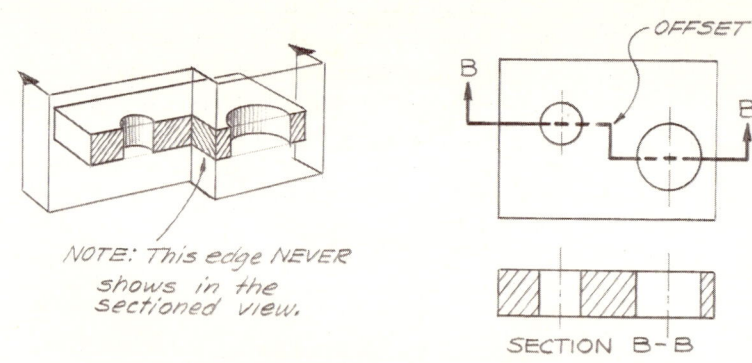

Figure 5.9 Sectioning—the offset section.

front	47	The front view is the actual sectioned view. This view is drawn as if the part of the object that lies in _____ of the cutting-plane line were removed completely.
section lines	48	To change the visual appearance of the "cut" part, a gray tone is added by drawing a series of _____ _____.
angle; evenly	49	Section lines are always drawn at a(n) _____ to the horizontal and vertical object lines. They are parallel lines, closely but _____ spaced.
darker	50	Spacing of section lines is not standardized. They should be close enough together to give a distinct gray tone to contrast with the white background of the drawing paper. The shorter the spacing between section lines, the _____ the tone.
even (uniform)	51	Section lines are thin lines—about the same weight as centerlines. Equal spacing between lines yields a(n) _____ gray tone.
cutting	52	Note in Figure 5.9 that the offset *is not* shown in the front view even though we see an edge view of the offset plane which should result in a "hard" straight line. It is a standard convention to omit lines caused by offsetting the _____ plane.
one-quarter	53	Figure 5.10 shows another type of section—the *half section*. This is used for simple objects and objects that have a symmetry about a centerline. Although called a half section, only _____ _____ of the object is removed.

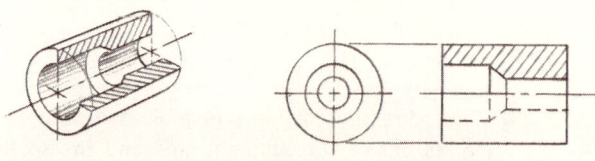

Figure 5.10 Sectioning—the half section.

OBJECT ORIENTATION, AUXILIARY VIEWS, AND SECTIONAL VIEWS   93

54 In the simple cylindrical part of Figure 5.10, the cutting-plane line and the labeling of the section has been _____. This is standard procedure in drawing simple parts where the manner of taking the section is obvious without labeling.

omitted

55 Note that the two views of Figure 5.10 are the _____ and the right-_____ views. A sectioned view can be any one of the six principal orthographic views or any auxiliary view.

front; profile

This has been a brief discussion of full, offset, and half sections and some of the standard procedures involved in making sectioned drawings. A more detailed account of this subject will be taken up in Chapter 13.

SUMMARY—ORTHOGRAPHIC PROJECTION

Chapters 3, 4, and 5 have presented the fundamental concepts of orthographic projection—the first of the two important forms of technical drawing. You will be using these concepts through Chapter 28, either to make orthographic drawings or to read orthographic drawings.

■ **56** Sketch 20 is a summary exercise on picturing objects in the orthographic form. Four objects are shown pictorially. Some dimensions are given and the basic H:W:D proportions are given for each object. Corner marks for some of the orthographic views are superimposed upon the ¼-in. grid to help you to get started. The viewing direction for at least one of the principal views is given by a viewing arrow on the pictorial drawings.

Sketch the orthographic views called for. Include all hidden lines. Note that object A calls for an auxiliary view and object D is marked for a full section view.

Check your solutions.

chapter 6 *proportion and scale*

WHERE ARE WE GOING?

Our plan in this programmed book is to present the theory upon which the two forms of technical drawing are based. The two forms are *orthographic projection* and *pictorial drawing*. Chapters 3, 4, and 5 covered the basics of *orthographic projection*. (There is more to come in later chapters.) We learned (1) how to project any or all of six principal orthographic views, (2) how to project simple auxiliary views, and (3) a few of the more important conventions of industrial drafting practice. Any one orthographic view shows two of the three principal dimensions of H, W, and D. Any two or more adjacent views show all three dimensions and comprise an orthographic drawing.

Before proceeding to the subject of *pictorial drawing* (Chapters 7 through 11), let us take a look at the subjects of *proportion* and *scale*—their meanings and their relationships with each other.

INTRODUCTION

In the exercises of Sketch 20, four objects were presented to be drawn in orthographic views. These objects could be of any size ranging from something that could be held in your hand (a transistor) to a small house. A transistor would have to be enlarged to give a drawing of reasonable size. A house would have to be reduced to get it on the drawing paper. Both drawings, however, must show the true proportions of the object to communicate correct information.

PROPORTION AND SCALE

DEFINITIONS

Proportion

Proportion, and we are talking about geometric proportion, is defined as "the equality of ratios." In mathematics, an equality is shown as two terms on either side of an equal sign. Thus, an equality of ratios could be $a:b = c:d$ where a, b, c, and d could represent anything such as numbers, distances, areas, volumes, etc. In Figure 6.1, we will let them represent straight lines of different lengths ($a = 2$ units, $b = 3$ units, $c = 4$ units, and $d = 6$ units). If we substitute numbers for letters, we obtain $2:3 = 4:6$ which is arithmetically correct since 4/6 can be reduced to 2/3.

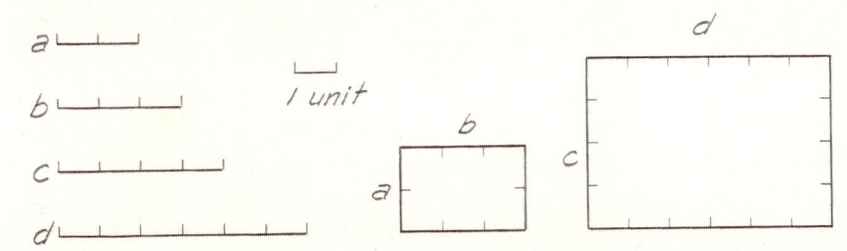

Figure 6.1 Two rectangles of equal proportions but of different scales.

Scale

Scale is defined as "relative dimensions without a difference in proportion." It is the proportion in dimensions between a drawing of an object and the actual object. Let's let lines a and b represent the two sides of a rectangle and c and d be two sides of another rectangle (Figure 6.1). It is obvious that the rectangles are not the same size, but we know they are the same proportion since we know that $a:b = c:d$. Each rectangle could represent a postage stamp (1 in. by 1½ in.) or the outline of a farm (2 miles by 3 miles) because $1:1½ = 2:3 = 4:6$. To actually represent one of these, we must define the units of measure, i.e., state the scale of the drawing.

PROPORTION VERSUS SCALE

In graphics, of the two concepts, proportion and scale, proportion is the more important. We can change scale at will, but if the proportions of the drawing are not correct, the drawing is not correct. This fact will become more evident in our study of pictorial drawing, when we are concerned with visual proportion as well as geometric proportion.

proportion	1	In the study of graphics we are concerned with both scale and proportion. Of the two, _____ is the more important.
equality	2	Proportion is defined as a(n) _____ of ratios.

96 ORTHOGRAPHIC PROJECTION/UNIT 2

H; W; D

3 A ratio is a fixed relation between quantities. If we wish to draw a specific object, we know that the object has a fixed relation between its three principal dimensions of _____, _____, and _____.

two

4 This fixed relation between H, W, and D is considered to be the proportion of the object. Such a proportion between three quantities is written H:W:D. Each one of the six principal orthographic views of an object displays _____ of the three dimensions, H, W, and D.

relation

5 These two dimensions represent a fixed _____ whether they are H and W, W and D, or H and D.

proportions

6 No matter what size we draw the object, we must maintain the _____ between dimensions.

Two-dimensional Proportions

ratio

7 A 3 × 5 card is a standard stationery item for keeping records or taking notes. The term 3 × 5 is a(n) _____ of H:W (or W:H depending upon how we look at the card).

inches; inches

8 The implication of the name "3 × 5 card" is understood to mean a rectangular card whose dimensions are 3 _____ by 5 _____.

3:5

9 The dimensions 3 in. and 5 in. are actually scale measurements of the two dimensions. We could convey the idea of a 3 × 5 card, through drawing, to any scale we wished as long as we maintained the proportion of ____:____.

same

10 Figure 6.2 shows a 3-in. by 5-in. rectangle at A. Two other rectangles are shown at B and C. At B the dimensions are c × d (c = ¾ in., d = 1¼ in.) and at C they are e × f (e = 4 in., f = 6⅔ in.).
It is obvious that the three rectangles are of different size, but are they of the same or different proportion? _____

3:5
(or ¾:1¼ or 4:6⅔)

11 If they are of the same proportion, then a:b = c:d = e:f = _____. (Supply the two numbers.)

ratios

12 You can prove for yourself that, arithmetically, 3:5 = ¾:1¼ = 4:6⅔. The dimensions a through f are different, but their _____ of a:b, c:d and e:f are equal.

scale

13 Each of the three rectangles is an accurate representation of a 3 × 5 card as long as we know what the _____ of each drawing is.

PROPORTION AND SCALE 97

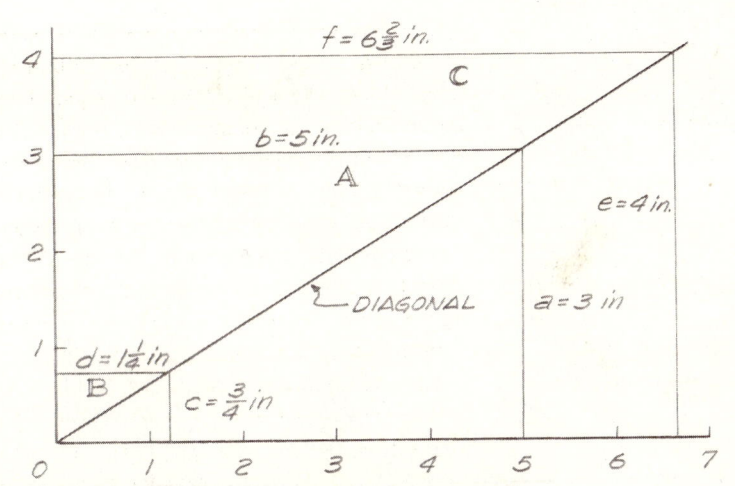

Figure 6.2   Three rectangles with the proportions of 3:5.

14 Assuming that A, B, and C represent a drawing of a 3 × 5 card, the scale of drawing A is 3 in. = _____ (or _____ scale); at B the scale is ¾ in. = _____; and at C the scale is 4 in. = _____ in.

(A) 3 in. (full);
(B) 3 in.; (C) 3 in.

In stating the *scale of a drawing*, it is standard procedure to let the first number represent the dimension of the drawing and the second represent the corresponding true-scale dimension of the object. It is also standard procedure to make one of these numbers a single unit of the measuring system (1 in., 1 ft, 1 cm, etc.). Thus the previous three scales should be stated:
  (A) $1'' = 1''$ or full scale
  (B) $¼'' = 1''$ (one-quarter scale)
  (C) $1'' = ¾''$
More of this later.

proportion;
scale

15 And so we see that *scale* is a *relative quantity*. For drawing purposes, _____ is more important than _____.

enlarged; reduced

16 The Introduction to this chapter pointed out that selection of the scale of a drawing is dictated by either (1) obtaining a picture of reasonable size for a(n) _____ scale or (2) fitting the picture on the paper for _____ scales.

common

17 In Figure 6.2, note the straight line that is the common diagonal line to all three rectangles. This diagonal line will be a common diagonal to all three rectangles *only if the rectangles are proportional*. Figure 6.3A shows three rectangles of different proportions. They do not have a(n) _____ diagonal.

98 ORTHOGRAPHIC PROJECTION/UNIT 2

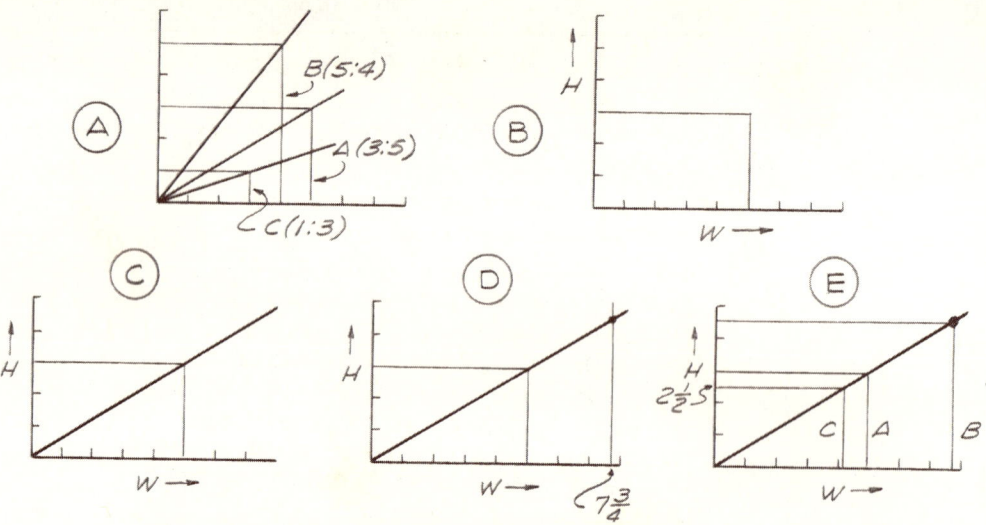

Figure 6.3 Proportioning rectangles using a common diagonal.

3:5	18	The fact that proportional rectangles have a common diagonal can be used to enlarge or reduce a drawing and keep it in proportion to the original. Figure 6.3B shows a rectangle of proportion H:W = ___:___.
diagonal	19	Suppose we wish to enlarge the 5-unit dimension to 7¾ units and draw a proportional rectangle. The first step (Figure 6.3C) is to sketch a(n) _____ in the original (3:5) rectangle.
diagonal	20	Then at 6.3D we find the dimension of 7¾ units on the W scale and from this point sketch a vertical line that intersects the _____.
proportional; common diagonal	21	Finally, at 6.3E, sketch a horizontal line through this point of intersection and complete the rectangle. The two rectangles are _____ because they have a(n) _____ _____.
reduced	22	Reducing the size of the original can be accomplished in the same way. In Figure 6.3E, rectangle C is a proportional but reduced-size version of A. The 3-unit dimension has been _____ to 2½ units.
■	23	In Sketch 21A two scales (H and W) are given. Sketch the following: (1) A rectangle of H = 3¼ and W = 4. (2) A proportional rectangle whose W dimension is 7½. (3) A proportional rectangle whose H dimension is 1¾. Check your solution.

Three-dimensional Proportions

Our purpose here is to learn to picture three-dimensional objects not just two-dimensional objects.

PROPORTION AND SCALE 99

24 Actually, a 3 × 5 card has three dimensions of H, W, and D (or thickness). It should be called a 3 × 5 × 0.008 card. Eight one-thousandths of an inch (8/1000 or 0.008) is the card thickness. The three-dimensional proportions of the card can be stated as ____:____:____.

3:5:0.008

25 Eight one-thousandths is about one-half of 1/64 in. and is so small as compared to, say, the 3-in. dimension that it is neglected in drawing a 3 × 5 card. Consider, however, the box shown in Figure 6.4. It has proportions of H:W:D = ____:____:____.

3:5:1½

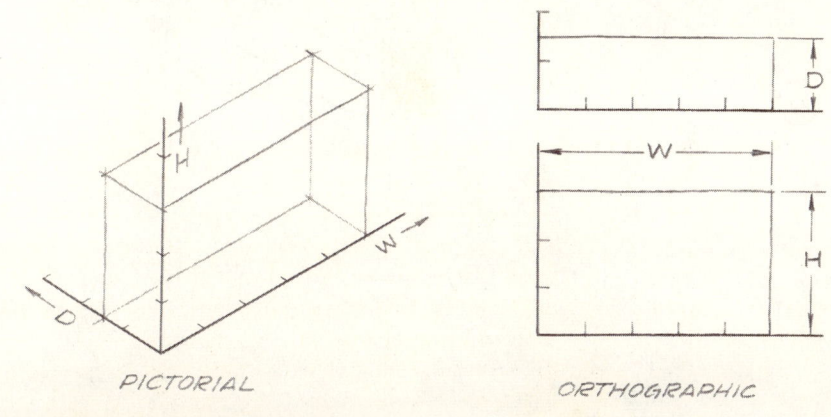

Figure 6.4 A rectangular box.

PICTORIAL

ORTHOGRAPHIC

26 Figure 6.5 shows three objects drawn within this same box, (A) a bar of soap, (B) a building, and (C) an air-control zone over an airport. It is obvious that these three objects are of different _____, but since they are shown in the same box, they are of the same _____.

scale (size);
proportion

27 A scale is shown for each drawing. Using these scales we see that the actual dimensions of the objects are:

	H	W	D
(A) Soap (in.)			
(B) Building (ft)			
(C) Control zone (mi)			

	H	W	D
(A):	3	5	1½
(B):	60	100	30
(C):	6	10	3

Figure 6.5 is convincing evidence that in technical work you should learn to think in terms of proportion rather than scale. The scale of a drawing becomes relatively unimportant and is usually dictated by convenience or whim. Let's now take a look at the problem of proportioning a three-dimensional rectangular box.

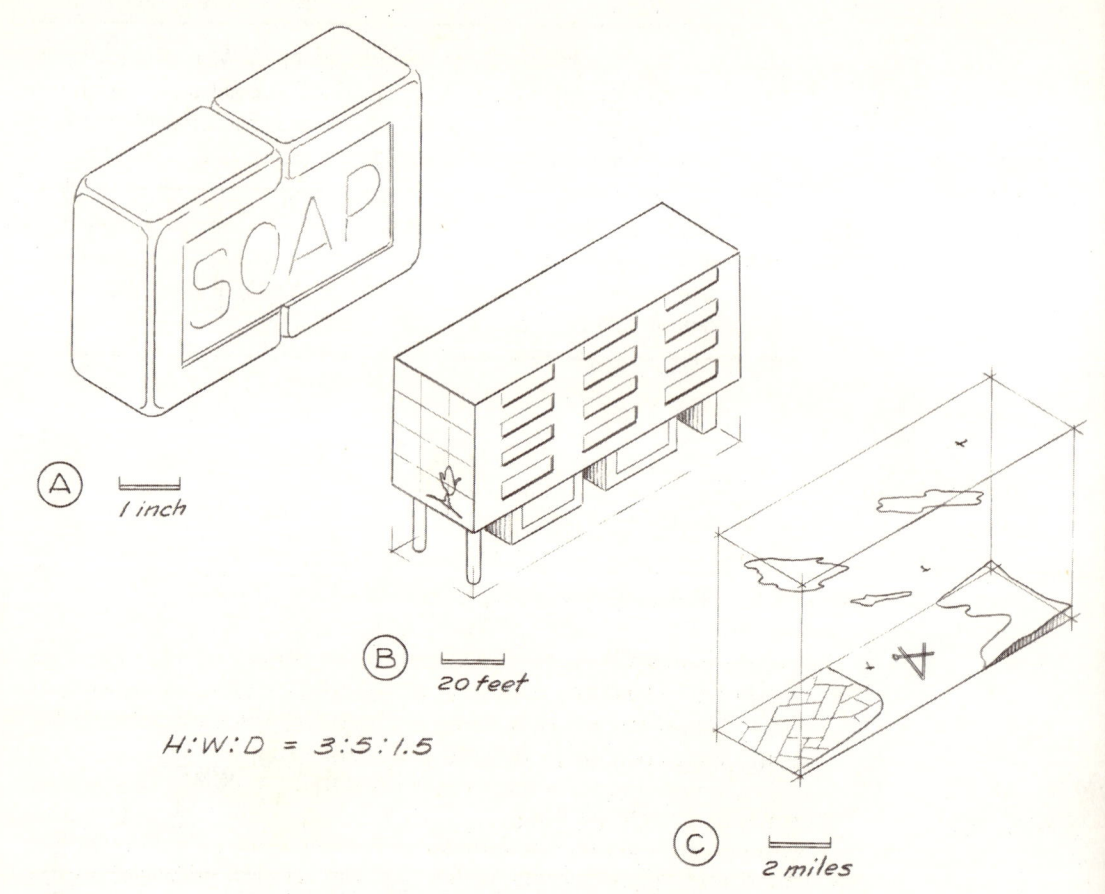

Figure 6.5 Three objects of the same proportions but different scales.

NOTE: Throughout this and other chapters, we concentrate the discussion and exercises on rectangular boxes. In Chapter 3, you will remember, we showed that any object, no matter how complex, can be enclosed in a single rectangular box of dimensions H, W, and D. The drawing of detail within the box becomes much easier once the basic proportions are determined (the enclosing box is drawn).

Figure 6.6A is a pictorial drawing showing the geometric relation between three boxes of different size but with the same proportions of H:W:D.

body

28 Note that a *body diagonal* has been sketched from point A at the lower front corner to point B at the upper rear corner. AB is a body diagonal because it passes diagonally through the _____ of the object.

common

29 Point A is common to all boxes. The body diagonal AB is also _____ to all three boxes.

PROPORTION AND SCALE 101

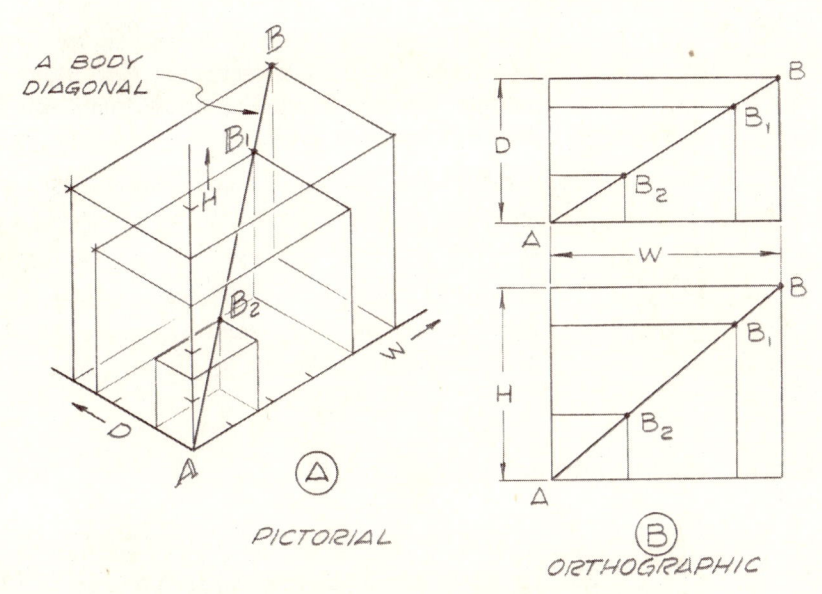

Figure 6.6 Three-dimensional proportions—the body diagonal.

30 Figure 6.6B shows the top and front orthographic views of the boxes of 6.6A. Note that each view shows that the rectangles representing the boxes are proportional because they have a(n) _____. Each view is similar to what we saw when we proportioned a 3 × 5 card.

common diagonal

31 We saw in Figure 6.3 and Sketch 21A that, given a rectangle, we can draw an enlarged- or _____-size rectangle of the same proportion as a given rectangle by knowing one of the new dimensions and utilizing the diagonal.

reduced

■ **32** Sketch 21B shows a box of proportions H:W:D = 3:4:2. Suppose we wish to change this to a proportional box whose height (H) is four units. The following steps direct you in sketching this new box. The procedure will be to use the diagonals of the two exposed vertical faces and then to check the solution by confirming that the two boxes have the same *body diagonal*.

STEP 1: Mark a point (A') at 4 units (the new height) on the H scale. Sketch two lines from this point, line A'B' parallel to line AB and line A'C' parallel to line AC. Check your construction.

(NOTE: Solutions are given for each step in this series.)

STEP 2: Sketch the diagonal of the front face (3 × 4) of the original box. Extend it to an intersection with line A'B'. Similarly, sketch the diagonal of the side face (3 × 2) and find an intersection with line A'C'.

STEP 3: Complete the front face of the new box by sketching a vertical line from the intersection on line A'B' down to line AB (extended). Do the same from the intersection on A'C' down to line AC.

102 ORTHOGRAPHIC PROJECTION/UNIT 2

STEP 4: Complete the top of the new box by sketching a line parallel to A'C' starting at the intersection on A'B'. Sketch a line parallel to A'B' starting with the intersection on A'C' until you find an intersection of these two lines (call it D').

STEP 5: The new box is now complete. Check your construction by sketching a body diagonal AD in the original box. When extended, it should pass through point D'.

MORE PRACTICE: On the same figure, sketch a box whose width (W) is 3 units. This will be a reduced-size box. The basic procedure is the same. Check your solution.

SCALE, SCALES, AND SCALING

drawing;
object

33 Earlier we showed that we can state a scale for a drawing with an equality involving two numbers (1" = 1", ¼" = 1", etc.). Customarily, the first number represents the dimension on the _____ and the second represents the corresponding true-scale dimension of the _____.

scale

34 We have seen how important it is to state the scale of a drawing. Figure 6.5 illustrated this. The bar of soap, the building, and the air-control zone were all drawn to the same size. To obtain full dimensional information from these drawings, we must know the _____ of each.

proportion

35 The word "scale" as a noun has two distinct and important meanings. The first we have already encountered. Scale is relative dimensions without a difference in _____.

MEASURING WITH A SCALE

The second use of the word "scale" as a noun has reference to the actual device used in making a measurement. A scale is a measuring device—a ruler. In a moment we will be discussing three types of scales:
    mechanical engineer's scale
    architect's scale
    civil engineer's scale

yes

36 The word "scale" as a verb means to make a measurement. To scale a drawing means to measure it. You scale a drawing or an object with a scale. Confusing? Yes _____. No _____.

37 The selection of an appropriate scale for any drawing usually depends upon two factors:
   (1) The drawing should be of a(n) _____ size.
   (2) The drawing must _____ on the drawing paper.

(1) reasonable (readable)
(2) fit

The choice of an appropriate scale is up to the person making the drawing. The technique of making a scale drawing can be aided by various scales (measuring devices) available to the engineer and the draftsman.

We mentioned earlier three types of scales: the mechanical engineer's, the architect's, and the civil engineer's. Figure 6.7 shows these three types and the various separate scales available on each.

These scales are named for the professional groups who have most use (but by no means exclusive use) for each. The objects that each group must draw dictate the scales they will use.

MECHANICAL ENGINEER'S (ME) SCALE
— 8 EQUAL DIVISIONS

$\frac{3''}{4} = 1''$ Others: $\frac{1''}{8} = 1''$ $\frac{3''}{8} = 1''$ $1'' = 1''$ (Full Scale)
              $\frac{1''}{4} = 1''$ $\frac{1''}{2} = 1''$ $2'' = 1''$ (Double Scale)

ARCHITECT'S (A) SCALE
— 12 EQUAL DIVISIONS

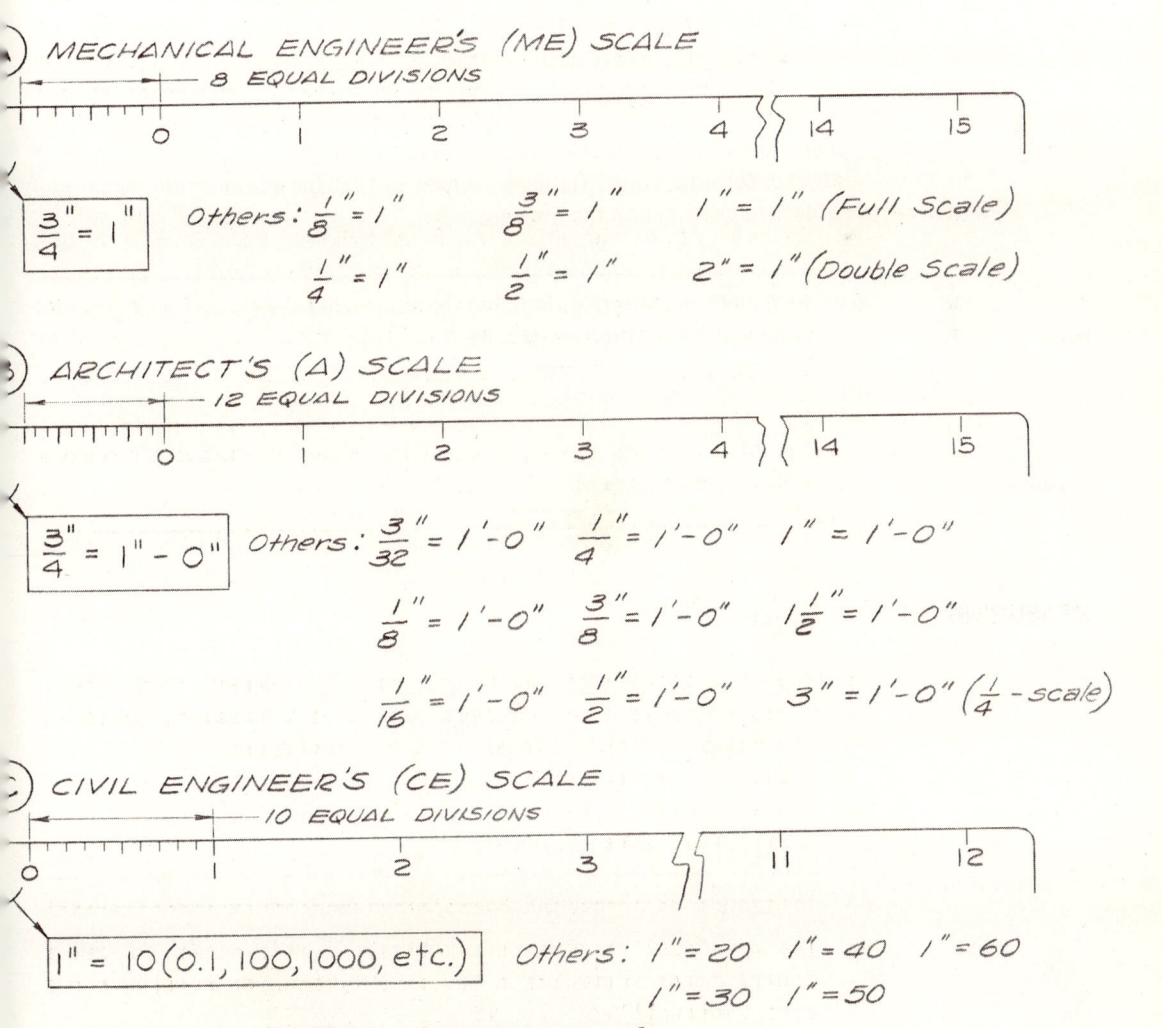

$\frac{3''}{4} = 1'-0''$ Others: $\frac{3''}{32} = 1'-0''$ $\frac{1''}{4} = 1'-0''$ $1'' = 1'-0''$
              $\frac{1''}{8} = 1'-0''$ $\frac{3''}{8} = 1'-0''$ $1\frac{1''}{2} = 1'-0''$
              $\frac{1''}{16} = 1'-0''$ $\frac{1''}{2} = 1'-0''$ $3'' = 1'-0''$ ($\frac{1}{4}$-scale)

CIVIL ENGINEER'S (CE) SCALE
— 10 EQUAL DIVISIONS

$1'' = 10$ (0.1, 100, 1000, etc.)   Others: $1'' = 20$ $1'' = 40$ $1'' = 60$
                                              $1'' = 30$ $1'' = 50$

Figure 6.7 Standard drawing scales.

104   ORTHOGRAPHIC PROJECTION/UNIT 2

size

38 The scale of a drawing is determined by the actual _____ of the object to be drawn.

inch

39 A mechanical engineer, in general, works on objects from very small size to approximately the size of an automobile or large machine tool (thousandths of an inch to about 25 ft). Figure 6.7A shows that his scales relate both multiples and fractions of an inch to the whole _____ and provide for both enlargement and reduction.

fractions;
reduction

40 An architect, in general, works on objects from the size of a room to an entire building or group of buildings (10 ft to about 500 ft). In Figure 6.7B we see that his scales relate _____ of an inch to the foot. They provide for _____ of the original only.

inch

41 A civil engineer, in general, works on objects from the size of a building or bridge to roads hundreds of miles long. His scales (Figure 6.7C) relate the _____ to decimal (10) and multiple decimal (20, 30, etc.) quantities. The units, whether inches, feet, miles, etc., are defined by the needs of the drawing.

With these three sets of scales it is possible to make an accurate drawing of any conceivable object (large or small) so that the drawing will be of reasonable size or will fit on the drawing paper.

inch

42 In the ME scale of Figure 6.7A, the first major division on each scale is subdivided into fractions of a(n) _____.

$1/64$, $1/32$,
$1/16$, $1/8$,
$1/4$, and $1/2$

43 Standard subdivisions of an inch are ____, ____, ____, ____, ____, and ____.

$1/2$ in. = 1 in.
$4 3/4$ in.

44 Thus each of the ME scales allows direct measurement in whole inches and fractions of one inch. Figure 6.8A shows the distance between two hole centerlines measured to a scale of _____ = _____. The measurement is _____.

inches

45 In the A (architect's) scale of Figure 6.7B, the first major divisions on each scale are subdivided into _____.

feet; inches

46 Inches are the only standard divisions of one foot. Thus each of the A scales allows direct measurement in _____ and _____.

11'-6";
$1/4$" = 1'-0"

47 In Figure 6.8B a small building is shown along with a scale. The width of the building is _____. The scale of the drawing is _____ = _____.

PROPORTION AND SCALE 105

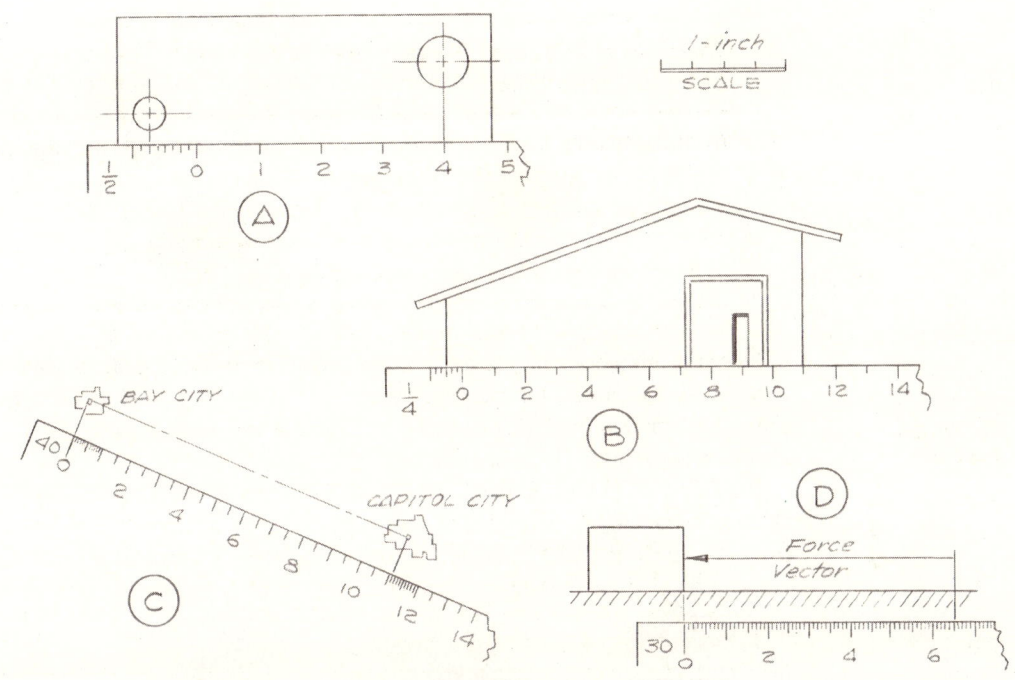

Figure 6.8 A measuring exercise.

10 per inch; 20 per inch	**48** In the CE scales of Figure 6.7C, each inch is divided into decimal (or multiple decimal) units. On the 10-scale there are _____ divisions per _____; on the 20-scale _____ divisions per _____; etc.
5280; mile	**49** These scale numbers of 10, 20, 30, 40, 50, and 60 can represent any units we wish. For example, the 10-scale can be 10 in. per inch, 10 ft per inch, 10 mi per inch, 100 ft per inch, 1000 mi per inch, etc. At a scale of 1 in. = 1000 ft, a line 5.28 in. long would represent _____ ft or one _____.
11.2 (at 1 in. = 4 mi) 112 (at 1 in. = 40 mi) 1120 (at 1 in. = 400 mi)	**50** Figure 6.8C shows a section of a map with the centerline of a proposed freeway between two cities. A 40-scale is also shown. The distance between the cities is _____ miles at a scale of 1 in. = _____ mi. (Choose your own multiple of the basic 40.)

Engineers find the CE scale useful in measuring quantities other than inches, feet, miles, etc. In the subject of vector geometry, they solve force, velocity, and other types of problems by graphic means. A force can be represented by a straight line of definite length and definite direction. The direction is shown by adding an arrowhead to the line. (See Figure 6.8D.)

106 ORTHOGRAPHIC PROJECTION/UNIT 2

6.5
(at 1 in. = 3 lb)
65
(at 1 in. = 30 lb)
650
(at 1 in. = 300 lb)

51 Figure 6.8D is a force diagram showing how much force is required to push an object along the ground. The length of the line represents the value of the force in pounds. A CE scale is shown. The force is _____ lb at a scale of 1 in. = _____ lb. (Choose your own scale multiplier.)

In many instances, especially in the precision-machining processes, measurements are stated in decimal divisions of an inch. Common subdivisions are 10 units per inch (1/10 or 0.1); 100 units per inch (1/100 or 0.01); and 1000 units per inch (1/1000 or 0.001 in.).

Drawings of precision parts need not carry the same precision. Accuracy to the nearest one-hundredth of an inch (0.01) is all that can be expected from a pencil drawing. Indeed, greater accuracy is useless since drawing paper can change that much from the effects of temperature and humidity variations.

3.25 in.

52 Figure 6.9A shows two hole centerlines and a CE 10-scale. What is the dimension between the holes to the nearest one-hundredth (0.01) inch? _____

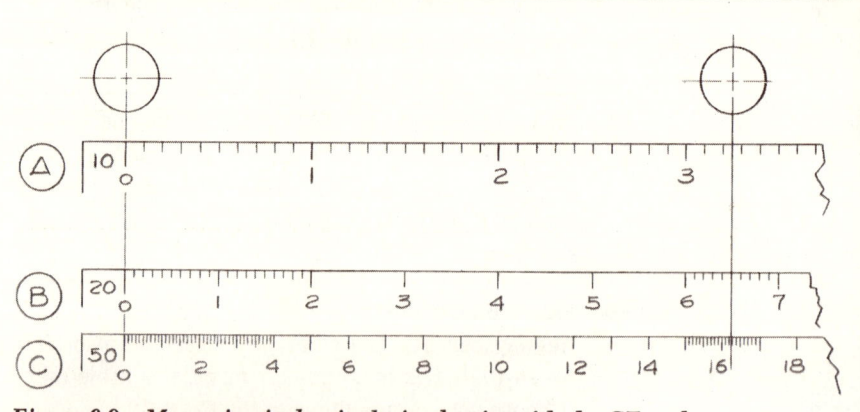

Figure 6.9 Measuring inches in decimal units with the CE scale.

estimate
(guess)

53 With the 10-scale it was necessary to _____ the second decimal position (0.05).

(20): 6.5
  or 65;
(50): 16.25
  or 162.5

54 Figure 6.9B and C shows the same dimension with a CE 20-scale and a CE 50-scale. The readings on these scales are (20) _____ and (50) _____.

55 To convert these to the correct inch dimension (3.25 in.) it is necessary to divide the first reading by _____ and the second reading by _____ .

2 (or 20);
5 (or 50)

56 Aside from the need to convert the reading, it can be seen from Figure 6.9 that the most accurate means of measuring to the nearest 0.01 in. is by using the _____ scale.

CE 50

A table of decimal equivalents of common fractions of an inch is very useful to the student of graphics. You should memorize the basic fractional units (1/64, 1/32, 1/16, 1/8, 1/4 and 1/2) and all the 1/8 and 1/4 units (3/8, 5/8, 3/4, and 7/8). A ready reference to these will come in handy when converting fractions to decimals and vice versa. Such a table will be found on the inside back cover of this book.

PROPORTIONING DISTANCES WITH A SCALE

57 The various scales shown in Figure 6.7 are used mainly for _____ distances. However, they can also be useful in *proportioning distances*.

measuring
(scaling)

58 Figure 6.10A shows two parallel lines 2¼ in. apart. Suppose we wish to divide the space between the lines into 13 equal spaces. Dividing 2¼ in. by 13, we get 0.173 in. per space. There is no scale that will give us 0.173-in. units directly. However, 13 × ¼ = 3¼. Thus, if as in the figure we position a standard inch scale with _____ on one line and _____ on the other, we may make a mark at every ¼-in. division and obtain 13 equal spaces in 2¼ in.

0; 3¼

59 We have proportioned a 2¼ in. space into 13 equal spaces. Suppose we wish to divide a 4-in. line (AB) into seven equal divisions (Figure 6.10B). Choosing ½ in. as our measuring unit, we lay off _____ ½-in. units (3½ in.) along a line *at any angle* to the given 4-in. line. Note that the new line starts at one end of the given line. (Point 0 at point A.)

seven

60 Next we connect points 7 and B with a straight line. Drawing lines from each point on line 0-7 that are _____ to line B-7, we find that line AB is proportioned into seven equal spaces as desired.

parallel

Another example of using a scale to proportion distances is shown in Figure 6.10C. We wish to place four marks on line CD such that the first is ¼ of the way from C to D; the second, ⅓; the third, ⅝; and the fourth, ⅞.

108 ORTHOGRAPHIC PROJECTION/UNIT 2

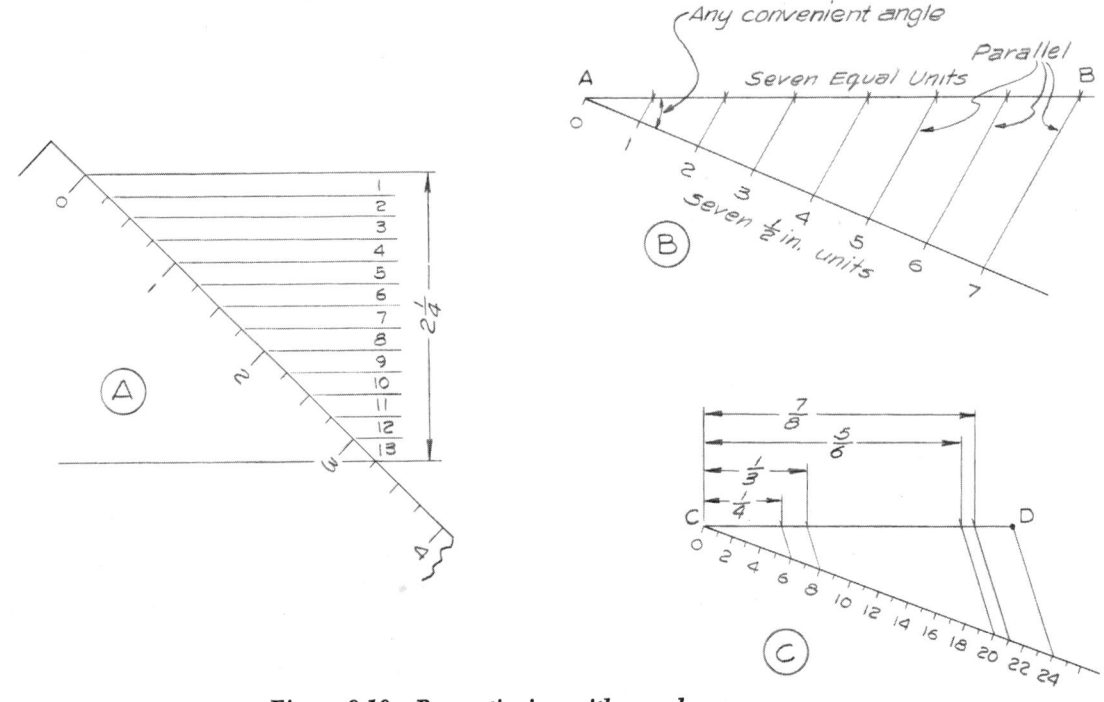

Figure 6.10 Proportioning with a scale.

$6/24$; $8/24$; $20/24$; $21/24$	**61** The first step is to change the fractions to a common denominator. The least common denominator is 24. Thus $1/4 =$ _____/24; $1/3 =$ _____/24; $5/6 =$ _____/24; and $7/8 =$ _____/24.
24	**62** The problem now is to divide *CD* into 24 equal spaces and then count off 6, 8, 20, and 21 spaces to get our desired divisions. Using the method of Figure 6.10B, we choose $1/8$ in. as a convenient basic unit and lay off _____ $1/8$-in. spaces on the sloping line.
6; 8; 20; 21	**63** Next we connect point *24* and point *D* with a straight line. Counting up from *C* _____, _____, _____, and _____, we draw four lines parallel to line *24-D*.
proportional	**64** The intersections of these lines with *CD* give us the desired spacings _____ to $1/4$, $1/3$, $5/6$, and $7/8$ of the total length of *CD*.

This chapter has pointed out the importance of maintaining correct proportions to obtain a correct drawing of an object. Scale has been presented as a relative quantity that varies with our drawing needs. Of the two, proportion is more important than scale to success in drawing.

UNIT 3 *pictorial drawing*

A BACKWARD AND THEN FORWARD LOOK

The goal of this programmed text is to present the subject of technical drawing to all who have need for knowledge of the subject. Let's pause here and consider two questions:

Where have we been?
Where are we going?

Flow Chart II answers these questions graphically. It is a graphic outline of the complete text. The shaded area covers the topics presented in the previous six chapters. The clear area shows the topics ahead: those remaining in Part I—Graphic Description of Objects, and those which comprise Part II—Graphic Solution of Problems. Numbers in circles are chapter numbers.

Chapters 7, 8, and 9 present an intuitive approach to sketching in the perspective form. Chapter 10 provides a geometric basis for perspective with a brief introduction to projected perspective. Finally, Chapter 11 presents the topics of isometric and oblique drawing. These are geometric approximations of true perspective drawing.

In the following presentation of pictorial sketching from an intuitive viewpoint, we are more concerned with how we see three-dimensional objects than with geometric rules and procedures. We have already discovered that any three-dimensional object, no matter how complex, can be enclosed in a rectangular box. Our procedure in the next three chapters will be to:

(1) Sketch the basic box in perspective (Chapter 7)
(2) Sketch detail within the box (Chapter 8)
(3) Learn to make perspective sketches of different sizes and of different object orientations (Chapter 9)

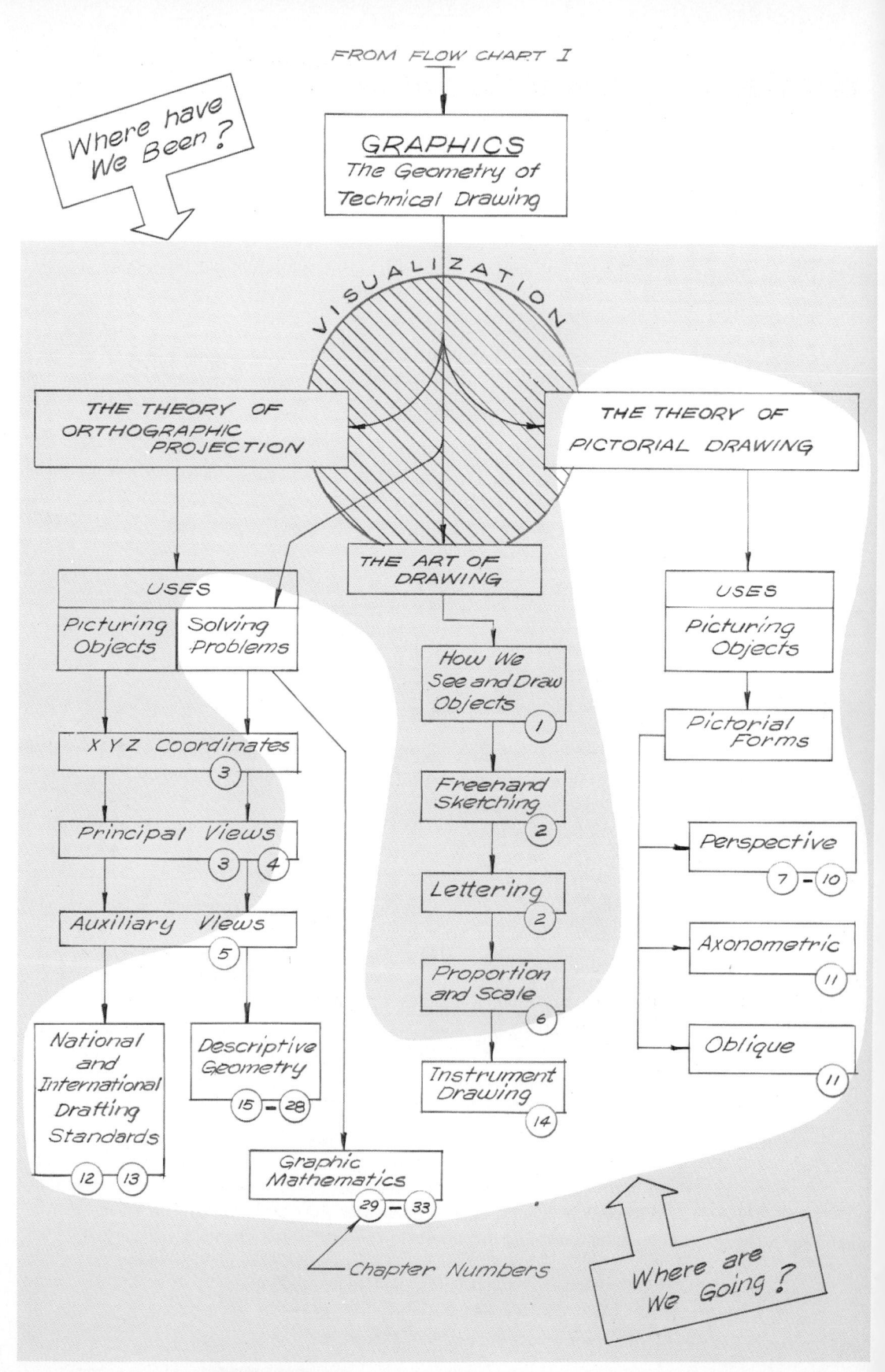

FLOW CHART II

chapter 7   *the theory of perspective*

INTRODUCTION

Perspective is defined as the art or science of *drawing natural objects as they appear to the eye.* It is the means by which we can picture a three-dimensional object on a two-dimensional drawing surface in a *realistic manner.*

In our study of perspective, we will take into account the laws of perspective and the effects of foreshortening as we learn how to make a drawing that is true to life. Our goal is to achieve photographic realism in a drawing. If we had the object at hand, taking a photograph would be easier than making a drawing. In technical work, however, many drawings must precede the point in the development of a design when we have a complete and accurate model. For this reason, the making of realistic drawings becomes an important aspect of the communication of technical information.

We will use the word "realistic" to mean *visually* correct and proportionate. Later, in Chapter II, we will look at another commonly used form of pictorial drawing, *isometric.* Here we will find a form that is *geometrically* correct and proportionate. Isometric, however, is not realistic (it contains distortion) because the visual proportions are not correct.

SKETCHING THE BASIC BOX

Let's use a familiar object to represent a rectangular box. Figure 7.1 shows photographs and sketches of a common building brick. They are presented in the two important forms of technical drawing: (A) orthographic projection and (B) pictorial (perspective) drawing.

112 PICTORIAL DRAWING/UNIT 3

---

1 Which of the four pictures is the most realistic representation of the brick to you? *(Check one.)*

(A) Photograph _____; sketch _____.

(B) photograph

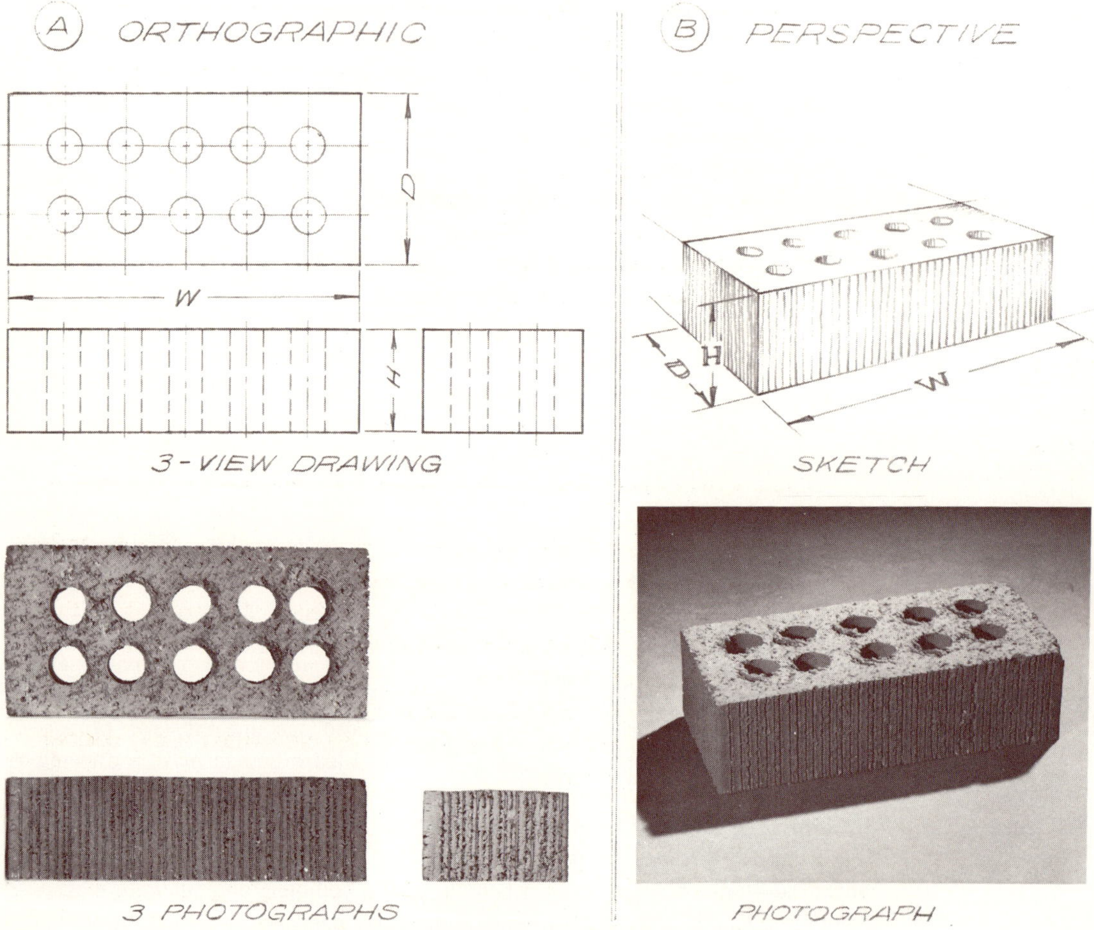

(B) Photograph _____; sketch _____.

Ⓐ ORTHOGRAPHIC　　　　　　Ⓑ PERSPECTIVE

3-VIEW DRAWING　　　　　　*SKETCH*

3 PHOTOGRAPHS　　　　　　*PHOTOGRAPH*

Figure 7.1 Photographs vs drawings—a common building brick.

---

2 Most people would agree that the photograph in B is the best and most realistic view of a brick. The photographs in A surely give the idea of a brick, but we know that, in real life, we would never see a brick as

three (3)　　_____ separate views.

---

3 Any one of the three orthographic views is familiar to us. Taken singly, however, any one gives us only _____ piece(s) of dimensional

two　　information about the brick.

---

4 Thus, the second most realistic view of the brick in Figure 7.1A and B

sketch in B　　is _____ (your own words).

THE THEORY OF PERSPECTIVE 113

height *H;*
width *W;*
depth *D*

5   This is true because, even though the sketch does not give us all the textural detail of the brick, it does give us an impression of the basic proportions of _____, _____, and _____.

6   If you wished to convey realistic information about a brick to a number of people who had never seen a brick, what means would you use? (Rate the following 1, 2, 3, 4, and 5 in order of (1) most effective, (2) second most, etc.)
    (A) A written paragraph _____
    (B) A set of orthographic photographs _____
    (C) An orthographic drawing _____
    (D) A perspective sketch _____
    (E) A recorded verbal description _____
    (F) A pictorial photograph _____

(A): 5
(B): 3
(C): 4
(D): 2
(E): 6
(F): 1

The answers to *A* and *E* may be arbitrary. It would depend upon whether your audience knew the written and spoken version of the language you used.

But enough about the brick. We used it here as a familiar example of the basic box. Our point is that knowing how to make a perspective sketch is the next best thing to having a photograph.

The goal of this chapter is summarized in Figure 7.2—the development of an ability to sketch a basic box of any given *H:W:D* proportions in any desired position and orientation in space.

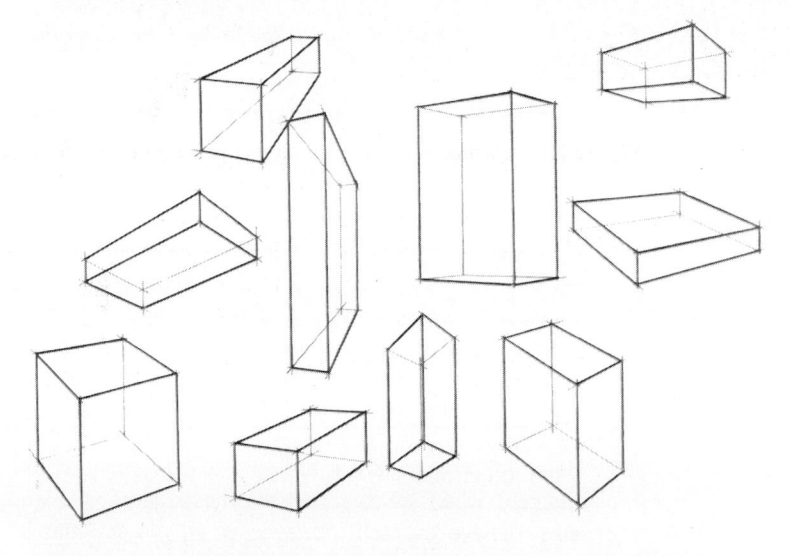

Figure 7.2   Goal: Sketch a rectangular box of any given H:W:D proportions in any desired position and orientation in space.

114 PICTORIAL DRAWING/UNIT 3

details (the object)

7 We concentrate on the basic box because we know that once it is drawn to correct H, W, and D proportions, our problem is reduced to adding _____ inside the box.

eye

8 Perspective is the art or science of drawing natural objects as they appear to the _____.

see
(view, observe)

9 This implies that we must know how we _____ objects.

vanishing point

10 To understand how we see an object in our field of vision we must first understand three concepts. These are shown in Figure 7.3. Two intersecting planes, one vertical and one horizontal, appear to stretch out to a single point called a(n) _____ _____.

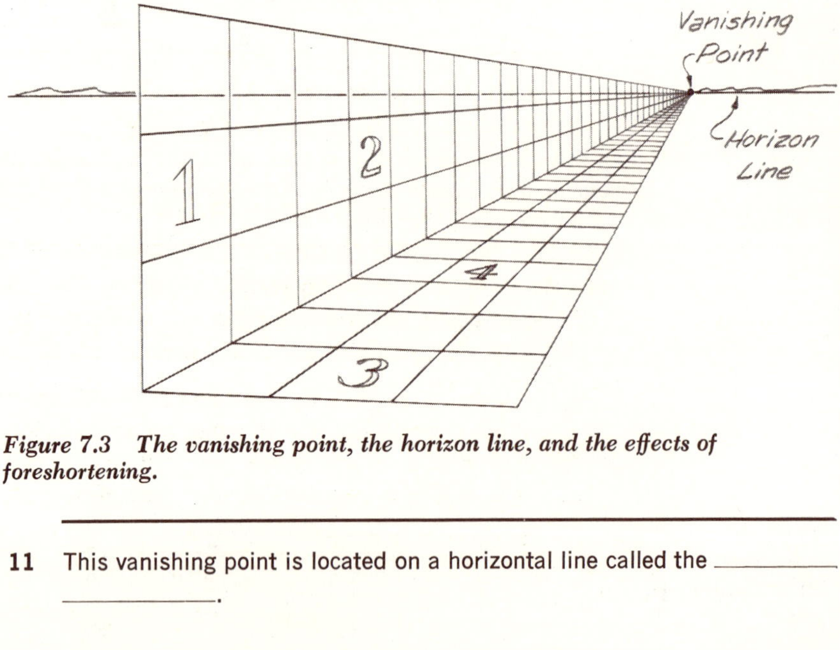

Figure 7.3 The vanishing point, the horizon line, and the effects of foreshortening.

horizon line
(horizon)

11 This vanishing point is located on a horizontal line called the _____ _____.

Square 1; Square 3

12 Both planes have a square grid on them. In each set which of the numbered squares *appears* larger:
  Square 1 or Square 2? _____
  Square 3 or Square 4? _____

appear (seem)

13 Square 2 and Square 4 are said to be *foreshortened*. Believe us for the moment when we say that all the squares of Figure 7.3 are of equal size. Believe yourself, however, when you say that S2 and S4 _____ to be smaller than S1 and S3.

THE THEORY OF PERSPECTIVE 115

SKETCHING IN PERSPECTIVE

To understand perspective you must understand the role played by three concepts:
(1) The horizon line
(2) The vanishing points
(3) Foreshortening

To sketch any one of the boxes of Figure 7.2, you must first make some decisions. Let's turn to the type of chart a computer programmer uses as he programs the computer to make decisions for him. Figure 7.4 is a logic diagram designed to show the decision steps in making a perspective sketch of a rectangular box. The diagram may look terribly complicated, but we will discuss it bit by bit. We also will make further elaborations on the three concepts as we need them.

First we must understand the diagram format. It consists of three elements:
(1) Rectangular boxes
(2) Diamond-shaped boxes
(3) Connecting lines

do

14 Examine a few of these elements. The rectangular boxes are *action* instructions. They tell you what to _____.

questions

15 The diamond-shaped boxes pose _____.

yes (or) no

16 The answers to the questions represent the decisions that must be made. These answers are either _____ or _____.

17 You use the diagram by beginning at START and following the connecting lines until you come to either an action box or a question.
    (1) If it is an action box, you _____.
    (2) If it is a question diamond, you make _____
    _____ (your words).

(1) do as instructed;
(2) a decision (answer the question)

18 The answer you give leads you on through the diagram. The first element after START in Figure 7.4 is _____ (your words).

a question (a question diamond)

Do as instructed. This is an invitation to skip the rest of Chapter 7. It is designed to let you evaluate whether you need the discussion that follows.

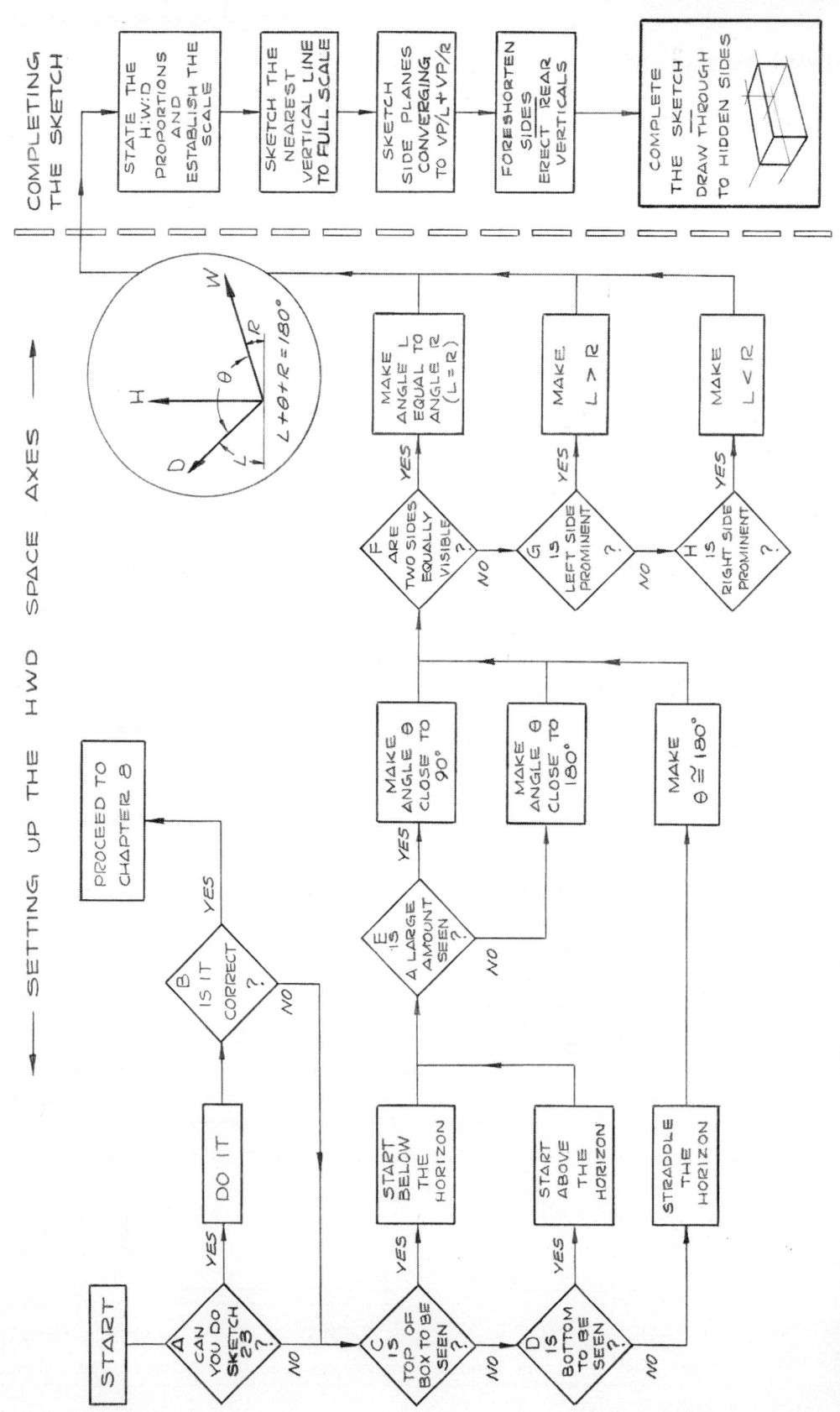

Figure 7.4 A logic diagram showing how to set up a perspective sketch of a particular box in space.

THE THEORY OF PERSPECTIVE 117

CRITERION TEST

Sketch 23 is called a criterion test. Read it carefully and decide (diamond A) whether you can do it. If "yes," proceed to do it. If "no," proceed on to the next question box. If you do the three problems in Sketch 23, evaluate your solutions *carefully* with the solutions given (diamond B). Do the two agree *closely?* If "yes," go on to Chapter 8. If "no," follow the diagram. It takes you back to the main stream.

---

19 Keep in mind the goal toward which we are working: To be able to sketch the basic box in
      (1) any desired position and orientation in space
      (2) to any given H:W:D proportion.
These two goals bring up two questions which must be answered before we start:
    How do we wish to _____ the box?
view (see);   What is the _____ of the box?
size (proportions)

---

20 Our answer to the first question tells us how to set up three space axes. Our answer to the second question tells us what _____
measurements   to make on the three axes.

---

21 When we think about H, W, and D of a rectangular box, we can imagine three mutually perpendicular lines. We have seen such a system before. In Chapter 3 we used the *xyz*-coordinate space _____ to
axes   build a frame of reference for orthographic projection.

THE HORIZON LINE

22 We will use the space axes in perspective sketching. The main decision-making part of the logic diagram of Figure 7.4 is labeled:
    Setting up the _____ space axes.
HWD

---

23 Once the axes are set up, we have defined graphically how we are _____ the box. Then we can lay off the H, W, and D dimen-
viewing   sions so that the box will be of correct visual _____.
(looking at);
proportions

---

24 Diamond C of the logic diagram (reproduced in part in Figure 7.5) asks
top   us whether our sketch will show the _____ of the box.

---

25 If this is what we desire, we are directed to start the sketch below the
horizon line   _____ _____.

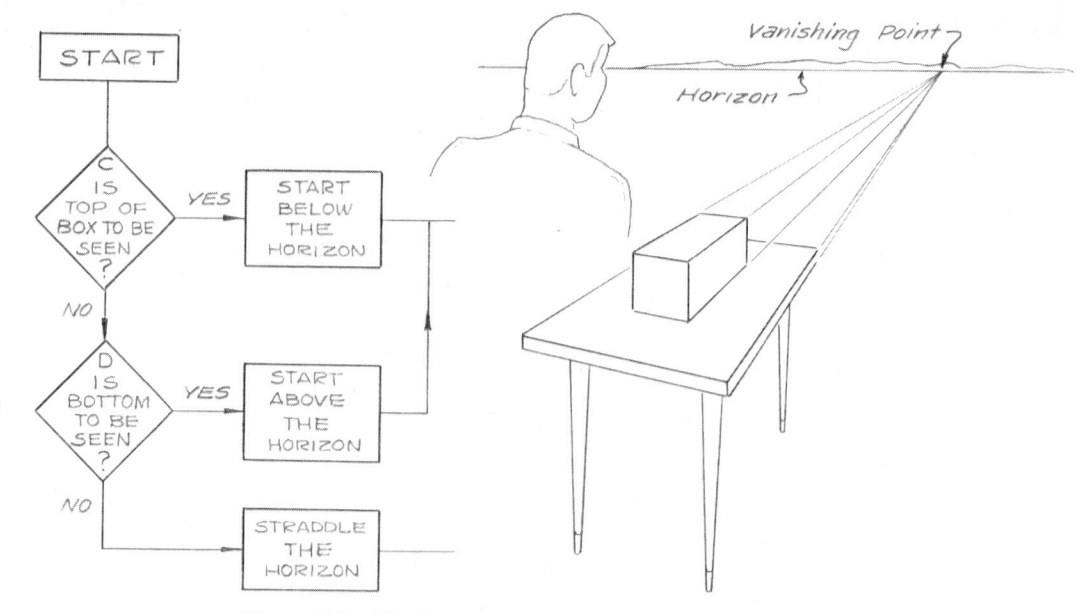

Figure 7.5 The horizon and the vanishing point.

above

26 Let's assume now that we do not wish to see the top of the box. Our next question then asks if the bottom is to be seen. If the answer is "yes," we are told to start the sketch _____ the horizon line.

straddle

27 Suppose we wish to see none of either the top or bottom of the box. There is no further choice so we are told that our sketch must _____ the horizon line.

We can ask here, "What is the horizon line?" To understand how we actually see an object in our field of vision, we must understand the role of the horizon line and vanishing points in perspective.

THE VANISHING POINTS

horizon

28 Consider Figure 7.5. We see an observer (let's consider him to be ourselves) looking at a rectangular block resting on a small table. The lines forming the sides of the block and the table *recede* to a vanishing point on the _____.

ground

29 We know the horizon as the apparent junction of earth and sky. In geometric terms, it is an infinite plane passing through the eyes of the observer and perpendicular to the vertical at the observation point, that is, parallel to the _____ plane. (See Figure 7.6.)

parallel

30 This latter definition implies that the ground plane and the horizon plane never meet since they are _____.

THE THEORY OF PERSPECTIVE 119

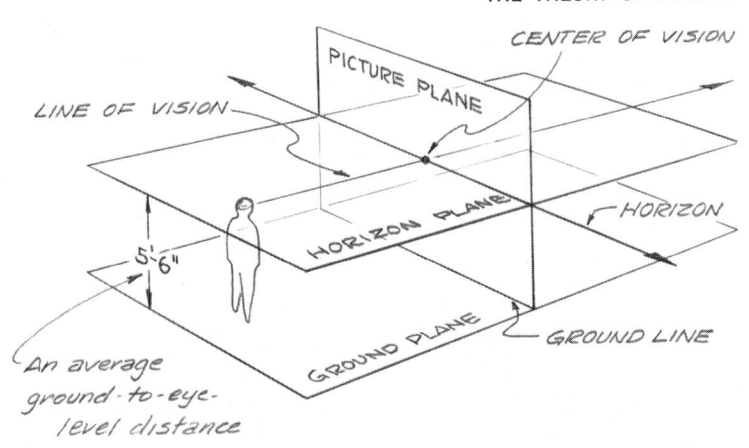

Figure 7.6 Geometric relations between the ground plane, the horizon plane, and the picture plane.

infinity	**31** From a geometric point of view, we assume that parallel lines and planes meet at _____.
two (2)	**32** Infinity is meaningless when considering our ability to see our horizon (junction of earth and sky). Consider Figure 7.7A. It shows the geometric relationship between a human being standing on a "flat" surface and observing the real horizon. His extent of vision is about _____ miles.

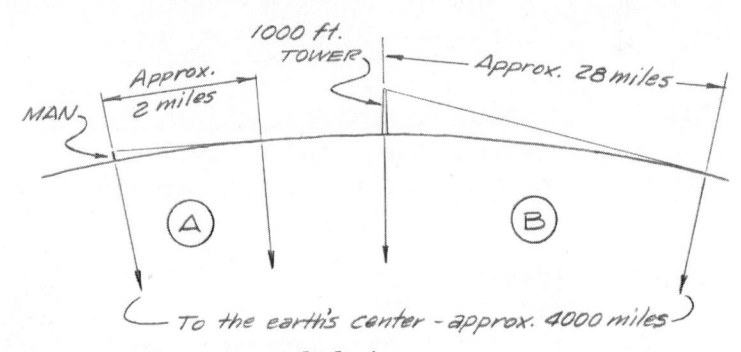

Figure 7.7 The real meaning of the horizon.

28	**33** To gain a view of more of the earth's surface, the observer would have to elevate his position with respect to the surface. Atop a 1000-ft tower (Figure 7.7B) his line-of-sight distance would be about _____ miles.

The important consideration is that the horizon line is *always at the observer's eye level* no matter where he is positioned, below ground, above ground, or standing on the ground plane.

120 PICTORIAL DRAWING/UNIT 3

	34	Look back at Figure 7.5. The horizon is drawn through the observer's eye level. The block on the table is below the horizon, and therefore, he is looking _____ the block (seeing the top and two sides). (*Choose one: Down on/up at.*)
down on		

	35	Figure 7.8A is the same as Figure 7.5 except that other objects have been added (a line of telephone poles, a road, a railroad track, a billboard sign, and an airplane). Horizontal lines on these objects that are parallel to the sides of the block (see Figure 7.8B) all recede to the same _____ _____ on the _____.
vanishing point; horizon		

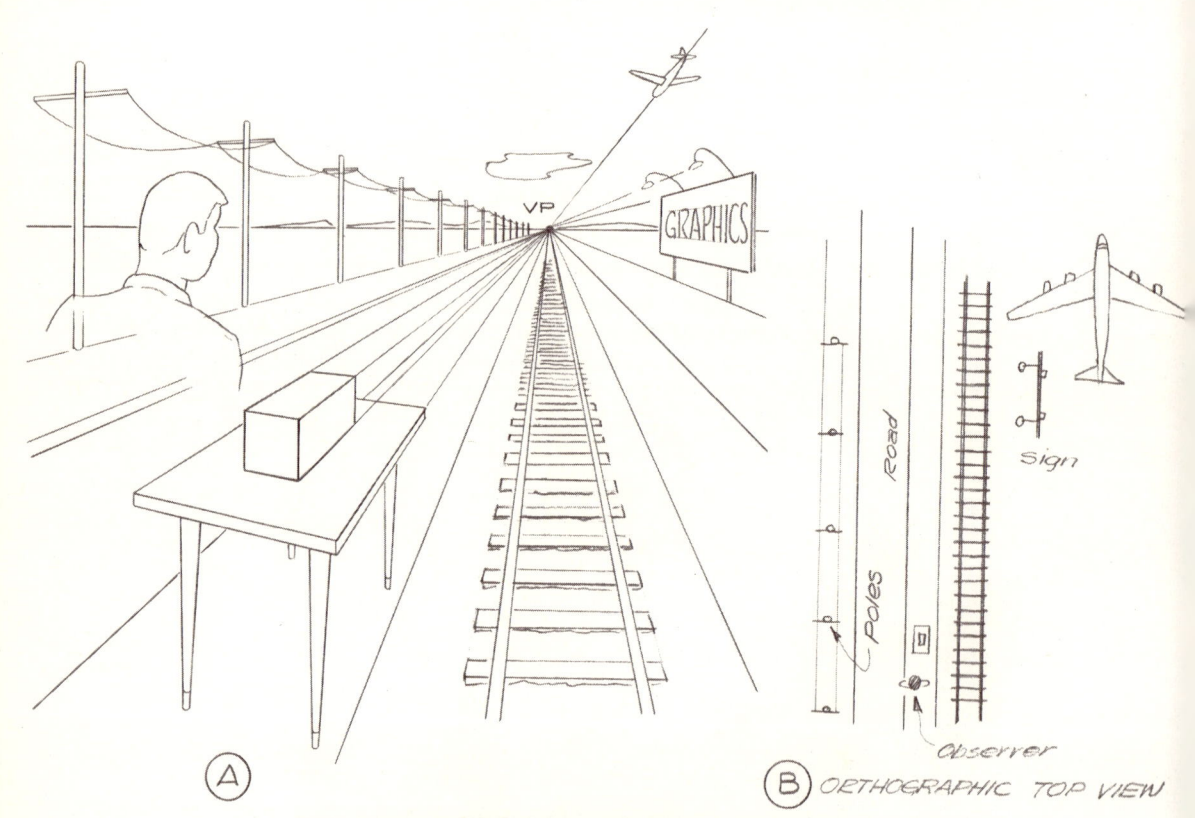

Figure 7.8 Horizontal, parallel lines recede to a single vanishing point (VP) on the horizon line.

	36	We may write the *first rule of perspective:* All *parallel, horizontal* lines on an object recede to a single vanishing point on the horizon. Lines through the tops and bottoms of the telephone poles are, in reality, parallel, horizontal lines and therefore recede to a vanishing point on the _____.
horizon		

	37	The projected line of flight of the airplane meets the single vanishing point; therefore, we know that the plane is flying _____ to the railroad tracks.
parallel		

THE THEORY OF PERSPECTIVE 121

First Rule of Perspective

We can generalize the first rule of perspective by stating:

All parallel lines on an object recede to a single vanishing point.

We have omitted from our first statement the facts that (1) the lines are horizontal and (2) the vanishing point is on the horizon. The implication of this omission is that only horizontal parallel lines have a vanishing point on the horizon. Figure 7.9 shows that vertical and sloping parallel lines do have single vanishing points, but they are not on the horizon. Figure 7.9 also illustrates how the horizon is used as a base line in deciding how we wish to view an object (bird's-eye view, gnat's-eye view, or normal view).

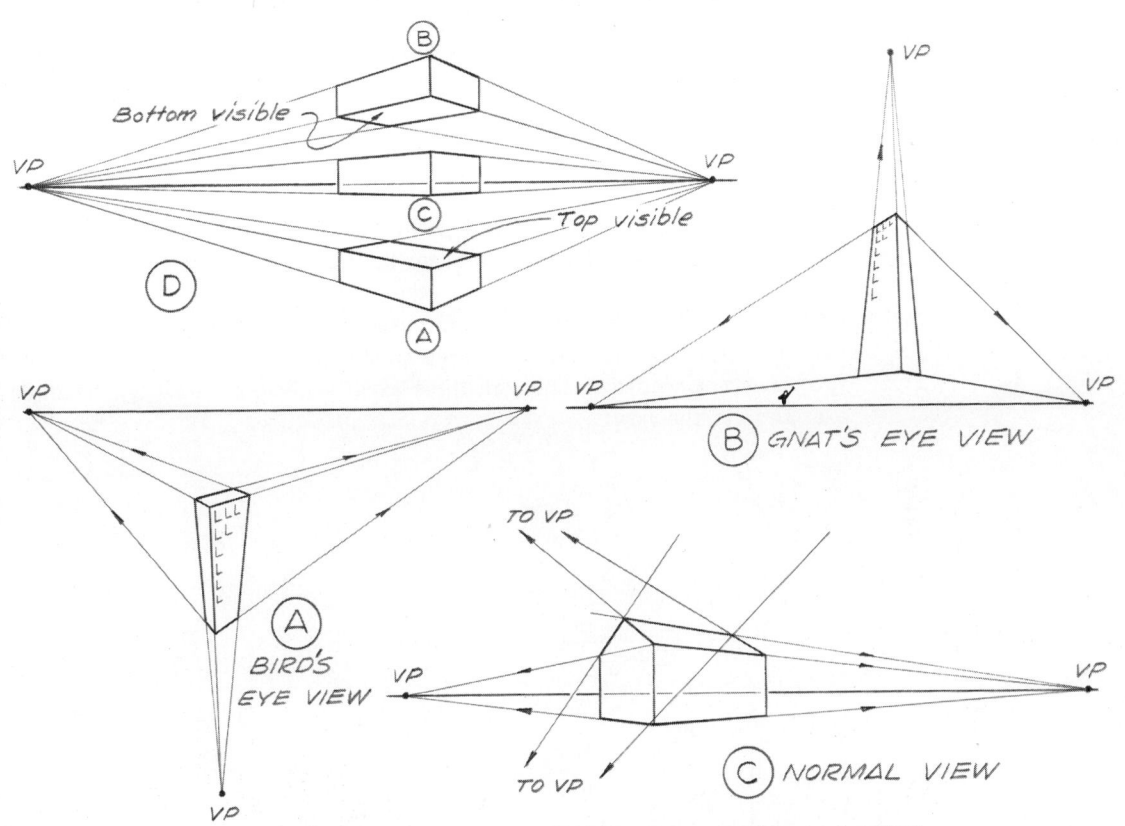

Figure 7.9 Every set of parallel lines has a vanishing point (VP).

	38	No matter where an observer stands with respect to the ground plane (on the ground, below ground, or in the air) the horizon is *always* at his _____ _____ .
eye level		
	39	If the observer's eyes are above the object being viewed (Figure 7.9A), he is looking down at the object. To sketch the object, he would start sketching the space axes _____ the horizon.
below		
	40	Conversely, he would start the sketch above the _____ if he is viewing an object much higher than he is. (Figure 7.9B.)
horizon		

122 PICTORIAL DRAWING/UNIT 3

at the horizon

41 If the main part of the object is at the observer's eye level, his sketch must start _____ (your words). (See Figure 7.9C.)

top, bottom, or neither top nor bottom

42 This reasoning relates correctly with the three action blocks of the logic diagram (Figure 7.4 and 7.5) which were the instructions for our decision, in relation to Figure 7.9D, to show the _____ _____ of the box (your words).

■ **43** Sketch 22A is an exercise in positioning the horizon line. Three scenes are presented. Based upon the foregoing discussion of the horizon and the vanishing points, sketch the approximate position of the horizon in each scene. Check your solutions.

Two-point Perspective

Before proceeding through the rest of the logic diagram, let's settle on two-point perspective as a standard form for the rest of this chapter. Figure 7.10 shows this form in relation to two other forms, one-point and three-point perspective. The system of space axes used in graphics consists of one vertical axis (H) and two horizontal axes (W and D). In orthographic drawing, two views are needed to show all three axes. In pictorial drawing, the three axes show in one view.

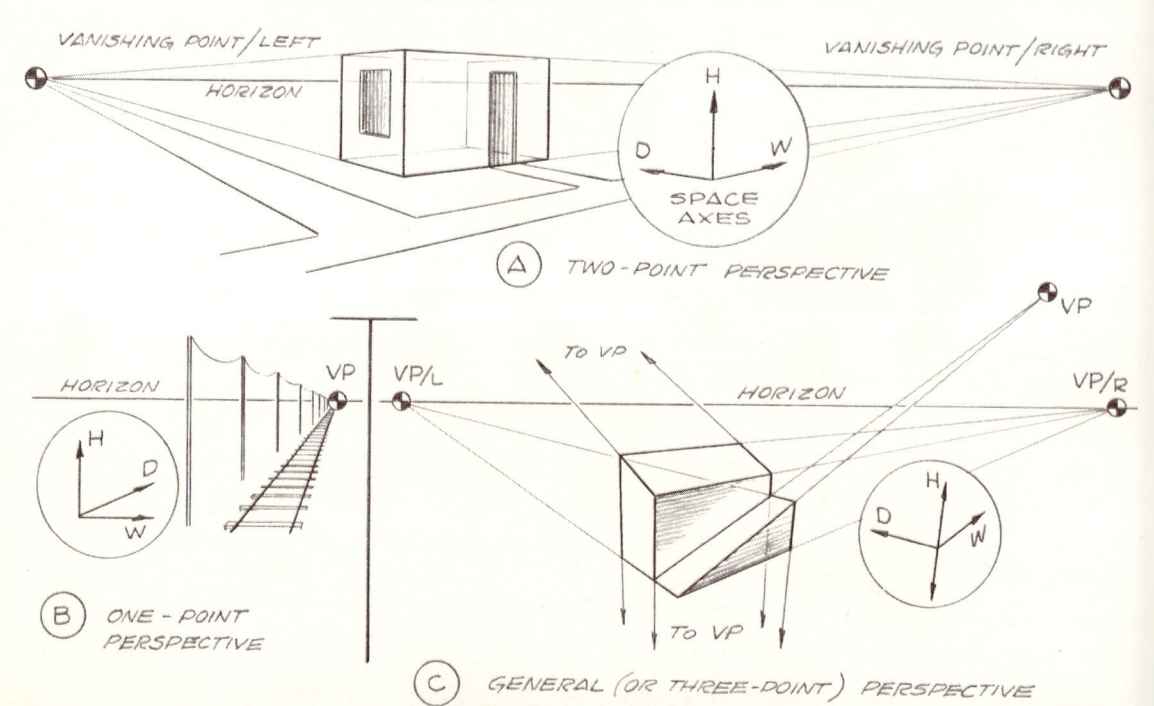

Figure 7.10 Perspective systems. Two-point perspective is the accepted form for technical drawing.

THE THEORY OF PERSPECTIVE 123

parallel;
parallel

44 The box enclosing any object is merely a system of three sets of _____ lines. Each set is _____ to one of the space axes.

vertical

45 Figure 7.10A illustrates two-point perspective. Lines parallel to the H axis are _____ lines.

right, left

46 Lines parallel to the W and D axes recede into space and converge on two points, the _____ and _____ vanishing points (VP).

horizontal

47 VP_R and VP_L are on the horizon because the W and D axes are _____ lines.

One-point Perspective

Figure 7.10B is a one-point perspective. The H axis is vertical, one of the horizontal axes (D) recedes to a single VP on the horizon, and the W axis is parallel to the horizon (VP_R at infinity). There is little use for this form in technical drawing. Architects use it to show room interiors when they wish to picture three walls, the ceiling, and the floor. It is sometimes used in illustration for dramatic effects.

Three-point Perspective

Figure 7.10C is a three-point perspective. This is the truest form in that lines parallel to the H axis converge on a third VP. Two-point perspective neglects this third VP and makes all H lines vertical and parallel. Distortions introduced by this assumption are negligible for reasonably sized objects.

Setting the Horizontal Angle θ

visible (seen)

48 Back to the logic diagram. Figure 7.11 reproduces the next part. Assuming that we wish to see some of either the top or bottom of the object, we are asked, (at diamond E) "Is a large amount of the top or bottom surface to be _____?"

90°; 180°

49 Both the "yes" and "no" answers lead to instructions which give a range for setting the horizontal angle θ between the two horizontal space axes, W and D. The range is between _____ degrees and _____ degrees.

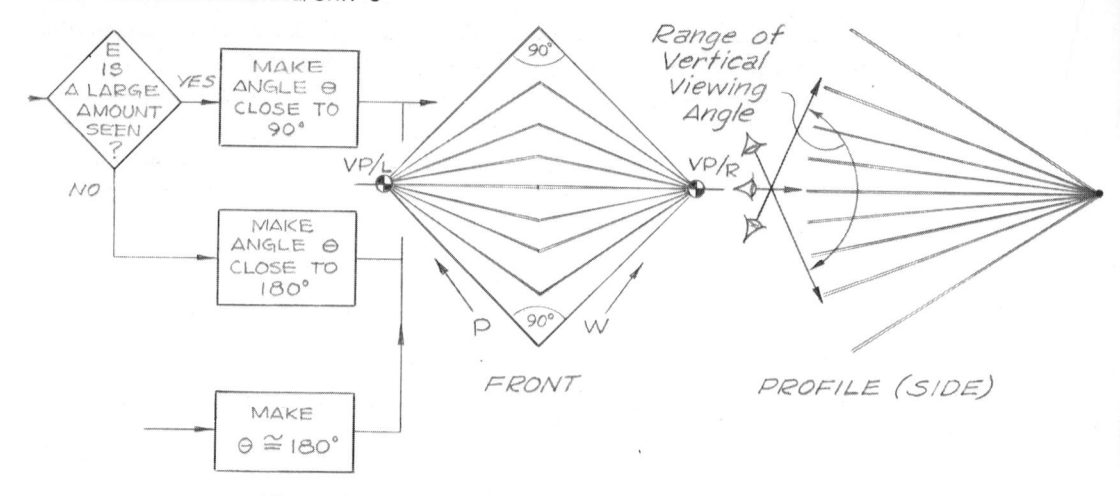

Figure 7.11 *Controlling the amount of the top or bottom of the box which is visible.*

50. If angle θ is closer to 180° than to 90°, we will see _____ (your words) of the top or bottom of the object. If θ is closer to 90°, we will see _____ (your words) of the top or bottom.

(180°): little
(a small amount);
(90°): a lot
(a large amount)

51. The diagram of Figure 7.11 shows the geometry involved. We see a fan-shaped series of perspective horizontal planes created by sketching a series of W and D axes to two vanishing points (VP_R and VP_L) on the horizon.

   Starting from the bottom of the diagram and working up, the angle between the W and D axes:
   (1) Is _____ degrees at the bottom.
   (2) Becomes _____ (smaller/larger).
   (3) Is _____ degrees at the horizon.
   (4) Becomes _____ (smaller/larger).
   (5) Finally ends up at _____ degrees.

(1) 90°;
(2) larger;
(3) 180°;
(4) smaller;
(5) 90°

52. Placing the box above the horizon exposes the _____ to view. With the box placed below the horizon, the _____ is seen.

bottom; top

53. The horizon is always at our eye level. This doesn't mean, however, that we cannot be looking up or down at an object. (See Figure 7.11—side view.) If we are looking up at the object, the object is _____ the horizon. If we are looking down, it is _____ the horizon.

above; below

THE THEORY OF PERSPECTIVE 125

viewing

54 Figure 7.11 shows the upper and lower *limits* (90°) for setting up the W and D axes according to the amount of the top or bottom we wish to see. It is a geometric translation into perspective of our actual vertical _____ angle if we were sketching the box from real life.

55 "Why is 90° the lower limit for angle θ?" The simple answer is that our sketch would become distorted (unrealistic) if we set the angle between the two receding axes at less than 90°.

vanishing points
(VP)

Consider Figure 7.12. It shows a circle whose diameter is the distance between the _____ (your words).

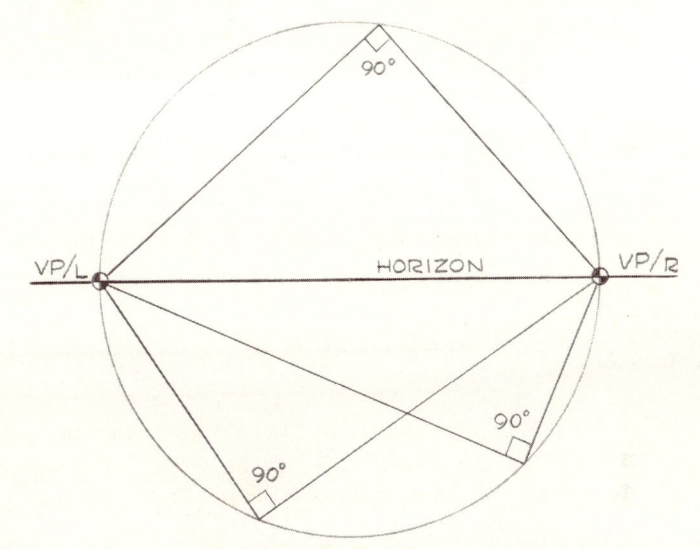

Figure 7.12 The lower limit of angle θ is 90° for a distortion-free sketch.

As a first guide, any box drawn outside of this circle will be distorted.

This puts a lower limit of 90° on angle θ. From plane geometry we know that if the diameter of a circle is one side of a triangle and the intersection of the other two sides lies on the circle, the angle between the two sides is 90° and the triangle is a right triangle whose hypotenuse is the diameter of the circle.

90; 180

56 Thus the limits of angle θ are _____ degrees minimum and _____ degrees maximum (when the base of the box is on the horizon).

57 We have not looked at one action block of the portion of the logic diagram of Figure 7.11. This is the instruction of what to do if we wish the box to straddle the horizon (neither top nor bottom of the box visible). The instruction is: Set θ _____ 180°.

$\cong$
(approximately
equal to)

58 It is reasonable to expect angle θ to be very close to 180° when the box straddles the horizon. Referring again to Figure 7.11, side view, we see that at this condition, the box is at the observer's _____.

eye level

59 In summing up this discussion of angle θ, we can say that a change from $\theta = 90°$ to $\theta = 180°$ is a change in the vertical position of the object in space which moves the object _____ our horizontal eye level. (*Choose one:* closer to/farther from.)

closer to

We will see more of how the selection of angle θ helps to set up the space axes for a successful perspective sketch. For the moment, let's complete the decision-making portion of our logic diagram. Remember that we are still concentrating on the problem of sketching a rectangular box in perspective.

60 After deciding how much, if any, of the top or bottom of the box is to be seen, we are asked next (in Figure 7.13) about the visibility of the _____ of the box.

sides

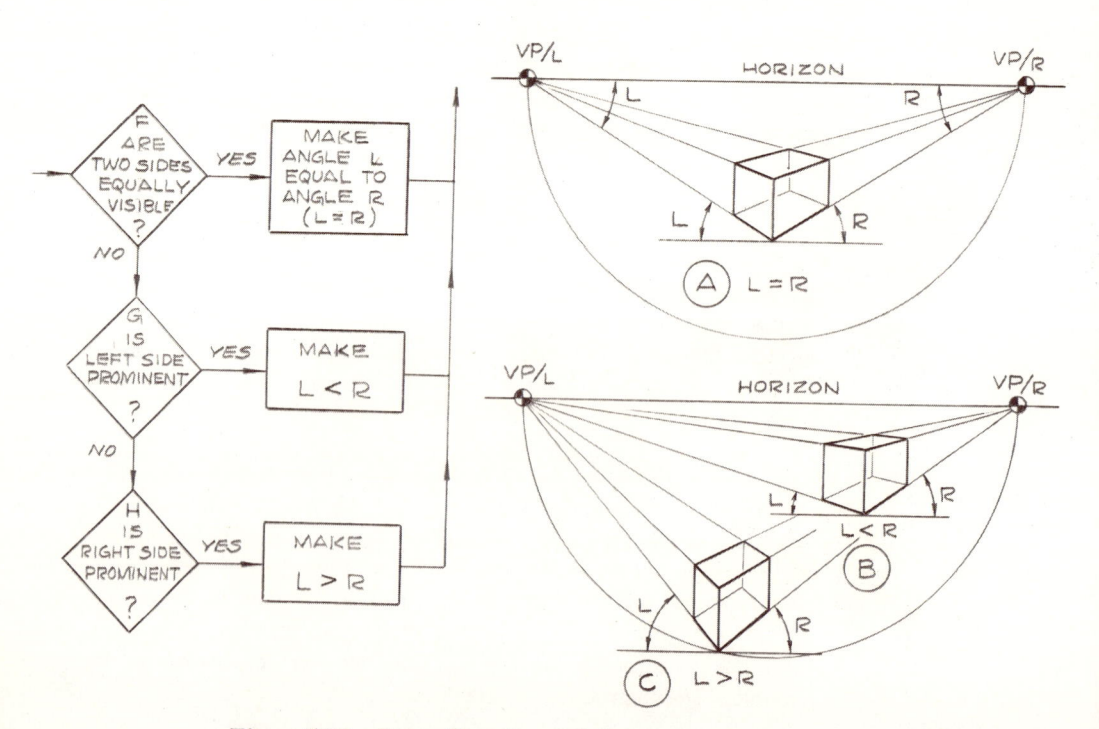

Figure 7.13 Controlling the visibility of the two exposed sides of the box.

THE THEORY OF PERSPECTIVE 127

=
(equal)

61 This introduces two new angles, L and R. If we wish both sides to be equally visible, we are told to make L _____ R.

the horizon
(a horizontal line)

62 This is shown in Figure 7.13A. Angle L is the angle that the *left* receding axis makes with _____.

equally

63 Angle R is the angle that the *right* receding axis makes with the horizon. In 7.13A L = R so that the two sides of the box are seen _____. (Don't be confused by the fact that the right side is larger than the left side. This is the nature of the box we are sketching.)

(1) < (is smaller than);
(2) > (is larger than)

64 Figure 7.13B and C shows the other two possible conditions:
(1) When the left side is displayed more prominently than the right, then L _____ R.
(2) When the right side is turned more toward our line of vision, then L _____ R.

Figure 7.14 shows how angle θ relates to angle L and R. The sum of the three, $L + \theta + R = 180°$. This will always be true. We haven't said anything about the actual sizes of angles L and R, only their *relative* sizes. In both examples of Figure 7.14, angle L is smaller than angle R which means that the left side of the object will be more prominent than the right. Angle θ is close to 90° in the set of axes above the horizon and close to 180° in the set below. In both instances, the sum of the three angles is 180°.

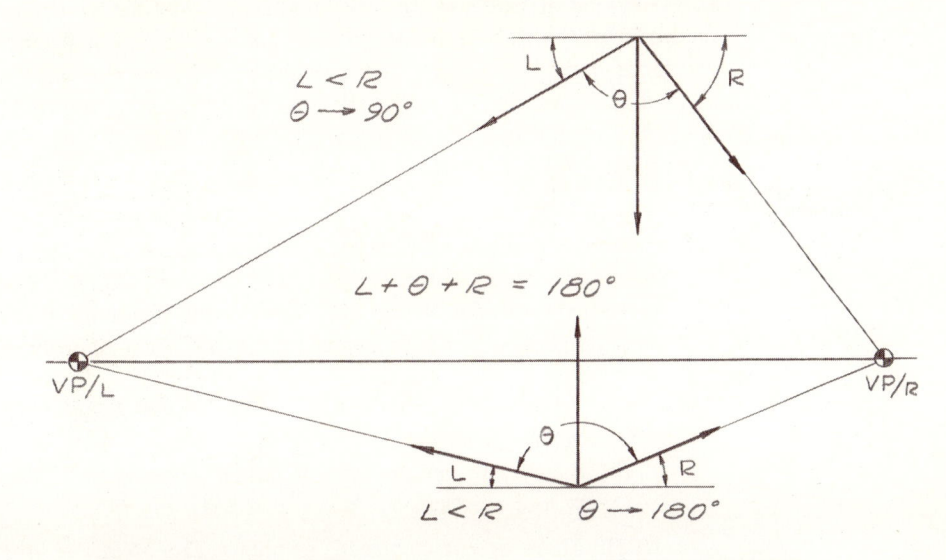

Figure 7.14 The relative sizes of angles L, θ, and R are important, knowing that $L + \theta + R = 180°$.

128 PICTORIAL DRAWING/UNIT 3

vanishing points
(VP$_L$; VP$_R$);
horizon

65 The relationships between these three angles control the position of the HWD space axes within a circle whose diameter is equal to the distance between _____ _____. The center of the circle is on the _____.

vertical

66 Angle θ controls the _____ position of the axes. (Choose one: horizontal/vertical.)

horizontal

67 Angles L and R control the _____ position of the axes.

(1) axes closer to VP$_L$ (left of center);
(2) axes centered;
(3) axes closer to VP$_R$ (right of center)

68 State simply and in your own words the horizontal position of the space axes within the limiting circle for the following conditions:
(1) Angle L large, angle R small: _____
(2) L = R: _____
(3) Angle L small, angle R large: _____

Distortion of a perspective sketch results if we set the horizontal position of the axes too close to either VP. Figure 7.15A shows an additional constraint placed upon the horizon-line geometry. The diameter of the limiting circle has been divided in quarters. Vertical lines have been drawn through points M and N, the midpoints of the two halves of the diameter. The shaded area is the zone of distortion. A distortion-free perspective sketch should be set up within the unshaded area.

This criterion for distortion-free perspective sketches seems to be quite confining, especially when we examine the figures on these pages. Normally, we make a sketch of reasonable size on standard paper (8½ × 11 or larger). For sketching at this size, *the VP must be at least 30 to 40 in. apart.* This naturally precludes having the VP available on the drawing paper unless we use large paper on a large board or table. Once having established angles θ, L, and R, we can learn to estimate the position of the VP by sketching lines that are not parallel, but converge (come together) toward an imaginary point. Figure 7.15B illustrates this idea.

■ 69 Sketch 22B is an exercise in setting up space axes for a sketch of a rectangular box. A horizon VP$_L$, VP$_R$ and the outline of the zone of distortion are given. Set up HWD axes for the following:
(1) A box with much of the top visible; right side more prominent than left.
(2) A box with a small amount of the bottom showing; right and left sides equally prominent.
(3) A box with neither top nor bottom showing; left side prominent. Check your solutions.

NOTE: There are no "correct" solutions. The solutions given meet the above criteria. Yours should be close to these.

THE THEORY OF PERSPECTIVE **129**

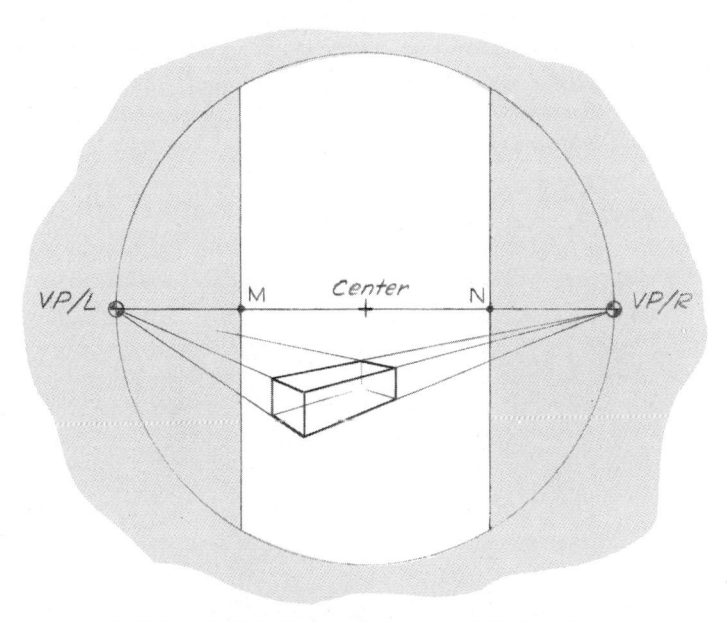

Figure 7.15A Further constraints on positioning the space axes for a distortion-free perspective sketch.

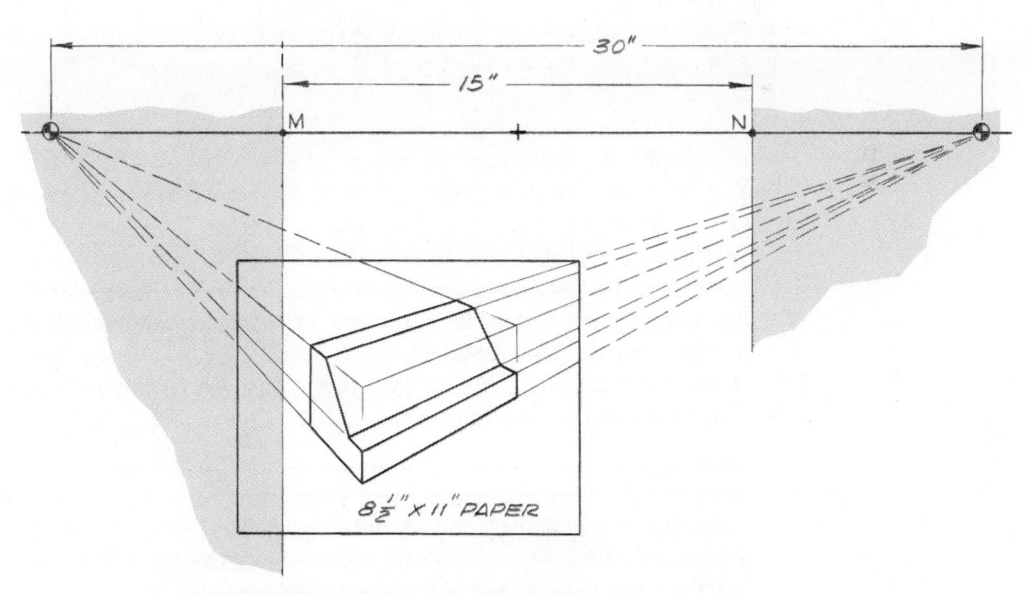

Figure 7.15B Perspective sketching involves estimating the positions of the vanishing points.

130 PICTORIAL DRAWING/UNIT 3

FORESHORTENING

Earlier we mentioned three concepts important to perspective sketching. They are:
  (1) The horizon line
  (2) The vanishing points
  (3) Foreshortening

We have discussed the first two. The concept of foreshortening will be taken up in this series. First let's set up the sketch to the point where we must consider foreshortening.

Sketching the Box

We are now ready to finish the box. All "yes/no" decisions have been made. Figure 7.16 shows the end of the logic diagram. It is a series of action blocks leading to the completed sketch.

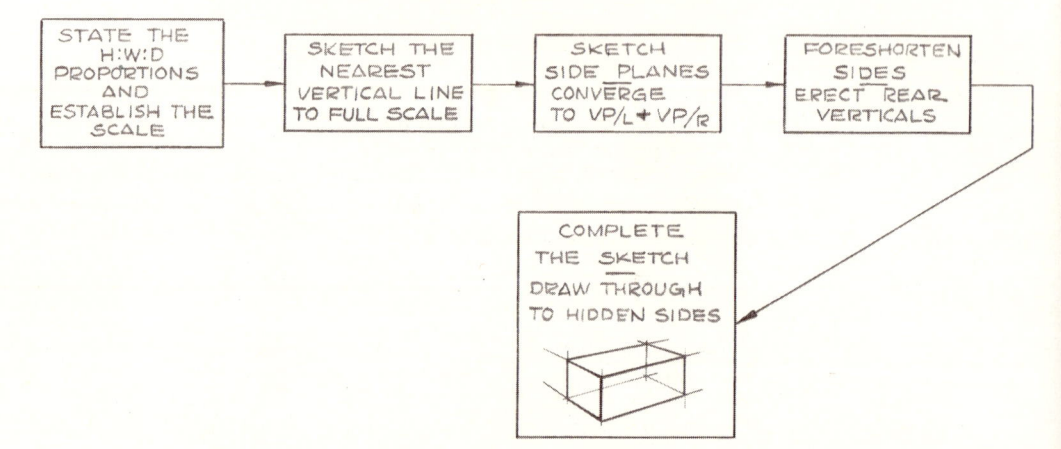

STATE THE H:W:D PROPORTIONS AND ESTABLISH THE SCALE → SKETCH THE NEAREST VERTICAL LINE TO FULL SCALE → SKETCH SIDE PLANES CONVERGE TO VP/L + VP/R → FORESHORTEN SIDES ERECT REAR VERTICALS → COMPLETE THE SKETCH — DRAW THROUGH TO HIDDEN SIDES

Figure 7.16 The final steps in the logic diagram—sketching the box.

---

70 We must know *what* we want to draw. This would be the object specifications. Information needed is:
SIZE: The ____:____:____ proportions of the enclosing box.
POSITION: How we wish to _____ the object.

H:W:D;
view (see, look at)

■ **71** The following directions will set up the planes of the two visible sides of the box. Work on the first set of axes you drew in Sketch 22B. The *position specification* was top and right side prominent. We will add the *size specification* that H:W:D = 3:6:4 (in ¼-in. units).

NOTE: We will assume that the largest dimension is always the *W* dimension. This is an arbitrary choice. *W* and *D* can be interchanged according to the nature of the object.

Measure or estimate the *H* dimension (¾ in.) on the vertical from the intersection of the *W* and *D* axes. Sketch lines from the top and bottom of this vertical line to both VP_L and VP_R. Check your solution.

THE THEORY OF PERSPECTIVE 131

These two triangles represent the planes of the visible sides of the box. Our next step is to determine the W and D measurements along the axes and sketch the vertical lines that define the rear limits of the sides. These sides are at an angle to our line of sight and are therefore *foreshortened*. Because of this we cannot make full-scale measurements along the W and D axes.

---

finger

72 Look at Figure 7.17. You see a man pointing his _____ at you (at me, us, etc.).

---

no

73 Actually he is pointing his entire arm directly at the observer. The arm carrying his hand and right index finger is almost nonexistent in the drawing. Does this mean that we can assume that he has no right arm (or, at least, a very short one)? _____.

---

profile (side)

74 The pictorial drawing is correct. We know instinctively that the arm is there, but since it lies along (is parallel to) our line of vision we are seeing an end view. Figure 7.18 is a _____ view of the same scene. In it we see the full arm.

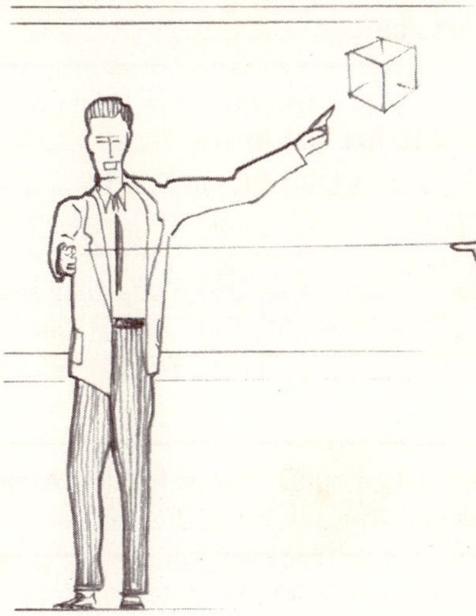

Figure 7.17   Foreshortening.    *Figure 7.18   Profile view.*

The man's right arm in Figure 7.17 is said to be *foreshortened*. Foreshortening is the apparent visual reduction in size of a line or a surface as it (1) is turned away from our line of vision or (2) is moved farther away from our point of vision.

132 PICTORIAL DRAWING/UNIT 3

75 Consider Figure 7.19. It illustrates both types of foreshortening. A truck and trailer combination is sketched in four different positions as we might view it from the side of the road. At 7.19A we see an orthographic side view.

x; y

The length of the trailer body is _____ and the height of the top above ground is _____.

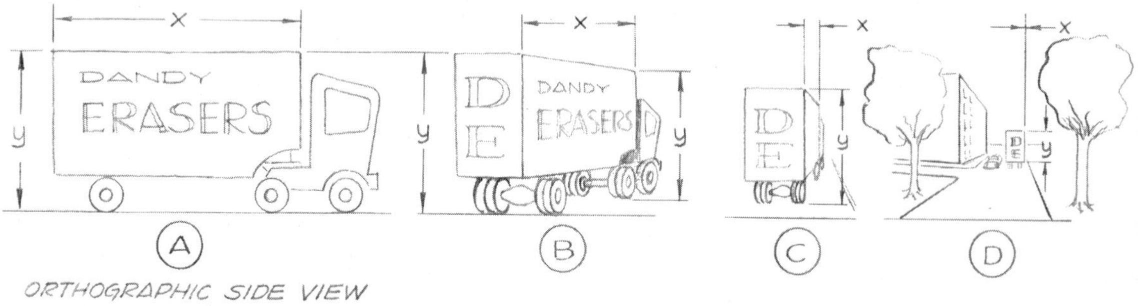

Figure 7.19 Lines and surfaces are foreshortened as they are (1) turned away from our line of sight and (2) moved farther away from our observation point.

76 In Figure 7.19B, the truck has started around a corner, turning away from our line of sight. The length x now appears to be _____ than full length.

shorter (smaller)

77 In 7.19B we are now looking obliquely at all lines of x dimension. They are said to be _____.

foreshortened

78 Dimension y, the height of the trailer body, shows two values in Figure 7.19B. It is full size at the _____ of the trailer body and something less than full size at the _____. (Choose one: front/rear.)

rear; front

79 The front of the trailer body is farther away from us than the rear of the body. Because of this, all vertical lines of dimension y appear to diminish in size from rear to front. The lines farthest away are said to be _____.

foreshortened

80 Figure 7.19C and D shows further scenes in the sequence. In 7.19D the dimension x is _____.

zero (0)

81 The truck is now far down the side street and dimension y is drastically foreshortened.

Dimension y would decrease to zero when the truck _____ from sight.

disappears

We know intuitively that the dimensions of the trailer body do not change size as the truck turns 90° to our line of sight and proceeds away from us. But the effects of foreshortening make them appear to change, and we would actually see them change if we were observing on the scene. Perspective sketching involves drawing objects as we actually see them. Therefore, the effects of foreshortening must be introduced into a perspective sketch.

The effects of foreshortening are ever present. The person with an untrained eye may not be aware that the dimensions of an object are constantly changing as he moves with respect to the object or the object moves with respect to him. The scene in Figure 7.19D is perfectly natural since the size of the truck is correct in relation to the size of objects in *its* immediate surrounding (trees, building, auto, etc.). It would be a real shock if the truck diminished in size and these objects did not.

82 Figure 7.20 shows another familiar scene that illustrates the two types of foreshortening. Foreshortening is the apparent visual reduction in size of a line or a surface as it is (your words):

(1) _____ our *line* of vision.
(2) _____ our *point* of vision (observation point).

(1) turned away from
(2) moved away from (or farther from)

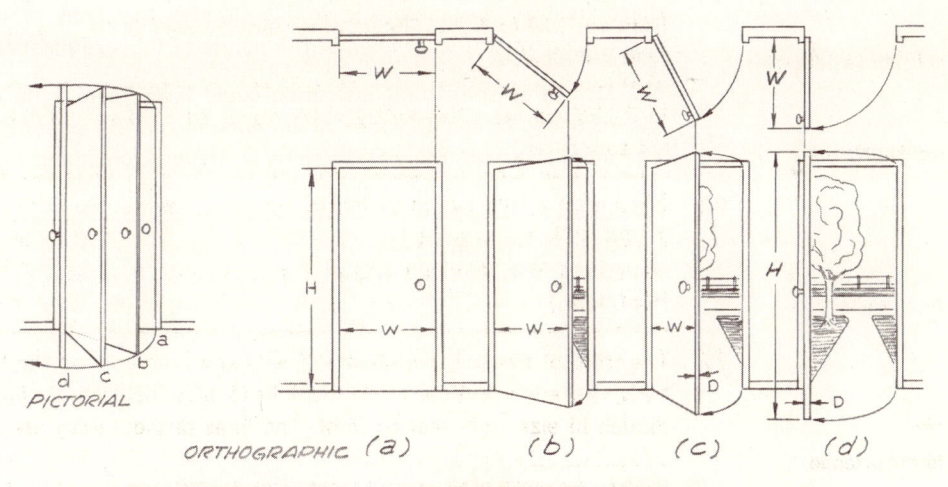

Figure 7.20 The H, W, and D dimensions of a door appear to change as it swings toward us.

83 A line or surface at any distance from our observation point appears to be of *maximum* relative size when it is _____ to our line of vision.

perpendicular

84 As the line or surface is turned away from the perpendicular to our line of vision, it passes through an infinite number of positions, each presenting a different foreshortened view. The size reaches a minimum when the line or surface is _____ to our line of vision.

parallel

134 PICTORIAL DRAWING/UNIT 3

Let's finish the sketch of the 3:6:4 box which we started a moment ago, taking into account the effects of foreshortening. In Sketch 22B, we have two triangles that represent the *planes* of the two visible sides of the box.

cannot

85 Remember that we started the sketch by measuring three full units in the H direction. We _____ measure off six full units in the W direction and four full units in the D direction. (*Choose one:* can/cannot.)

foreshortened

86 The sides of the box are not perpendicular to our line of vision and, therefore, are _____.

An Intuitive Method of Foreshortening

■ **87** Figure 7.21 shows an intuitive method for foreshortening receding lines and surfaces. Make the following constructions on Sketch 22B using Figure 7.21 as a guide.

On a horizontal line through the base of the H line measure 6 units to the right and 4 units to the left. Sketch *circular* arcs to transfer these full-size measurements to the bottom side lines of the sketch. Erect light, vertical trial lines at the points of full-size measurement.

NOTE: The 6- and 4-unit measures could have been made directly on the receding W and D lines. When working on grid paper, it is easier and more accurate to count divisions on the horizontal and then transfer the lengths to the receding lines.

FORESHORTENED RECTANGULAR SIDES MUST <u>LOOK</u> <u>LIKE</u> THE CORRECTLY PROPORTIONED RECTANGLES

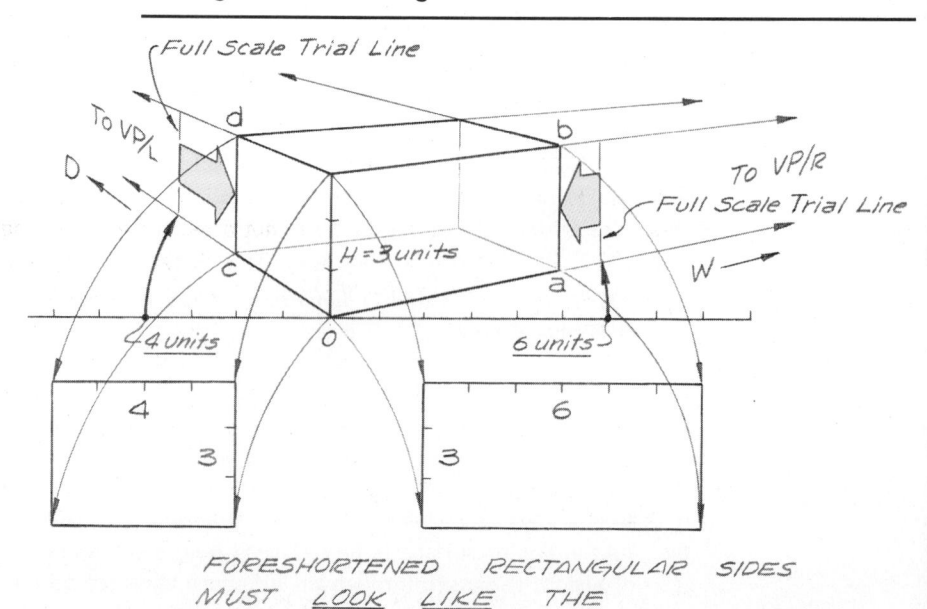

Figure 7.21 An intuitive method of foreshortening the receding sides of a box.

THE THEORY OF PERSPECTIVE   135

no

88 These vertical trial lines cut off the receding planes at full-scale W and D measures. Do the resulting rectangular perspective planes *look like* a 3 × 6 rectangle and a 3 × 4 rectangle? Yes _____; no _____.

89 Your answer should be "no" because the rectangular planes really do not look right. If you cannot see that they don't look right, you should know by now that the sides cannot be drawn _____ because they are foreshortened.

full scale
(full size)

■ **90** Move the right and left vertical lines in by sketching light trial lines *until you think the sides look like* 3 × 6 and 3 × 4 rectangles. Reference to Figure 7.21 may help you on this.

■ **91** From the tops and bottoms of these vertical lines (points a, b, c, and d of Figure 7.21), sketch lines to the correct VP. Add the hidden back vertical line (light construction). Darken all visible lines. The sketch is complete. It should look like a box of H:W:D = 3:6:4. The back hidden lines were sketched in lightly. It is all right, even desirable, to leave light construction lines on a technical sketch. They help to give you a feel for the solid, three-dimensional object.

The foregoing method of foreshortening was called an *intuitive* method. It required you to decide when the proportions of a rectangle in perspective *looked like* the desired proportions. This is not easy, but you can train your eye to become skilled in proportioning visually just as you can train your hand to become skilled at manual tasks. Skill comes through practice.

Basic-square Method of Foreshortening

Another method of foreshortening surfaces in perspective sketching is one using a square as a basic unit and then multiplying the square graphically to find the extent of the surface desired.

Figure 7.22 shows the principle behind the method. In 7.22A, the 3 × 6 rectangle is shown in an orthographic view. It has been marked off in an 18-square grid. In 7.22B the figure is not a rectangle but a long thin triangle that is 3 units high at the wide end. This is the same type of triangle we obtain when we draw parallel horizontal lines in perspective. Both sketches were drawn by the same method.

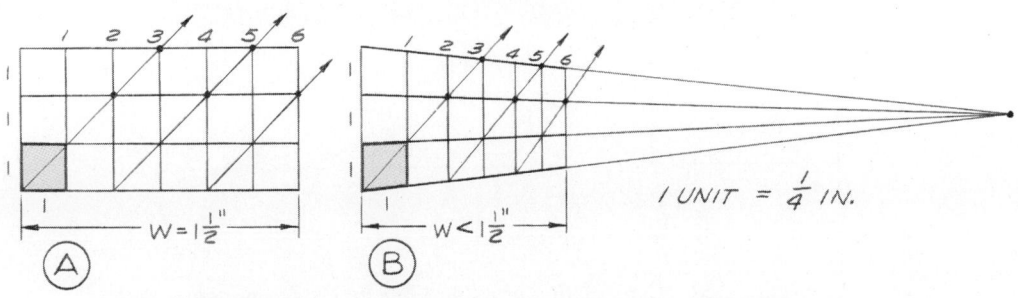

Figure 7.22   The basic-square method of foreshortening receding planes.

136 PICTORIAL DRAWING/UNIT 3

three (3)

92 The height *H* was divided off into _____ units.

93 Horizontal lines were drawn to divide the area into three equal horizontal strips. In Figure 7.22B, the lines are horizontal perspective lines meaning that they _____ (your words) a single point, the vanishing point.

converge on
(recede to,
vanish to)

94 Vertical line No. 1 was sketched one unit from the left, full-scale vertical line, creating a vertical column of squares. Vertical lines No. 2 and 3 were sketched through points which are the intersection of the horizontal lines and a(n) _____ constructed through the bottom square in the first column.

diagonal

95 This created a 9-square array. Two more diagonals were constructed until the full, 18-square array was achieved. In Figure 7.22A this is an accurate 3 × 6 rectangle. Note, however, that each column of "squares" in 7.22B is slightly narrower than the preceding column. This gradual narrowing is caused by the fact that the horizontal lines are _____.

converging

96 Because the horizontal lines of 7.22B are converging, both the height and width of the squares becomes smaller as we move from left to right. Each square is foreshortened and thus the entire _____ is foreshortened.

rectangle

Figure 7.23 shows this method applied to the problem of Figure 7.21. The results are the same. The basic-square method still involves one intuitive step. The first square constructed must be foreshortened visually *to look like a square* in perspective. In Figure 7.22 we omitted this step by making the first column exactly one unit wide. As we will see in a moment, this can be done if angle *L* or *R* is small

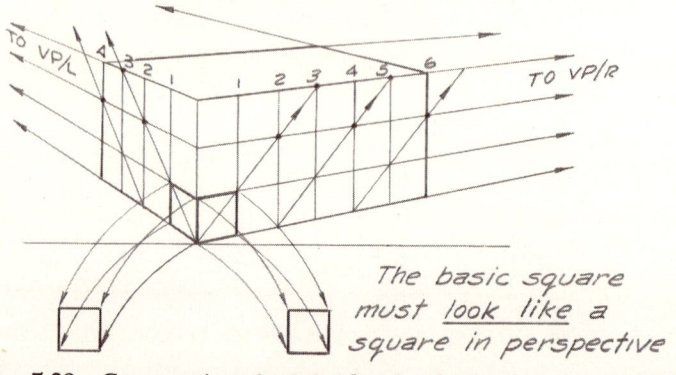

Figure 7.23   Construction of a 3:6:4 box by the basic-square method.

THE THEORY OF PERSPECTIVE 137

■ **97** In Sketch 22B, you sketched the space axes for a box with neither top nor bottom showing and with the left side prominent. On these same axes, sketch a box with H:W:D = 2:3:2. Use the basic-square method with ½ in. = 1 unit. Check your solution.

In effect, we have reached the goal of this chapter—to sketch a rectangular box of any size and in any position or orientation in space. Let's pick up a few loose ends.

98 Consider Figure 7.24A. It shows a horizon and a phantom observer holding a cube at a position below the horizon. The horizon is always located at the observer's _____ _____ no matter where the observer is positioned with respect to the ground plane.

eye level

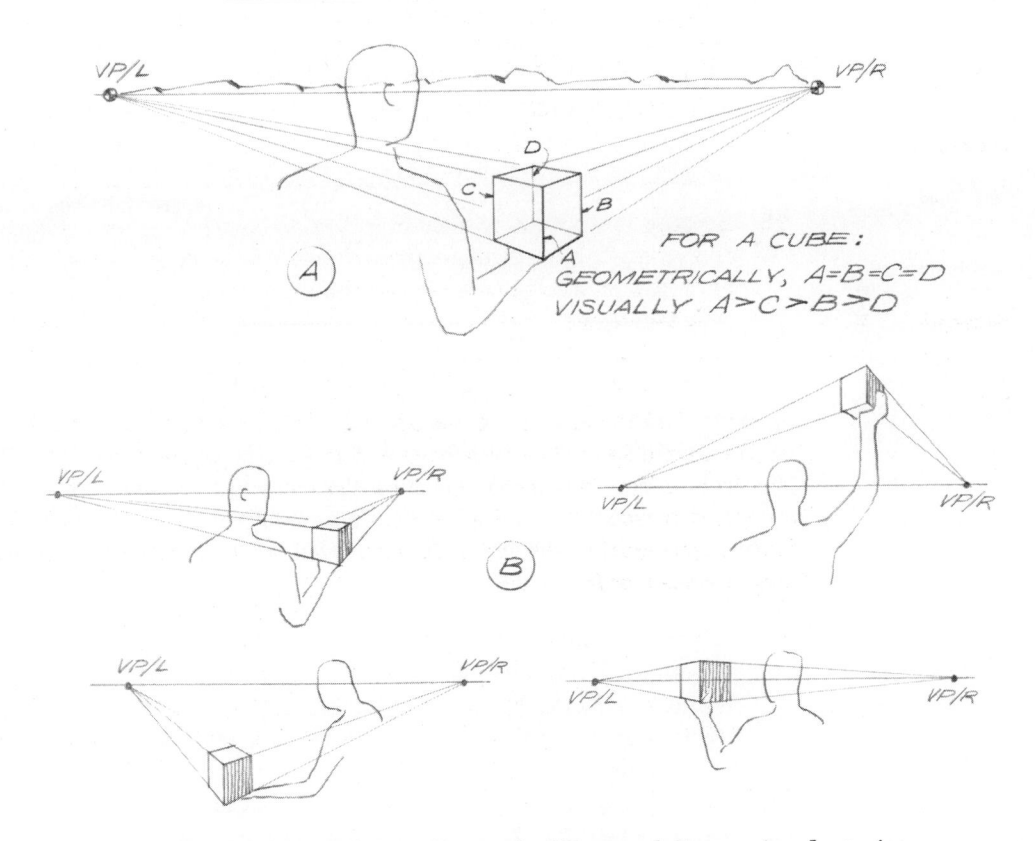

Figure 7.24 Lines receding to the VP control distance foreshortening.

99 Holding the cube below the horizon exposes the _____ and _____ _____.

top; two sides

138 PICTORIAL DRAWING/UNIT 3

VP_L; VP_R

100 The two sets of horizontal, parallel lines on the cube are shown vanishing (receding) toward two points, _____ and _____, on the horizon.

vanishing point (VP)

101 Notice that the vertical edges of the cube are parallel lines. For most of our work in technical drawing we will neglect the vertical _____ _____ and keep vertical lines parallel.

horizontal

102 Figure 7.24B shows the cube in other positions with respect to the horizon. It is as if you fastened elastic bands to the front edge of the cube and the other ends to the VP. As you move the cube around in front of you, the bands are the straight lines that determine the _____ edges (both top and bottom) of the cube.

shorter

103 Since these lines converge back to the VP, three of the vertical edges will be shorter than the front edge. In Figure 7.24A, vertical edges B, C, and D are _____ than vertical edge A.

vanishing points (VP)

104 This obeys the rule of foreshortening that states, "Object lines become smaller as they are moved farther away from the observer." We know that lines $A = B = C = D$. (It's a cube). The horizontal lines receding to the _____ _____ govern the amount that these lines are shortened.

What about the other rule of foreshortening that states "An object line becomes shorter as it is turned away from the observer's line of vision"? The angle between a plane surface and our line of vision can vary from 90° (perpendicular) to 0° (parallel). We control this in our sketch by the choice of angles L and R. The cubes of Figure 7.24 show this effect in the amount of the two sides or the top of the cube that shows as they are moved with respect to the observer.

picture plane

105 Figure 7.25 is a pictorial drawing of an observer looking at an object located behind a(n) _____ _____. You will remember that in Chapter 1 we discussed how we see and draw objects and we set up a visualization scheme similar to Figure 7.25.

a'; b'; c'

106 Figure 7.25B is a three-view orthographic drawing of the scene. In the top view we see the edge view of the picture plane and can thus find the intersection of projectors drawn from the observer to points on the object. Three such projectors are shown to points, a, b, and c on the object. The images of these points on the picture plane are _____, _____, and _____.

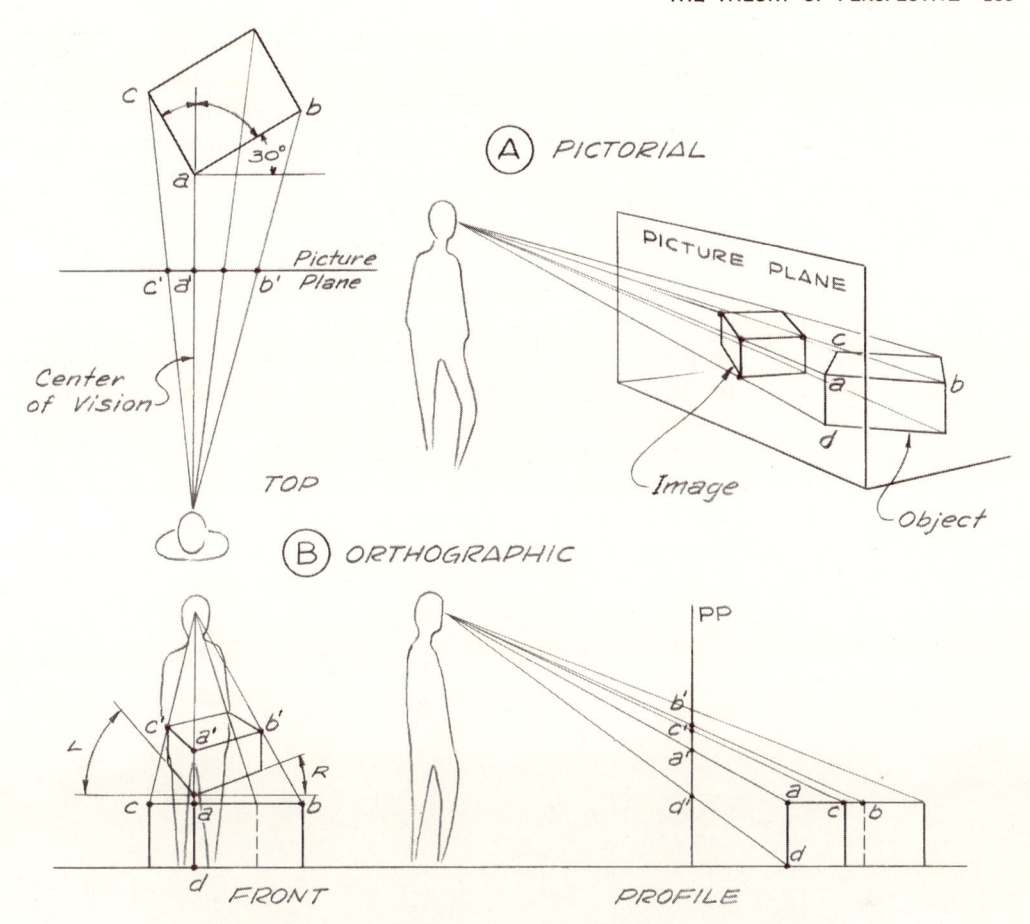

Figure 7.25 Viewing angle controls angular foreshortening.

	107	From the geometry of Figure 7.25B, it is evident that side *ab* and side *ac* on the object are both _____ than their projected images *a'b'* and *a'c'* on the picture plane. (*Choose one: shorter/longer.*)
longer		
	108	It is further evident that the amount of foreshortening of *ac* is greater than that of *ab*. This is true because *ac* makes a _____ angle with the observer's line of vision than does side *ab*. (*Choose one: larger/smaller.*)
smaller		
	109	Vertical lines on the object are foreshortened depending on their vertical position with respect to our line of sight. In the profile view of Figure 7.25B, the nearest vertical edge *ad* is reduced in length at the picture plane. The projected image is the line _____.
a'd'		
	110	If the object had been positioned so that vertical line *ad* was in the picture plane, then the projected image *a'd'* = _____.
ad		

140 PICTORIAL DRAWING/UNIT 3

In the work you did on Sketch 22B you made the H axis measurement full size. The assumption here was that this line, the nearest vertical edge of the box, was in the picture plane. This is an important assumption in that it gives us one measurable line on a drawing that is otherwise not measurable. Making this line full scale neglects the vertical foreshortening of this one line, but for reasonably sized objects and reasonable viewing angles, this omission is negligible.

---

111 Figure 7.26 is a projected perspective drawing of a rectangular box ($H:W:D = 1:2:1$). A full discussion of this drawing will come in Chapter 10, but it is useful to see the geometry now. A top view of the box is oriented so that line *ad* makes an angle of 30° with the picture plane and line *ae* lies in the _____ _____ .

picture plane

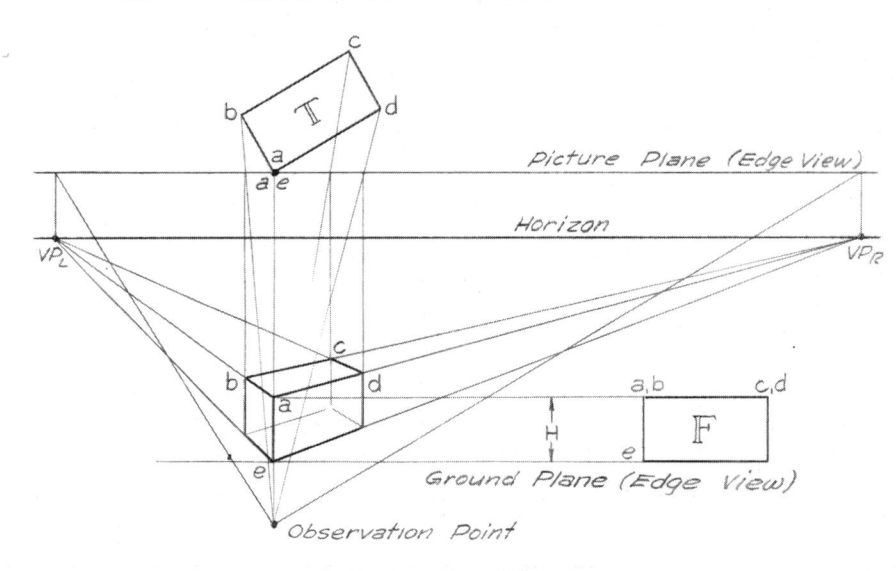

Figure 7.26 A projected perspective drawing.

---

112 Notice that a front view is positioned on the ground line. This view supplies the height dimension *H* which is not available in the top view. Since the front (nearest) edge *ae* of the box is in the picture plane, this line is the only _____ _____ line in the perspective drawing.

true-length
(full-scale, full-size)

---

113 Any line that lies in the picture plane is a true-length line. All others are _____ .

foreshortened

---

114 Figure 7.26 shows how the foreshortening is controlled by (1) the intersection of projection lines with the picture plane for _____ object lines and (2) the convergence of horizontal lines to the two vanishing points for _____ object lines.

horizontal;
vertical

THE THEORY OF PERSPECTIVE 141

Perspective Sketching without Vanishing Points

Up to now we have kept our objects small and used vanishing points that were on the printed page. Earlier, we said that, to make a sketch of reasonable size on a standard sized sheet of paper (8½ × 11), the vanishing points should be at least 30 to 40 in. apart. This precludes having the VP available. The methods developed in this chapter are designed to be used to make perspective sketches without vanishing points on the paper.

The geometry is shown in Figure 7.27. The lower half of the circular zone of distortion is shown in Figure 7.27A. The distance between VP is 32 inches. An 8½ × 11 sheet of paper is drawn to scale. All construction outside of the 8½ × 11 borders is dashed to indicate that it was not actually drawn. Figure 7.27B shows the 8½ × 11 sheet to an enlarged scale. This is what we actually draw. We must imagine where the VP are and sketch lines that converge in their direction.

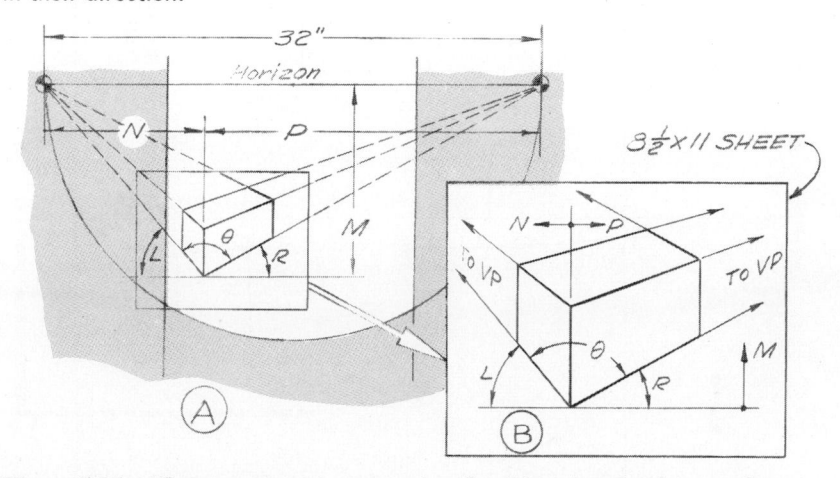

Figure 7.27 The geometry of perspective sketching showing how angles L, θ, and R control the sketch.

---

angle θ	**115**	The selection of angles, L, θ, and R give guideposts as to the location of the horizon and the VP. Study the three dimensions M, N, and P on Figure 7.27A. What geometric parameter controls the magnitude of dimension M _____ (your words).
small; large	**116**	When angle θ is close to 180°, dimension M will be _____. M will be _____ when θ is close to 90°. (Choose one: small/large.)
angles L and R	**117**	Distances N and P position the vertical axis horizontally. N + P equals the distance between VP. What parameter(s) control(s) the magnitudes of these two dimensions? _____ (your words).
(1) <; (2) >; (3) =	**118**	Complete the following: (1) When L > R, then N _____ P. (2) When R > L, then N _____ P. (3) When N = P, then L _____ R.

Keeping these relations in mind will help in freehand perspective sketching.

Sketching converging lines

To sketch converging lines effectively, use the technique of gazing beyond the plane of the paper (as in a vacant stare). This technique was developed for sketching parallel lines in Chapter 2. By removing your direct attention from the lines, you see both (or all lines) that are converging together. You will find it easier to bring them gradually together intentionally rather than making them parallel or diverging. Be careful not to make them converge too fast. This will bring the VP in close to the object and distort the sketch.

SUMMARY

Figure 7.28 is a summary sketch for Chapter 7. It shows five cubes oriented in space. All the geometry developed in this chapter is shown.

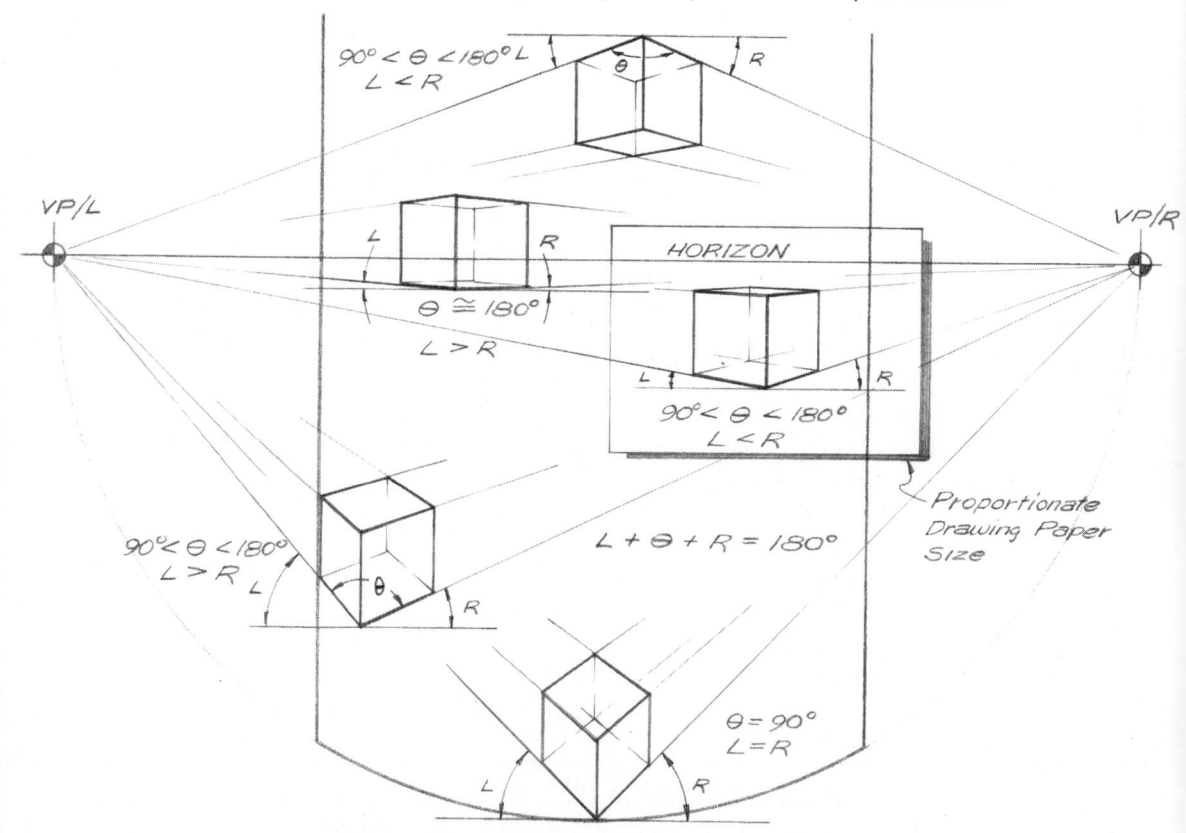

Figure 7.28 Summary sketch—a cube in five different positions in space.

■ **119** Sketch 23 is a criterion test for Chapter 7. You are asked to sketch three specific boxes. The position and size specifications and drawing scales are given. Do not use a horizon line or vanishing points. Check your solutions.

---

chapter 8 perspective sketching

ADDING DETAILS

Once we have sketched the basic box of *H:W:D* proportions, we wish to add details inside of the box and complete the sketch. This chapter presents:
Units of measurement
Shading (to enhance detail of objects)
Multiple boxes
Sloping lines and surfaces
Circles
Circular arcs
Irregular curves

Units of Measurement

1 In discussing all forms of pictorial drawing, we stress the construction of a cube. The arrangement of the space axes is of prime importance in starting a pictorial sketch. On a cube, the lengths of these axes are _____ and thus allow us to start off with a fundamental unit of measure in three directions.

equal

You should always think in terms of a unit cube in making a pictorial sketch. This is not a new concept. We use basic units when considering one- and two-dimensional concepts—so why not do the same in three dimensions? Figure 8.1 illustrates our use of basic units.

143

144 PICTORIAL DRAWING/UNIT 3

100

2. Figure 8.1A illustrates a one-dimensional unit. We are measuring the distance between two cities on a map. The basic unit is a measure of linear distance. It is one inch as far as the map is concerned, but is _____ miles as far as actual distance on the face of the earth is concerned.

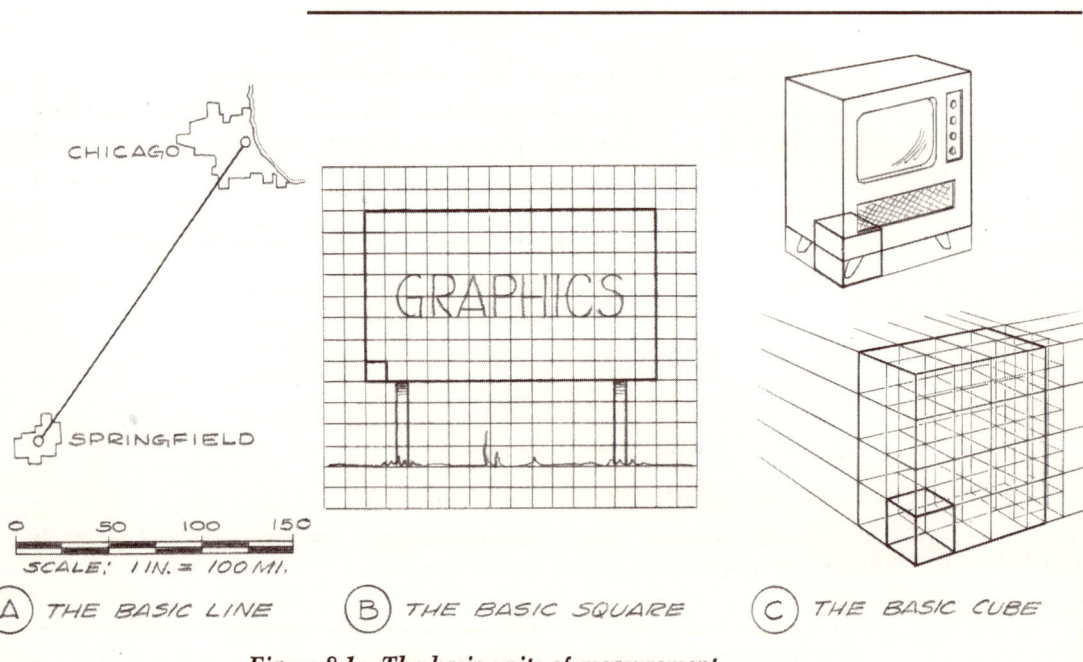

(A) THE BASIC LINE   (B) THE BASIC SQUARE   (C) THE BASIC CUBE

Figure 8.1 The basic units of measurement.

8:14

3. Figure 8.1B shows the two-dimensional basic unit. It is a unit of area. To maintain scale and proportion in two directions we use a basic square. Grid paper is a convenient device to supply the basic squares automatically. The basic units in each direction can be set as desired. The signboard shown has proportions of H:W = ____:____.

proportions

4. We need not define the unit of measure unless we wish to know the exact size of the sign. We can make an accurate sketch just knowing the _____.

H: 48;
W: 42;
D: 24

5. Figure 8.1C extends this thinking to three dimensions. Here the basic unit is the cube. The entire object, a TV console, was sketched by starting with one basic cube and then multiplying and subdividing it in three directions. If we use a 12-in. cube, the overall dimensions of the TV console are H = _____ in., W = _____ in., and D = _____ in.

PERSPECTIVE SKETCHING 145

In perspective sketching we must proportion visually. In Chapter 11 we will discuss the axonometric and oblique forms of pictorial drawing, in which we proportion geometrically. Geometric proportioning is easy. The pictorial forms that permit it are only approximations to reality. Learning to proportion visually through perspective sketching will result in a realistic picture and will develop your powers of visualization—a talent useful to all of graphics.

Shading

6 Chapter 2 introduced the subject of shading. It gave a gray scale showing how we can develop gray tones between _____ and _____ using a parallel-line pencil technique. The gray scale is reproduced here in Figure 8.2.

white, black

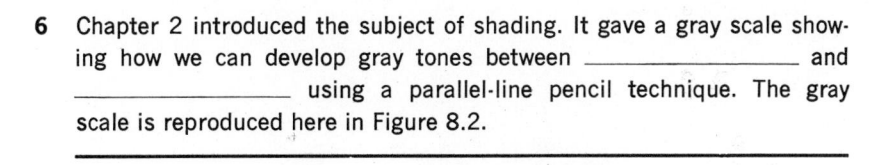

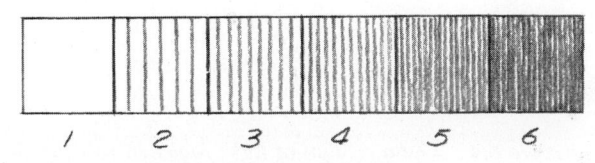

Figure 8.2 The white-to-black gray scale.

Shading can be used on pictorial drawings as a powerful means of bringing out the shape of the object. Shading enhances the three-dimensional quality that you work hard to achieve in your line work.

7 Shading is never added to orthographic drawings except in the one case of sectioning. In taking a section through an object, you deliberately remove part of the object to show _____ detail. (See Figure 8.3.)

interior

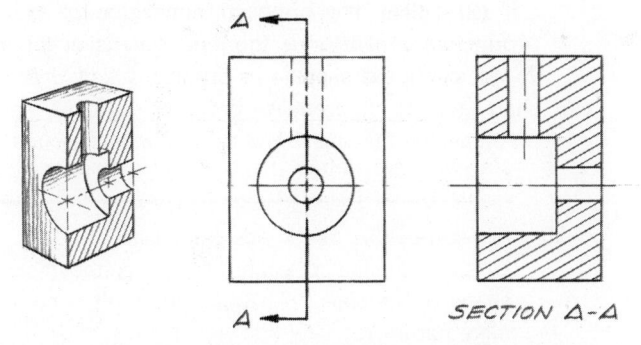

Figure 8.3 The use of a gray tone to show a sectioned view.

8 In this case the cut portion is shaded by drawing evenly spaced, parallel lines to give it a _____ tone. The person reading the drawing immediately interprets this as a sectioned view.

gray

To understand shading, we must first understand how a light source illuminates an object. For simplicity's sake, we will consider only a single light source shining on the object.

Figure 8.4 shows a rectangular box illuminated by a single source coming from different directions:

    (A) Over our (the observer's) left shoulder
    (B) Over the right shoulder
    (C) From straight above
    (D) From above and behind

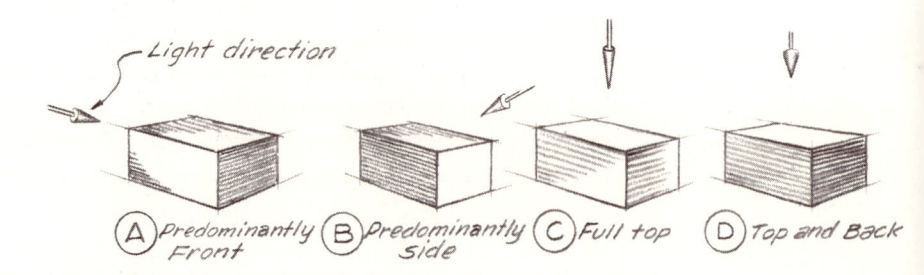

Figure 8.4 The direction of light rays controls the shading of an object.

9 The three exposed faces each have different tones of shading on them. The darkest tone is always on the side _____ from the light source. The other faces are shaded in _____ tones.

away (opposite); lighter

We will standardize on a light source coming over our left shoulder. Note that in all cases the shade lines:

    (1) Are parallel lines (in perspective they must converge toward a VP in order to represent parallel lines).
    (2) Follow the principal dimension of the face of the object. The principal dimension is the long dimension for rectangular faces.

Do not cover every plane or surface solidly with shade lines. The purpose of shading is to differentiate between adjacent planes on the object or between planes on the object and the white background.

10 In Figure 8.5A, the shade lines run _____ to the principal dimension of each face. This technique tends to break up the shape of the object. At 8.5B, this fault is corrected and the object looks more natural.

opposite (perpendicular)

11 Do not attempt to "color" a pictorial drawing. With a few quick strokes, add some shading either at the edge between two adjacent _____ on the object or between a plane and the _____.

planes; background

PERSPECTIVE SKETCHING 147

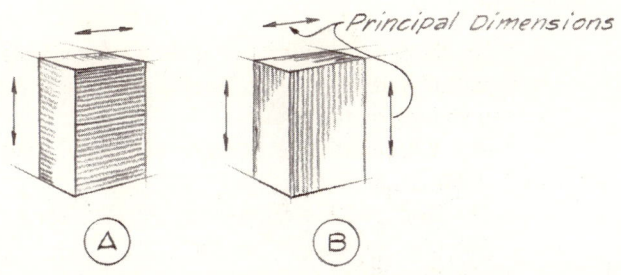

Figure 8.5 Direction of shade lines usually follows the principal (long) dimension of each face.

---

background;
white

12 This differentiation between planes should be done on one of the planes only. There should be a definite light-to-dark change at the intersection. (See Figure 8.6.) The figure also shows a white object. In this case the dark tone is added to the _____ leaving the object _____.

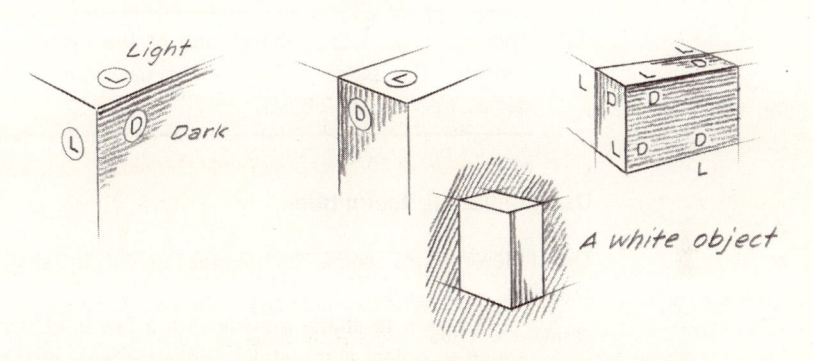

Figure 8.6 Line shading should achieve a contrast of light and dark.

---

dark (shaded)

13 A white object never appears all-white unless it is flooded with light from all directions. We are considering only a single light source coming over the left shoulder, therefore at least one face will be _____. (See Figure 8.6.)

---

dimension
(direction)

14 Figure 8.7A shows a pyramid. Note that the shade lines follow the principal _____ of each face and not the intersection lines of adjacent plane surfaces.

---

parallel

15 Shade lines on flat surfaces should always be _____ and evenly spaced.

148 PICTORIAL DRAWING/UNIT 3

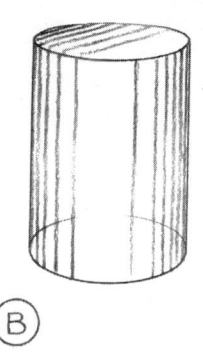

Figure 8.7 Shading a curved surface calls for graded line shading.

---

16 If the spacing between shade lines changes, the surface will tend to appear curved. In fact, changing the spacing is the criterion for shading a curved surface.

Figure 8.7B shows a shaded vertical cylinder. The shade lines follow the principal direction of the surface (parallel to the axis) and are _____ spaced to give a rounded effect. (*Choose one:* evenly/unevenly.)

unevenly

---

17 The _____-hand side of the cylinder is darkest because we consider the light to be coming over our _____ shoulder as we view the object.

right; left

Useful shading techniques

The following are some techniques useful in shading pictorial technical sketches:

(1) Learn to shade quickly with a few bold strokes. Do not attempt to render an object completely; this is the work of the artist.

(2) Use a soft dull pencil so you can get a thick, expressive line.

(3) Vary the pressure on the pencil to change from light to dark.

(4) Keep in mind the direction of the light source. This will show you which side is darkest.

(5) Use bold expressive line work. Lines should be free—not tight and precise. *Don't scribble.* Work quickly, but deliberately.

(6) Keep shading simple. Too much can ruin a good line drawing.

(7) Use shading to differentiate between adjacent surfaces or between a surface on the object and the background.

---

18 Remember that a good pictorial sketch is a combination of correct line construction and effective line shading. Of the two, the _____ _____ is more important.

line construction

■ 19 Sketch 24 shows some examples of shaded objects of a variety of shapes. Some line drawings of these same objects are also given. Try your hand at applying freehand line shading to these drawings.

PERSPECTIVE SKETCHING 149

This brief discussion on shading has been placed here to encourage you to use simple line shading on your pictorial sketches. In the next few sketch exercises, concentrate first on correct line construction. Once you are satisfied with the construction of the sketch, add some very simple line shading and see if it doesn't improve the appearance of the sketch.

Multiple Boxes

Just as we can sketch a single box that encloses the complete object, so also can we sketch a number of small boxes that *enclose important details* of the object. Consider Figure 8.8. A motion-picture projector is sketched in both orthographic and perspective form. The orthographic drawing is complete.

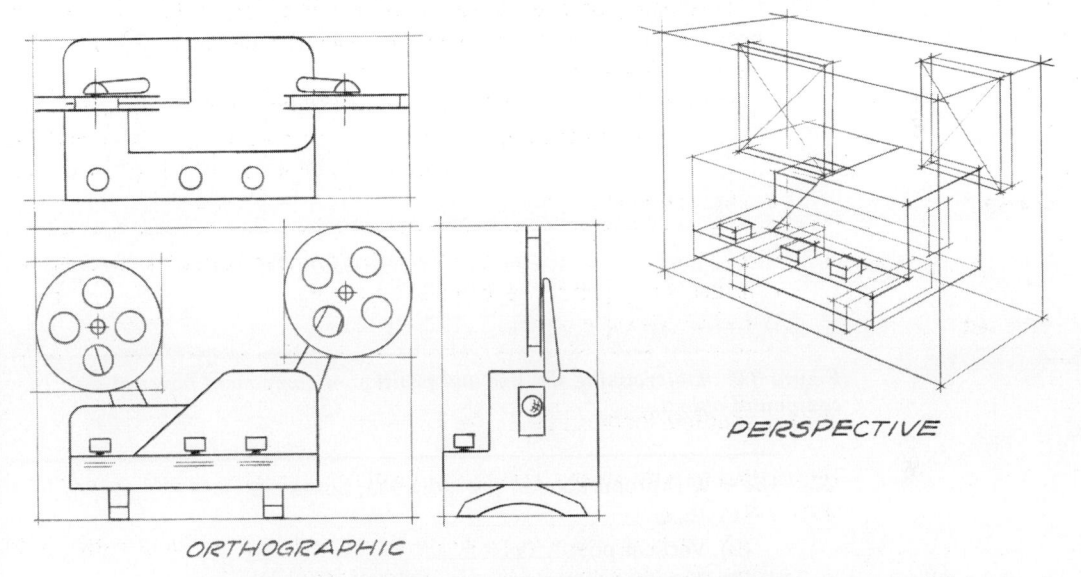

Figure 8.8 Multiple-box construction, the foundation of a successful perspective sketch.

boxes

20 The perspective sketch is incomplete in that all the details of the machine are presented as _____.

And yet the object is beginning to be recognizable. We can begin to see the film reels even though they are square flat prisms rather than flat cylinders. Note also that many parts are drawn through even though the parts are not visible. This is especially true of the boxes enclosing the leg structures. Drawing through is an excellent way to learn to visualize three-dimensional objects.

The perspective sketch of Figure 8.8 shows the construction necessary to make a good finished sketch. In this preliminary form, the important point is to get all the boxes positioned correctly within the main box and to draw through the object. Knowledge of hidden details is helpful in sketching the visible details.

7:7:5

21 Figure 8.9 is a simpler object. It is a combination of three rectangular boxes whose overall H:W:D = ____:____:____.

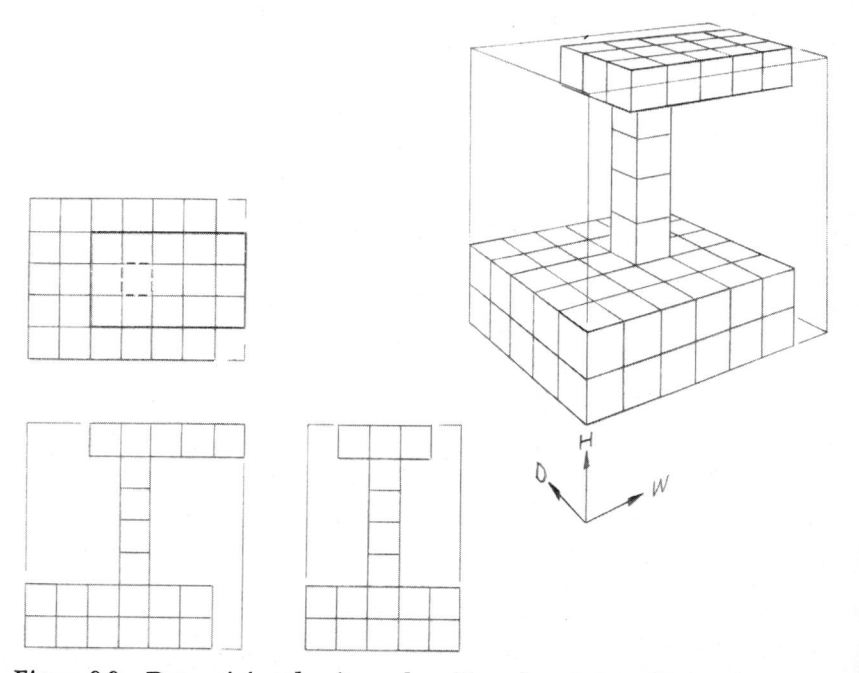

Figure 8.9 Determining the size and position of component boxes of a compound object.

22 The H:W:D proportions of the individual boxes are:
(1) Base ____:____:____.
(2) Vertical post ____:____:____.
(3) Top ____:____:____.

(1) 2:6:5
(2) 4:1:1
(3) 1:5:3

23 We can find the position of the one-unit-square post by coordinate means. Measuring in the W- and D-coordinate directions from the nearest corner of the top surface of the base, we see that the post sets at W-coordinate ____ and D-coordinate ____.

(W): 4; (D): 3

24 The post connects with the top box. Again measuring from the nearest edge of the top box, this connection is at W-coordinate ____ and D-coordinate ____.

(W): 2; (D): 2

25 It was easy to read the drawings of Figure 8.9 and to describe the component parts. We were interested in the coordinate information that described (1) the _____ of each box and (2) the _____ of each with respect to some reference point.

(1) size (proportions, dimensions);
(2) position

■ 26 Sketch 25 is an exercise in proportioning and positioning the parts of a compound block. $H:W:D = 6:8:4$. Use ¼-in. units. Steps to follow are:

(1) Determine how you wish to view the object, set up the *HWD* axes, and sketch the main box according to procedures of Chapter 8. This construction is given in Sketch 25.

(2) Divide the block into smaller boxes. There are four distinct boxes including one "negative" box, a cutout.

(3) Position these by coordinate means.

(4) Finish all visible lines with bold expressive-line treatment.

REMEMBER: Sketch lightly at first. Sketch through to get a feel for the entire block. Check your solution.

Sloping Lines and Surfaces

Up to now we have been concerned only with lines on objects that were parallel to one of the *HWD* axes. Let's look now at sloping, or oblique, lines. Figure 8.10 shows an object presented earlier.

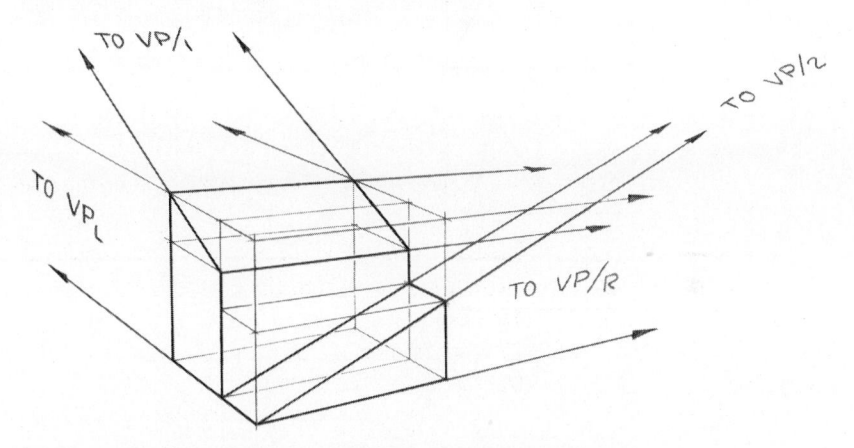

Figure 8.10 Sloping lines are found by coordinate means.

horizon

27 In addition to the standard vanishing points, VP_L and VP_R, there are two more to consider, VP_1 and VP_2. We know that VP_L and VP_R are located on the _____.

no

28 Are VP_1 and VP_2 located on the horizon? Yes _____; no _____.

horizontal

29 No, they are not, because the lines converging to these two points are not sets of _____ parallel lines.

HWD axes

30 Finding VP_1 and VP_2 would be a tedious job. The sketch of Figure 8.10 was made without worrying about the new vanishing points. An examination of the sketch will show that the *ends* of each sloping line are located on lines that are parallel to the _____ (your words).

152 PICTORIAL DRAWING/UNIT 3

ends

31 Therefore, as the construction of Figure 8.10 shows, the preliminary work just blocked in three sub-boxes and disregarded the sloping lines. Once this construction was complete, the sloping lines were found by using coordinates to locate the _____ of the lines.

converge

32 Once the ends were located, the sloping lines were sketched. If the preliminary construction is done accurately according to the rules of perspective sketching, the sloping lines will truly _____ to their respective vanishing points.

(W): 3; (D): 2

33 Figure 8.11 is an example of using coordinate methods to find sloping lines in perspective. The object is an oblique pyramid. In 8.11B the basic box has been sketched. The apex of the pyramid is located on the top surface at coordinates W_____ and D_____.

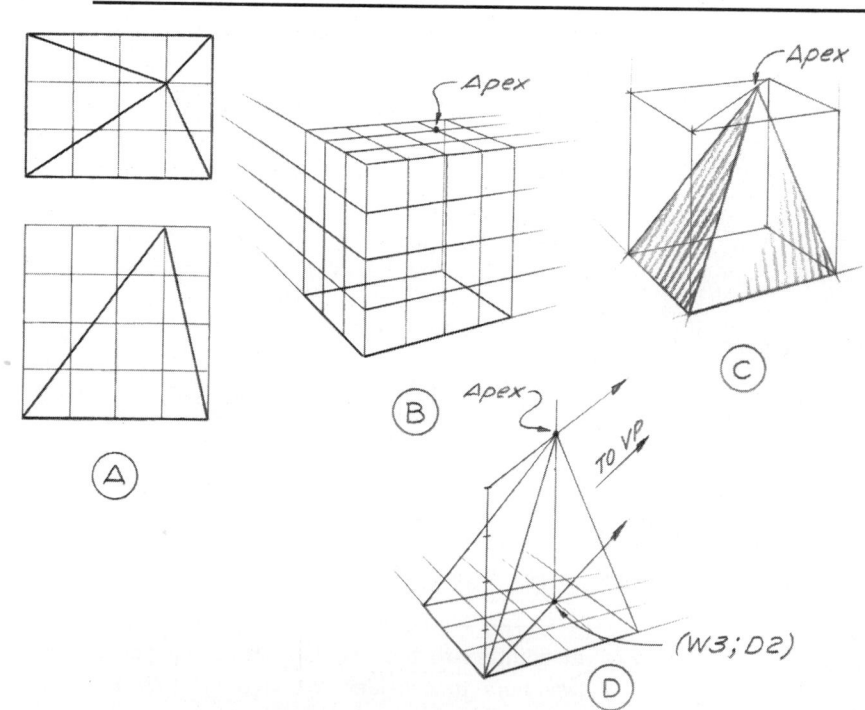

Figure 8.11 Finding the apex of an oblique pyramid.

base

34 Once the apex is found, the sketch is finished by connecting this point with the corners of the rectangular base as in 8.11C.

Figure 8.11D shows another method involving less construction. Here the W and D coordinates of the apex were found on the _____ of the pyramid.

no

35 A vertical line was then erected at point W3, D2. Could the apex be found by measuring four full units along this line? Yes _____; no _____.

nearest

36 The only full-scale line on a perspective sketch is the _____ vertical line. All others are foreshortened.

37 In the case of Figure 8.11D, this is the front corner of the box. A quick way of finding the foreshortened height of the apex is shown in the sketch. A line was sketched on the base from the front corner through the coordinate point of the apex. A vertical was erected at the front corner and measured full scale (4 units).

A line sketched from the top of this vertical through the apex of the pyramid would lie in the top of the box and be parallel to the first line sketched across the base.

Therefore, this top line was sketched in perspective and was made to

converge

_____ on the first (base) line.

38 The intersection with the first vertical is the apex of the pyramid and

foreshortened

gives the _____ height of the apex.

Figure 8.12 shows an object with a number of sloping surfaces. Procedure for sketching such a form is as follows:
(A) Sketch basic box to correct proportions and desired viewing position.
(B) Plot ends of sloping lines by coordinate means.
(C) Finish off with bold line work and simple line shading.

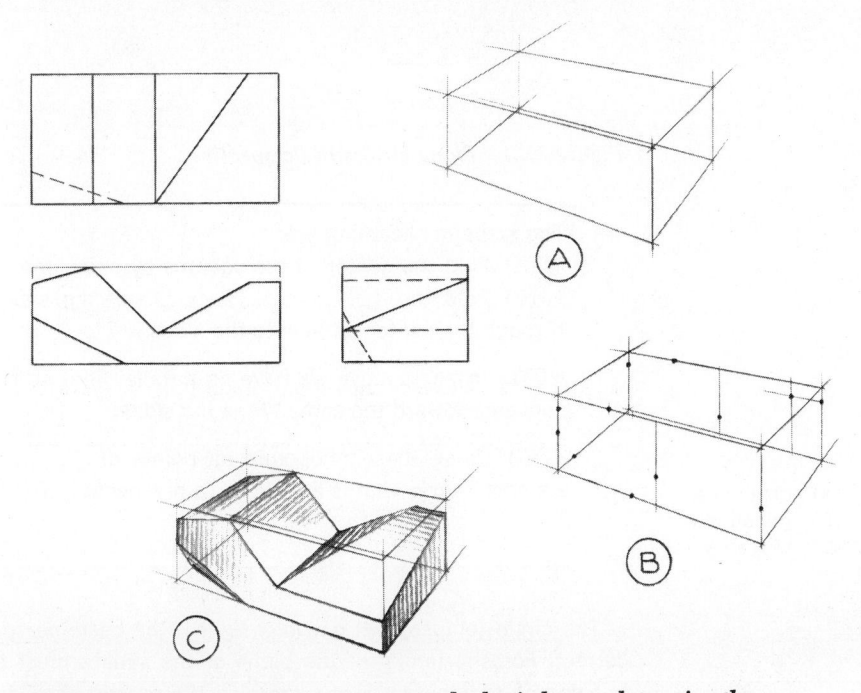

Figure 8.12 Using coordinate points on the basic box to determine the ends of sloping object lines.

154 PICTORIAL DRAWING/UNIT 3

■ **39** Sketch 26 is an exercise in sketching objects with sloping surfaces. Follow the procedure of Figure 8.12 and make a finished, shaded perspective sketch of the block shown. The enclosing box and three sub-boxes are given. Check your solution.

Circles

40 Circles, circular arcs, and curves are common details of technical objects. We must learn how to sketch them in perspective. We learned in Chapter 2 that, when viewing a circle obliquely, we see a(n) _____.

ellipse

41 Figure 8.13 shows a perspective cube with perspective circles (ellipses) sketched on the three exposed faces. The ellipses were developed by first finding five points by coordinate means. They are:
(A) The _____ of the square surface.
(B) The _____ of the four sides.

(A): center
(B): midpoints

The Squared Circle

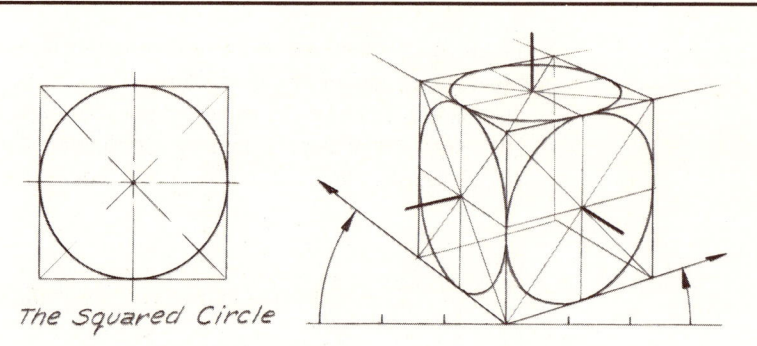

Figure 8.13 Plane circles in perspective.

42 The steps in sketching are:
(A) Find the center of the squares by sketching _____.
(B) Find the _____ of the sides by sketching lines through the center parallel to the sides.

NOTE: In perspective we have no parallel lines so that these lines must converge toward the same VP as the sides.

(C) Using these midpoints as points of _____ sketch a smooth curve that is an ellipse (a perspective circle).

(A): diagonals;
(B): midpoints;
(C): tangency

The circle (ellipse) will not look right if the basic perspective square is not correct. Foreshortening of the plane of the square must be estimated with a fair degree of accuracy. Let's examine the properties of an ellipse to see how we can aid the sketching problem.

PERSPECTIVE SKETCHING 155

major, minor

43 Figure 8.14A shows one method of describing an ellipse—that is, by showing its _____ and _____ diameters.

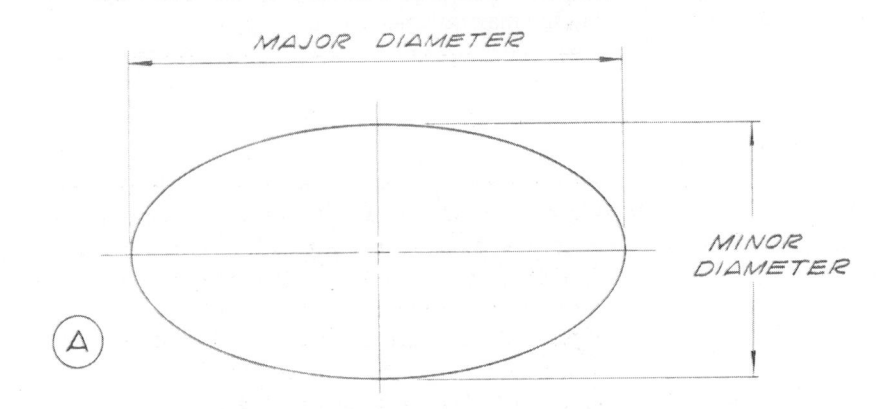

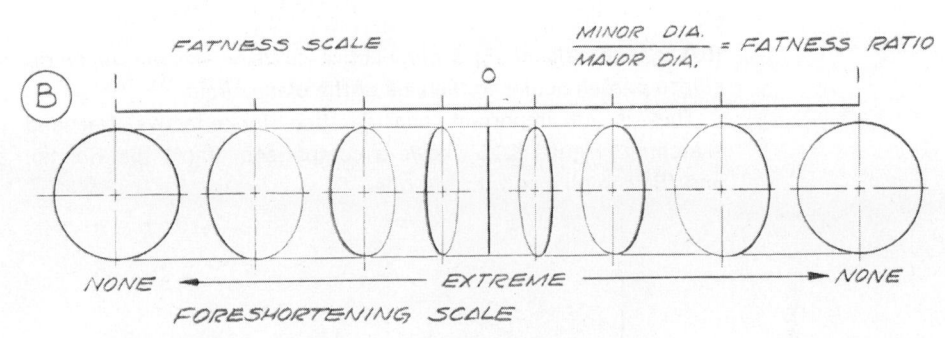

Figure 8.14 A foreshortened plane circle is an ellipse.

straight line

44 Figure 8.14B shows that a circle can be viewed from a full circle (major dia. = minor dia.) to a(n) _____ _____ (minor dia. = 0). The "fatness" of an ellipse is the ratio of its minor diameter to its major diameter.

(1) one (1);
(2) zero (0)

45 An examination of the ellipse-fatness scale of Figure 8.14B shows that for:
    (1) a fat ellipse, the value of the (minor dia.)/(major dia.) ratio approaches _____, and for
    (2) a thin ellipse, the value of the ratio approaches _____.

diameter

46 From Figure 8.14B we can see that the major diameter of an ellipse is always equal to the _____ of the circle represented by the ellipse.

156 PICTORIAL DRAWING/UNIT 3

perpendicular

47 Consider Figure 8.15. It is a repeat of Figure 8.13 but without the box and construction lines. Axes perpendicular to the three circular planes have been emphasized. Note that each axis is _____ to the major diameter of the ellipse.

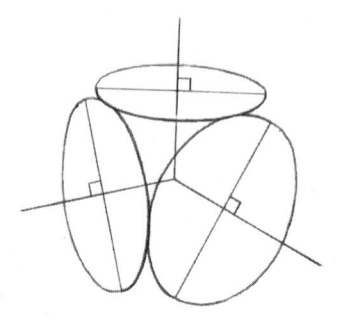

Figure 8.15 A useful rule: Make the major diameter of the ellipse perpendicular to the axis of the plane circle.

RULE: When sketching a perspective circle, make the major diameter of the ellipse perpendicular to the axis of the plane circle.

This is an important consideration for effective freehand perspective sketching. Figure 8.16 shows a comparison of (A) the violation of the rule and (B) compliance with the rule.

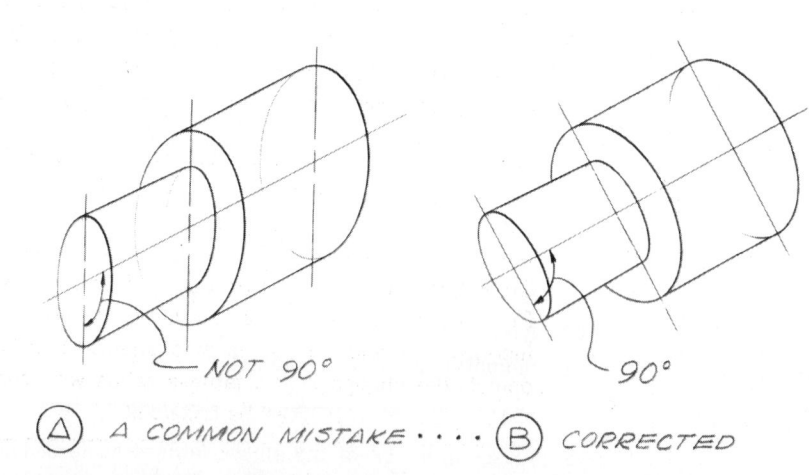

(A) A COMMON MISTAKE · · · · (B) CORRECTED

Figure 8.16 Realism in sketching cylinders is achieved by making the major diameters of the ellipses perpendicular to the axis of the cylinder.

ratio

48 Note also in Figure 8.16 that the cylindrical objects were sketched without the familiar enclosing-box construction. An understanding of the preceding rule permits the sketching of ellipses without complicated construction. Just follow the rule and have an idea of how "fat" you want the ellipse to be—"fatness" is the _____ of minor diameter to major diameter.

49 A perspective drawing will show parallel ellipses (the two ends of a cylinder as in Figure 8.17A) as having slightly different fatness ratios. Comparing ellipses M and N, we see that M is thinner than N and that the major diameter of N is _____ than that of M (the effect of the convergence of the sides of the cylinder).

shorter

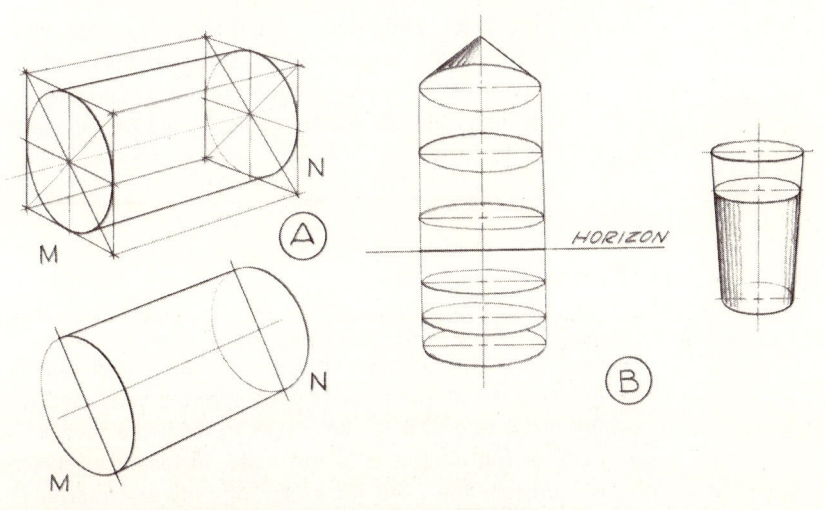

Figure 8.17 Analyzing the fatness ratio of ellipses.

50 Ellipse N is a fuller ellipse (more nearly a circle) because it is farther away from us than is M. Being farther away, it is turned more toward our line of sight. Theoretically, if the cylinder were infinitely long, the far ellipse would be a true _____.

circle

51 Figure 8.17B shows other objects that illustrate this. A farm silo is shown. At the horizon, the ellipse fatness ratio is ____:____.

0:1

52 On either side of the horizon the minor diameter increases in size as the distance of the plane of the circles from our horizontal line of sight increases. A water glass is also shown. Even in this relatively small object, the change in the fatness ratios with change in distance below the _____ can be observed.

horizon

Notice that in Figures 8.16 and 8.17, the ellipses are *drawn through*, that is, the complete ellipse is sketched even though only one-half is visible. This is an especially useful technique to obtain accurate ellipses.

■ **53** Sketch 27 is an exercise in sketching perspective circles. In 27A, three perspective squares are given. Sketch circles in these squares by using the coordinate method of Frame 42. Check the accuracy by sketching the axis perpendicular to the plane of each circle to see if it is also perpendicular to the major diameter. Check your solution.

■ 54 In Sketch 27B sketch four ellipses to represent 1½-in.-diameter circular disks at the ends of the four rods (axes) shown. An example is given. Make the ellipses thin, medium, or fat as called for. Evaluate your own ellipses.

■ 55 In Sketch 27C, sketch a horizontal cylinder without making the enclosing box. An example of the cylinder enclosed in a box is shown. *HWD* space axes for your sketch are given. Consider the W axis to be the axis of the cylinder. Refer to Figure 8.17A. Shade the sketch. Check your solution.

Circular Arcs (Partial Circles)

Figure 8.18 shows an object with a rounded end and a one-quarter circle notch removed. Partial circles are sketched in exactly the same manner as full circles. The coordinate-point method was used to make the sketch. The full circle is drawn for both arcs to illustrate that the arcs are truly just portions of full circles. It is fortunate, in technical sketching, that most objects will present the need for sketching only one-quarter circles (90° of arc) and one-half circles (180° of arc). Seldom will any object display an odd subdivision of 360° of arc.

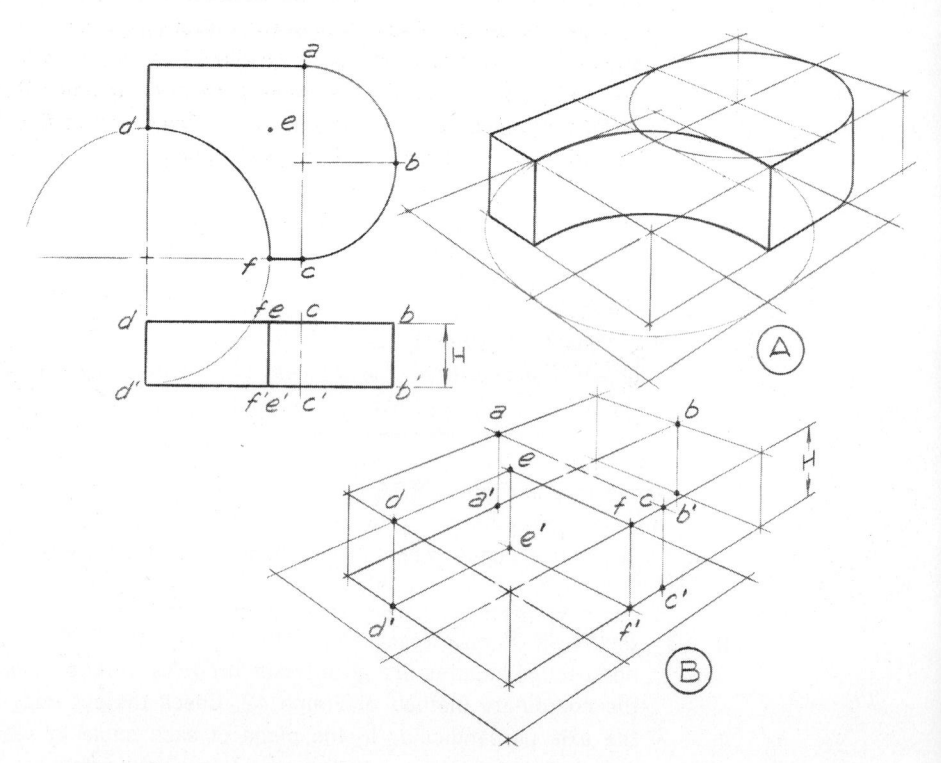

Figure 8.18 Construction for sketching partial circles in perspective.

PERSPECTIVE SKETCHING 159

56 Figure 8.18B shows the basic construction used to make the sketch of 8.18A. Using coordinate means, the following points were needed to sketch the circular arcs on the top surface:

(A) Semicircle; points _____, _____, _____.

(B) Quarter-circle; points _____, _____, _____.

(A) a; b; c
(B) d; e; f

57 Once found, these points were dropped down a distance H to the corresponding construction lines on the _____ surface. (Points a', b', c', d', e', and f'.)

bottom

58 Here is a good example of the advantage of drawing through (sketching object lines on the hidden side). A successful sketch must have both sets of curves drawn accurately. The curves representing the circular arcs on the _____ and _____ surfaces of the object are sets of parallel curves.

top (and) bottom

Figure 8.19 shows some objects having partial circles. The required quarter-circle and semicircle are blocked in and then a smooth curve (partial ellipse) is sketched. Foreshortening affects the size and shape of the enclosing squares.

Note in Figure 8.19D that foreshortening on the right-hand receding axis is quite extreme—and yet the end of the object looks like a semicircle.

The major-diameter method cannot be used in most cases because the full major diameter is not available. However, the circular holes through the objects can be found by this method. It is relatively easy to find the center of the hole and then sketch an axis through this point.

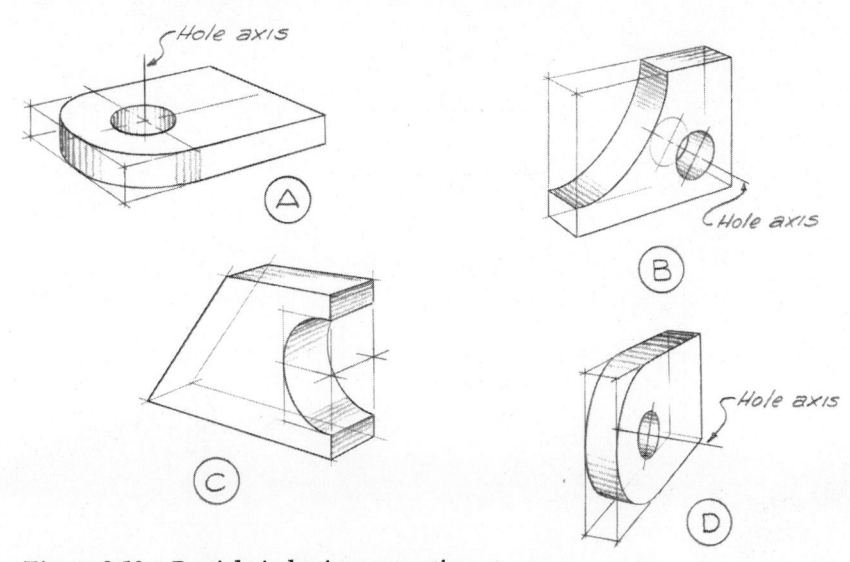

Figure 8.19   Partial circles in perspective.

■ 59 Sketch 28A is an exercise in sketching partial circles. The object, presented orthographically has three partial circles: (1) a semicircular vertical member, (2) a quarter-circle rib, and (3) a corner notch that is less than one-quarter circle. In the case of the latter curve, sketch the full quarter-circle and then use only the needed portion in finishing the sketch.

The complete box construction is given for a perspective sketch. Sketch the three partial circles. Finish the entire sketch with bold, expressive line work and simple line shading. Check your solution.

Irregular Curves

Irregular curves are defined as curves which are not circles or circular arcs. Figure 8.20A shows a plate cam, a device to translate linear motion in one direction (x) into controlled linear motion in another direction (y). The top surface of the cam is an irregular curve—actually, two parallel curved lines.

Irregular curves are sketched in perspective by first plotting coordinate points on the curve and then connecting the points with a smooth, curved line. An orthographic, true-shape view of the curve is needed to determine first the coordinates of selected points.

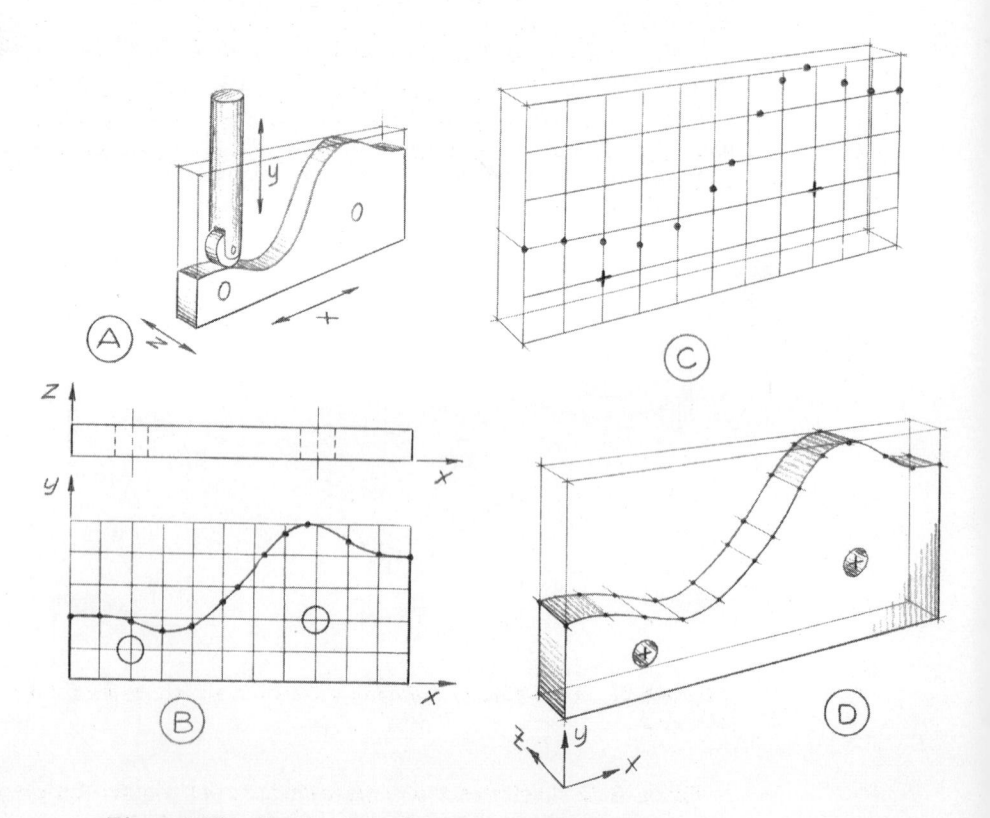

Figure 8.20 Plotting irregular curves in perspective.

PERSPECTIVE SKETCHING 161

grid

60 Figure 8.20*B* is an orthographic sketch of the cam. A ¼-in. _____ has been superimposed upon the front view.

x (and) y

61 The grid is included to aid in finding coordinate points. It automatically supplies measurements in the _____ and _____ directions.

intersections

62 Thirteen points have been marked at _____ of the curved line and *x* (or *y*) grid lines.

63 In Figure 8.20*C*, the enclosing box and the grid have been sketched in perspective form. The 13 points on the curve have been transferred to the perspective grid. This defines the curve on the _____ surface of the cam.

front

z (or depth)

64 The finished sketch is shown at Figure 8.20*D*. The curve on the rear surface of the cam was obtained by moving each of the 13 points backward (in the _____ direction) a distance equal to the thickness of the cam.

Figure 8.21 shows the same cam plate sketched in perspective but to an enlarged scale. This illustrates one advantage of the grid method. Enlargements and reductions of complicated drawings can be made easily merely by adjusting the size of the basic grid square. If the two grids are accurate, correct proportioning is accomplished automatically. This technique applies equally well to enlarging or reducing a drawing within one drawing form and to translating from orthographic to pictorial as illustrated in Figures 8.20 and 8.21.

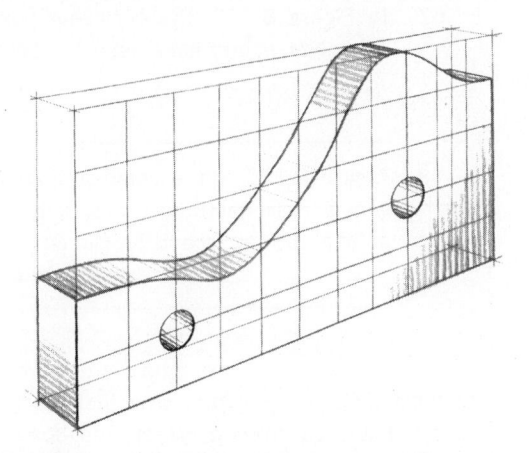

Figure 8.21 Enlarging or reducing a sketch is made easy by the use of a grid.

Figure 8.22 illustrates the same technique but without the use of a grid. An odd-shaped object is shown orthographically in 8.22A.

162 PICTORIAL DRAWING/UNIT 3

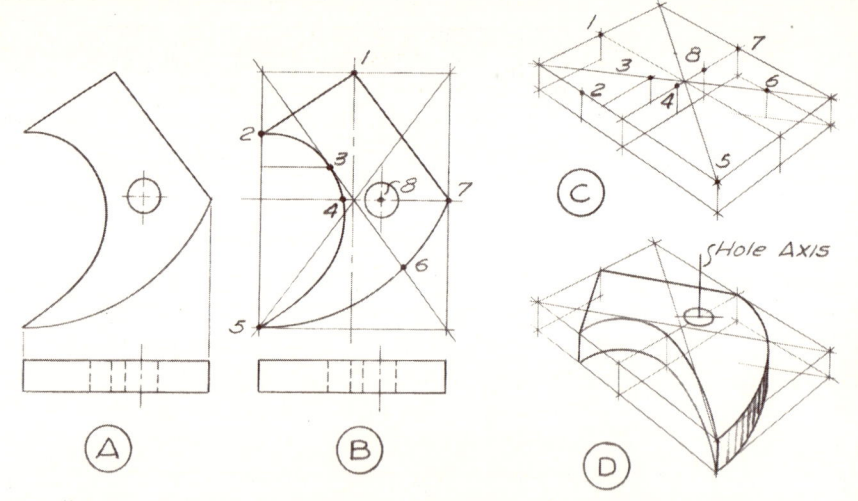

Figure 8.22 Using diagonal lines and centerlines to locate coordinate points.

diagonals	65	In Figure 8.22B, the top view has been boxed in. The center of the rectangle representing the box has been found by sketching _____.
center of circular hole	66	Horizontal and vertical centerlines have been sketched through this center. Eight points have been marked at intersections of the object outline and (1) the enclosing box, (2) the diagonals, and (3) the centerlines. Point 8 defines the _____ (your words).
top and bottom surfaces	67	In Figure 8.22C, the enclosing box has been sketched in perspective. The eight points have been transferred to this sketch on both _____ _____ (your words).
through	68	Figure 8.22D is the completed sketch. The eight points are enough to give a reasonably accurate sketch. As with all pictorial sketching, we gain a feel for the three-dimensional quality of the part by sketching _____ to the hidden side.

In the last two examples, we have been dealing with essentially two-dimensional curves. It is true that there has been a second identical curved line displaced from the first, the combination yielding a three-dimensional object.

(two-)dimensional	69	However, in both examples, the plotting of coordinate points on any one plane curve has been a two-_____ problem.

70 Figure 8.23 shows an object requiring three-dimensional coordinate plotting to make a perspective sketch.

As with any perspective sketch the first step is to sketch the enclosing box. $H{:}W{:}D = $ ____ : ____ : ____.

5:6:4

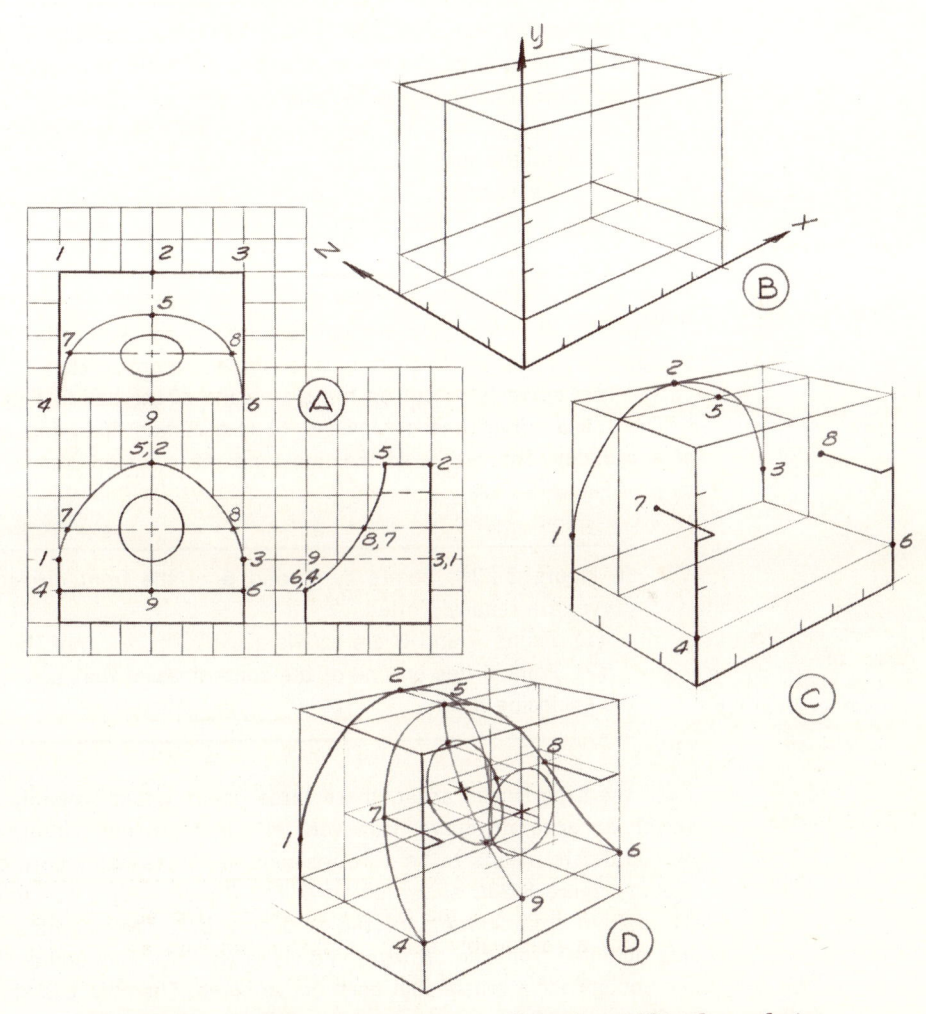

Figure 8.23   Using xyz coordinates to plot points within the enclosing rectangular box.

71 Figure 8.23B shows the box construction. Note that the total box is subdivided into three boxes. The object has a rectangular base of proportions ____ : ____ : ____.

1:6:4

The portion of the object above this rectangular base can be drawn in perspective by sketching two curved lines:

(1) The rear curved line (defined by points 1, 2, and 3) is a plane curve, that is, it lies in a plane. This is called a single-curved line.

(2) The front curved line (defined roughly by points 4, 5, and 6) is a double curved line. It is a space curve because it does not lie in a single plane.

72 The rear plane curve (123) shows either as a straight line or as a true shape in each of the three orthographic views of Figure 8.23A. Check the appropriate boxes in the table.

	Top (T)	Front (F)	Profile (P)
Straight line	_____	_____	_____
True shape	_____	_____	_____

straight line: T and P; true shape: F

Points 1, 2, and 3 are enough to allow us to sketch the plane curve in perspective since we know the true shape (approximately circular) of the curve. This curve is drawn on the rear plane of the box in Figure 8.23C.

The space curve (line 4, 7, 5, 8, 6) is a curved line lying on the surface of a cylinder. The profile of the line (and the cylinder) shows in the profile view of Figure 8.23A.

73 In Figure 8.23C, points 4, 5, and 6 of the front curved line (space curve) are readily plotted.
(1) Points 4 and 6 are at two _____ of the base box.
(2) Point 5 lies on one of the centerlines of the _____ of the enclosing box.

(1) corners; (2) top

To the experienced artist, these three points would probably be enough to sketch an accurate curve. However, let's find two more intermediate points by coordinate means to be sure we achieve a reasonably correct perspective sketch. Figure 8.23C shows points 7 and 8 plotted by x,y,z coordinates from the bottom front corners of the enclosing box. The coordinate information was obtained from the orthographic views (8.23A). Five points (4, 7, 5, 8, 6) are enough for a reasonable perspective curve. The object is shown complete in Figure 8.23D.

Note that the hole is not an ellipse since it is formed by two intersecting cylinders. A four-point approximation was used to find the horizontal and vertical limits of the hole. Lines (5,9) and (7,8) represent the hole centerlines on the curved surface. Projections from an ellipse sketched on the front face of the enclosing box back to these lines showed the limits of the actual hole.

■ 74 Sketch 28B presents an exercise in plotting and sketching irregular curves. The orthographic views and the enclosing box of an object are given. Finish the sketch by finding the curves. Use expressive line work and simple line shading. Check your solution.

■ 75   Sketch 28C is a dot-to-dot drawing similar to the one you did in Sketch 2. The points are not numbered. The object should be familiar to you. Sketch the object by joining the points with straight or curved lines, remembering:

(1) To seek the outline first by connecting outside points.

(2) To sketch straight lines between points far apart and a smooth curved line between sequences of points close together.

Check your solution.

SUMMARY

The object of Sketch 28C is the same object shown in Figure 8.23. It has been reoriented in space.

Sketches 1 and 2 introduced the dot-to-dot concept and presented the idea that the purpose of this text is to help you learn to draw by showing you where to place the points (dots). Compare Figure 8.23D and Sketch 28C. Except for orientation, the objects are identical. In 8.23 we plotted the points by coordinate means and then joined them in correct sequence to create the perspective sketch. In Sketch 28C we were given the points. The fact that there are many more points is immaterial since we could have plotted all of them in 8.23.

You should now have an idea of how to plot points on a two-dimensional surface so that they represent a pictorial sketch of a three-dimensional object.

The change in object orientation between these two drawings represents another facet of perspective sketching with which you should be familiar. This subject is presented in Chapter 9, which follows next.

chapter 9 perspective sketching—size and orientation

Figure 9.1 summarizes our work in perspective sketching. We have learned how to sketch the enclosing box and how to use coordinate points to establish detail inside of the box. Chapter 9 takes a look at the problem of sketching objects to a size relative to their surroundings and in any orientation relative to their surroundings.

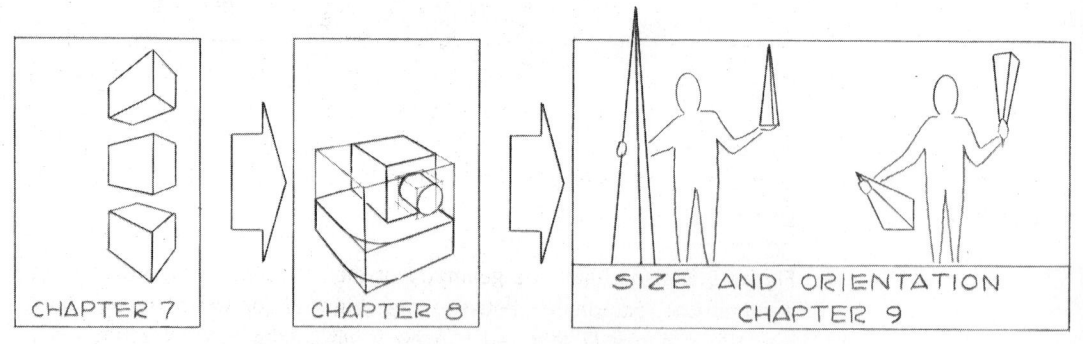

Figure 9.1 A graphic summary of the subject of perspective sketching.

OBJECT SIZE

We have dealt so far with individual objects drawn to any size we wish. We have not attempted to relate an object to its surroundings. Let's look now at the problem of sketching an object of any size in its true-to-life setting. There are two questions to be answered:

(1) How much do we see when we direct our gaze in any given direction?

(2) How can we establish a scale that will let us sketch objects to a true relative size?

PERSPECTIVE SKETCHING—SIZE AND ORIENTATION 167

15

1 Figure 9.2 shows that a person with normal eyesight has a *cone of acute vision* within which he can see things clearly and in detail. The sides of this cone make an angle of _____ degrees with the axis, called the *center of vision*.

Figure 9.2 Our cone of acute vision.

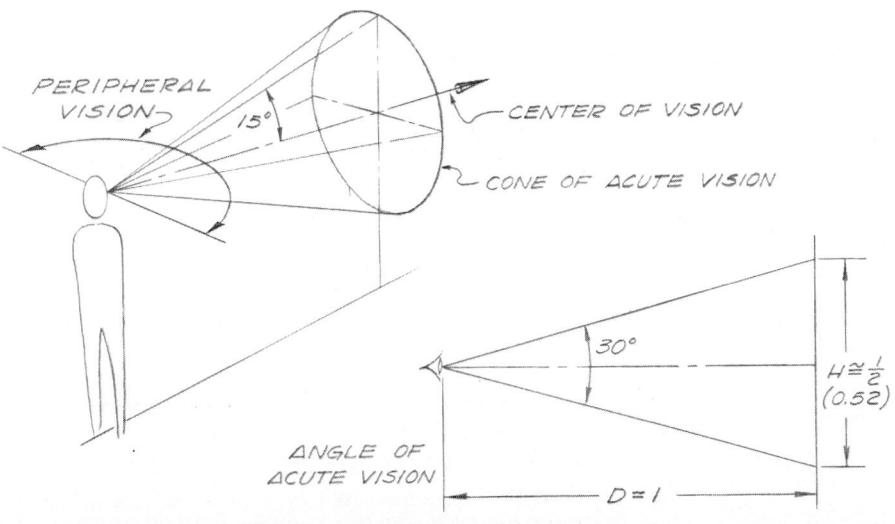

30

2 This means that our angle of acute vision is _____ degrees.

cone (or angle)

3 We actually are aware of much more than the part of the view contained in the 30° cone. Most people have 180° peripheral vision. We are aware of what is in the 180° "cone" but cannot examine details within this area unless we bring the details within the _____ of acute vision.

½

4 Figure 9.2 also shows the geometry of the 30° cone. If we wish to find a numerical relationship between the height H (or width) of an object and the distance D required to view it within the cone of acute vision we can set up a ratio of $D/H = 1/$_____.

two

5 Stated as a simple relationship, this ratio becomes $D = 2H$, or the viewing distance required to keep an object within the cone of acute vision must be at least _____ times the height (or width) of the object.

12 ft; 12 in.;
100 ft

6 Thus if we wish to view a 6-ft-tall person, we must be _____ feet away. For a 6-in.-high box we must be at least _____ in. away and for a 50-ft building, _____ ft away.

168 PICTORIAL DRAWING/UNIT 3

Figure 9.3 shows how we change lenses on a camera to overcome this geometry and still get the *same size* image on the film. In 9.3A, a standard-lens camera photographs a 6-ft man. In (B) we are too close to include the entire figure so we change to a wide-angle lens. In (C) we cannot get close enough to get a large image so we use a telephoto lens.

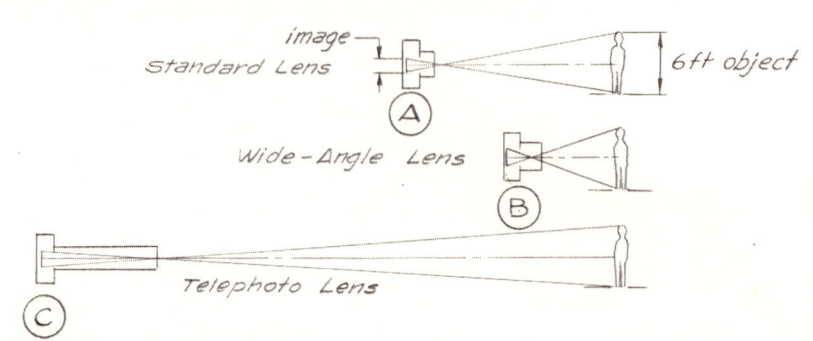

Figure 9.3 We change lenses in a camera to change the angle of acute vision.

7 What would we get on the film if we used a standard-lens camera at positions B and C of Figure 9.3? (Your words):
   At B _____
   At C _____

(B) part of the subject;
(C) small image, much background

Relating these ideas to perspective sketching, we can learn that placing our observation point too close to the object will cause distortion and placing it too far away will cause loss of detail.

The second question raised was: How, when picturing more than one object, can we establish a relative scale for our drawing? There are occasions in technical sketching when we wish to show a group of objects in relation to their surroundings or to each other. We want the scale of each object to give a true-to-life picture. This means that the scale must be such that all objects in the picture are in their correct relative proportions.

8 You will remember that the horizon is a horizontal line that appears at our eye level as we gaze horizontally into space. The average ground-to-eye-level distance for humans is _____ ft.

5½ (5′6″)

9 We can use this fact to establish a scale in any part of a perspective sketch even though the scale changes constantly due to the effect of foreshortening. Figure 9.4A shows Mr. Average Human standing erect on the ground plane. Because the horizon passes through his _____ _____, we know that the distance A is _____ ft.

eye level;
5½ ft.

PERSPECTIVE SKETCHING—SIZE AND ORIENTATION 169

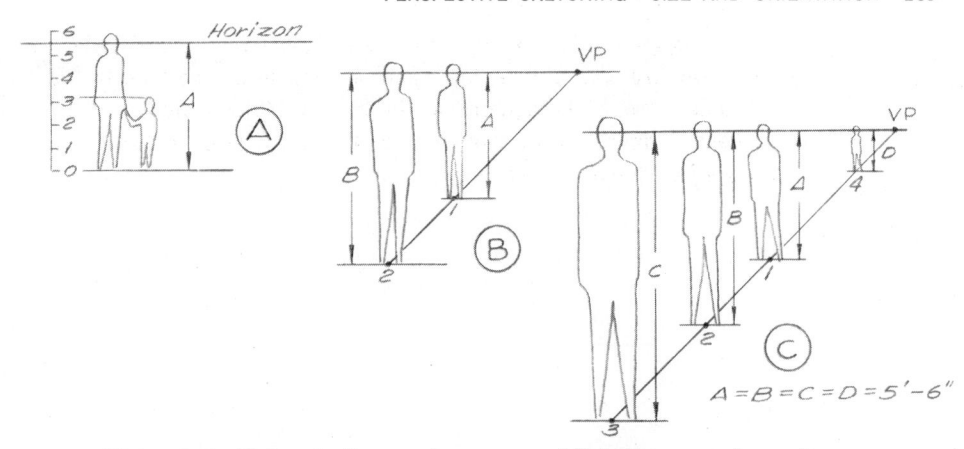

Figure 9.4 Using the human figure to establish distances above the plane of the ground.

scale	10	A 3-ft-tall child is standing next to the man. We can sketch him in place by placing his feet at the same level as the man's feet and then estimating 3 ft using the 6-ft man as a(n) _____.
larger (taller)	11	Suppose another 6-ft man entered the scene but was closer to us (see Figure 9.4B). We can sketch the newcomer by putting his eye level on the horizon but his feet must be farther down than distance A because he is closer to us and, therefore, _____ in apparent size B.
vanishing point	12	Since both men are the same height, we actually know that $A = B = 5'6''$. Sketching a line through points 1 and 2, we establish a ground line that gives us a(n) _____ _____ on the horizon.
5.5	13	We know that any point on this ground line is actually _____ ft. below the horizon.
different	14	Figure 9.4C shows two more average-height figures placed on this line. Each figure is a(n) _____ measured size on the drawing, but appears to be of the correct visual height considering its respective distance from our observation point.

Let's use this knowledge to create a scene (pictorial not behavioral). It will include:
    (1) an average-height man who is standing in front of
    (2) a 5-ft-high lathe (machine tool). At his side is
    (3) a 30-in.-high table (top, 30″ × 30″). On the table is
    (4) a 10-in.-high cylinder and behind the lathe is
    (5) a 7-ft-high computer (the lathe is automatically controlled).

170 PICTORIAL DRAWING/UNIT 3

14

15 We will view the scene from 15 to 20 ft away. Our approximate equation $D = 2H$ shows that we must be at least _____ ft away in order to draw the 7-ft computer without distortion.

5½ ft.

16 Figure 9.5 shows this scene. First, a horizon was sketched. A human figure was added to establish a ground line (use a simple stick figure as shown if you cannot draw the human form). The table was added knowing that the distance from point a (bottom of front leg) to the horizon is _____.

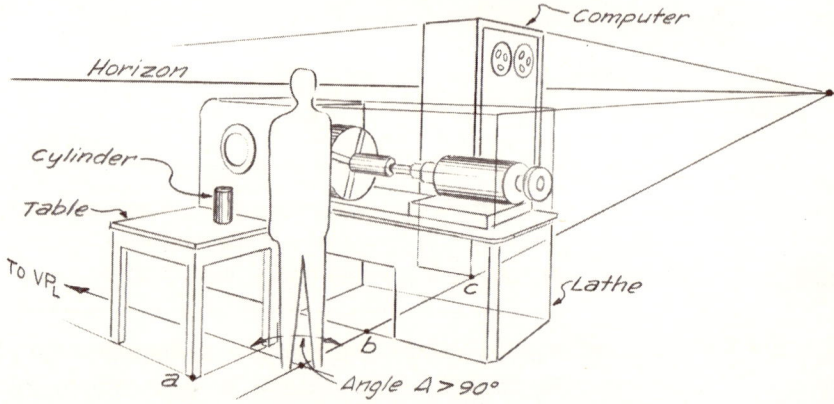

Figure 9.5 Keeping a group of objects in relative proportion.

cube

17 An estimate of 30 in. (2½ ft) was made to find the top of the front leg. The basic box for the table is a(n) _____.

The lathe is behind the human figure. A point b was selected on the ground line a short distance behind the man. Here again, the distance from point b to the horizon is 5½ feet. A distance of 5 ft was estimated and the enclosing box and then some details of the lathe were sketched.

The computer cabinet is behind the lathe. Again, on the ground line, a point c was selected a short distance behind the lathe. This step shows the advantage of sketching the hidden lines when making the box construction. Point c is 5½ ft below the horizon so we again have a scale for determining the 7-ft height of the computer.

The sketch was completed by adding the cylinder on top of the table. The 10-in. height was proportioned from the 30-in. height of the table.

far apart;
$\theta > 90°$

18 The scene of Figure 9.5 could have been sketched larger or the point of view could have been changed by adjusting angles L and R. To make a larger sketch, the human figure would have to be made larger to start with.

Based upon our knowledge of perspective sketching, the important considerations are to keep the vanishing points _____ _____ and make sure that angle θ _____ 90 degrees.

PERSPECTIVE SKETCHING—SIZE AND ORIENTATION 171

5 ft.

19 Figure 9.6A shows the same scene as viewed from a stair landing 5 ft above ground level. The horizon is still at our (the observer's) eye level. The eye level of the human figure in the picture, however, is now _____ below the horizon.

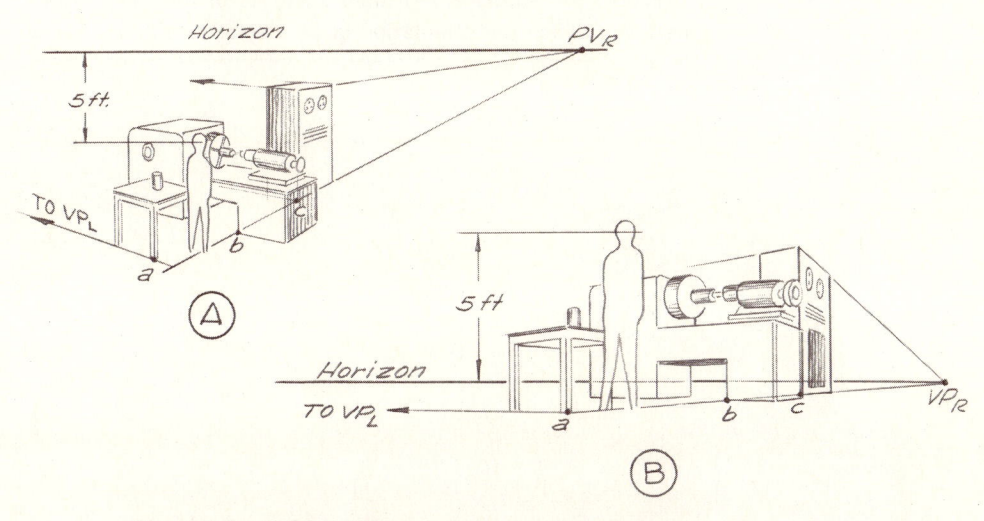

Figure 9.6 Taking different vertical points of view.

6 in.

20 In Figure 9.6B we see the scene as if we were coming up a flight of stairs and our feet were still 5 ft below the ground level. This places the horizon just _____ inches above ground level at any point in the picture.

■ **21** Sketch 29 invites you to try your hand at creating a scene. The ingredients are (1) a man of average height; (2) a 36-in.-high TV console; (3) a 7-ft doorway; and (4) a 2'6" child. Arrange these objects in any manner you wish using the techniques of Figures 9.5 and 9.6. The answer sketch shows three of many possible arrangements of the objects.

OBJECT ORIENTATION

Figure 9.7 is a perspective sketch of a sheet-metal part called an adjusting bracket.

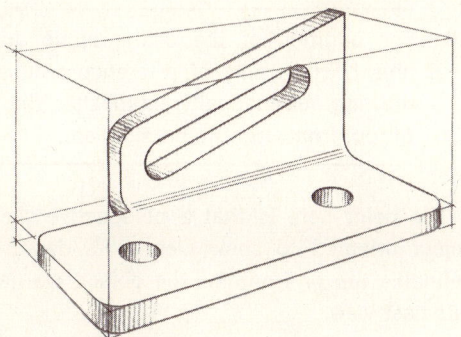

Figure 9.7 Adjusting bracket.

172 PICTORIAL DRAWING/UNIT 3

22 Is the part right side up, upside down, on its side, or don't you know? _____.

I don't know

23 There is no way of knowing which is the top, bottom, or side of the bracket. It is an unfamiliar object.

Figure 9.8 shows seven more sketches of the same object. Do these help you define the orientation of the object? Yes _____; no _____.

no

Figure 9.8 Seven other possible space orientations of the adjusting bracket.

24 It is still an unfamiliar object and we have no way of orienting it with respect to the familiar space directions of top, bottom, right-hand side, etc. Do we care? Yes _____; no _____.

no!

25 It really doesn't matter how the object is oriented as long as the sketch conveys all pertinent information about its details and proportions.

Figure 9.9 is an orthographic drawing of the bracket. Considering this drawing, which of the eight sketches of Figures 9.7 and 9.8A to G portray the object as being right side up? _____

9.8C

26 The addition of the orthographic drawing defines a top view of the object. The frame of reference around which we build an orthographic drawing automatically establishes the familiar space _____ of top, front, right-hand side, etc.

directions

The point here is that a perspective sketch of an object should have the object oriented to show clearly its details and proportions. If there is no definable top or front for the object, we have complete freedom in selecting this best view.

PERSPECTIVE SKETCHING—SIZE AND ORIENTATION 173

We would sketch a familiar object (car, human, house, etc.) in its natural position so as not to defy logic or the laws of gravity.

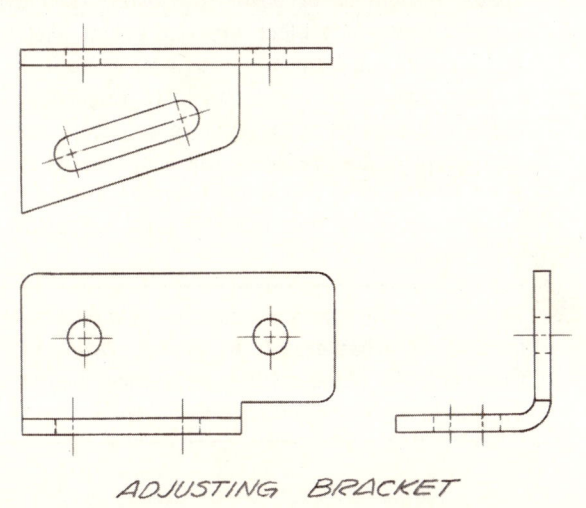

Figure 9.9 An orthographic projection automatically establishes the familiar space directions of top, front, side, etc.

27 An object such as the adjusting bracket of Figures 9.7 and 9.8 should be oriented in a sketch to show best the object's _____ and _____.

details;
proportions

28 Rate the eight sketches of the bracket using G for good, F for fair, and P for poor.
Fig. 9.7 _____; Fig. 9.8A _____, B _____
Fig. 9.8C _____, D _____, E _____, F _____, G _____

9.7 Good
9.8A Fair
  B Good
  C Good
  D Good
  E Poor
  F Good
  G Poor

29 There is one danger in sketching a single object from many angles. One or more of the views may turn out to be a different object unless you are careful with your visualization. Are all eight sketches of the bracket correct? Yes _____; no _____. If your answer is "no," which is incorrect? _____.

no;
D is incorrect

30 At a casual glance, all eight sketches look like a correct representation of the bracket. A closer examination of Figure 9.8D shows that the angular portion containing the slotted hole is _____ _____ (your words).

on the wrong side of the rectangular portion.

174 PICTORIAL DRAWING/UNIT 3

■ **31** Sketch 30 is an exercise in changing the orientation of an object in space. A number of boxes are given, sketched in perspective. Orthographic views of a block are also given. Sketch the block in the boxes changing its top-front orientation for each.

Your goal in this exercise is to select views that show the block's detail and proportion best. Once you start plotting coordinate points to add detail to the box, you will soon see whether you are developing a good view or not. If you start one poorly oriented view, finish it just to prove to yourself that it really does not display the block well. Check your solutions.

■ **32** Sketch 31 is a summary exercise in perspective sketching covering the material of Chapters 7, 8, and 9. Orthographic views are given for three objects. Make a freehand perspective sketch of each. The enclosing box has been drawn for object (A).

chapter 10   projected perspective

This chapter completes the discussion of perspective drawing. It is a short account of a system by which we can make an accurate perspective drawing of an object by projecting from known orthographic views of the object. The subject is included here more to demonstrate further the geometry of perspective than to present a rigorous and useful tool. Most students of graphics will find the projection techniques too tedious and time-consuming for their purposes. The techniques of the intuitive methods for perspective sketching of Chapters 7, 8, and 9 and the approximate methods of pictorial drawing of the next chapter, Chapter II, should be developed as useful tools of communication.

BASIC CONSTRUCTION

Figure 10.1 is a projected perspective of a rectangular box whose dimensions are in the ratio of $H:W:D = 1:2:1$. This is the complete construction required to make a perspective drawing of the box. It is essentially a *superimposed top and front view* of a physical arrangement of (1) object, (2) observer, (3) picture plane, and (4) ground plane. Figure 10.2 is a perspective sketch of this physical arrangement.

1. The object, placed on the ground plane, is viewed by an observer through a picture plane. The _____ of his line-of-sight projectors with the picture plane define a true perspective view of the object.

intersections

176 PICTORIAL DRAWING/UNIT 3

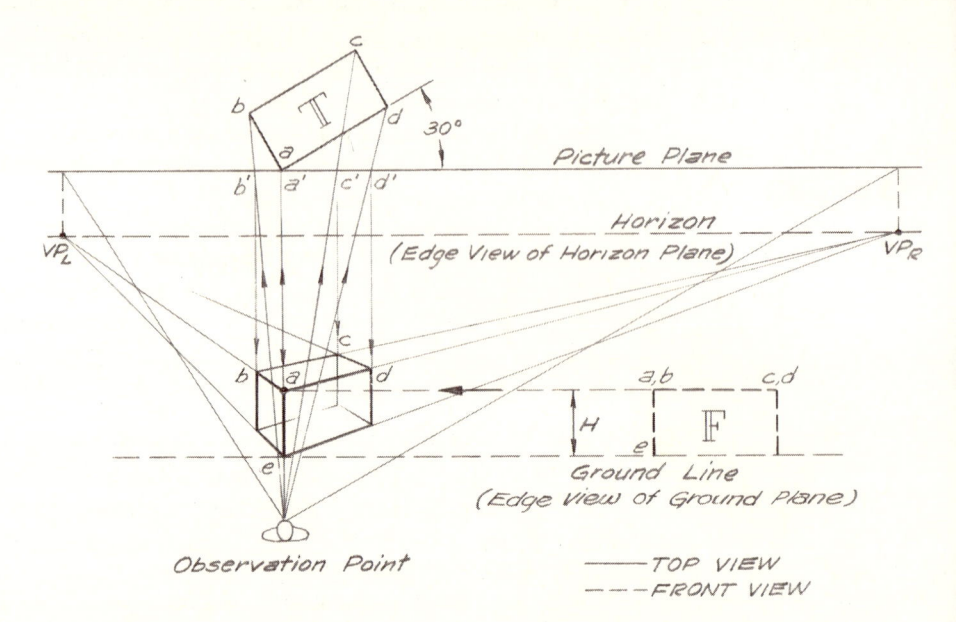

Figure 10.1 A projected perspective drawing of a box, H:W:D = 1:2:1.

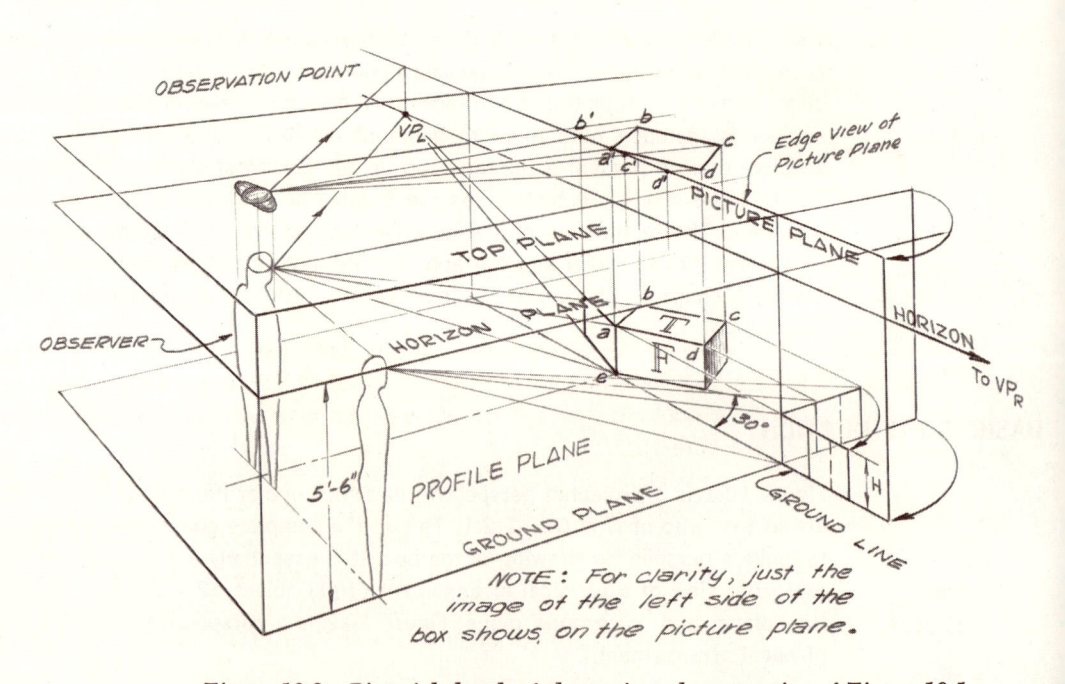

Figure 10.2 Pictorial sketch of the projected perspective of Figure 10.1.

PROJECTED PERSPECTIVE 177

2 We stated that Figure 10.1 is made up of a superimposed top and front view of the scene of 10.2. In 10.1, solid lines represent the top view and dashed lines represent the front view.

30

The block is positioned behind the picture plane with its front vertical edge in the picture plane. The front plane of the block makes an angle of ____ degrees with the picture plane.

3 Considering just the top view for the moment (simplified in Figure 10.3), we see that four line-of-sight projectors have been drawn from the observation points to the four corners (*a*, *b*, *c*, and *d*) of the block. Since the picture plane shows as an edge, we can locate the intersections of these projectors. These are points ____, ____, ____, and ____.

a', *b'*, *c'*, and *d'*

4 In Figure 10.2, these points are projected vertically downward into the front view. As we have seen earlier, these projectors control the foreshortening of the _____ dimensions of the object. (*Choose one*: vertical/horizontal.)

horizontal

5 The foreshortening of the vertical dimensions is controlled by the convergence of horizontal object lines to the left and right _____.

vanishing points

6 Vanishing points for horizontal lines always lie on the horizon. In Figure 10.3 we see that the vanishing points were found by first drawing lines from the _____ point parallel to the sides of the top view of the object (side *ab* on the left and side *ad* on the right).

observation

7 The intersections of these two lines with the _____ _____ are found. From these points of intersection vertical lines are projected down to the _____ to yield the vanishing points (VP_L and VP_R).

picture plane; horizon

8 We have seen how the VP were obtained, but why was it done this way? Remember that we are still considering only the top view (solid lines) of Figure 10.3.

If the observer gazes in a direction *parallel* to line *ab* on the object, his line of sight will, theoretically, meet line *ab* at _____.

infinity

9 Our horizon is, again theoretically, at infinity and therefore VP_L is at infinity. However, we are only concerned with what we see on the picture plane. Therefore we cut off this line of sight before infinity. The cut-off point is at the _____.

picture plane

An image of the horizon lies on the picture plane (see Figure 10.2) but, of course, we do not see it in the top view since the picture plane shows as an edge. Thus we must project the point of intersection onto the horizon

178 PICTORIAL DRAWING/UNIT 3

10 Now look at the front view of Figure 10.3. (Remember that the two views, top and front, are superimposed.) The front view (dashed lines) consists of the _____, the ground line (actually an edge view of the _____ plane), and a front orthographic view of the box.

horizon;
ground.

11 The horizon and ground line have a definite relationship dictated by the way we are viewing the object. This was discussed in Chapter 9. In the present example, the distance between the ground line and the horizon is _____ feet.

5½ (5'6")

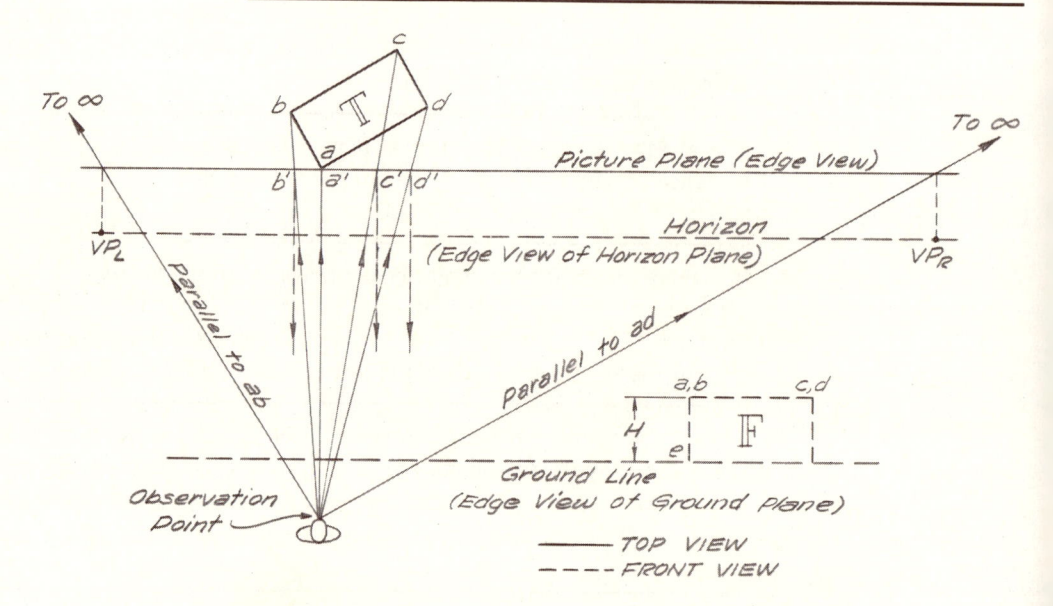

Figure 10.3 Finding the vanishing points in projected perspective.

These two lines may be placed anywhere with respect to the picture plane or the observation point. They are in a completely different view. It is only the *relation between them* that is important.

The front view of the block resting on the ground plane exists for only one reason—and that is to supply the vertical dimension H to the construction. The pictorial of Figure 10.2 suggests that this should be the right profile view taken directly from the particular object orientation as shown by the top view. This is not necessary since both the front and profile views supply the dimension H and the standard front view is much easier to draw—especially with complicated objects.

Let's finish the perspective of Figure 10.1. We have vertical projectors from points a', b', c', and d' which control the foreshortening of horizontal lines on the object.

PROJECTED PERSPECTIVE 179

12 One line on the object is not foreshortened, that is, it is true scale. It is the front vertical edge and is not foreshortened because it _____

lies in the
picture plane
_____ (your words).

13 Thus, as shown, we can project the dimension H across from the front view of the object until it intersects the vertical line from point _____
a'
which describes the front vertical edge of the block.

14 From the top and bottom of this line we can draw lines to both vanishing points and describe the perspective planes of the front and side of the object. The limits of the two sides are found at the intersections of
b' and d'
the vanished lines and vertical projectors from points _____ and _____.

15 Drawing a line from point d (on the perspective) to _____ and from point b to _____ will finish the back sides and locate the back upper
VP_L; VP_R
corner at c.

Note that the projector from c' was not needed. However, having it there is a good check on the construction. As a matter of fact, it is often possible when making a projected perspective to find only a few points on the object by the procedures discussed here and then find the rest by a series of projections back and forth between the VP and points on the object. Finding one point often locates a construction point for two more.

Measuring Lines

Consider another example in Figure 10.4. The object could be a modern bus-stop shelter 8 ft high × 12 ft wide × 8 ft deep. As in the last example, we start with a top orthographic view oriented behind the picture plane and a profile view positioned on the ground line. This object and its placement differs from the last example in that the object has an irregular profile and has no part of it in the picture plane.

16 The fact that the shape is irregular can be overlooked at the moment. We will enclose the object in a rectangular box (H:W:D = 8:12:8), make
details
the perspective drawing of the box, and then add _____ later.

17 Note that the horizon is drawn through the object (see left profile view)
5'6"
indicating that the shelter is higher than _____.

18 The observation point and the object orientation (top view) have been selected to give about equal views of the front and side. We will not see the _____ or the _____ of the shelter because it
top, bottom
straddles the horizon.

180 PICTORIAL DRAWING/UNIT 3

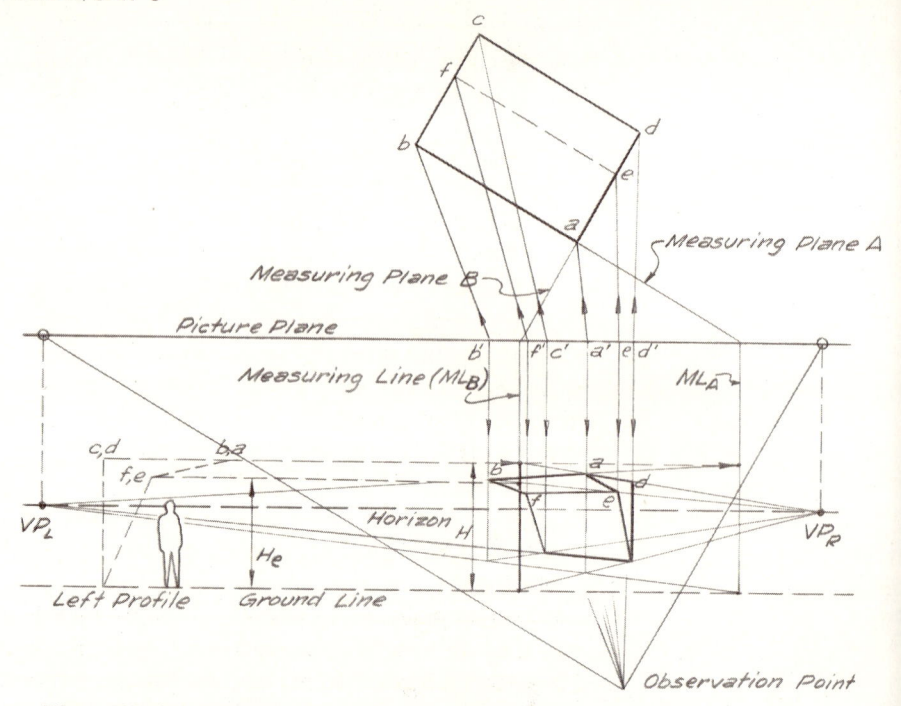

Figure 10.4 A projected perspective using measuring lines.

---

19 The only problem in finding the perspective of the enclosing box is that there will be no true scale lines since no part of the object lies in the picture plane. Figure 10.5 shows how we can overcome this.

We can pass *measuring planes* through the vertical faces and extend them until they intersect the picture plane. These lines of intersection are called *measuring lines* and, since they lie in the picture plane, are _____ _____ lines.

true (full) scale

20 In the top view of Figure 10.4 measuring planes A and B appear as _____ (your words). Their intersections with the picture plane appear as single _____.

edges
(straight lines);
points

Projecting down from these points we have measuring lines (ML_A and ML_B) on which we may make true scale measurements and then, using the VP, project back behind the picture plane where the object actually is.

Construction of the enclosing box in Figure 10.4 proceeds in the same manner as that of Figure 10.1 except for the use of the measuring lines to transfer full-scale measurements to the perspective. Figure 10.5 shows how the true height of the shelter eventually became the foreshortened height of the perspective. Once the enclosing box was found in Figure 10.4, only points e and f had to be plotted to complete the drawing.

There are usually many paths to a given point on a projected perspective drawing. Familiarity with the basic procedures will provide a facility in finding points. Remember that you are searching for points in space and must be able to visualize the spatial relationships between points on the object.

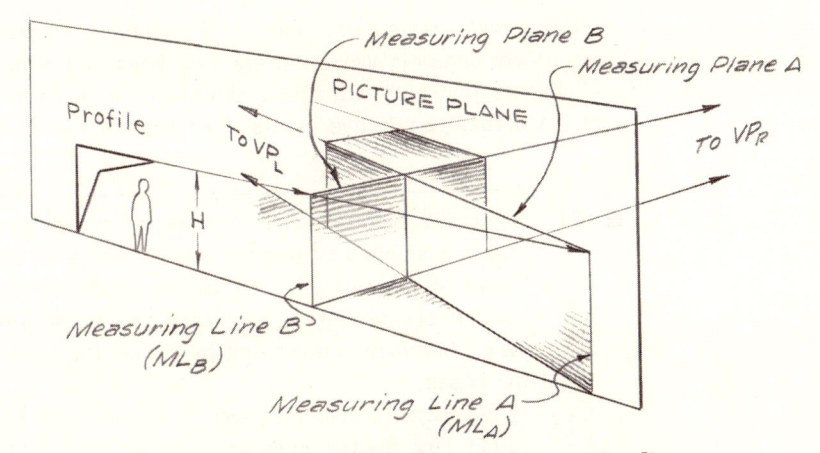

Figure 10.5 The concept of measuring planes and measuring lines.

Curved Lines

Circles and curved lines are handled in the same manner for projected perspective as they are for other pictorial forms. The procedure is:

(1) Establish a few select points on the circle, arc, or curve.
(2) Find these points in the perspective drawing by projection techniques.
(3) Draw a smooth curve through the points.

Study Figure 10.6. It is a complete projected perspective of an L-shaped block with a semicircular top on the vertical member. The semicircles have been defined for projection purposes by five equally spaced points. Measuring planes have been passed through the front and back vertical faces so that the

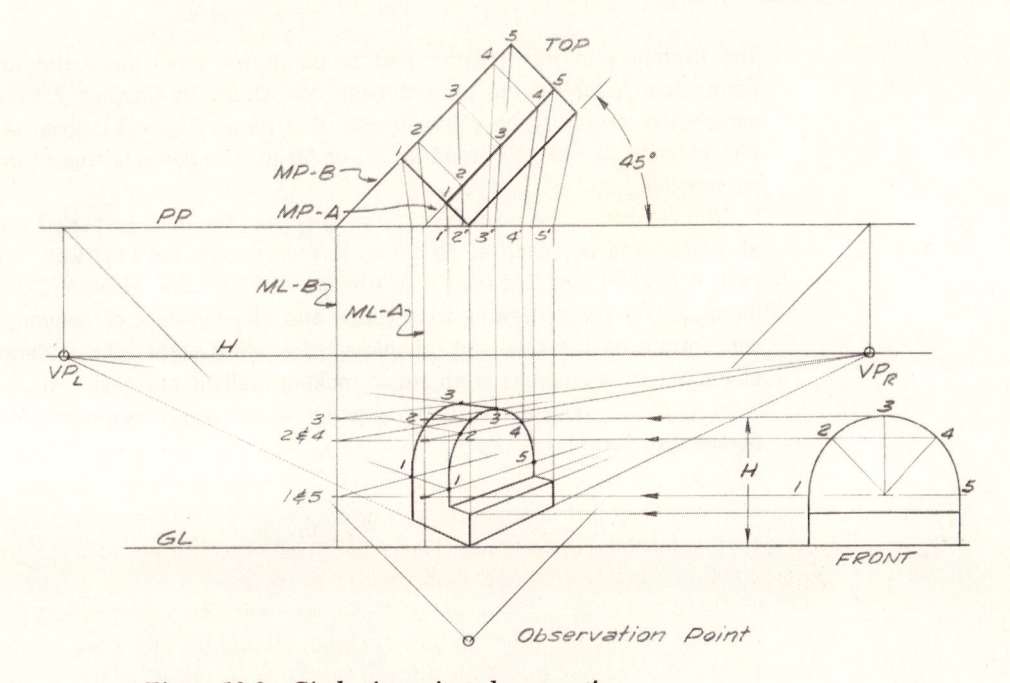

Figure 10.6 Circles in projected perspective.

height measurements for each of the five points can be projected from the front orthographic view to measuring lines and thence into the perspective. They are matched with direct sightings and projections from each point to yield actual perspective points. Two smooth curves completed the drawing.

■ 21   Sketch 32 is an exercise in projected perspective. The object is a small house positioned behind the picture plane as shown in the top view. A profile view is available for height measurements. The ground line and horizon are set. We are viewing the house from above (a bird's-eye view). Complete the drawing by finding the projected perspective view of the house.

Suggested procedure (refer to Figures 10.1 through 10.5 as needed):
(1) Find the two vanishing points.
(2) The house is behind the picture plane so a measuring line is needed. (A measuring plane and measuring line is given.)
(3) Find the complete enclosing box.
(4) Find the rectangular box below the sloping roofs.
(5) Find the ridge-pole line, ef.
(6) Complete the sloping lines. The drawing is finished. Check your solution.

NOTE: Projected perspective is normally done carefully, with instruments. Try this exercise freehand. Use a straight edge to draw lines to the VP if you wish.

A NOTE ON SIZE

The examples in this chapter had to be drawn small since the vanishing points had to be on the printed page. We stated in Chapter 7 that, for a perspective sketch to be of reasonable size on an 8½ × 11 sheet of paper, the distance between VP must be 30 or 40 in. The same is true of projected perspective.

This completes the discussion of perspective. You have just tried your hand at a projected perspective. To some, this may seem the best way to make a perspective drawing because it is precise and to scale. However, if you are interested in using drawing as a quick and effective tool of communication, concentrate on the freehand techniques discussed earlier. An understanding of projection techniques is an aid to making realistic sketches.

Practice! If you are a doodler, practice doodling geometric shapes in the perspective form.

chapter 11   other pictorial drawing forms

Figure 11.1 shows a cube sketched in all of the pictorial drawing forms useful in graphics. We have just presented the subject of perspective sketching. This chapter is concerned with all the others shown: axonometric (embracing isometric, dimetric, and trimetric) and oblique (including cavalier and cabinet).

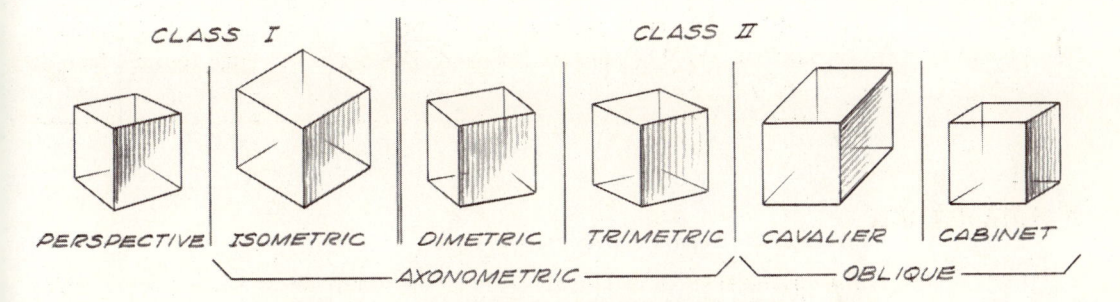

Figure 11.1   The classification of pictorial drawing forms.

The pictorial forms are classified as to types and degrees of usefulness. Class I we consider to be the most useful. Perspective drawing has true-to-life realism. Isometric drawing, the most useful of the axonometric group, has a simple geometric base but contains visual distortion.

Class II forms have limited application to technical drawing and are useful only in certain specific cases.

ISOMETRIC DRAWING

Isometric drawing is a pictorial form based upon the perspective form. It is easier to draw since we neglect the effects of foreshortening. The result is a pictorial form which is geometric and measurable but visually distorted. We accept the distortion in favor of measurability.

The word "isometric" comes from the prefix *iso* meaning "the same" and *metric*, "measure." The implication is that isometric is a drawing in which all three space axes (*x*, *y*, and *z*) are drawn to the "same measure" or the same scale.

184 PICTORIAL DRAWING/UNIT 3

In Chapter 7 we spent considerable time learning the relationship between angles L, θ, and R and how their selection affected the sketch. In isometric drawing, we can state immediately that $L = R = 30°$ and $\theta = 120°$. There are no variations. Note that $L + \theta + R = 180°$.

same

1 Isometric is one of three forms under the general heading of *axonometric drawing*. The other two are dimetric (two axes the same scale and one different) and trimetric (all three axes at different scales). Isometric implies that all three axes are the _____ scale.

perpendicular (normal)

2 Let's analyze isometric drawing by considering the cube. Figure 11.2A shows how we have been viewing the cube (especially in orthographic projection). Our line of sight is _____ to the front face.

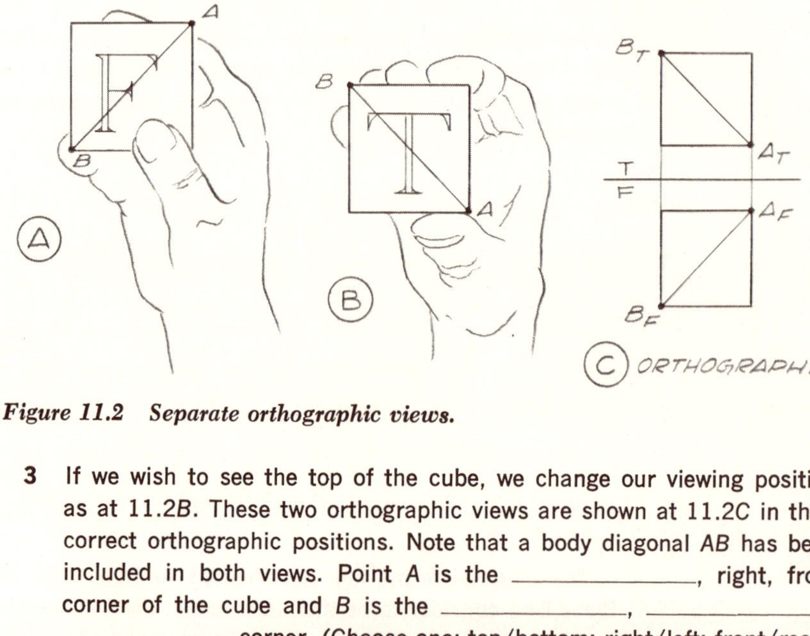

Figure 11.2 Separate orthographic views.

(A): top;
(B): bottom, left, rear (any order)

3 If we wish to see the top of the cube, we change our viewing position as at 11.2B. These two orthographic views are shown at 11.2C in their correct orthographic positions. Note that a body diagonal AB has been included in both views. Point A is the _____, right, front corner of the cube and B is the _____, _____, _____ corner. (*Choose one:* top/bottom; right/left; front/rear.)

profile (side); foreshortened

4 Suppose from the position shown in Figure 11.2A we rotate the cube about a vertical axis to the position shown in Figure 11.3A. We have now exposed the right _____ face of the cube. Dimensions W and D are not shown full scale. They are said to be _____.

full

5 The height H is _____ scale.

W and D (by trigonometry) are reduced in size by $\sqrt{1/2}$ or 0.71 of full scale. The body diagonal AB now appears as a vertical line in the front view and coincides with the intersection of the front and right profile planes of the cube.

Let's now tilt the cube forward until the *body diagonal AB appears as a point*. (See Figure 11.3B.) The angle of tilt needed to accomplish this is actually 35°16′. This action foreshortens the dimension H to 0.816 of its true-scale value (cosine 35°16′ = 0.816).

We find that the other two axes, W and D, now appear as angular lines and are also foreshortened to 0.816 times their true-scale lengths.

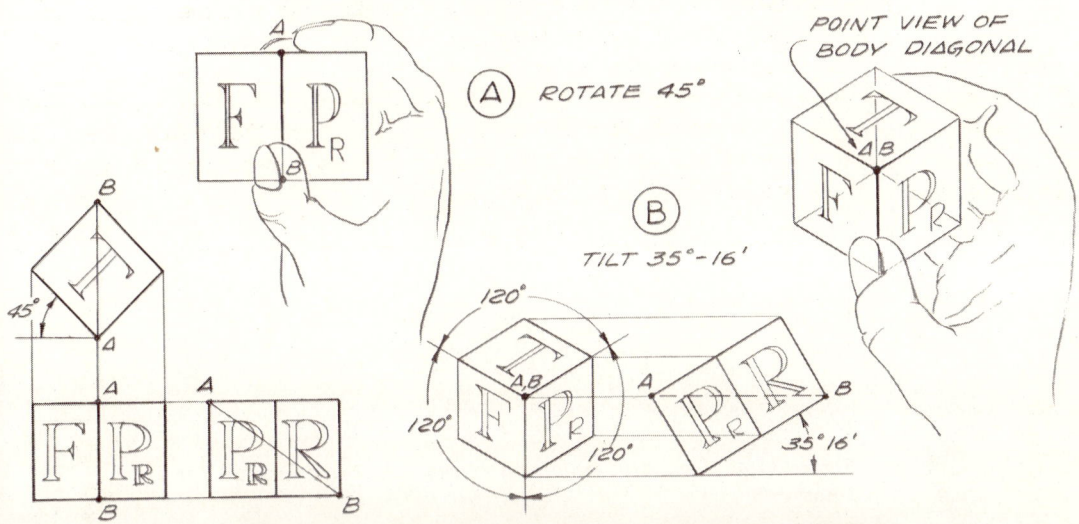

Figure 11.3 The orthographic basis for isometric projection.

equally (the same)	6 Thus by orienting the cube so that the body diagonal shows as a point, we have created a front view in which the three principal axes are foreshortened _____.
120	7 Note the angular relationship of the three axes in Figure 11.3B. They are equally spaced about point A meaning that their angular spacing is _____ degrees.
vanishing points (VP)	8 We have not considered the second effect of foreshortening, namely that objects or parts of objects farther away from us are smaller. Thus each line on the cube is foreshortened the same amount and is parallel to one of the principal axes that show H, W, and D. (There is no convergence toward _____ _____.)
120; foreshortened; parallel	9 This is an *isometric projection*. We have not used projection techniques to draw it. These will be shown later in descriptive geometry. It has three characteristics that we can use to duplicate the drawing directly: (1) The three space axes (representing H, W, and D) are symmetrical about a point and make angles of _____ degrees with each other. (2) The three axes are _____ equally, being 0.816 times full scale. (3) Any line on the object that is _____ to one of the three axes has the same foreshortening.

Figure 11.4 shows the principal orthographic views of a 1-unit cube (11.4A) and an isometric projection of the same cube (11.4B). But we said earlier in this chapter that we would neglect foreshortening in isometric drawing. Let's do just that.

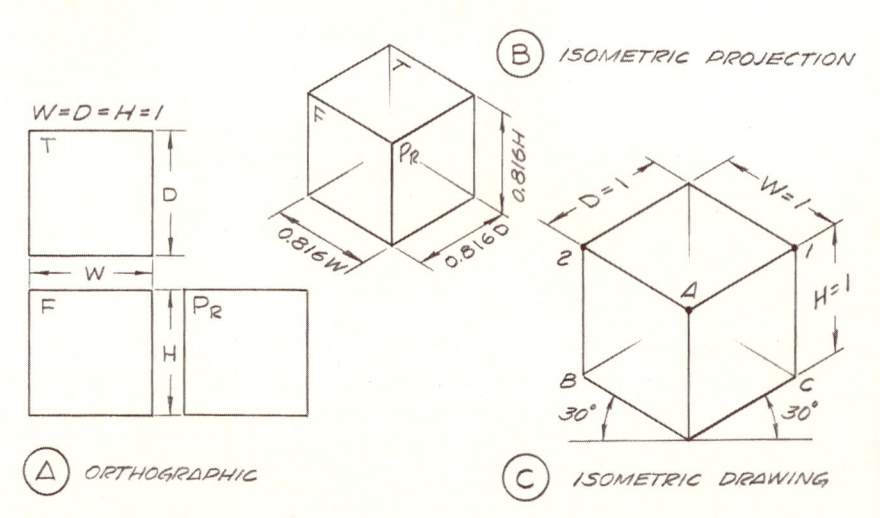

Figure 11.4 Isometric projection versus isometric drawing—for a 1-unit cube.

---

10 Figure 11.4C is a sketch of the 1-unit cube with foreshortening of the three axes omitted. In other words, we have made full-scale measurements along these three axes. This is called an isometric _____ as opposed to 11.4B which is an isometric projection.

drawing

---

11 Procedure for making an *isometric drawing* of a cube is as follows (refer to Figure 11.4C):
(1) Set up three axes, one vertical and the other two at angles of _____ degrees with the horizontal.
(2) Make _____ _____ measurements of H, W, and D along appropriate axes (H is always vertical but D and W can be either right or left as desired).
(3) From the ends of the axes (points A, B, and C) draw lines _____ to the appropriate axes. This gives intersections (points 1 and 2) from which additional lines may be drawn to complete the cube.

30;
full-scale;
parallel

---

12 Note in Figure 11.4C that the isometric drawing of the cube appears larger than both the orthographic views at 11.4A and the isometric projection at 11.4B. This is as expected since, in 11.4C, we have _____ foreshortening.

neglected
(omitted)

OTHER PICTORIAL DRAWING FORMS   187

13 Freehand isometric sketching can be assisted greatly by the use of an isometric grid. Such a grid is shown in Sketch 33. It consists of three sets of parallel lines; one _____ and the other two at _____ degrees to the horizontal.

vertical;
30

The spacing of the parallel lines is such that we have a standard scale measurement along the three principal directions. The actual perpendicular distances between adjacent parallel lines is of no consequence since we do not measure in these directions.

■ 14 Sketch 33 gives dimensioned orthographic views of a simple L-shaped block. A ½-in. isometric grid is also given. Make an isometric drawing of the block on the grid according to the following steps.

(1) First determine the H:W:D proportions of the enclosing box. Expressed in ¼-in. units, they are H:W:D = 4:8:6.

(2) Using point X on the grid as the lowest point of the drawing, erect the dimension H (1"). Next measure W and D from point X along the sloping axes. W = ⅜ and D = ⅝. W may be sketched to the left or to the right as desired.

(3) Now that H, W, and D are laid out, complete the enclosing box using the grid lines as guides. Do not concern yourself with details of the object until the box is drawn. Sketch in the hidden lines at the back lightly to gain a feeling for the solid object.

(4) The bottom portion of the object is really another box with the proportions H:W:D = 3:8:6. Starting at point X sketch this box. Check your solution to this point.

(5) The step in the object is merely another box with proportions H:W:D = 1:3:6. Starting at the farthest rear corner of the total box, sketch this small box on top of the base box.

(6) The construction is now complete. Finish the sketch by going over the important (visible) lines with bold, expressive strokes. Make the closest lines (around point X) darker. Leave hidden lines as light construction lines. Check the final solution.

The block of Sketch 33 could have been oriented in space in a number of ways. The goal in making a pictorial sketch should be *to show the object to the best advantage.* Figure 11.5 shows isometric sketches of the block in some other positions with comments. All are correct isometric drawings.

The decision is yours. Your point of judgment should be "Does the drawing show the details of the object clearly?" Often a few quick, reduced-scale sketches (Figure 11.5) will help you select the best view.

188 PICTORIAL DRAWING/UNIT 3

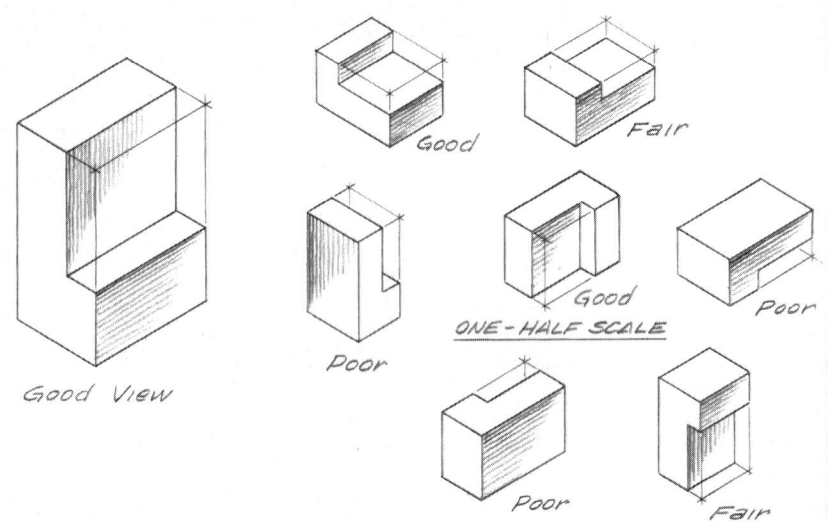

Figure 11.5 An object may be oriented in many ways.

parallel	**15**	The L-shaped object was easy to draw because it had no curves, circles or angular lines. All lines fell on the isometric axes or lines _____ to the isometric axes.
line (edge)	**16**	Thus, full-scale measurements could be made along every _____ of the block.

Non-isometric Lines

rectangular box	**17**	Consider the pyramid of Figure 11.6. Following the procedure of Sketch 33, we would start an isometric sketch of the pyramid by sketching a(n) _____ _____ of proportions $H:W:D = 5:3:3$. (See Figure 11.6B.)

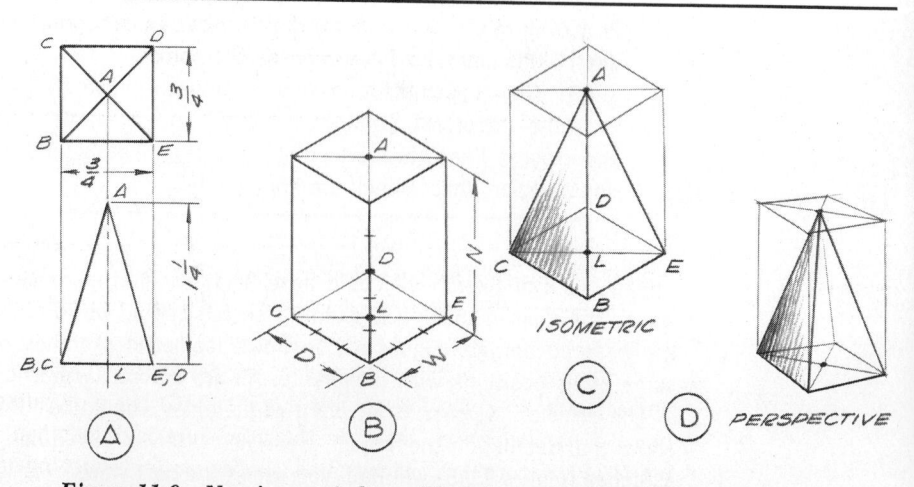

Figure 11.6 Non-isometric lines. Find the end points by coordinate means and join them with a straight line.

OTHER PICTORIAL DRAWING FORMS 189

true-scale

18 Lines *AB, AC, AD,* and *AE* are not parallel to any of the three isometric axes. Therefore, _____ _____ measurements cannot be made along these lines. These lines are called *non-isometric lines*.

H (height)

19 Lines *BE* and *CD* are parallel to the *W* dimension and lines *CB* and *DE* are parallel to the *D* dimension. No lines on the surface of the pyramid are parallel to the _____ dimension.

20 However, note that the altitude *AL* of the pyramid is a vertical line that joins the apex *A* with the center point of the base of the pyramid. This line can be measured since it is parallel to the *H* isometric direction.

Because of the geometry of the object, locating point *A* allows us to finish the object by sketching *AB, AC, AD,* and *AE*. (Figure 11.6C.)

non-isometric

These four lines are called _____ _____ lines.

At Figure 11.6D you see a perspective sketch of the pyramid. The sketch is included here to point out that it is a *better picture* of the pyramid than is the isometric sketch. The extreme symmetry of both the pyramid and the isometric system detracts from the pictorial sketch. The perspective system offers an opportunity to change the point of view.

21 From Figure 11.6 we have learned that the method for finding non-isometric lines on an isometric sketch is as follows:

(1) Find the two points that represent the _____ of the line.

ends;
straight line

(2) Join these points with a _____ _____.

■ **22** Sketch 34 is an exercise in sketching non-isometric lines. The object is a combination of straight and sloping surfaces. On the isometric grid provided, make an isometric sketch of the object, remembering to:

(1) Select an advantageous viewpoint—one that shows the detail of the object best.

(2) Start with the basic box, $H \times W \times D$.

(3) Find the end points (coordinates) for all non-isometric lines, and then sketch straight lines between the points.

(4) Check your solution.

Isometric Circles

ellipses

23 In pictorial drawings, circles become _____.

In Chapter 2, Freehand Sketching, we developed a method for sketching a circle inside an enclosing square (Figure 2.16). We then tilted the square to make a parallelogram, repeated the procedure and sketched an ellipse (Figure 2.19). Figure 11.7 shows that procedure for sketching isometric circles (ellipses) is the same. Figure 11.7A shows a circle on the front face of a

190 PICTORIAL DRAWING/UNIT 3

cube. Figure 11.7B is an isometric sketch of the cube. To transfer the circle to the isometric cube we proceed as follows:

(1) Find the center of the parallelogram that represents the front of the cube by sketching two diagonals.

(2) Then we find the midpoints of the sides of the parallelogram by sketching two lines through the center parallel to the sides of the parallelogram.

(3) This gives us four points of tangency that we use as guides in sketching a *smooth curve* (ellipse) which is the isometric circle required. (Figure 11.7C.)

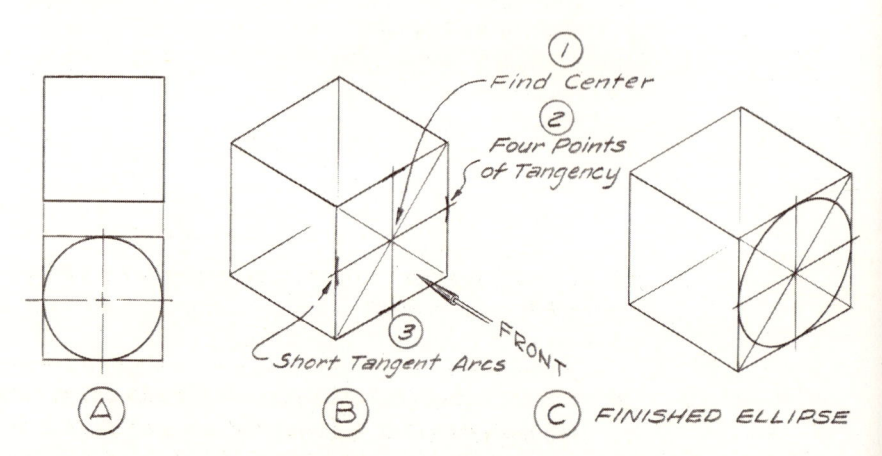

Figure 11.7   Steps in sketching an isometric circle.

■ 24   Sketch 35A is an isometric sketch of a 1½-in. cube. Sketch isometric circles on the three exposed (visible) surfaces of the cube. Check your solution.

Instrument Isometric Circles

parallelogram
(rhombus)

25   Because of the symmetry of the isometric system, a good approximation to an ellipse can be drawn with a compass. Consider Figure 11.8A. It shows a(n) _____ ABCD.

square

26   If we say that Figure 11.8A is an isometric sketch, then the parallelogram ABCD (actually, a rhombus, an equal-sided parallelogram) represents a(n) _____ in isometric form.

27   In Figure 11.8B, the diagonals of the rhombus have been drawn and the midpoints, E, F, G, and H, of the four sides have been plotted.

Triangle ABD is an equilateral triangle. A line has been drawn from point E, the midpoint of side AB, to D. Line ED is the altitude of triangle ABD and therefore, the angle between line ED and side AB is a(n) _____-degree angle.

OTHER PICTORIAL DRAWING FORMS 191

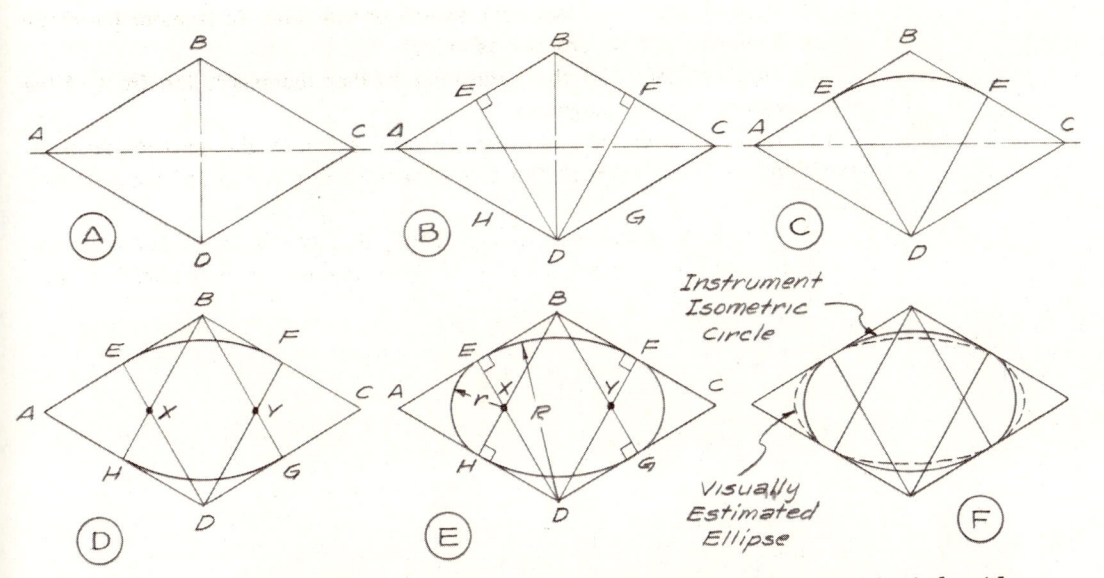

Figure 11.8 Steps in constructing an approximate isometric circle with a compass.

tangent	28. Similarly, the angle between FD and side BC is also 90°. Therefore, if we use point D as a center and ED (or FD) as a radius we can draw a circular arc (Figure 11.8C) that will be _____ to the sides AB and BC at points E and F.
one-fourth (¼, 25%)	29. This arc represents _____ of the complete ellipse.
B; G (and) H	30. Figure 11.8D shows the opposite quarter arc drawn. Here point _____ was the center of the arc and _____ and _____ are the tangent points.
X; Y	31. All that remains is to close the ends of the ellipse. Note in Figure 11.8D that the intersection of lines ED and HB has located a point _____ and the intersection of FD and GB has located a point _____.
XE = XH = YF = YG	32. Points X and Y are the centers for the two end arcs using a radius of _____ = _____ = _____ = _____.
perpendicular	33. These arcs will be tangent at the midpoints (E, F, G, and H) of the sides of the isometric square for the same reason that the long arcs were. That reason is that lines ED, FD, GB, and HB are _____ to the sides.

The complete ellipse is shown in Figure 11.8E. This is not a true ellipse but just an approximation. Figure 11.8F shows a comparison of an instrument-drawn ellipse and a sketched ellipse. The instrument ellipse is accurate enough to look realistic especially in light of its simplicity of construction.

Figure 11.9 shows a cube with ellipses constructed on the three visible faces to illustrate that the system works for all three principal isometric planes. The system involves finding four centers and two radii, *R* and *r*, for each plane.

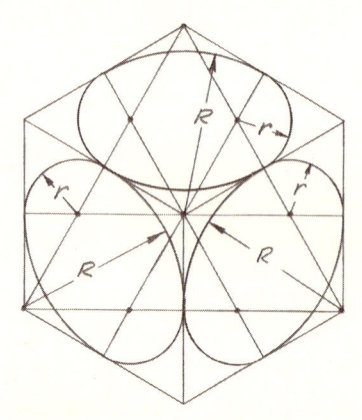

Figure 11.9 Isometric circles—instrument constructed.

■ **34** Sketch 35*B* shows a 1-in.-diameter cylinder that is 1½ in. high. Make an isometric sketch of the cylinder on the grid provided according to the following steps:

(1) Start the sketch by drawing the enclosing box (prism). The proportions are $H{:}W{:}D = 1\frac{1}{2}{:}1{:}1$ in 1-in. units or 6:4:4 in ¼-in. units.

(2) Now sketch isometric circles on the top and bottom surfaces of the prism (box). These should be identical ellipses.

(3) The two vertical sides of the cylinder are represented by two straight lines that are tangent to the ellipses at their extreme ends (ends of the major diameters). Sketch these lines. The sketch is now complete except for the finish line work.

(4) For practice, you might try sketching the cylinder lying on its side as shown in the solution sketch.

REMEMBER: In all sketching exercises, do your preliminary construction lightly and then finish the important lines with darker, expressive pencil strokes. Use a soft pencil so that both line qualities can be made with the same tool.

■ **35** Sketch 35*C* shows a flat plate with a 1-in.-diameter hole through it. The hole may be considered to be a negative cylinder. Make an isometric sketch of the plate. Check your solution. Treat the hole in the same manner as the cylinder of Sketch 35*B*. Part of the ellipse on the bottom of the object will not be visible. Leave this part as light construction line.

Figure 11.10 shows the plate in other positions. Note that one of the positions is a bottom view; that is, the object is above our line of sight and we are looking at the bottom of the object. This can be done easily in the isometric system by reversing the W and D axes. No matter what position you choose, the sketching problem is the same.

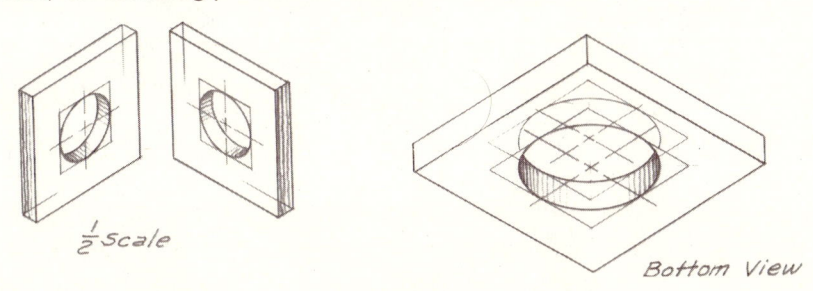

Figure 11.10   Different views of the object of Sketch 35C.

Arcs or Partial Circles

36 Partial circles are handled in the same manner as full circles. Consider the object of Sketch 36A. It has basic proportions, H:W:D = 2:9:6 (¼-in. units). One end is rounded with a _____-in.-radius semicircle and one corner at the other has a quarter-circle notch of _____-in. radius.

¾-in.;
½-in.

■ **37** On the isometric grid of Sketch 36A make an isometric sketch of the object. Steps in the construction are as follows:

(1) Start by sketching the enclosing box. (Given.)

(2) To sketch the semicircular end, we must find points of tangency A, B, and C on the top surafce and A', B', and C' on the bottom. Similarly, we need D, E, D', and E' for the quarter-circle. Find these points and sketch the appropriate partial parallelograms.

REMEMBER: Full-scale measurements can be made only on the isometric axes or lines parallel to these axes. Check your solution.

(3) Now, lightly sketch the circular arcs on both the top and bottom surfaces. The construction is complete. Finish the sketch using expressive line treatment. Be careful of arc visibility. *Don't forget the short vertical line at the semicircular end* that represents the profile of the cylindrical shape. It must be tangent to the two semi-ellipses. Check the final solution.

Figure 11.11 shows the object of Sketch 36A oriented in different ways. Again, the drawing problem is exactly the same in each case. The choice of position is up to you. Some positions do not show the nature of the quarter-circle notch well and should be avoided.

You should recognize that the arcs in partial isometric circles are made up of one, two, or three of the quarter-arcs that make up a full ellipse. Therefore, we should be able to construct these with a compass as was shown in Figure 11.8.

194 PICTORIAL DRAWING/UNIT 3

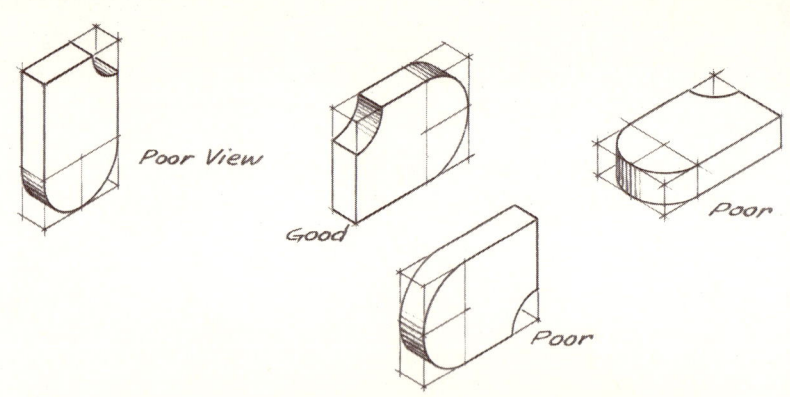

Figure 11.11 Always select a view which shows the details of the object best.

---

tangent

38 The procedure there was to find two sets of centers from which we can construct arcs that are _____ to the sides of the enclosing isometric square (rhombus) at the midpoints of the sides.

39 Consider Figure 11.12. It shows a perspective sketch and an orthographic sketch of a block with a(n) _____-circular notch on the front, a(n) _____-circular notch on the right rear corner and a less-than-one-quarter notch on the left rear corner.

semi-(circular);
(one-)quarter-

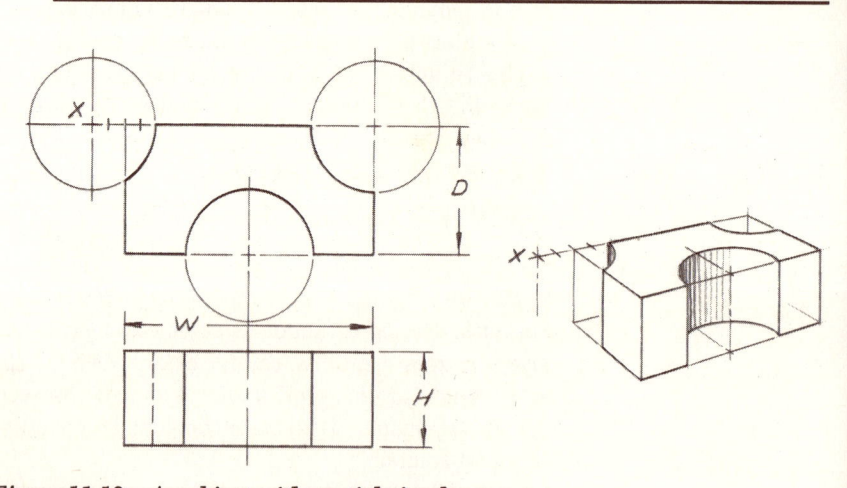

Figure 11.12 An object with partial circular arcs.

---

40 Figure 11.13A shows the instrument construction required to draw the semicircle on the top surface of the block. Note that it was necessary to construct the _____ isometric square in order to find the two centers, A for radius R and B for radius r.

full
(whole,
entire)

OTHER PICTORIAL DRAWING FORMS 195

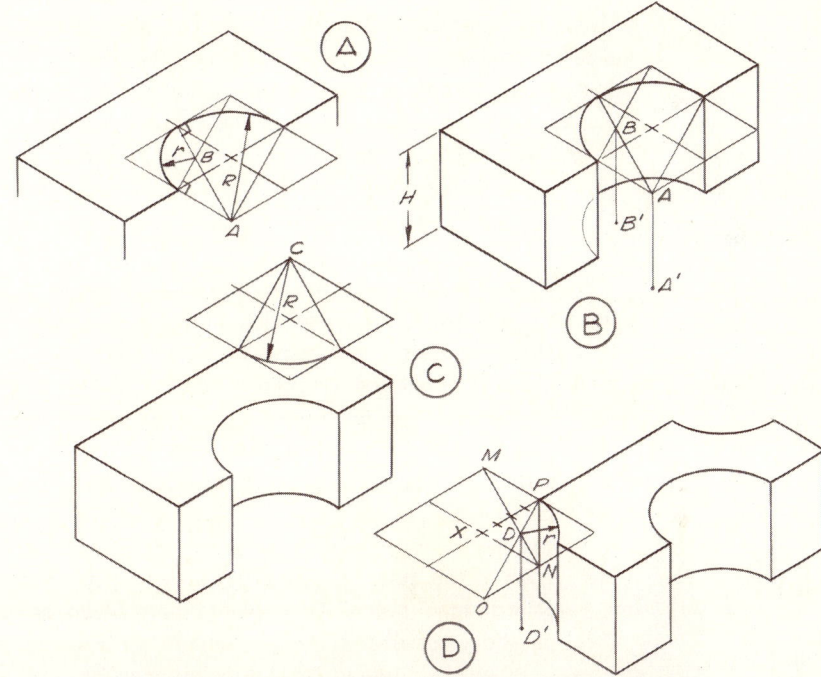

Figure 11.13   Partial isometric circles constructed with a compass.

	41	Using points A and B as centers, we were able to draw the two arcs to form the semicircle on the top surface. In Figure 11.13B, the arc center points A' and B' were found for the semicircle on the bottom of the block. It was not necessary to reconstruct the enclosing parallelogram, but merely to construct vertical lines from points A and B and measure
height (H)		down a distance equal to the _____ of the object along each line.
	42	Figure 11.13C shows the construction required to find the arc center C for the quarter-circle arc at the right rear corner of the block. Again,
full (entire)		the _____ enclosing parallelogram had to be drawn.
	43	Using C as a center and with radius _____, we can draw an arc to depict the corner notch on the top surface in the isometric form. It is not necessary to find the arc on the bottom surface because that arc
R; visible (seen)		is not _____.
	44	The notch at the left rear corner is not a full quarter-circle. Construction, however, is the same. The circle center X is offset _____ units from the left edge of the block along the extension of the rear
two (2)		horizontal line.
	45	Knowing the radius of the arc and using the center X, we can construct an enclosing parallelogram (Figure 11.13D). The circle radius, XP =
four (4)		_____ units.

196 PICTORIAL DRAWING/UNIT 3

The arc center D is located by finding the intersection of lines MN and OP. By using a radius r, the arc is drawn to its intersections with the rear and side horizontal lines on the top surface of the block.

A small portion of the corresponding arc on the bottom surface will be visible in the isometric drawing. The center for this arc D' is found by dropping a vertical from point D and measuring down a distance H to point D'.

Irregular Curves

points (dots)	46	In Chapter 2, Freehand Sketching, we saw that irregular (noncircular) curves could be sketched if we have enough _____ along the curve to define the irregular shape.
smooth	47	The problem in isometric sketching is, first, to find the points that describe the curve and then to join the points with a(n) _____ curved line.
far; close	48	Look at the object in Sketch 36B. It is a flat slab of metal with a parabolic-shaped notch. Nine points have been selected to describe the parabolic curve. When the curve is flat, the points may be _____ apart, but when the curve is sharply changing, the points must be _____ together.
centerline	49	Note that the points are symmetrically located on either side of a vertical _____ through points 0 and 5. A parabola is a symmetrical curve. We will save drawing time by selecting symmetrical points as shown.
■	50	On the grid provided (Sketch 36B), make a full-size isometric sketch of the parabolic plate. Follow instructions in the next few steps: (1) As before, start with the basic enclosing prism (box) knowing that H:W:D = 10:8:2 (¼-in. units). The isometric box is given. (2) Find the vertical centerline on the front face of the prism. The nine points must be plotted by using coordinate measurements. In a coordinate measuring system, we must have an origin from which we can make measurements. We will select point 0, the intersection of the vertical centerline and the top, front, horizontal edge of the prism. (3) Points 1, 5, and 9 lie on the coordinate lines. Find points 1 and 9 by measuring three units on either side of point 0 along the top, front edge. Find point 5 by measuring eight units down from point 0 on the centerline. (4) We now find the other points by measuring in both coordinate directions. Points 2 and 8 are three units down from the top. Sketch a line in the W direction that is three units below point 0. The distances

from the centerline to points 2 and 8 are not necessarily even integer units. Transfer these by estimation or mark them on the edge of a piece of paper and then transfer them directly. Check your solution.

(5) Repeat this procedure for points 3 and 7 and 4 and 6. All nine points have now been transferred to the isometric form. Lightly sketch a smooth curve through the nine points.

(6) We must now find the corresponding points on the rear surface. We could repeat the procedure used for the front surface, but let's consider another method. On the object, each of the nine points on the rear surface is exactly two units ($\frac{1}{2}$-in.) *behind* the corresponding points on the front. Sketch a line through each point in the D direction and estimate a two-unit distance on each.

(7) This procedure defines the nine points on the rear surface. Sketch a light curve through these points.

NOTE: Always include the hidden side of an object in sketching. If you make your construction light and your finish lines dark, the light construction will not detract from the sketch and you will gain the advantage of seeing through the object.

(8) All points have now been plotted. Finish the visible lines with carefully sketched but bold and expressive lines.

The selection of the coordinate origin 0 in the last example was arbitrary. The prime consideration in selecting the point is to make the drawing job as easy as possible. Figure 11.14A shows a few points plotted with the origin at one corner of the prism. Figure 11.14B has the origin completely off the object. As you can see, this is a possibility, but an inadvisable one in that it complicates visualization and drawing.

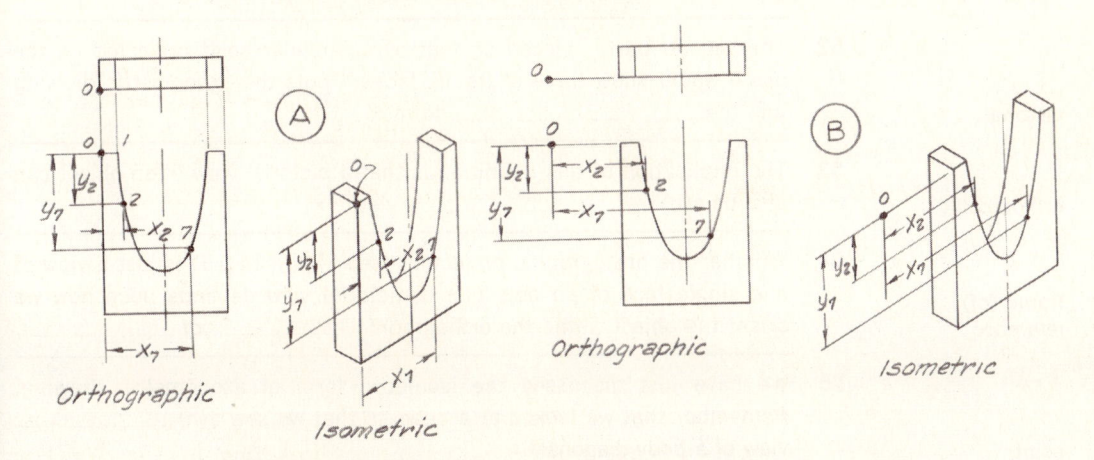

Figure 11.14   The origin of the coordinate system may be located at any point.

198 PICTORIAL DRAWING/UNIT 3

Summary of Isometric Considerations

This is the end of the discussion of isometric drawing. Remember the following points. They will be of use in the coming discussion on dimetric, trimetric, and oblique drawing.

(1) True-scale measurements can be made only on the isometric axes or lines parallel to these axes.

(2) Foreshortening is neglected. Isometric is a parallel system so there is no reduction in full-scale size of lines placed farther away from the observer.

(3) No angular measurements can be made. A non-isometric line is drawn by finding the coordinate points which describe the ends of the line.

(4) Circles and curves must be sketched by finding points (by coordinate methods) that lie on the curved lines.

DIMETRIC AND TRIMETRIC DRAWING

The introduction to this chapter showed a classification chart of the pictorial forms of technical drawing. The three forms called *isometric*, *dimetric*, and *trimetric* fall together under a general classification of axonometric drawing. By definition, axonometric drawing is the depiction of an object by means of perpendicular projectors on a single plane (the picture plane) which is placed so that a rectangular solid projected on the plane shows three of its six faces.

orthographic	51	The first part of this definition, "the depiction of an object by means of perpendicular projectors on a single plane," we have already encountered as the definition of _____ drawing.
pictorial	52	The second part, "placed so that a rectangular solid projected on the plane would show three of its six faces," puts the axonometric drawing into the _____ form.
orthographic	53	The implication of this definition is that a pictorial view of an object can also be a(n) _____ view.
frame (of) reference	54	Whether the orthographic projection (see Figure 11.15) is just a view of one single face of an object or a pictorial view depends upon how we orient the object within the orthographic _____ of _____.
point	55	We have just discussed the isometric form of axonometric drawing. Remember that we looked at a cube so that we saw a(n) _____ view of a body diagonal.

OTHER PICTORIAL DRAWING FORMS 199

Figure 11.15 The result of orthographic viewing depends upon the orientation of the object behind the picture plane.

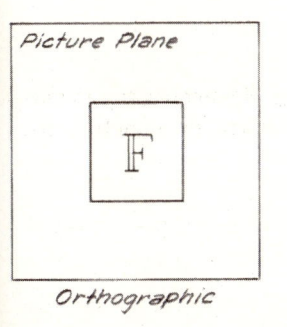

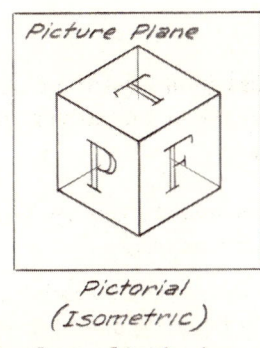

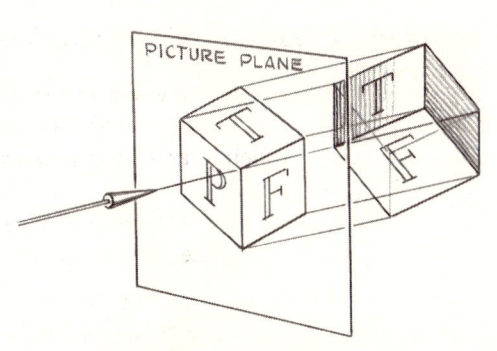

Figure 11.16 The basis for the isometric pictorial drawing form.

perpendicular	**56** Figure 11.16 shows this method of viewing. The projectors that define the image on the picture plane are _____ to the picture plane. The special orientation of the object gives us the pictorial (isometric) form.
120	**57** Isometric drawing has extreme symmetry. An isometric *projection* shows equal foreshortening (0.816 of full size) on the three axes. In an isometric *drawing*, we neglect foreshortening and make full-scale measurements on all axes. The three space axes are _____ degrees apart.
two (2); three (3)	**58** How do the dimetric and trimetric forms of axonometric projection differ from isometric? Isometric means "the same measure," that is, the same measure (or scale) is used on all three space axes. This comes about because all three exposed faces are inclined at the same angle to the picture plane. By the same reasoning, dimetric means "_____ measures" and trimetric means "_____ measures."
	In the dimetric form, two faces of the object have equal angles of inclination to the picture plane (see Figure 11.17) and the third face has a different angle. Thus two *scales* are used, one for the two faces (directions) that have equal inclination, and one for the face of different inclination.
axes	**59** In the trimetric form, all three faces have different angles of inclination with the picture plane. Thus three separate scales are used to measure distances along the three space _____.
scales; ¾	**60** In the dimetric cube of Figure 11.17, the top and front faces have equal inclination to the picture plane and thus the vertical and the right-hand axes have the same _____. The left-hand axis is foreshortened by _____ of full scale.

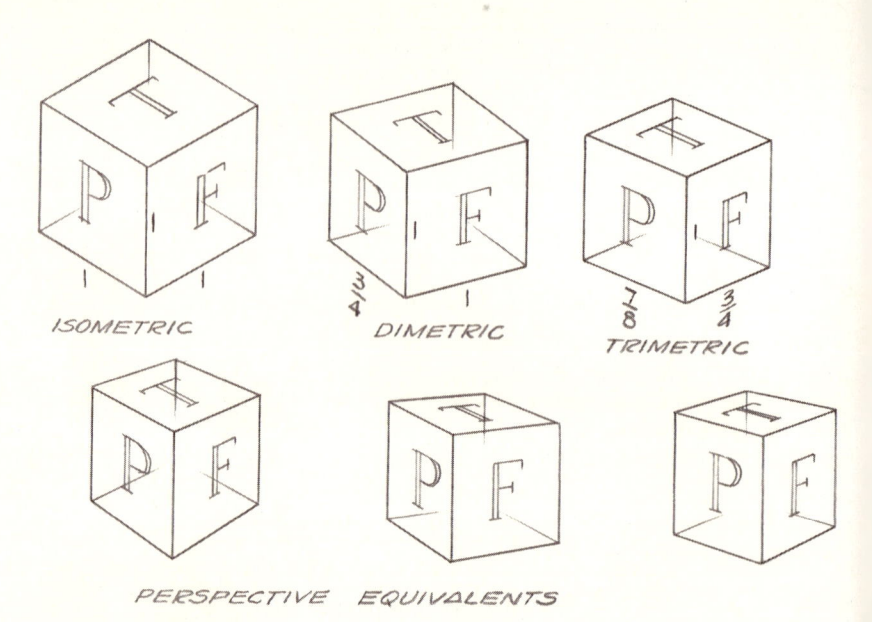

Figure 11.17 The axonometric pictorial forms and their perspective equivalents.

---

unequal (different)

61 The trimetric cube of Figure 11.17 shows three _____ scales on the three space axes. This means that each axis is foreshortened to a different scale.

---

perspective

62 Note in Figure 11.17 that the change from isometric to dimetric to trimetric changes the visual appearance of the cube much in the way that we changed the appearance of an object by selecting different values for angles L, θ, and R in _____ sketching.

The real value of the dimetric and trimetric pictorial forms lies in their ability to give us a geometric and measurable form of pictorial drawing that is *visually more accurate* and more pleasing than the isometric form. Figure 11.18 shows an object sketched in the three axonometric forms.

The object of Figure 11.18 is a model of a body-centered cubic crystal. The compound cesium chloride (CsCl), has a body-centered-cubic molecular structure. Chemists and chemical engineers are interested in being able to sketch molecular structures of elements and compounds. Notice that the isometric form is quite inadequate because of its extreme symmetry. The crystal appears to be just an interesting geometric design.

The dimetric and trimetric forms begin to show the three-dimensional character of the object. However, unless you must rely on a measurable pictorial form, perspective sketching can give the same advantages and will produce sketches with photographic realism if done correctly.

OTHER PICTORIAL DRAWING FORMS 201

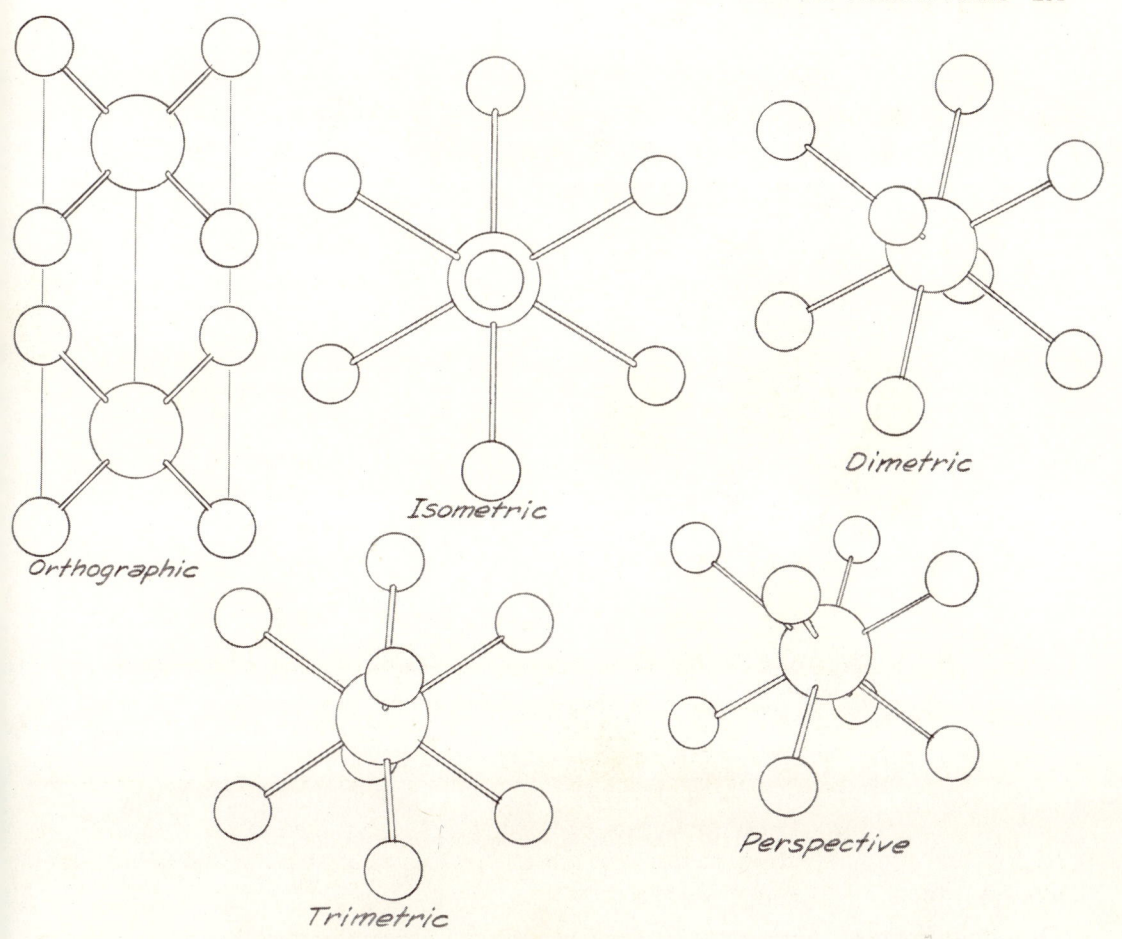

Figure 11.18 The extreme symmetry of isometric can be detrimental to the communication of information.

axonometric	**63** Figure 11.19 is a comparison of the three _____ forms in an orthographic-viewing system. A cube is shown pictorially positioned behind a picture plane and a front orthographic view of the plane is drawn.
point	**64** In isometric drawing (Figure 11.19A) there is essentially only one way to orient the object behind the front reference plane (picture plane). To obtain the isometric we must always see a(n) _____ view of a body diagonal.
same	**65** We may expose other faces of the cube (rear, bottom, etc.) by changing its orientation but we would always get exactly the _____ drawing.

202 PICTORIAL DRAWING/UNIT 3

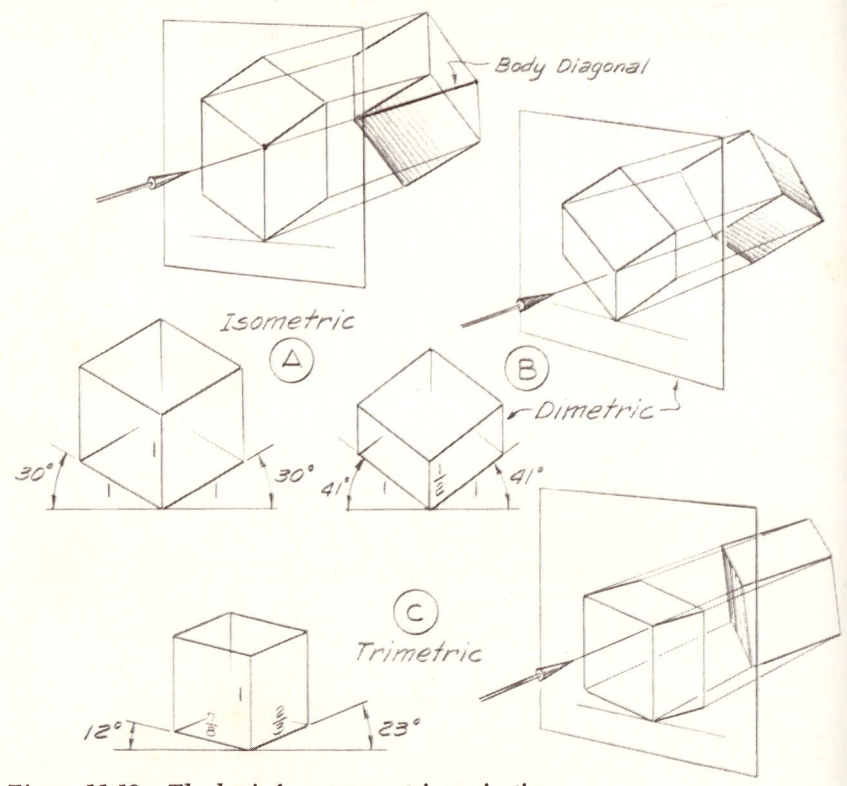

Figure 11.19 The basis for axonometric projection.

Dimetric Forms

½	66	The dimetric projection of Figure 11.19B shows one possible view within the dimetric system. The vertical axis has been foreshortened to _____ the scale of the other two axes by tilting the cube forward.
41	67	This angle of tilt puts the two receding axes at _____ degrees with the horizontal.
1:½:1	68	In a projected dimetric view (as Figure 11.19B) all three axes are foreshortened. But, since two carry equal foreshortening, we can say that the ratio of foreshortening of the space axes is left (L): vertical (V): right (R) = _____:_____:_____.
1:1:1	69	Do not confuse this ratio of foreshortening with the ratio of dimensions of the object (H:W:D). Since we are illustrating a cube, this latter ratio is _____:_____:_____.
right; ½	70	Figure 11.20A shows a dimetric drawing of a 1-in. cube. The natural foreshortening of the dimetric *projection* has been neglected in the dimetric *drawing*. One axis, the _____ (left/right) receding axis is foreshortened to _____ full size.

OTHER PICTORIAL DRAWING FORMS 203

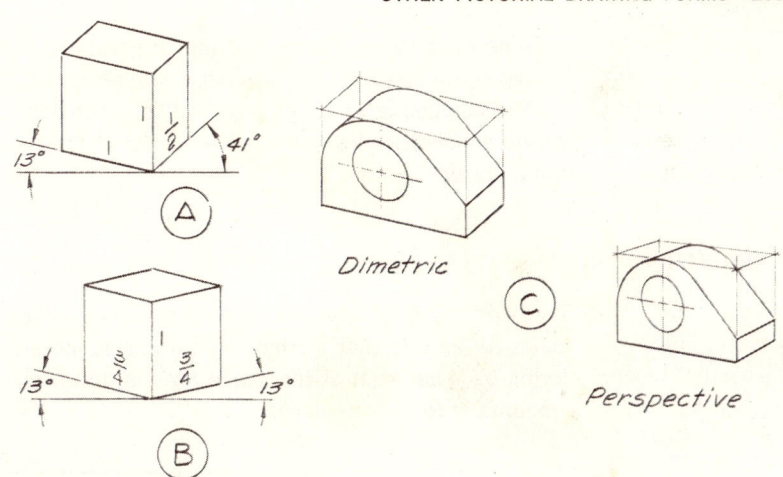

Figure 11.20 Dimetric drawing.

1:1:½	71	The ratio of foreshortening is *L:V:R* = ____:____:____.
vertical; right	72	Comparing the dimetric cubes of Figures 11.19B and 11.20A, we see that they are identical except in orientation. The foreshortened axis has been changed from the _____ position to the _____ receding position.
dimetric	73	This gives us another way to look at the cube (another point of view) while still maintaining the geometry of the _____ system.
¾:1:¾	74	There are other orientations and other foreshortening ratios in the dimetric system. Figure 11.20B shows a dimetric drawing of a cube with foreshortening ratio of *L:V:R* = ____:____:____.

Note that the angles of the receding axes with the horizontal have changed. This is natural since the cube had to be reoriented to obtain the new foreshortening ratio. The cube could have been drawn with the ratios *L:V:R* = 1:¾:¾ or ¾:¾:1. The angles would be different for each new ratio.

Summary of dimetric considerations

The technique of making a dimetric drawing is exactly the same as that of isometric with the one exception that the unequal foreshortening must be included. As a review, the steps in sketching a rectangular box are:

(1) Lay out the three space axes—one vertical and two receding at the correct angles according to the system desired.

(2) Make scale measurements along the three axes taking into account the *L:V:R* foreshortening ratio.

(3) Finish the enclosing box by sketching a series of parallel lines.

The sketching of circles, circular arcs, irregular curves, and non-dimetric lines (lines not parallel to any of the space axes) is the same as in the isometric form.

204 PICTORIAL DRAWING/UNIT 3

Dimetric drawing has limited use in graphics. A good understanding of the theory of perspective and of visual proportioning will be more useful to you than remembering the angles and ratios of the dimetric system. Figure 11.20C shows a dimetric drawing of an object and the perspective equivalent of the same object.

Trimetric Forms

Trimetric drawing has the same drawbacks as the dimetric form. Its only claim to fame is that a trimetric drawing is more realistic than its two sister forms because each of the space axes is drawn to a different scale (different amounts of foreshortening).

⅞:1:⅔

75 Look back at Figure 11.19C. The trimetric projection shown there has a foreshortening ratio of *L:V:R* = ____:____:____.

12; 23

76 The angle of tilt necessary to achieve this *L:V:R* ratio makes the angles of the receding axes _____ degrees for the left axis and _____ degrees for the right one.

picture plane

77 Figure 11.21A shows two other possible trimetric drawings of a cube. There are many combinations of axis foreshortening. Each one must be analyzed on the basis of the orthographic viewing system. The foreshortening ratio depends upon how the cube is oriented behind the _____ _____.

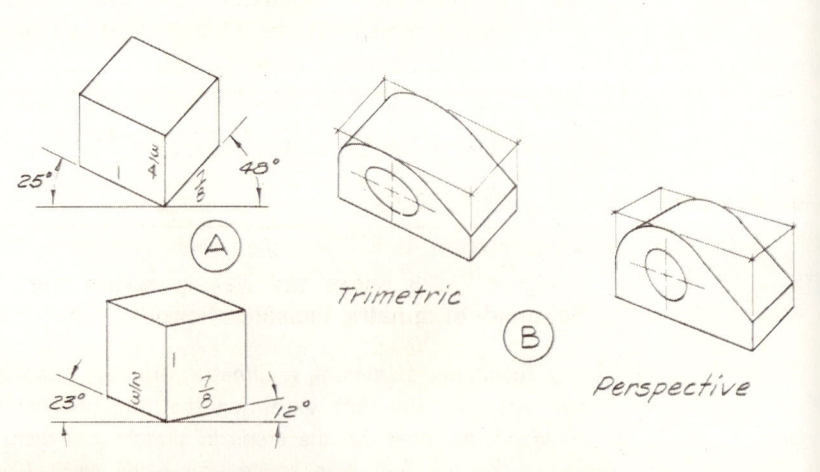

Figure 11.21 Trimetric drawing.

infinite

78 Since we have no restrictions on the foreshortening ratio such as those in isometric (all axes the same) and dimetric (two the same), there are a(n) _____ number of positions in which we can place a cube to obtain a trimetric drawing of it.

Evaluation of trimetric considerations

The problems of drawing, and especially sketching, in the trimetric form are severe. There are odd angles to calculate and measure and three different scales to calculate. Again, a sound knowledge of perspective sketching will prove more useful to both the student of graphics and the professional in making satisfying sketches.

Figure 11.21B shows an object sketched in the trimetric drawing form and also in the perspective form. A comparison of these two will show that the perspective is the more realistic and satisfying sketch.

OBLIQUE DRAWING

The introduction to this chapter showed another special pictorial drawing form that is of use in graphics. This is the oblique form.

axes

79 Consider Figure 11.22. It is an oblique drawing of a 1-in. cube. In all our discussions of pictorial sketching, we have started the drawing by setting up the three space _____.

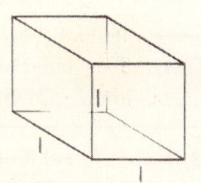

Figure 11.22 An oblique drawing of a cube.

90; 0 or 180

80 These are the vertical and two receding (horizontal) axes. Note that in Figure 11.22 one of the horizontal axes does not recede, but makes an angle of _____ degrees with the vertical (of _____ degrees with the horizontal).

Figure 11.23 shows the viewing system from which we develop the oblique drawing form. The name "oblique" comes from the fact that our line of sight is no longer perpendicular to the picture plane. We are viewing the cube through the picture plane at an angle other than 90°.

**in (on);
perpendicular**

81 Figure 11.23A shows two significant points of difference with the pictorial forms studied so far:
   (1) One face of the cube (the front) is _____ the picture plane or, at least parallel to it.
   (2) The arrow showing the normal orthographic line of sight has been changed in direction so that it is no longer _____ to the picture plane.

206 PICTORIAL DRAWING/UNIT 3

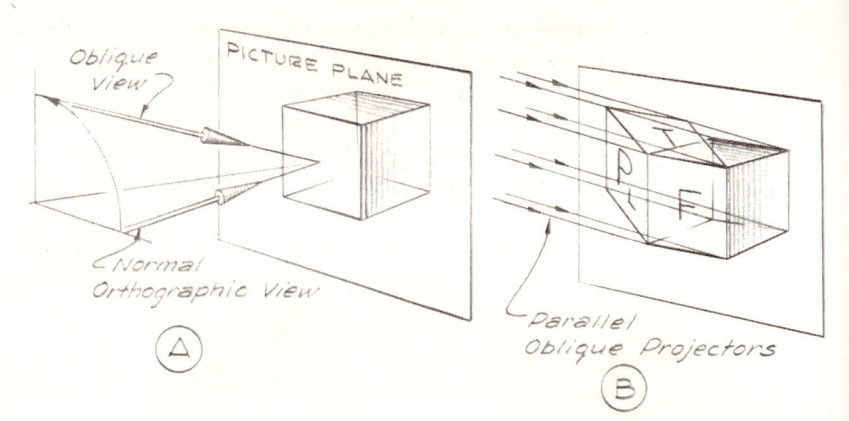

Figure 11.23   The basis for the oblique drawing system.

square	**82** If we viewed the cube of Figure 11.23A orthographically, we would see a(n) _____ that represented the front of the cube.
front, top, side	**83** However, viewing it obliquely, we see three faces of the cube, the _____, _____, and _____. (See Figure 11.23B.)
H; W; D	**84** This makes oblique a true pictorial form since we can show the ____, ____, and ____ of an object in one view.
hole	**85** It is significant that the front face of the cube is in (or parallel to) the picture plane. This fact makes the oblique form ideal for drawing certain objects. Consider Figure 11.24A. It is a cube with a horizontal _____ through it.

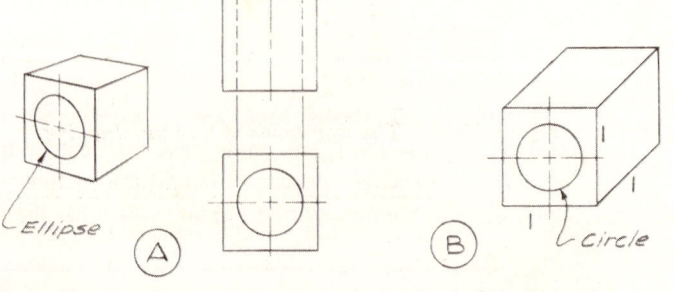

Figure 11.24   Circles can be drawn as circles rather than ellipses in the oblique system.

circle; ellipse	**86** Figure 11.24B is an oblique drawing of the cube. Note that the face showing the hole was selected as the front of the oblique pictorial and, because of the geometry of the axes, the hole was drawn as a(n) _____ rather than a(n) _____.

OTHER PICTORIAL DRAWING FORMS   207

From past discussions on the drawing of pictorial circles, we can see that it is a distinct advantage to be able to draw circles with a compass rather than plotting and sketching ellipses.

1:1:1	**87** Note that in Figure 11.24B the foreshortening ratio of the three axes is ____:____:____.
cube	**88** However, the object of 11.24B does not look much like a(n) _____. It looks more like a long box of square section.
recede; foreshortened	**89** Distortion is introduced by the fact that (1) the left-hand axis does not _____ into space and (2) the right-hand axis has not been reduced in scale (_____) proportionately to its direction.

Cavalier versus Cabinet Form

cabinet	**90** There are two kinds of oblique drawing, the *cavalier* and the *cabinet*. Figure 11.25 shows a cube sketched in both forms. The difference between the forms is in the foreshortening of the one receding axis. Which form looks more like a cube, cavalier _____ or cabinet _____?

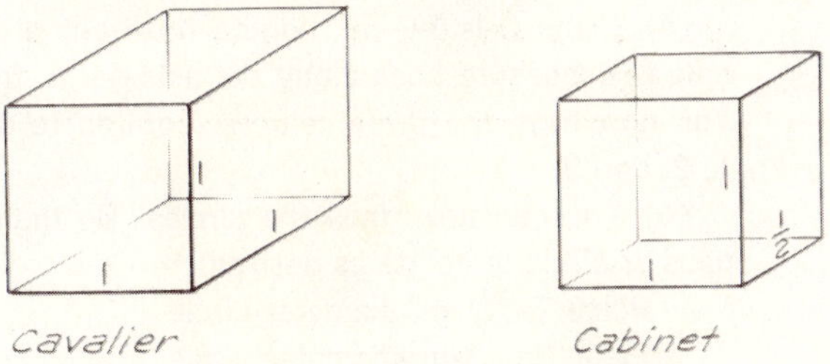

Figure 11.25   The two types of oblique drawing.

parallel	**91** The 1:1:½ foreshortening ratio of the cabinet form reduces the visual distortion. It still does not have the realism of perspective, but it is adequate and is a form that can be sketched quickly. Its primary advantage is that circular or odd-shaped details may be oriented _____ to the picture plane and *drawn in their true shape*.

The angle of the one receding axis with the horizontal may be of any desired value. It is common to use either 30 or 45° since these can be drawn easily with standard drawing instruments.

The receding axis may go off to the left or right and upward or downward. (See Figure 11.26.) If this axis is drawn downward, the sketch shows the front, bottom, and one side of the cube.

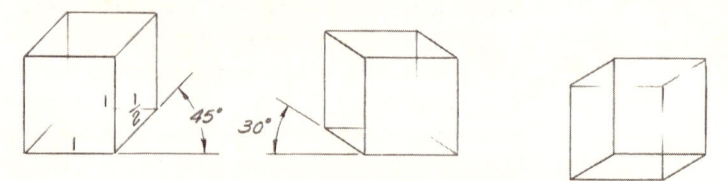

Figure 11.26 Variations in viewing position—cabinet drawing.

■ **92** Sketch 37 shows a cylindrical part in a two-view orthographic drawing. Make an oblique (cabinet) sketch of this part in the space provided. Use the procedures set forth in the next few steps:

(1) First, set up the three oblique axes. Remember that the receding axis may be right or left and at any angle desired.

(2) Now, sketch the enclosing box. It is 1.5 in. × 1.5 in. × 2.5 in. The front view shows a series of concentric circles. The planes of these circles, planes 1, 2, and 3 in profile view, should be parallel to the picture plane. This means that the two 1.5-in. dimensions are measured on the vertical axes and the 90° axis. We are making a cabinet drawing so that the measure on the receding axis is 2.5 ÷ 2 = 1¼ in.

(3) Next, find the axis of the cylinder. Sketch diagonals on the front and rear 1.5-in. squares to find their centers (or find the centers on the grid) and join these points with a centerline. This is the axis of the cylinder. Check your solution.

(4) Plane 2 is 1½ in. behind the front plane. Cut this dimension in half and measure back along the axis ¾ in. from the front center point. You now have the three centers required to draw the circles on planes 1, 2, and 3.

(5) You can now draw the circles. Do them freehand or use a compass or circle template as desired.
        Plane 1:   1-in.-diameter circle
        Plane 2:   1-in.-diameter circle
                     1.5-in.-diameter circle
        Plane 3:   1.5-in.-diameter circle

(6) Join these circles with straight lines parallel to the axis and tangent at the outer edges. These lines form the contours of the part. Finish off all visible lines with bold dark lines. Check your solution.

93 The solution to Sketch 37 shows what would have resulted if you had not made the planes of the circles parallel to the picture plane. It is distorted mainly because it seriously violates the cardinal rule of perspective circles which states that the major diameters of the ellipses must be _____ to the axis of the circles represented by the ellipses.

perpendicular

OTHER PICTORIAL DRAWING FORMS   209

94 The circles of this sketch were sketched using the same construction methods developed for perspective and isometric. Irregular curves are also handled in the same way. The procedure is:
(1) Find a few select points on the curves.
(2) Transfer these by coordinate methods to the oblique form.

smooth
(3) Sketch a(n) _____ curve through the points.

95 Lines that are not parallel to any of the oblique axes are also handled by _____ methods: Find the end points and then join

coordinate
them with a straight line.

Figure 11.27 shows an object sketched in both the cavalier and the cabinet forms. The cavalier sketch looks awkward and distorted when compared with the orthographic drawing. Forget this form, but remember that the cabinet form comes in handy for quick sketches of objects that have many odd-shaped or parallel circular elements.

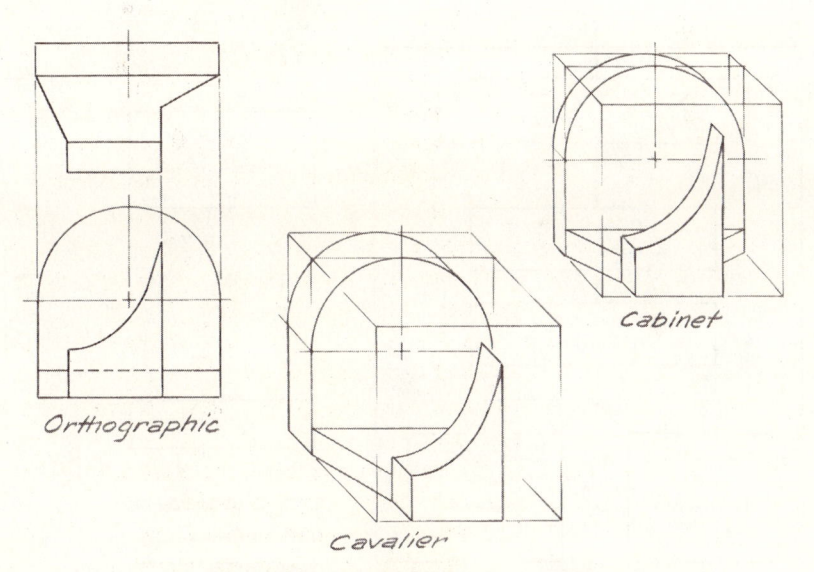

Figure 11.27   A comparison of cabinet and cavalier oblique drawing.

ADVANTAGES OF PERSPECTIVE SKETCHING

This completes the discussions of the theory of pictorial forms of drawing. *Perspective sketching* has been stressed as the most important form for the student and the professional to use in the communication of ideas and plans both with himself and with other people.

Perspective is useful because
(1) If properly drawn, a perspective sketch has photographic realism and thus transmits maximum information.
(2) It is flexible in that it permits us to take any point of view we wish, without need to remember a number of angles and foreshortening ratios.

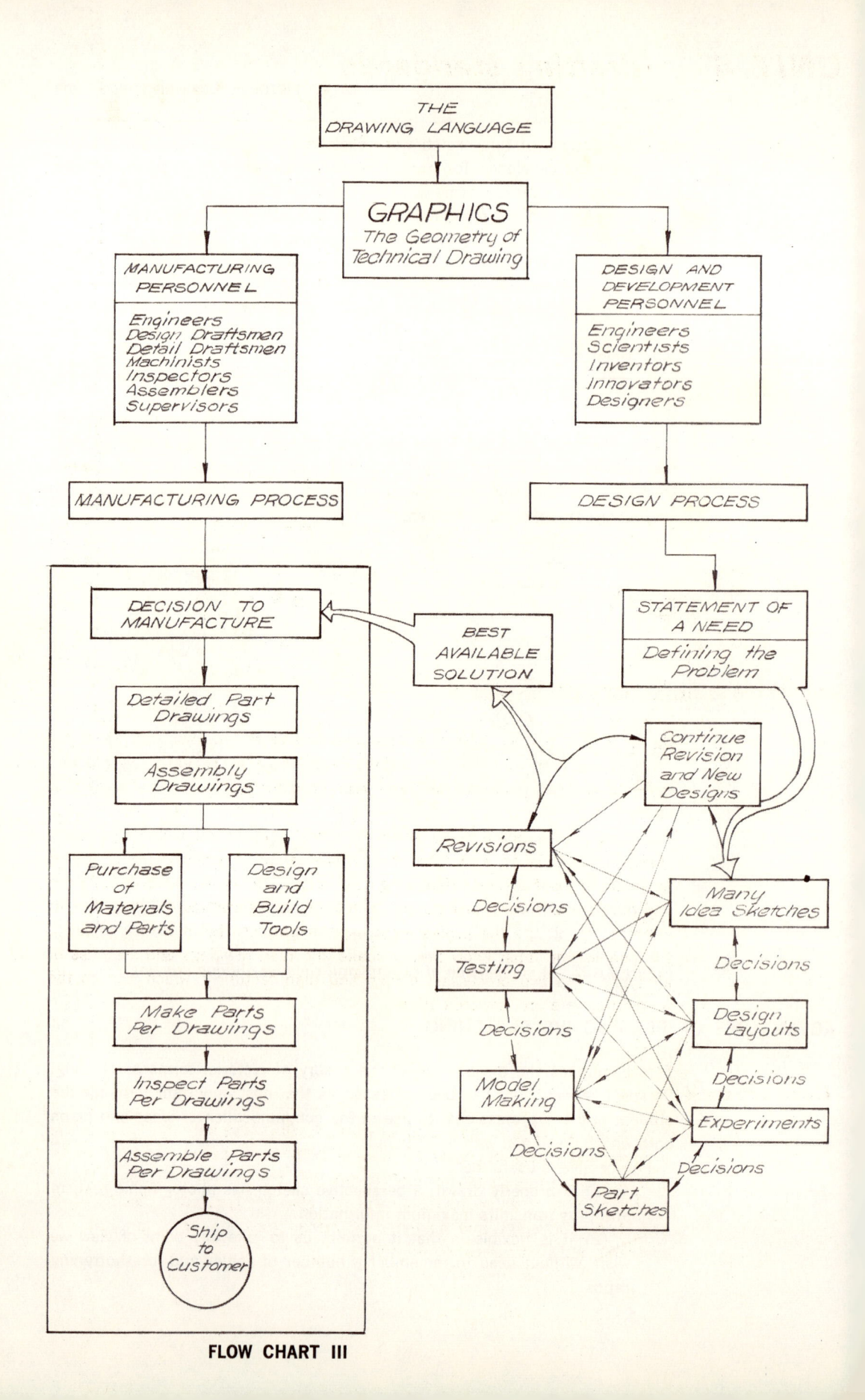

FLOW CHART III

UNIT 4   *drafting standards*

THE PROFESSIONAL USES OF GRAPHICS

So far in this text we have not attempted to answer the important question, "Why study graphics?" In one sense, Flow Chart III suggests an answer. Perhaps you can place yourself and your need for technical drawing on this chart.

Graphics, the language by which technical information is developed and communicated, is an important tool of the technical profession. Its use is universal throughout manufacturing and service industries, and in research and development work for industry, education, and government. Flow Chart III attempts to picture the professional uses of graphics by indicating some of the people who must know how to make and read drawings and sketches of all types. The two processes, design and manufacturing, which lead to the production of a product are outlined.

Design Process

The design process is loose and variable. As the chart indicates, progress in a new design may be made by flowing both forward and backward or by crossing over some steps. Parts of the process may recycle many times before a best solution is achieved (usually at the eleventh hour on a specific deadline date). Drawings and sketches of many types are used for analyzing problems and communicating information.

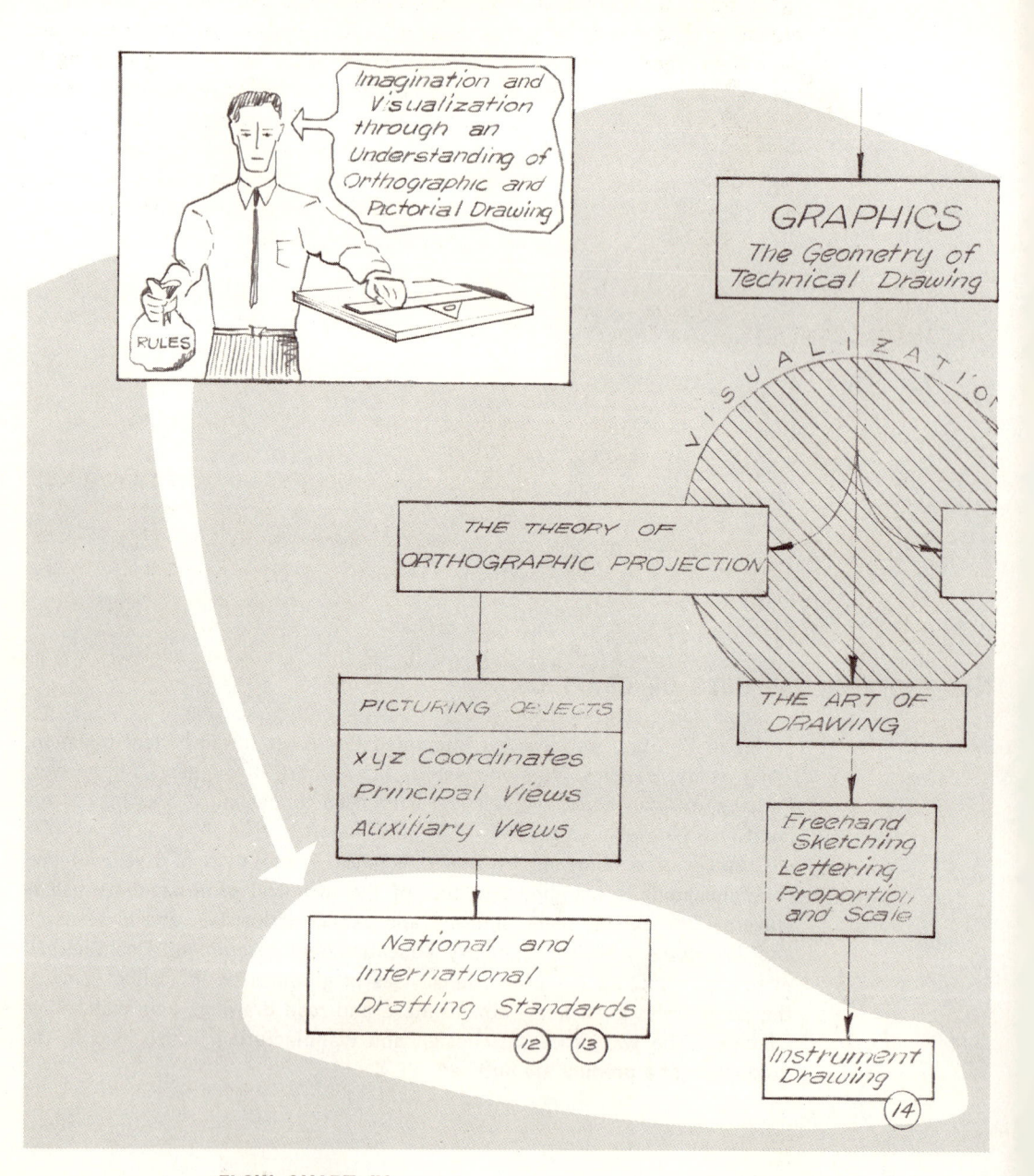

FLOW CHART IV

Manufacturing Process

In contrast, the manufacturing process is straight and rigid. It is a highly organized process involving thousands of people. The jobs of these people are dependent upon their getting accurate information. The *working drawing* is the primary means for disseminating accurate technical information to many people. Working drawings are made according to standards accepted the world over. Each person in the process has learned to read drawings according to these accepted standards. Disregard of the standards by anyone preparing drawings for production can cause serious breakdowns in communications. These standards are discussed in the next three chapters.

RULES AND CONVENTIONS—DRAFTING STANDARDS

Flow Chart IV is a portion of Flow Chart II (which precedes Chapter 7). It shows the topics for Chapters 12, 13, and 14, which comprise our study of the professional uses of graphics.

The inset figure illustrates the requirements for the successful professional use of graphics. If a student can visualize and use his imagination to portray a three-dimensional object or concept on a two-dimensional surface, all he needs to make effective use of this understanding is a knowledge of the tools of drawing and a bagful of rules that are standards of the technical world.

chapter 12   working drawings—dimensioning

INTRODUCTION

Chapter 12 is concerned with the skills required to communicate technical information to others through drawing. Flow Chart I, preceding Chapter 1, showed the skill and communication aspects of technical drawing. So far we have dealt mainly with personal communications—how to transmit physical reality to the drawing surface. We have concentrated on your personal needs to communicate visual and mental images to yourself and produce a drawing in one or more of several forms.

Flow Chart I shows two divisions of the skill aspect of technical drawing:

(1) Reading of drawings (interpreting skill)
(2) Making of drawings (performing skill)

Reading skill will be a by-product of our development of performing skill. The best way to learn to read a technical drawing is to learn to make one. We will concentrate, therefore, on making drawings to communicate precise technical information which other people need to know.

Who are these other people? The diagram of Figure 12.1 shows a line of communication from the originator of industrial products and processes to a wide variety of people in every country of the world. The means of communication is the *working drawing*.

Encircling the working drawing is an area of professional standards and conventional practices that ensure a clear communication channel. The vast number of people who need technical information have learned to read working drawings by these standards. You, then, must follow conventional practice closely or the communication link breaks down.

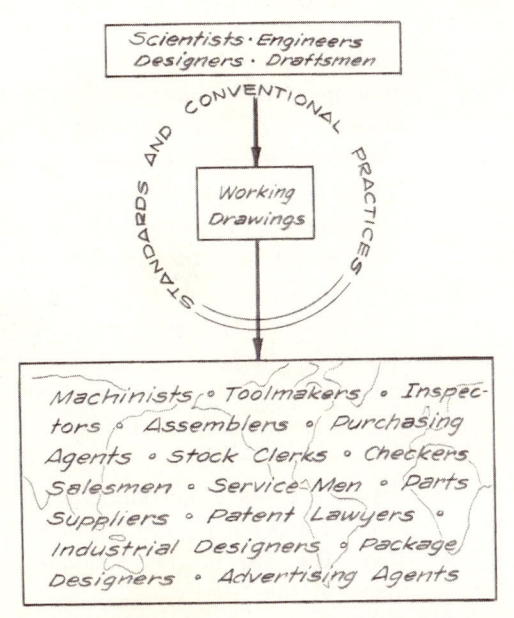

Figure 12.1 Who are the people who need to read technical drawings?

Standards imply rules—rules that say "Do this, not that." If you think about drawing a picture of anything that exists on the face of the earth and also anything that hasn't been thought of yet, it is impossible to imagine how any one set of rules could cover all problems. They can't. Rules throughout society are guide lines—guide lines that are subject to intelligent interpretation by responsible people. Follow the rules and standards presented in this chapter. When a situation arises in which you find it impossible or, at least, questionable to stick with the rules, remember your audience. They must understand what you have on your mind.

In the United States, the drawing standards are originated and controlled by the American Standards Association.* The Society of Automotive Engineers and the military organizations of the United States also have much to do with the setting of drawing standards. Most of industry conforms to these standards, staying as close as possible to the accepted procedures even when deviation from them becomes necessary. Many of the examples in this chapter are extracted from the American Standard Drafting Manual, Dimensioning and Notes (ASA Y14.5–1957).

Almost every country in the world has a similar standards association. In recent years much effort has been devoted to reaching agreement on international standards in many areas.

WORKING DRAWINGS

standards

1 The introduction to this chapter states that it is concerned with the _____ and conventions of making working drawings.

* Now the United States of America Standards Institute (USASI), 10 East 40th St., New York, N.Y. 10016

216 DRAFTING STANDARDS/UNIT 4

information

2 Working drawings are completely dimensioned and detailed drawings that are used to communicate technical _____ to other people.

standards

3 There is no room for ambiguity or impressions in working drawings. The people who must receive the information contained in the drawings have been trained to read them according to accepted drawing _____ _____.

communication

4 Therefore, these same standards must be used in the making of a drawing or the drawing's essential purpose, the _____ of information, will not be met.

eccentric shaft

5 Figure 12.2 is a working drawing. It is a collection of lines, letters (words) and numbers arranged according to standard form and procedure. The only unique features of Figure 12.2 are the orthographic views of the _____ _____ and its particular dimensions.

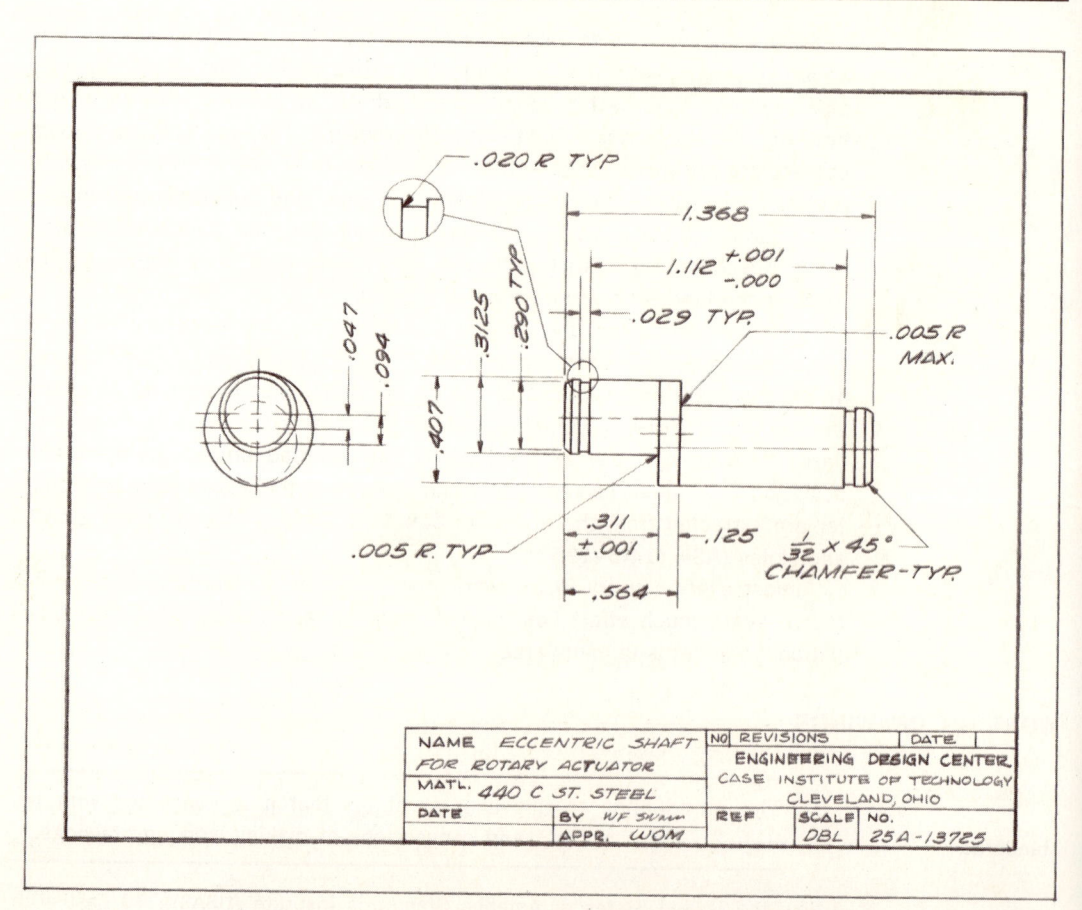

Figure 12.2 A working drawing. Specifically, a detail drawing.

WORKING DRAWINGS—DIMENSIONING   217

people

6 This is the information the drawing is intended to convey. It is drawn according to accepted standards so that any number of _____ will be able to receive the correct information quickly and efficiently.

This drawing must stand alone. The designer of the rotary actuator in which the eccentric shaft is used and the draftsman who made the drawing cannot go along to explain the part. A breakdown in communications through an inadequate or sloppy drawing carries serious implications. A minor error or omission on the drawing could cause long delays or, especially in mass-production industries, could turn thousands of parts onto the scrap heap.

standards

7 A working drawing is a collection of lines, letters, and numbers, arranged on a sheet of paper according to accepted _____ of form and procedure.

product (device)

8 Figure 12.3 shows the types of drawings that come under the heading of *working drawings*. Flow Chart III, preceding this chapter, showed the different steps in both design and manufacturing—the two essential processes required to transform an idea into a usable or a salable _____ .

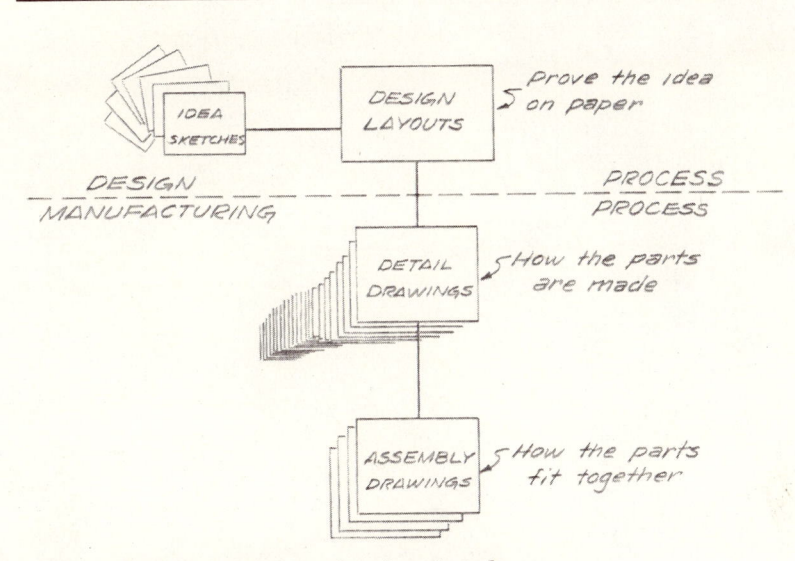

Figure 12.3   The role of working drawings in industry.

manufacturing

9 Working drawings belong to the _____ process which comes after the creative work involved in the design and testing of the idea.

detail; assembly

10 The two most important types of working drawings, as shown in Figure 12.3, are the _____ drawing and the _____ drawing.

218 DRAFTING STANDARDS/UNIT 4

dimensional

11 A detail drawing shows complete _____ details of a single part. Figure 12.2 is a detail drawing. Each part in an assembly of parts is manufactured and inspected according to its detail drawing.

assembled
(put together)

12 An assembly drawing shows how all the individual parts are _____ _____ to make the finished product. Figure 12.4A is an assembly drawing of a cantilever shaft assembly.

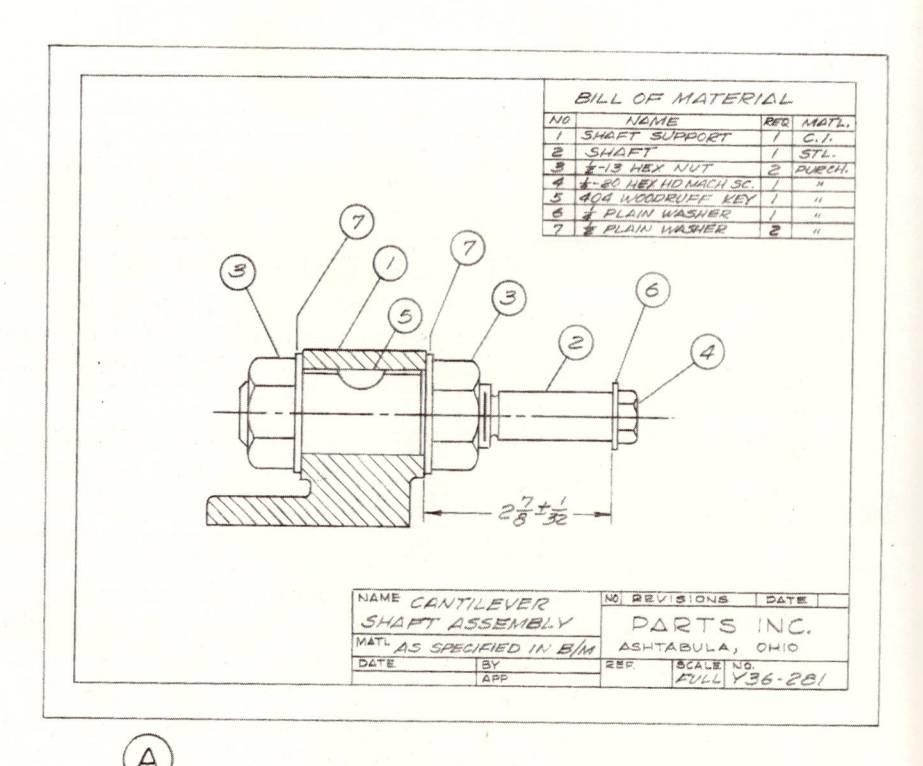

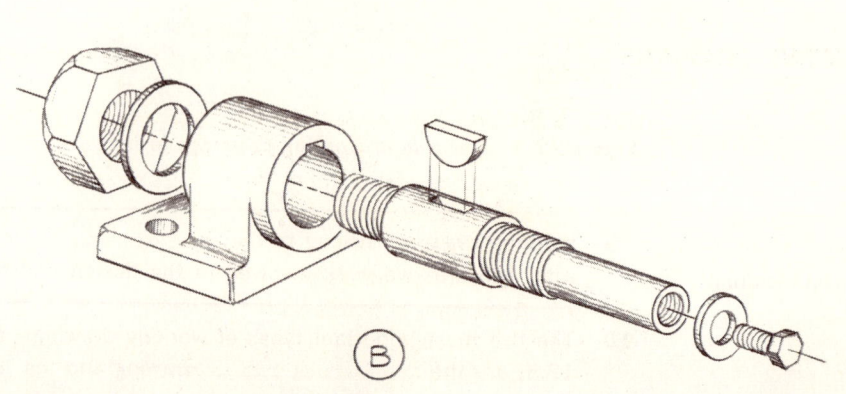

Figure 12.4   Assembly drawings.

WORKING DRAWINGS—DIMENSIONING 219

seven (7);
two (2); five (5)

13 The assembly is made up of _____ separate parts. How many of these must be manufactured? _____ How many are purchased as standard parts from other companies? _____

detail

14 Almost every product is an assemblage of manufactured and purchased parts. Manufactured parts must each have a(n) _____ drawing.

bill (of) material

15 Purchased parts need not have detail drawings unless the supplier must perform some special operation on the part before shipping it to the product manufacturer. All other purchased parts are specified by a listing called a(n) _____ of _____. (See Figure 12.4A.)

shaft

16 Note in Figure 12.4A that one dimension is included. It shows how far the _____ must protrude from the shaft support. This is a setting that must be performed in the assembly department.

The assembly is drawn in one orthographic view. A second or third view would not add much to this particular assembly since it is symmetrical about a single centerline and all seven parts are shown clearly.

section

17 The single orthographic view of part No. 1, the shaft support, is drawn as a full _____.

This is done to clarify the position of interior parts, such as part No. 5, the Woodruff key, a device to keep the shaft from turning in the support member.

Assembly drawings are traditionally orthographic drawings. Consider Figure 12.4B. It is a perspective sketch of the assembly in *exploded* form. The parts are separated and drawn individually but, by their position, it is obvious how the parts are assembled. Here is an example of how the realism of a pictorial drawing serves a function better than the precision of an orthographic drawing.

DETAIL DRAWINGS

Most of the drawing standards to be discussed here apply to detail drawings. The subject may be outlined broadly as follows:
    (1) Design of the drawing (general layout on a sheet of drawing paper)
    (2) Line specification (selection of lines)
    (3) Placement of lines
    (4) Selection of letters and numbers
    (5) Placement of letters and numbers

Let's proceed through these five items and then, in Chapter 13, discuss some conventional drawing practices and symbolism and the effects of precision on the making of working drawings.

Design of the Drawing

18 To start a drawing we must first have a sheet of paper. Drawing-paper sizes are standardized. Figure 12.5 shows the five lettered sizes. Starting with the A size (8½ × 11 in.—standard file size in the United States), the sizes expand in even multiples of the A size to B (____ × ____); C (____ × ____); D (____ × ____); and E (____ × ____).

B (11 × 17);
C (17 × 22);
D (22 × 34);
E (34 × 44)

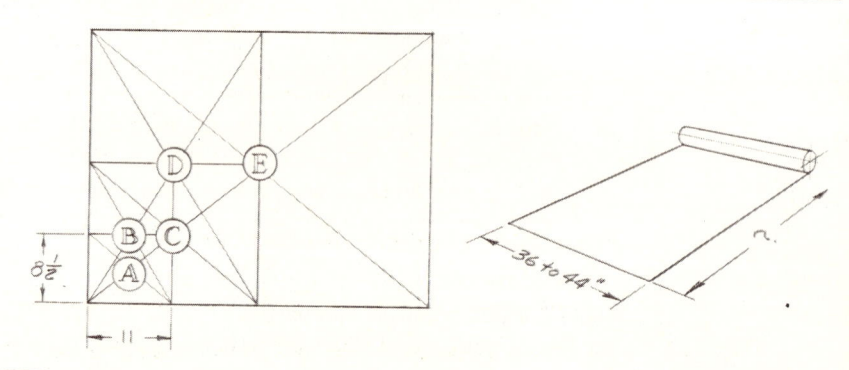

Figure 12.5 Standard drawing paper sizes.

19 Each size is a multiple of the basic 8½ × 11 (A) size to facilitate folding down to file size. For drawings larger than the E size, paper must be cut from a(n) ____. (Figure 12.5.)

roll

20 Figure 12.6 is a sketch showing the elements that constitute a working (detail) drawing. The object drawn is a simple rectangular box. The first, and most important, parts of any detail drawing are the ____ ____. (Area 1 in Figure 12.6.)

orthographic views

21 Chapters 3, 4, and 5 have discussed the choice of view to describe an object completely and unambiguously and the procedures for drawing these views. Area 2 is a clear area (a sort of no man's land) that ____ each orthographic view.

surrounds

22 Its purpose is to set the orthographic views off from the dimensions and notes which surround them. The figure shows that this area should be ____ in. to ____ in. wide if possible.

⅜ (to) ½

23 Area 3 is the space reserved for ____.

dimensions

24 This area is concentrated between views. Notice that it becomes narrower around the outside of the views. This indicates that it is good practice *not* to scatter dimensions but to concentrate them ____ views if possible.

between

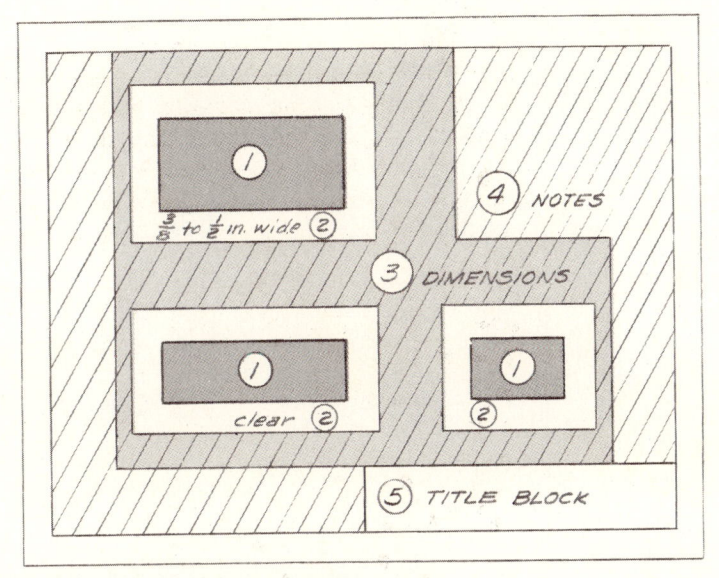

Figure 12.6   The design of a detail drawing.

notes

25 Area 4 comprises the rest of the paper area out to the printed border. It is reserved for _____.

26 Notes carry information and specifications that are not included on either the orthographic drawing itself or in the dimensions. This is also true of Area 5, the _____ _____.

title block

Every drawing should have a title block to show the (1) name of the part, (2) part number, (3) date, (4) scale of the drawing, (5) signatures of the persons who made and checked the drawing, (6) material of which the part is made, and (7) any other bits of information important to the completeness of the drawing.

The company name and adress is usually printed in the title block. Much of the title-block information is important record data that may prove useful to prove prior invention in patent disputes.

guide lines

27 In the foregoing discussion of the design of a working drawing we used the phrase "if possible" a few times. As with all rules and standard procedures, they should be considered _____ _____ and subject to interpretation according to the needs of the situation.

Figure 12.7 shows the detail drawing of Figure 12.2 with the design areas superimposed upon it. You can see that it conforms to the accepted pattern.

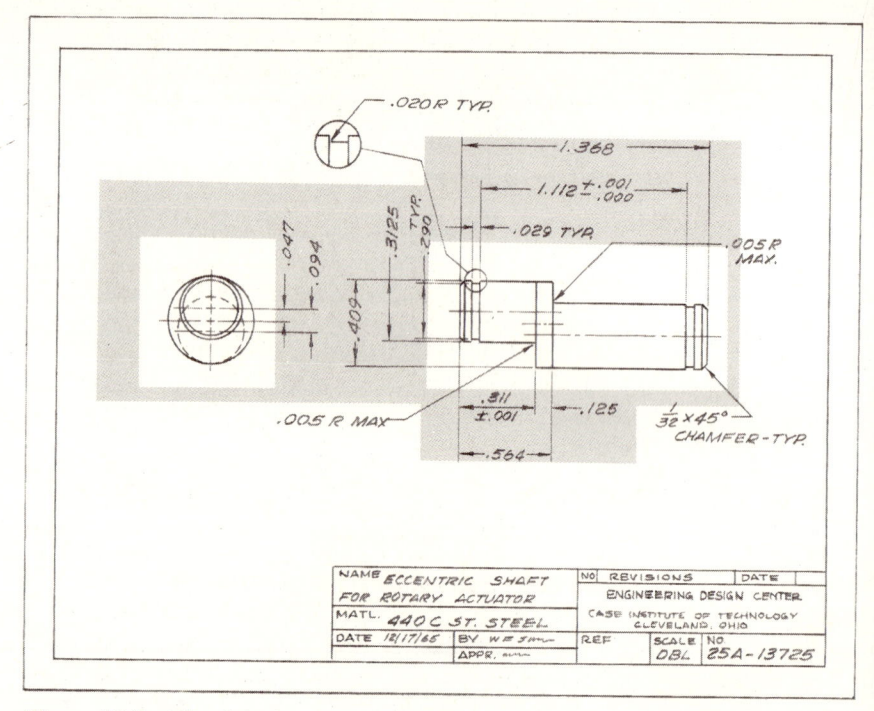

Figure 12.7 The drawing design pattern applied to the eccentric shaft of Figure 12.2.

Line Specification

The first question we might ask is, "What lines do we put on a drawing?" Figure 12.8 is an alphabet of lines showing the different types of lines used on a working drawing and their weights. Achieving these line weights on an instrument drawing using pencil involves a combination of (1) type of pencil point (sharp or blunt), (2) pressure on the pencil, and (3) grade of pencil lead. Some practice at drawing all three types will prove which combination of techniques is best suited to your own hand.

thick, medium, thin

28 In Figure 12.8 there are three separate weights of line used: _____, _____, and _____.

thickness

29 All these lines must be *dark* lines to achieve a reproducible drawing. They vary in _____ but not in density.

visible

30 The first six lines of Figure 12.8 are the most common since they appear on every detail drawing. First is the _____ line.

hidden

31 This is the line used to show all visible edges on the orthographic views. No. 2 is the _____ line.

WORKING DRAWINGS—DIMENSIONING   223

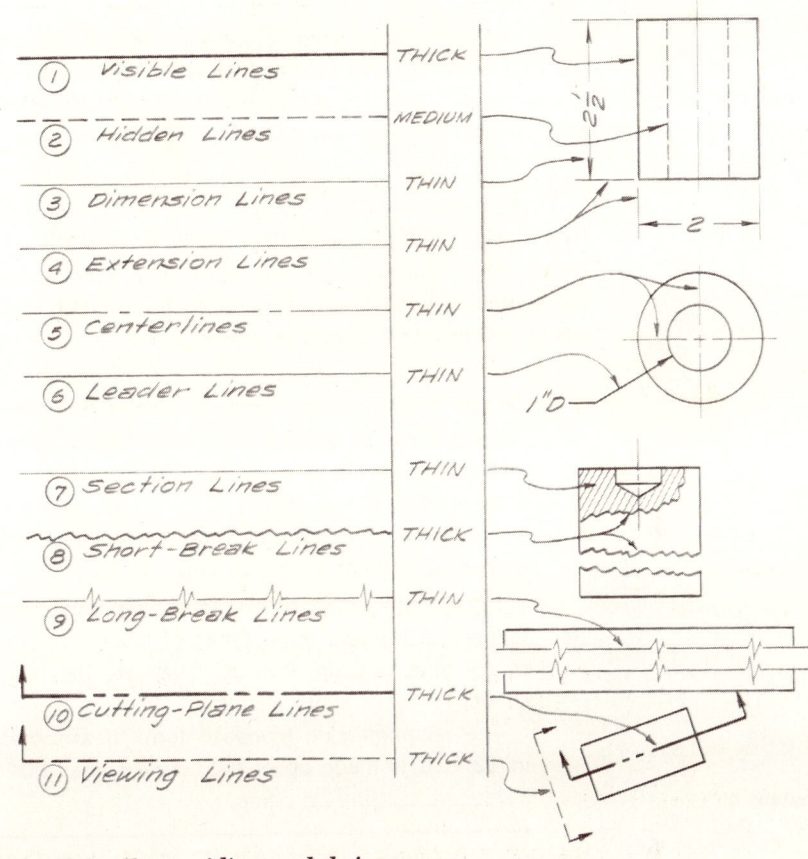

Figure 12.8   Types of lines and their uses.

equal; equal

32 The hidden line is a dashed line used to show invisible lines—lines that are part of the object but cannot be seen in the views presented. Note that the dashes are of _____ length and the spaces between dashes are all shorter but of _____ length.

dimension

33 No. 3 is the _____ line. It is the line that relates a dimension number to the actual distance on the part.

extend

34 No. 4 is the extension line. These thin lines _____ the actual distance on the part out across the no man's land to the dimension lines. They should project about ⅛ in. beyond the point of the arrowheads on the dimension line.

note

35 Centerlines (No. 5) have been discussed before. They are thin, alternate long-and-short-dashed lines that mark all circular centers and axes.

The leader line is No. 6. It is used to direct the information supplied by a(n) _____ to the appropriate detail on the drawing.

224 DRAFTING STANDARDS/UNIT 4

36 Section lines (No. 7) are thin. They are angular, parallel lines, evenly spaced and are used to show an intentional cutting apart of the object to reveal interior detail. Varying the spacing of section lines changes the _____ tone of the cut.

gray

There are two types of "break" lines, No. 8 for short breaks and No. 9 for long breaks. They are used to show an intentional breaking apart of the object and removal of the center portion. This is done on long parts where the center section carriers no needed information and drawing the full object would necessitate a large-size drawing.

37 Lines 10 and 11 are used to direct attention to the manner or position in which a particular view is presented. No. 10 is a cutting-plane line used to show where a(n) _____ is taken. No. 11 is a viewing line used to point out the direction of an auxiliary view that is not in its correct orthographic position.

section

38 Figure 12.9 shows other types of lines that are most important to the clarity of a working drawing. They are the two lines that form arrowheads.

The recommended standard form of arrowhead is shown at the left in 12.9. It is made up of two, curved lines that meet precisely at a(n) _____ line.

extension

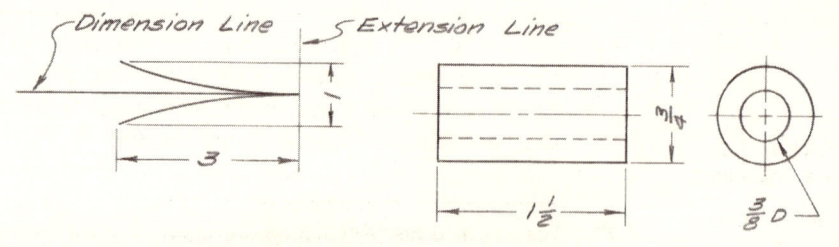

Figure 12.9 Arrowheads.

39 The length-width ratio of the arrowhead is ____:____.

3:1

An arrowhead is dark and thick and should be drawn freehand with a soft pencil (H or 2H), the same grade used for lettering.

■ **40** Sketch 38A is an exercise in sketching arrowheads. Some samples are given along with some blank stems. Try to duplicate the samples.

WORKING DRAWINGS—DIMENSIONING   225

Placement of Lines

We have seen *what* lines are used on a working drawing, now let's look at the recommended standards on where to place the various types of lines.

size (S), location (L)

41 Second to the orthographic views of an object, the dimensions are the most important features of a detail drawing. Figure 12.10 shows the two types of dimensions needed to describe an object precisely. They are _____ dimensions and _____ dimensions.

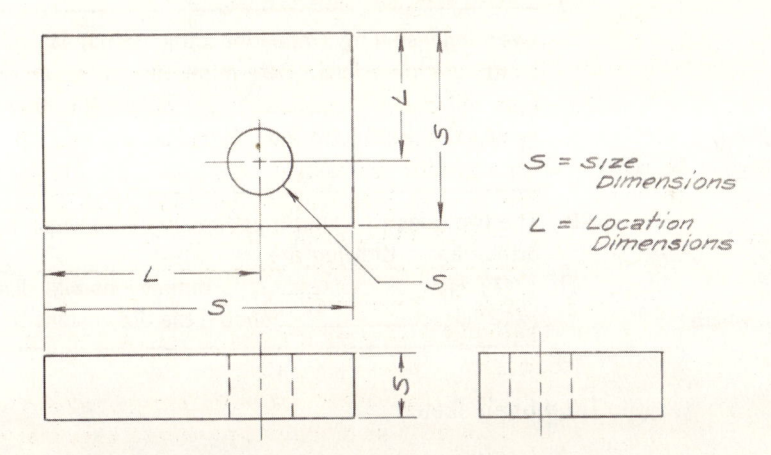

Figure 12.10 An object is described dimensionally with size and location dimensions.

H, W, and D; size (diameter)

42 Size dimensions show how large (or how small) the object or parts of the object are. In Figure 12.10 three of the S dimensions describe the ____, ____, and ____ of the rectangular block. The fourth S dimension tells the _____ of the hole.

where

43 The two L dimensions show _____ the hole is positioned on the block. All other details, such as the fact that the hole goes through the block, are made clear by the drawing itself.

■ **44** Sketch 38B is an exercise in recognizing size and location dimensions. An example has been worked out. Add an S or an L to each dimension according to your opinion of whether it is a size or a location dimension. Check your solutions.

⅜ (to) ½

45 An orderly appearance is the goal to work toward in a detail drawing. If the dimensions are placed in a logical and orderly manner, the drawing will be easy to read. Do not crowd dimensions. Figure 12.6 showed that there should be a clear area of about ____ in. to ____ in. around the orthographic views.

226 DRAFTING STANDARDS/UNIT 4

46 Only extension lines and leaders cross this area. Dimension lines should be separated from each other by at least ⅜ in. also. The numerical dimension is placed in a break about midway in the _____ _____.

dimension line

How many dimensions should be put on a detail drawing? The general answer is—enough dimensions so that the reader does not have to (1) make complicated calculations, (2) measure the drawing, or (3) make assumptions about the part.

47 Over-dimensioning (repeating dimensions) is just as much of a sin as omitting dimensions. Determine the minimum number of dimensions that will _____ define the object to a person who is seeing the drawing the first time.

completely (precisely)

48 The two questions that must be answered when adding dimensions to an orthographic drawing are
(1) How _____ dimensions will describe the object?
(2) _____ should the dimensions be placed?

many; where

Thumbnail sketch

An excellent practice to help answer these questions is to make a quick, thumbnail sketch of the drawing first. This is shown in Figure 12.11A. In making this sketch you can concentrate on the choice of dimensions and their placement without bothering with form. The finished drawing is shown at 12.11B. Here, the concentration is on the orderly appearance of the drawing.

The thumbnail sketch also helps you decide what size paper to use, how much space should be left between views for the correct placement of dimensions, and what notes are needed.

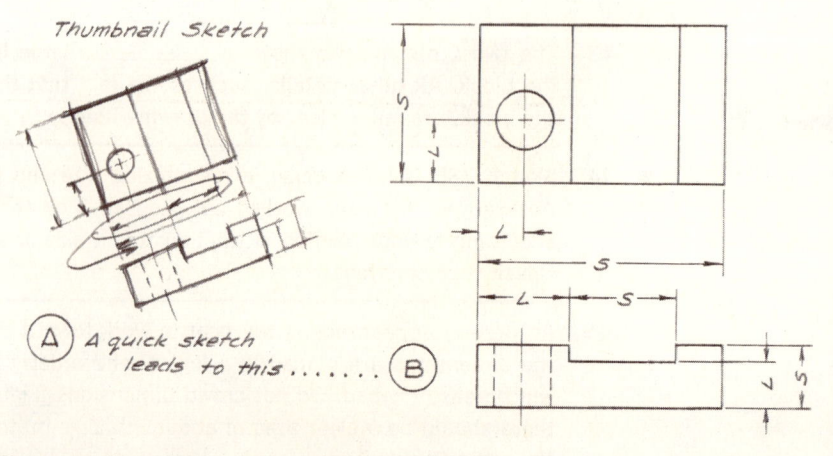

Figure 12.11 Make a thumbnail sketch first.

WORKING DRAWINGS—DIMENSIONING 227

The five basic 3-dimensional shapes

Most objects can be considered to be a combination of the following shapes:
- A. Rectangular prism
- B. Cylinder
- C. Cone
- D. Pyramid
- E. Sphere

---

circle

49 Figure 12.12 shows these basic shapes with their size dimensions correctly positioned. Note that the dimensions giving the diameters of a cylinder and a cone are placed on the view showing their profile rather than on the view that shows them as a(n) _____. This is standard procedure.

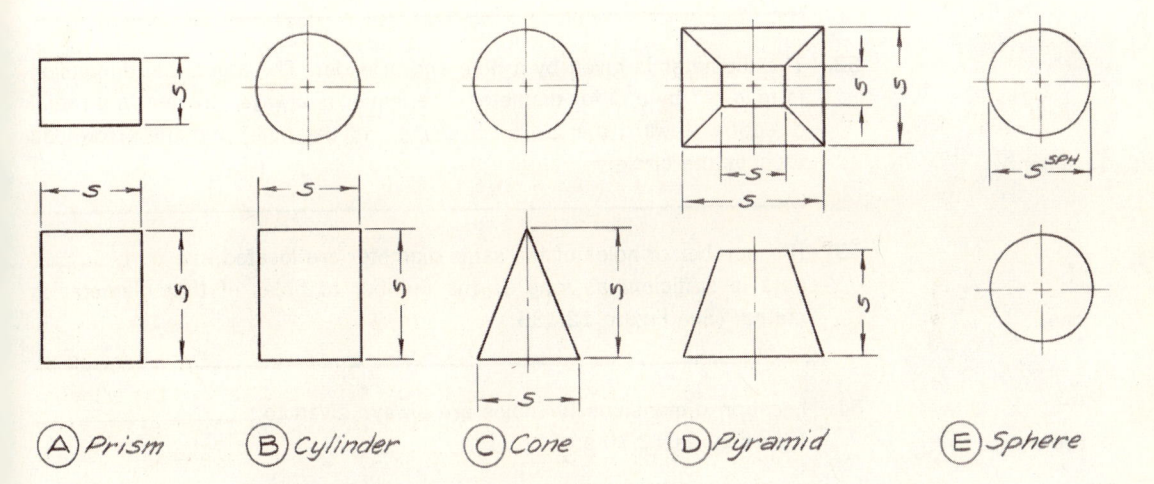

Figure 12.12 Approved dimensioning of the five basic 3-dimensional shapes.

---

centerline

50 In parts symmetrical about centerlines, such as the truncated pyramid at 12.12D, the centerline carries the description of the symmetry. In other words, the _____ does not have to be dimensioned to the sides of the object.

Dimensioning holes

diameter; circle

51 Figure 12.13 shows the correct dimensioning for holes. The size dimension always gives the _____ of the hole and is always directed to the view showing the hole as a true _____.

228 DRAFTING STANDARDS/UNIT 4

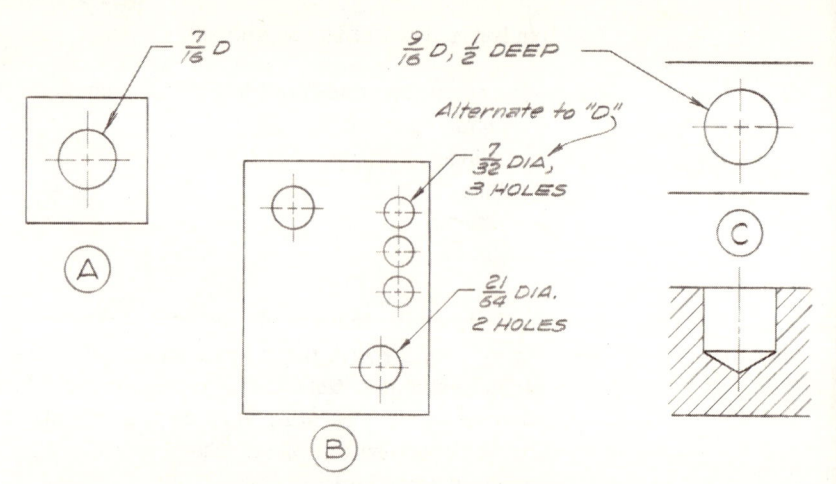

Figure 12.13 Approved dimensioning of hole size.

center	52 The diameter is given by a note and a leader. The numerical dimension is followed by a *D* for diameter. The leader is always directed in a radial direction (toward the _____ of the hole) and the arrowhead touches the circle.
one	53 If a number of holes of the same diameter are located in a part, _____ note is sufficient as long as the number of holes of that diameter is stated. (See Figure 12.13*B*.)
centerlines	54 Location dimensions for holes are *always* given to _____. (See Figures 12.10 and 12.11.)
depth; diameter	55 Blind holes (Figure 12.13*C*) are holes that do not go through a part. The _____ dimension of blind holes is given in a note along with the _____.

Dimensioning rounded ends and corners

radius; *R*	56 Figure 12.14 shows the recommended procedures for dimensioning corner radii and rounded-end parts. The dimensions are given by a note that includes the numerical _____ followed by the letter ____.
radius	57 The center of the radius must be marked. The leader is drawn to the center and the arrowhead is placed on the leader to show the exact _____ being specified.

WORKING DRAWINGS—DIMENSIONING 229

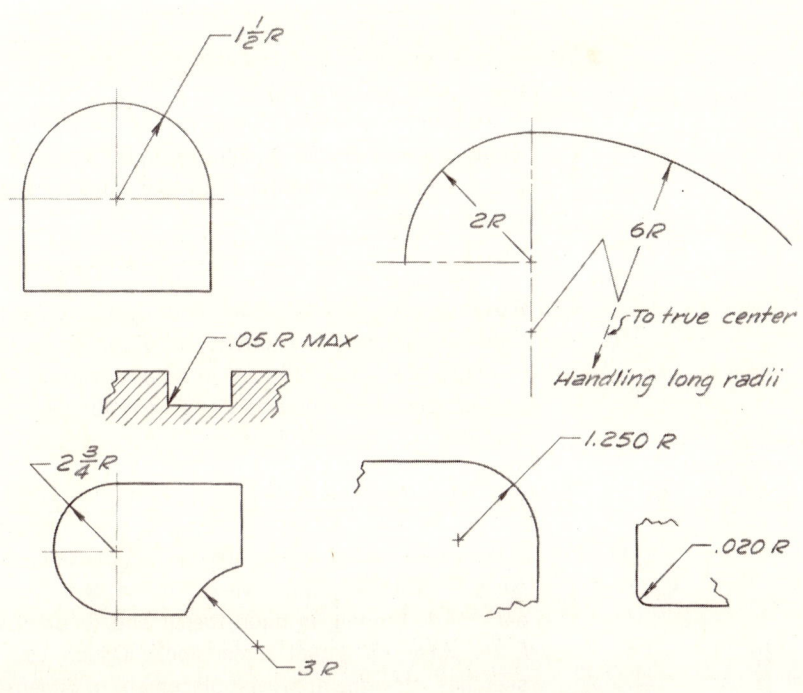

Figure 12.14 Approved dimensioning of radii and rounded ends.

58 Figure 12.15 shows two accepted methods for dimensioning a rounded-end part. At 12.15A the end radii are dimensioned with a note and the _____-to-_____ dimension is given.

centerline-to-centerline

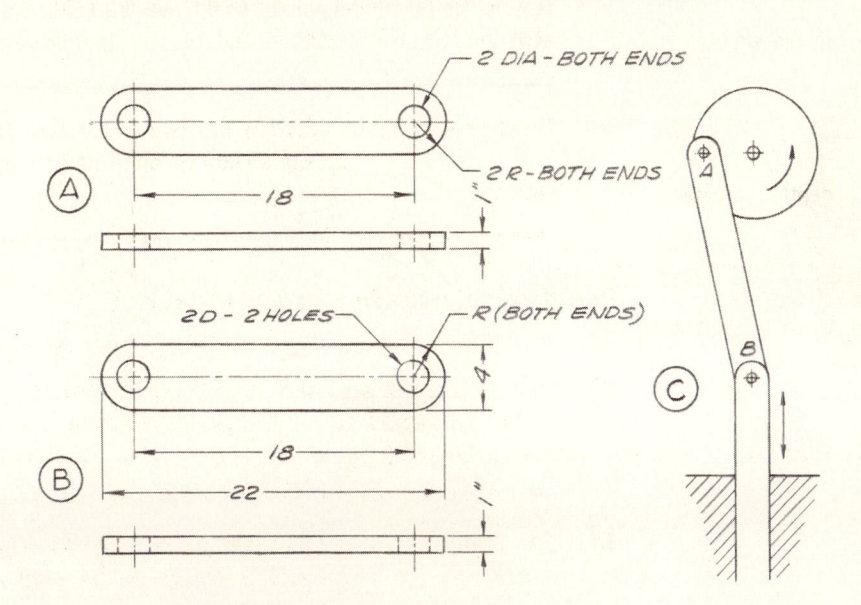

Figure 12.15 Dimensioning rounded-end parts.

230　DRAFTING STANDARDS/UNIT 4

59 This is approved dimensioning *even though calculations must be made* to determine overall length and width.

　　An alternative method is given at 12.15B. Here the overall length and width is dimensioned and the rounded ends are marked with just a(n) _____ to indicate that they are circular. The center-to-center dimension is still included.

R

---

60 **RULE:** Always dimension between lines and points having a definite relation as defined by the function of a part or its fit with a mating part.

　　Figure 12.15C shows a sketch of a simple drive mechanism. An important dimension that relates to the function of the mechanism is the distance, *AB*, between hole _____.

centers
(centerlines)

■ **61** Sketch 39A shows a simple sliding mechanism involving two mating parts. The orthographic views of the two parts are given. Dimension each part, keeping in mind the fit and function of the parts. Use *S* and *L* in place of actual dimensions. Check your solution. The answer sketches show the approved placement of dimensions.

Miscellaneous placement rules

62 **RULE:** Place dimensions on the view which shows the true shape of the feature being dimensioned. In Figure 12.16, it was necessary to draw a(n) _____ _____ to follow this rule.

auxiliary view

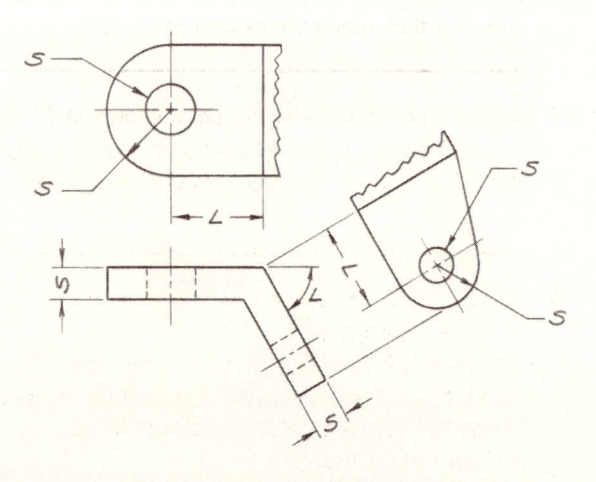

Figure 12.16   Place dimensions on true-shape views.

WORKING DRAWINGS—DIMENSIONING 231

extension	**63** **RULE:** Avoid running any dimension lines to visible (object) lines. Do not place dimensions on the orthographic view. Always use _____ lines and keep the first dimension at least ⅜ in. away from visible lines. (See Figure 12.17A.)

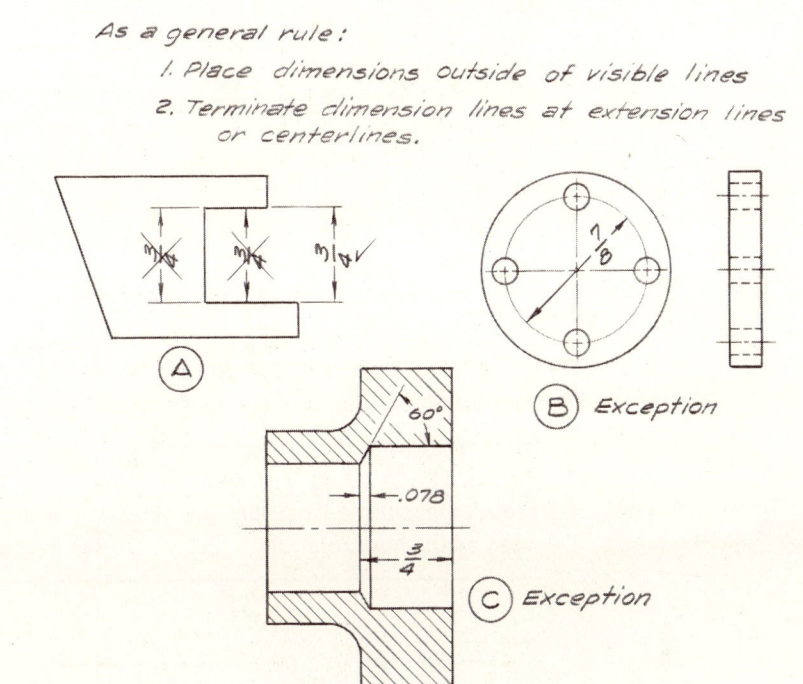

As a general rule:
 1. Place dimensions outside of visible lines
 2. Terminate dimension lines at extension lines or centerlines.

Figure 12.17 A general rule of dimensioning and some exceptions.

circle	**64** Figure 12.17B shows an exception to the last rule. The L dimension that locates the four holes on a circular centerline is better placed on the view that shows the centerline as a(n) _____.

removed (omitted)	**65** Another permissible exception is shown at 12.17C. If a dimension must be placed on a sectioned portion of the drawing, the section lines should be _____ around the dimension.

short; long	**66** **RULE:** Establish location dimensions from a single datum or reference surface. (See Figure 12.18.) Figure 12.18A also illustrates the rule that extension lines should never cross dimension lines. Pyramid the dimensions with the _____ dimensions close to the object and the _____ dimensions spaced successively outward.

232 DRAFTING STANDARDS/UNIT 4

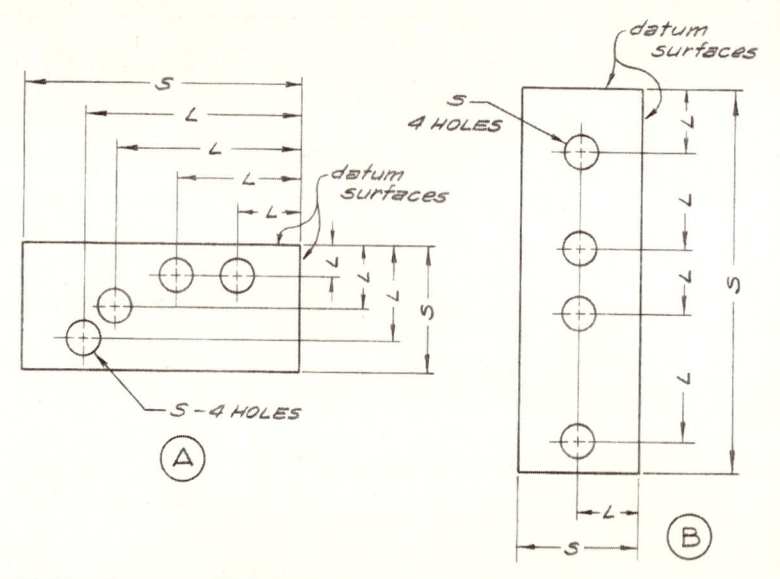

Figure 12.18 Locate multiple features from the same surfaces.

bar (line)

67 Figure 12.19 shows accepted procedures for placing dimensions in crowded positions. Note that the extension or dimension lines are never used as the horizontal _____ of the fractions.

Figure 12.19 Approved dimensioning for crowded spaces.

reference (REF)

68 To avoid the accumulation of tolerances, the sum of a series of dimensions should not add up to the overall dimension. *Leave one dimension in a series open.* (See Figure 12.20A.)

Figure 12.20B shows that it is permissible to include the last dimension in a series to avoid the necessity of calculating it on the part of the reader. However, it should be labeled as a(n) _____ dimension.

WORKING DRAWINGS—DEMINSIONING 233

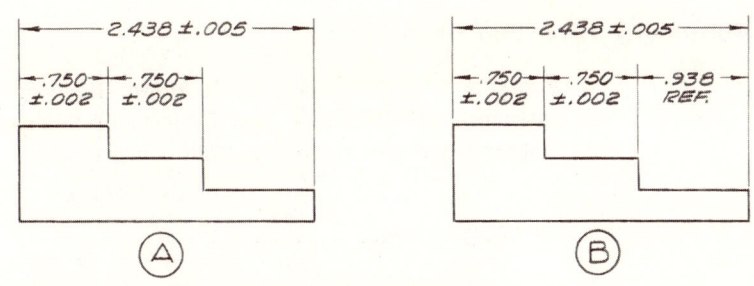

Figure 12.20 In a series, leave one dimension open.

orderly

69 Figure 12.20 illustrates also that dimensions should be kept in line wherever possible and not staggered. This is part of the problem of maintaining a(n) _____ appearance.

Placement of notes and leaders

Leaders should always be drawn at an angle to the surface or detail being described. Never make them horizontal or vertical. They may be confused with dimension or extension lines. The angle should be large (45° or greater) as shown in Figure 12.21.

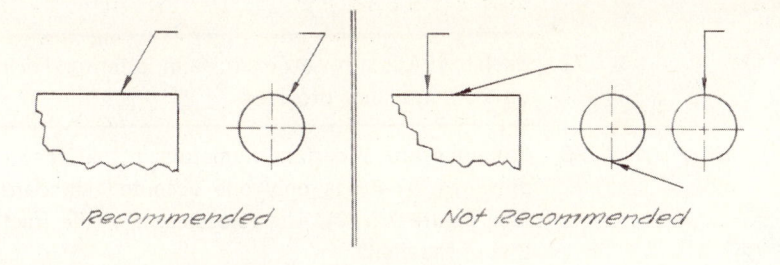

Figure 12.21 Approved construction of leaders.

touch

70 The point of the arrowhead should *always* _____ the surface or detail described by the note.

■ **71** We could go on endlessly listing specific rules for the placement of dimensions. Remember these two general rules:
  (1) Select dimensions for completeness of information without repetition.
  (2) Place dimensions for a logical and orderly appearance.
Sketch 39B is an exercise in selecting and placing dimensions. Dimension the three parts shown using S and L for the actual dimensions. Check your solutions with the answer sketches which are drawn according to accepted standards. Note that the optimum number of dimensions (including notes) is given for each part.

234 DRAFTING STANDARDS/UNIT 4

Selection of Letters and Numbers

Letters and numerals should be drawn freehand in standard engineering block form. A discussion and exercise in freehand lettering was given in Chapter 2. A review of the important points follows:

Freehand lettering

(1) Work freehand with a soft (H or 2H) sharp pencil.
(2) Use bold single strokes.
(3) Most lettering should be ⅛ in. high or less. Titles are ³⁄₁₆ in. high.
(4) Use inclined or vertical letters according to personal (or company) preference.
(5) Always use light guide lines.
(6) Letters in general are as wide as they are high. Guide lines control the height of the letter. You control the width. Exceptions to this rule are the letters I, J, M, and W.
(7) In composing words (or numbers with more than one digit) keep letters (or numbers) close together.
(8) The visual area between letters must be constant.
(9) Problem letters in composition are A, I, J, L, T, V, W, and Y. Problem numerals are 1 and 7.

■ 72 Sketch 40A is a review exercise in lettering. For practice, copy the notes on the guide lines provided.

---

two (2)

73 Fractions are important numerals in that they occur often on working drawings. There is only one accepted standard for drawing fractions (see Figure 12.22). The total height of the fraction is _____ times that of the digit.

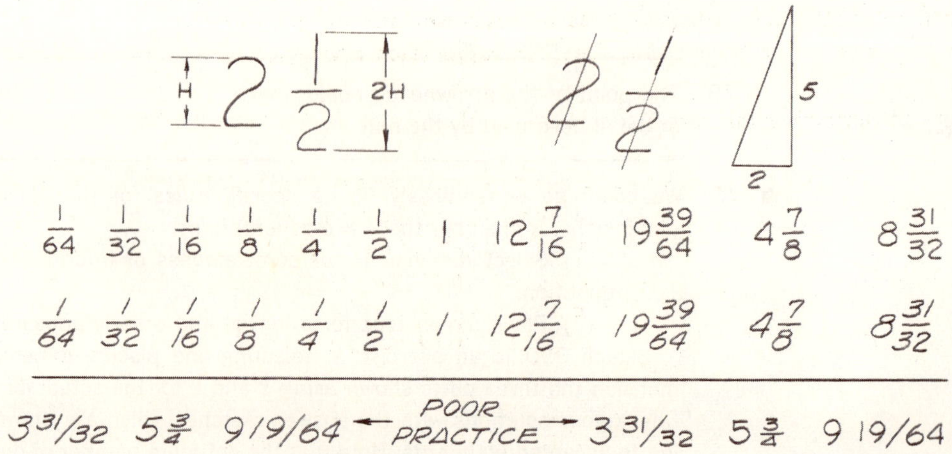

Figure 12.22 Accepted form for fractions.

WORKING DRAWINGS—DIMENSIONING

touch	74	The bar is always horizontal and centered (visually) on the height of the digit. The numerator and the denominator do not _____ the bar.
shorter	75	This means that the numerator and the denominator are actually _____ than the whole digit. (*Choose one: shorter/taller.*)
■	76	If you use inclined lettering, the entire fraction must be inclined. These rather rigid specifications need not be accompanied by measurements and guide lines for compliance to the rules. A little understanding and lots of practice will make fractions easy to draw according to standard. Sketch 40B is an exercise in drawing numerals and fractions.
feet; inches; inches	77	The inch-foot system is the standard for measurement in the United States. Therefore, every drawing should be dimensioned in _____, _____, and fractions or decimals of _____.
¹⁄₆₄; ¹⁄₃₂; ¹⁄₁₆; ⅛; ¼; ½	78	Fractions should always be reduced to their lowest form. Only the common fractions of an inch are acceptable. These are _____, _____, _____, _____, _____, and _____.
decimals	79	If a dimension cannot be expressed by a common fraction, then _____ of an inch should be used.

The standard usage of decimals is either the two-place decimal (0.28) or the three-place decimal (0.275). In rounding off decimals it is common practice to round off to an even number to facilitate halving the number if necessary.

All-decimal dimensioning is becoming popular in many industries. It avoids the awkward fractions of the inch-foot system. In fact, there is a strong movement (with some good arguments) to switch the United States standard to the metric sysem.

omitted	80	Figure 12.23A is a completely dimensioned drawing. Note that the inch marks (″) are _____. This is standard practice except where ambiguity might result (as in the case of 1″).
degrees; minutes; seconds; (°); (′); (″)	81	Angles are dimensioned in _____, _____, and _____. (See Figure 12.23B.) The symbols are always used and are (_____), (_____), and (_____).

236 DRAFTING STANDARDS/UNIT 4

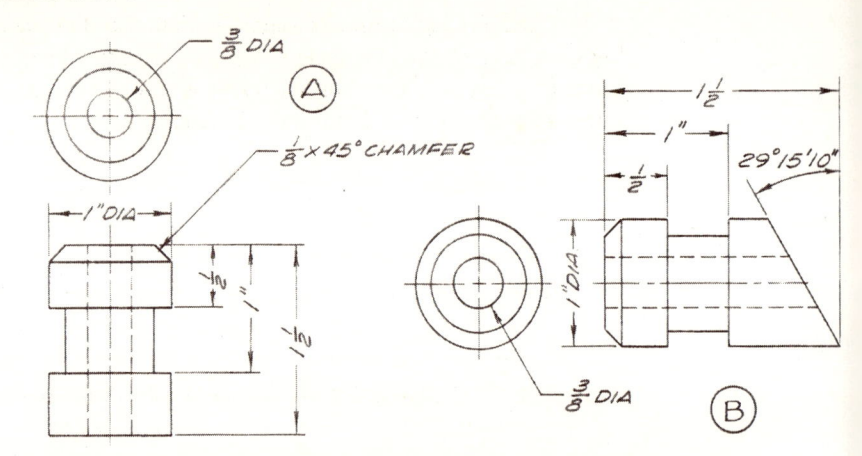

Figure 12.23 Inch marks are omitted except where ambiguity might result.

Placement of Letters and Numbers

bottom;
right(-hand)

82 There are two acceptable ways to place numerical dimensions in the break of the dimension line. (See Figure 12.24.) The *aligned system* (12.24A) has the dimensions in line with the dimension line. These are read either from the _____ of the drawing or the _____-hand side.

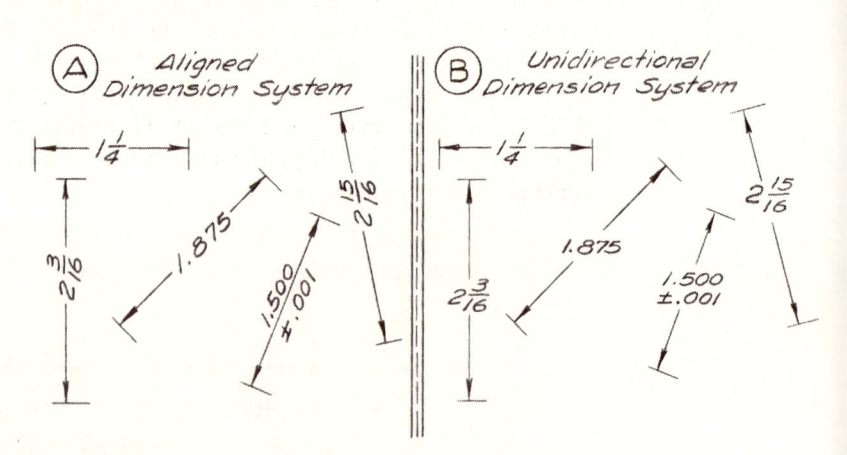

Figure 12.24 Two approved dimensioning systems.

bottom

83 In the *unidirectional system* (12.24B), all dimensions are horizontal and are read from the _____ of the drawing. The use of either system is optional with the individual or the company.

WORKING DRAWINGS—DIMENSIONING 237

The fraction bar is always separate from the dimension line. Figure 12.25 shows a use of the dimension line to separate two numbers. They are not parts of a fraction; they are the upper and lower acceptable limits of a single dimension. (More later about limit dimensioning.)

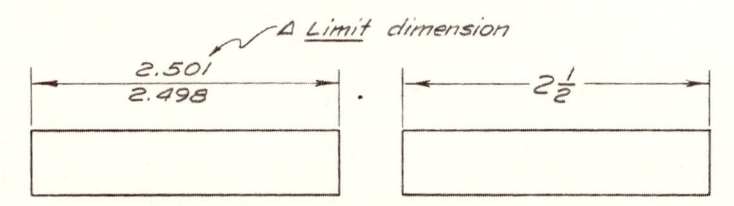

Figure 12.25 Dimension lines are never used as the bar of a fraction.

leader

84 All notes are horizontal. Notes are placed away from the views and the dimensions and a(n) _____ is used to direct the information in the note to the approprite detail.

■ **85** Sketch 41 is an exercise in the placement of dimensions and notes. The correct dimension lines have been started. Finish the lines and add the dimensions. Consider the grid squares to be ¼ in. Check your solutions.

This is the end of Chapter 12. You have seen examples of the many rules that control the production of a good working drawing. All rules of dimensioning are reasonable and logical if you keep in mind the fact that the drawing must be readily understood by a large number of people who have been trained to the rules. Any student who needs to understand these rules more fully is urged to purchase a copy of the American Standards Association Standard Drafting Manual, ASA Y14.5–1957. It may be obtained for a nominal price from the USASI, 10 East 40th Street, New York, N.Y. 10016.

The next chapter takes up a few of the conventional drafting practices which, when combined with good dimensioning, help to produce a clear and unambiguous working drawing.

chapter 13   working drawings—conventional practices

The many rules and standard practices which dictate the form of a working drawing are guide lines. They assure the originator of a drawing that other people will be able to obtain the exact information that he put into the drawing. But, as guide lines, they can be altered to serve special purposes. Some of these alterations have become conventional practices through years of unanimous acceptance. These conventions are either (1) violations of basic rules, or (2) accepted ways of drawing or labeling details that occur over and over on working drawings.

There are a few practices that are direct violations of the rules of orthographic projection. These violations are well-intentioned in that they help to clarify certain types of drawings.

SECTIONS

We have met one of these violations, the sectional view.

internal detail

1 A section is a deliberate cutting away of part of an object in one or more views to show _____ _____ .

2 The three types of sections that we have seen so far are shown in Figure 13.1. They are the _____ section, the _____ section and the _____ section.

full; half; offset

3 A cutting-plane line is included in a view adjacent to the sectioned view to show where the object was cut. A gray tone is added to the cut view by drawing evenly spaced, parallel _____ lines.

section

WORKING DRAWINGS—CONVENTIONAL PRACTICES   239

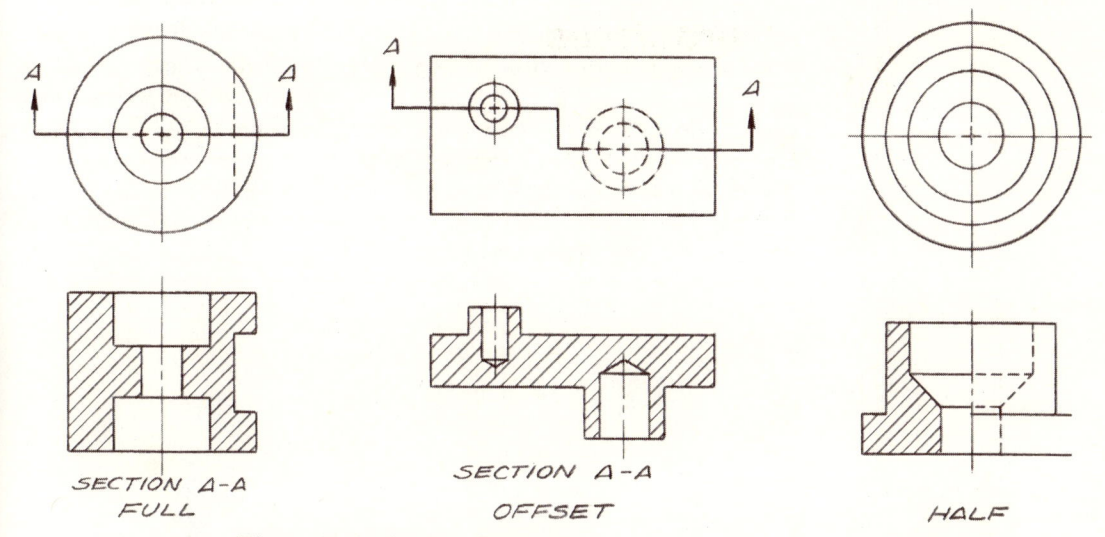

Figure 13.1 Section drawings.

cut (sectioned)	4 The cutting-plane line may be omitted if the symmetry of the object makes it obvious where the object was _____.

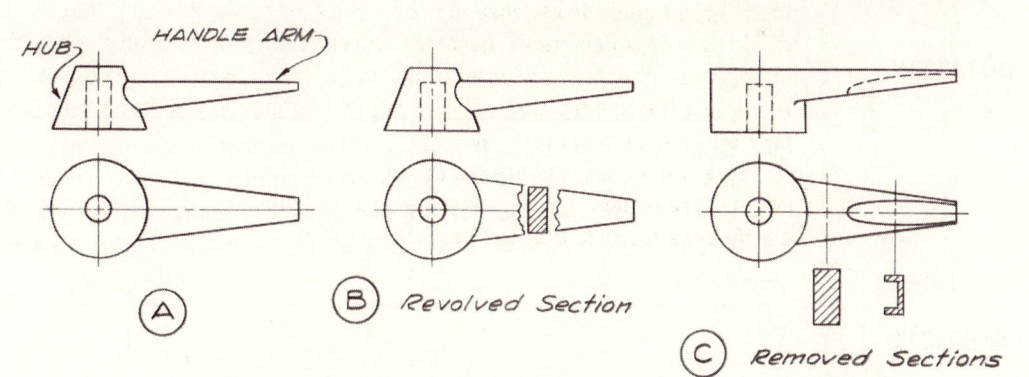

Figure 13.2 Revolved and removed sections to show cross-sectional detail.

shape	5 Figure 13.2 shows another type of section. The orthographic views of a handle are shown at 13.2A. The drawing is complete except that the cross-sectional _____ of the handle arm is not shown.
profile	6 A profile view would not clarify the drawing because of the tapered outline of the arm. Figure 13.2B shows a *revolved section* giving a typical cross-sectional shape of the arm. The section is drawn as if the arm were broken and a(n) _____ view taken at the break.

240 DRAFTING STANDARDS/UNIT 4

removed sections

7 Figure 13.2C shows another way to do this—especially if the arm does not have a uniform cross section along its length. These are called _____ _____ and any desired number may be taken to show detail and changes in cross section.

broken

8 Figure 13.3 shows a much used form of sectioning. It is a _____ section. This is used when only a small portion of the part needs to be sectioned to show interior detail.

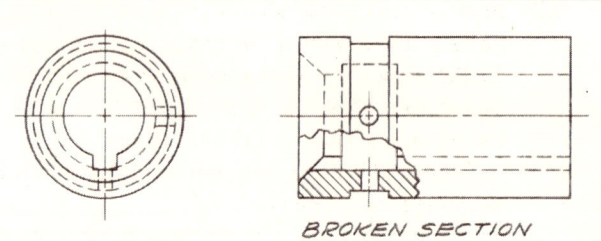

Figure 13.3 The broken section is used to show internal detail of only a small portion of an object.

The sectional views shown in Figures 13.1, 13.2, and 13.3 are the most commonly used. A section can be taken of any three-dimensional object. Its purpose is to clarify the drawing. To do this, it must be correctly drawn and clearly labeled so that the reader knows exactly where and how the part was cut.

ROTATION

foreshortened

9 Consider Figure 13.4. It is an angular arm with three bearings. Notice that the two views do not agree orthographically. This is conventional practice in drawing a part of this type. If the drawing were a direct orthographic projection, the right half of the arm in the top view would appear _____.

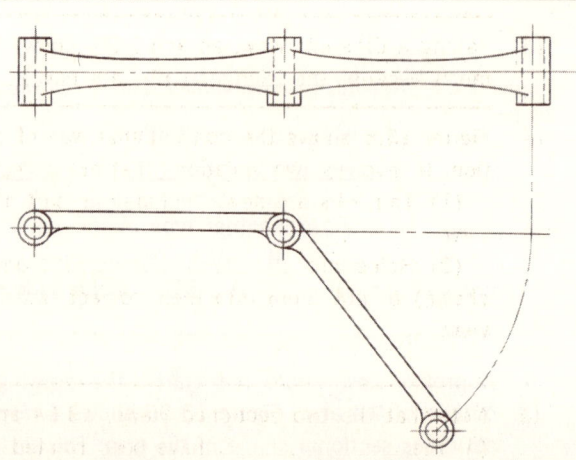

Figure 13.4 Rotation—an acceptable violation of orthographic projection rules.

WORKING DRAWINGS—CONVENTIONAL PRACTICES   241

front	**10** By rotating one half of the part into a straight line, the drawing of a foreshortened view is avoided. The _____ view still shows the correct angular character of the part.
trained	**11** Since this violation is an accepted conventional practice, people accustomed to reading working drawings will not be misled by it since they were _____ according to the same standards.
conical	**12** Another example is shown in Figure 13.5A. It is a bearing plate—a flat circular plate with a cylindrical hub strengthened by four ribs. Figure 13.5B shows the orthographic views including a full section. The sectioned view gives the wrong impression of the object and does not show the mounting holes in the base plate. Figure 13.5C is a perspective sketch of the object as it appears in the sectioned view. The object is _____ rather than cylindrical.

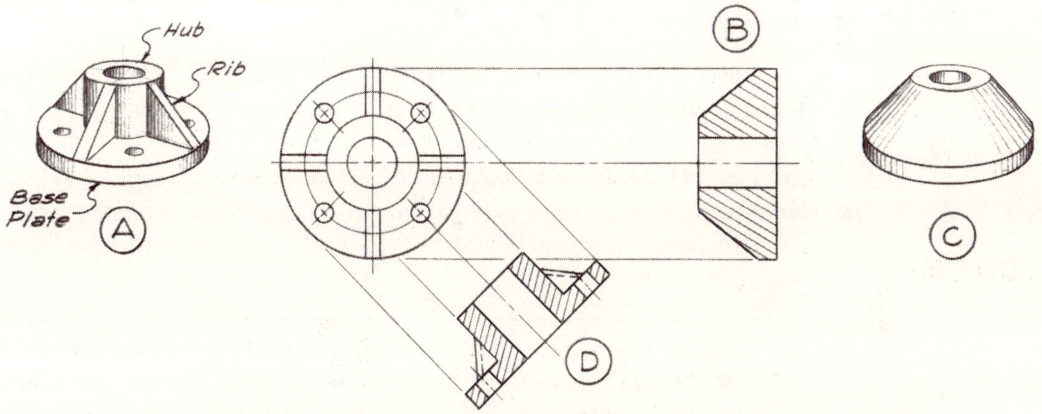

Figure 13.5 Problem: A true projection gives an awkward sectional view.

foreshortened	**13** Taking a different view, as at 13.5D, shows the holes but doesn't help much because the triangular ribs are now _____.
sectioned; rotated	**14** Figure 13.6 shows the conventional way of drawing this object in section. It involves two violations that are accepted practice: (1) Thin ribs are never included as part of the _____ view. (2) Holes are _____ onto the cutting plane (not shown) to put them into their correct radial position in the sectioned view.
(A) holes; (B) ribs	**15** Note that the two sectioned views, 13.6A and 13.6B, are identical. At (A) the _____ have been rotated onto the cutting plane and at (B) the _____ have been rotated.

242 DRAFTING STANDARDS/UNIT 4

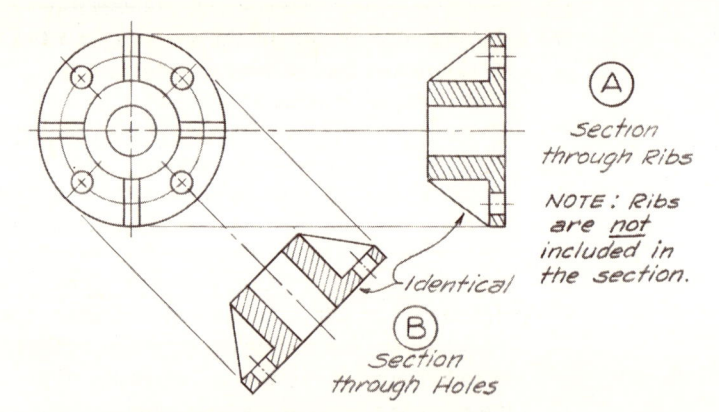

Figure 13.6 Solution: Use rotation on ribs or holes to give a realistic sectioned view and do not section the ribs.

radial

16 The holes (A) are rotated to put them in their correct _____ position with respect to the object centerline on the sectioned view.

foreshortened

17 The ribs (B) are rotated to avoid a(n) _____ view of the ribs.

■ 18 Sketch 42 is a sketch exercise on sectioning and conventional practices. Complete the views indicated according to accepted practice. Check your solutions.

PRECISION

We have discussed the making and dimensioning of working drawings. In the design of a product, care must be taken to ensure that parts fit together. They must either fit tightly so that they will not shift out of position or fit with just the right amount of clearance so that they will operate together freely.

These are problems of precision. A full discussion of precision is beyond the scope of this graphics book—it is more suited to a text and course on engineering design. However, to get a feel for precision and some of the dimensioning problems involved, let's take a look at the subject.

1-in. cube

19 Figure 13.7A presents a hypothetical situation to illustrate the meaning of manufacturing precision. We wish to make a(n) _____ _____ out of metal.

The table shows the various machining processes that would be suitable for the job. It also shows what we might expect in dimensional precision, cost, and surface finish from each process. These are three important interrelated factors that must be considered in making any part for any product. There are

WORKING DRAWINGS—CONVENTIONAL PRACTICES 243

many other considerations that are not shown in the table. Cost of capital equipment (machines), labor costs, and precision of shape are a few more. The balancing of all these factors is one of the many decision steps that must be made before an idea becomes a useful device.

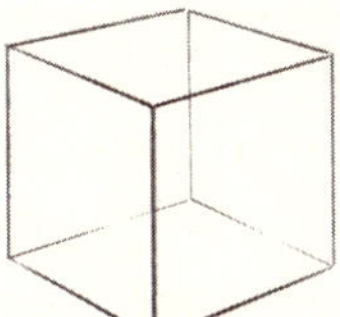

PROBLEM: MAKE A 1-INCH CUBE OF METAL

PROCESS	ACHIEVABLE TOLERANCE	COST INDEX	SURFACE QUALITY
Flame cut from 1-in. plate	± 1/16	1	Poor
Power Hacksaw (from 1" sq. bar)	± 1/32	1	Poor
Sand Casting	±.025	6	Poor
Die Casting	±.002	10	Good
Shaping	±.001	10	Fair
Milling	±.001	6	Good
Lathe (cutoff from 1-in. sq. bar)	±.001	3	Good
Surface grinding	±.0002	60	Excellent
Lapping	±.0001	120	Excellent

(A)

(B)

0 .0001 .001 .010 1/64 (.015) 1/32 (.031) (.0625) 1/16

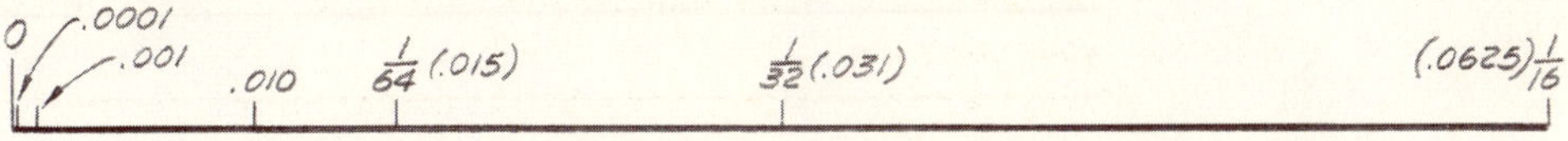

Figure 13.7 *The meaning of manufacturing precision.*

(±) 1/16;
(±) 0.0001

20 Study of the table shows a wide variation in dimensional precision with tolerances varying from ± _____ in. to ± _____ in.

1/16

21 If you have no feel for the meaning of these numbers, look at the expanded scale in 13.7B. A five-inch line represents _____ inch. To this scale, 0.0001 (one ten-thousandth of an inch) is almost unmeasurable.

1 to 120

22 The cost index varies from _____ to _____.

23 This means that, if the cube cost 10 cents to make by the flame-cut or power-hacksaw processes, it would cost $_____ by the lapping process.

$12;
improves

The surface quality _____ as cost and precision increase. (*Choose one:* improves/deteriorates.)

244 DRAFTING STANDARDS/UNIT 4

clearance;
interference

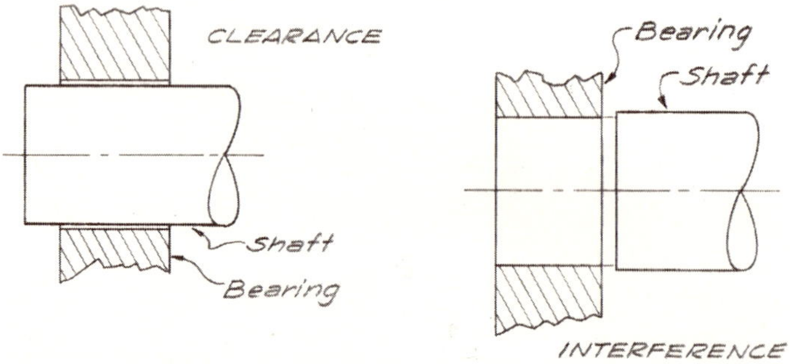

24 Figure 13.8 shows two kinds of fit between mating parts (a shaft and bearing in this case). They are the _____ fit and the _____ fit.

Figure 13.8   Clearance and interference fits between mating parts.

smaller; smaller

25 In a clearance fit, the shaft is intentionally made _____ than the hole so that the parts will go together easily. In an interference fit, the bearing hole is intentionally made _____ than the shaft so that the parts must be forced together. (*Choose one: larger/smaller.*)

TERMINOLOGY

nominal

26 Figure 13.9A relates these ideas to numbers. Some terminology of a part is in order. The nominal size of a part is the dimension used for general descriptions. The _____ size of both the shaft and the bearing hole of 13.9A is 1¼ in.

0.002

27 *Tolerance* is the *permissible variation* in a dimension. If we talk of a dimension of ½ ± 0.001 in., the tolerance is _____ in.

tolerance

28 The maximum limit of the ½-in. dimension is 0.501 and the minimum limit is 0.499. Thus, _____ is the difference between maximum and minimum limits.

hole; shaft

29 *Allowance* is the *intentional difference* in mating dimensions to permit a desired clearance or interference fit. Allowance is always the *tightest possible fit* permitted by the limits of the two mating parts.

Relating this to a shaft and hole, allowance is calculated by minimum _____ limit minus maximum _____ limit.

WORKING DRAWINGS—CONVENTIONAL PRACTICES 245

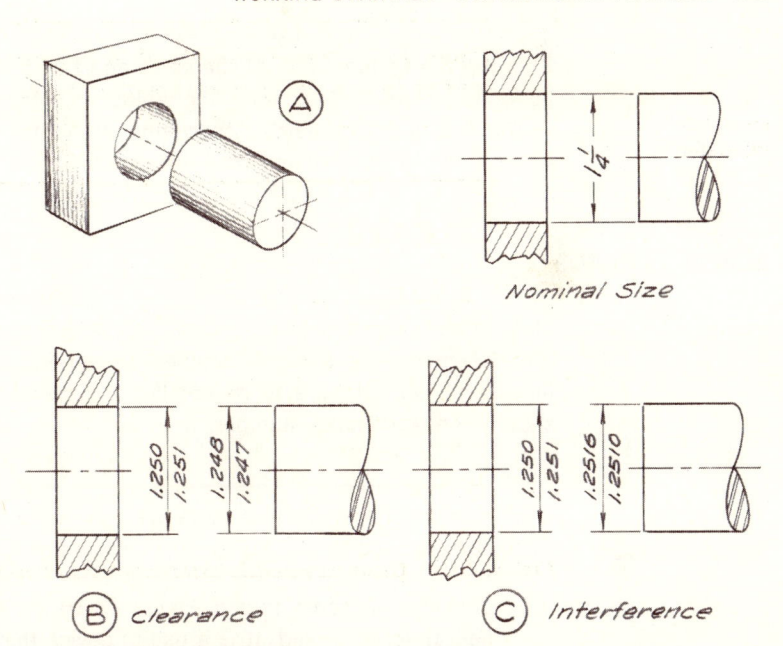

Figure 13.9 *Limit dimensioning.*

Clearance and Interference Fits

In a clearance fit, allowance is considered positive (+). Allowance is negative (−) in an interference fit.

---

30 Figure 13.9*B* and *C* shows complete limit dimensions for the nominal 1¼-in. shaft and hole for both clearance and interference fits.
  Clearance fit:
   (A) Shaft tolerance = _____ in.
   (B) Hole tolerance = _____ in.
   (C) Allowance (tightest fit) = _____ in.
   (D) Loosest fit = _____ in.
  Interference fit:
   (A) Shaft tolerance = _____ in.
   (B) Hole tolerance = _____ in.
   (C) Allowance (tightest fit) = _____ in.
   (D) Loosest fit = _____ in.

Clearance fit:
 (A) 0.001;
 (B) 0.001;
 (C) +0.002;
 (D) 0.004;
Interference fit:
 (A) 0.0006;
 (B) 0.001;
 (C) −0.0016;
 (D) 0.000

246 DRAFTING STANDARDS/UNIT 4

The exact numbers of the last example came from standard tables of limits. Knowing the nominal size of the parts and the desired fit, the tolerances can be found by looking them up in the tables.

The American Standards Association (ASA) has publications describing all aspects of limit and true-position dimensioning.

SCREW THREADS

Most manufactured products consist of designed parts and purchased parts. The designed parts must be detailed on working drawings. Purchased parts, in general, do not have to be detailed but merely listed by name or by the supplier's identification number.

fasten (hold)

31 Prominent among these parts are nuts, bolts, screws and other threaded devices used to _____ parts together.

rotated

32 Consider the screw thread. Figure 13.10 shows the profile of a standard helical screw thread. It is a useful device that moves in the direction of its axis when it is _____ about the axis.

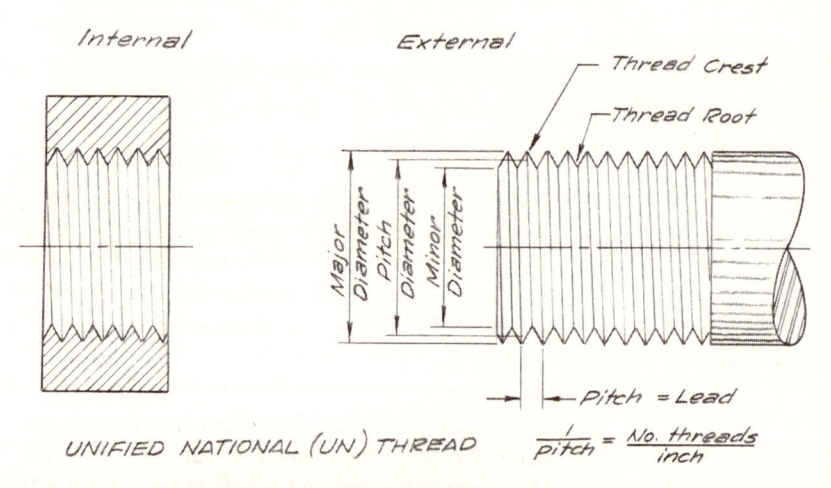

Figure 13.10   Internal and external screw threads drawing in the semiconventional symbol.

Screw thread form, shape, sizes, and tolerances are highly standardized. A designer specifies a machine screw by stating the desired diameter, length, and head style. He can expect to receive a quality product that meets the standards in all respects.

Two things that relate directly to working drawings are (1) how to draw screw threads and (2) how to specify screw threads. Figure 13.10 is a *semiconventional* representation of a Unified National internal and external thread. It is not a realistic form in that, being a helix, the line joining the crests of the threads should be a curved line rather than a straight line.

WORKING DRAWINGS—CONVENTIONAL PRACTICES   247

regular, simplified

33 The semiconventional symbol is hard to draw. Figure 13.11 shows two standard thread symbols that enjoy exclusive use in drafting. They are the _____ symbol and the _____ symbol.

Regular Symbol   *Simplified Symbol*

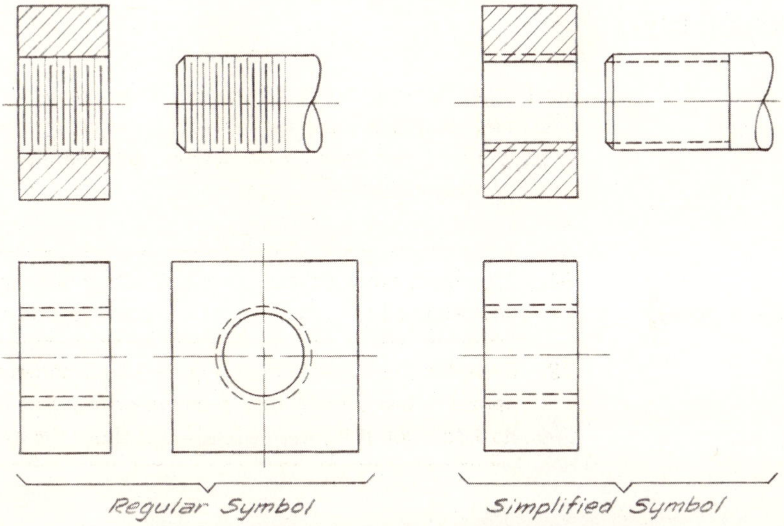

Figure 13.11 Approved drawing symbols for internal and external threads.

major; minor

34 The *regular* symbol is designed to give a visual impression of a screw thread. To draw the regular symbol, first draw the shaft or hole to nominal size. Establish the minor diameter (see Figure 13.10) by estimating a thread depth of about 1/16 in. (not critical). Draw alternate long thin lines (_____ diameter) and short thick lines (_____ diameter).

1/10 (0.1)

35 The spacing of these lines must be even but need not show the actual number of threads per inch. About 10 threads per inch is a useful guide. This means that the distance between adjacent thin lines is _____ in.

hidden (dashed)

36 The *simplified* symbol is easier to draw. Estimate the minor diameter (again use about 1/16-in. thread depth) and draw _____ lines for the desired length of thread.

specification

37 Threads are specified by a note of standard form and content with a leader directed to the symbol on the drawing. Figure 13.12 shows a typical thread _____ note.

nominal diameter

38 As shown in Figure 13.12, the first number stands for _____ _____.

248 DRAFTING STANDARDS/UNIT 4

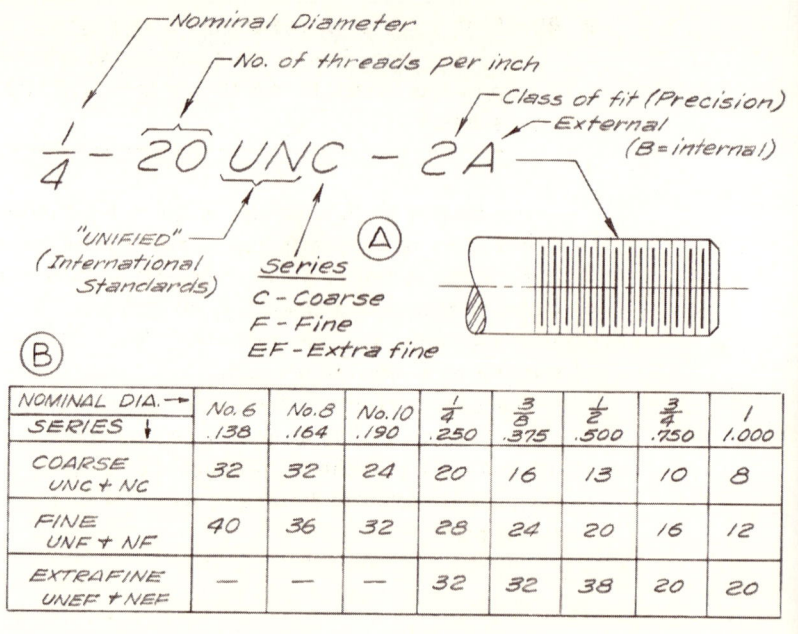

Figure 13.12 Thread specification.

---

threads per inch

39 The second number represents the number of _____ _____ _____ .

$\frac{1}{20}$ or (0.05 in.)

40 The pitch of a thread is the reciprocal of the number of threads per inch. Thus, for a screw of 20 threads per inch the pitch is _____ in.

---

series

41 The "UN" stands for Unified National which is the designation for a relatively new international standard thread form. The "C" denotes the thread _____ .

The coarse, fine, and extra fine series cover about all thread needs. The number of threads per inch for a given nominal diameter varies with each series. This is all information that is available in standard tables. A portion of a thread table is reproduced in Figure 13.12B.

The "2A" at the end of the note represents a class 2 external thread. 2B would be a class 2 internal thread. Class 1 is a loose-fitting thread, class 2 is a general-purpose thread, and class 3 is a precision thread.

The content and order of presentation of the thread specification note of 13.12A is standard. Any threaded shaft or hole on a detail drawing should be labeled with such a note.

■ **42** Sketch 43 is an exercise in thread-symbol drawing and specification-note writing. Sketch the symbols and add the note in the guide lines provided. Refer to the thread table (13.12B) as needed. Check your solution.

The content of this chapter is but a brief introduction to the topics presented. This is especially true of the discussions of limit dimensioning and screw-thread notation. A comprehensive account of these topics is beyond the scope of this programmed text on graphics. You are urged to consult the ASA standards and the many excellent conventional engineering drawing texts for more detailed accounts.

The brief account of the standard practices and conventional treatments presented here should at least make you aware of the existence of the highly detailed field of technical drafting. Our purpose is an understanding of the geometry of technical drawing so we will leave the drafting aspects of the field at this point.

chapter 14 drawing instruments and their use

INTRODUCTION

Chapter 14 deals with the art of drawing with instruments. Referring back to Flow Chart IV (preceding Chapter 12), we see that this is the last topic under the main heading, The Art of Drawing. We have discussed the theories of orthographic projection and pictorial drawing, concentrating these theories on the picturing of three-dimensional objects. Chapters 12 and 13 finished the topics under the heading of Picturing Objects by filling the bag of rules through discussions of dimensioning and other standard professional practices.

After this chapter, the rest of the text is devoted to the uses of graphics for solving problems. The two topics under this heading are Descriptive Geometry and Graphic Mathematics.

INSTRUMENT DRAWING

This programmed text has stressed freehand drawing throughout, and it will continue to stress freehand techniques. We are concerned for the moment, however, with understanding drawing theory. Drawing instruments are an aid to accuracy. We have been seeking accuracy of visual form and proportion but not necessarily accuracy of scale. This latter is achieved through carefully and precisely done instrument drawings.

Many students of graphics are aware of the tools of drawing and their basic uses. They have gained this information through an early preparatory course, personal interest in drawing, or perhaps a part-time job in a drafting office. Other students have never seen or handled drawing instruments.

DRAWING INSTRUMENTS AND THEIR USE   251

Chapter 14 is designed for both groups of students. It is divided in the following way:

Part 1: Frames 1 through 28 name all the modern instruments and discuss some of their important uses in technical drawing.
Part 2: Frames 29 through 60 give a more detailed description of instruments and their uses.

If you are among the students in the first group—those who know something about drawing instruments—proceed only to frame 28. If at that point, you are satisfied that you know and understand modern instrument usage, proceed directly to Chapter 15. If you have had no experience with drawing instruments, you will do well to proceed through Chapter 14 to the end.

In either case, remember that a book cannot transmit skill in the use of instruments. This you gain for yourself through constant practice.

DRAWING INSTRUMENTS

Part 1

Drawing instruments can be divided into five classes (see Figure 14.1):
   Class 1: Basic tools
   Class 2: Tools for drawing straight lines
   Class 3: Tools for drawing curved lines
   Class 4: Tools for measuring distances
   Class 5: Special tools

Class 1 instruments

1   The basic tools of class 1 are those needed to get started. First you need a flat, smooth surface on which to draw. This is the _____ _____.

drawing board

2   The most indispensable tools are _____ _____ (your words).

paper, pencils, and eraser

Paper. Technical drawings are usually made on high-grade translucent vellum paper. Copies can be made by exposing a special paper with a chemically treated surface to ultraviolet light through the original drawing. The paper must be translucent to let the light pass through to the copy paper. Pencil lines block the light. The ultraviolet light causes changes in the chemical surface of the copy paper, which is then developed in liquid ammonia for a blueprint (white line on blue background) or in vaporous ammonia for a white-print (blue or black line on a white background).

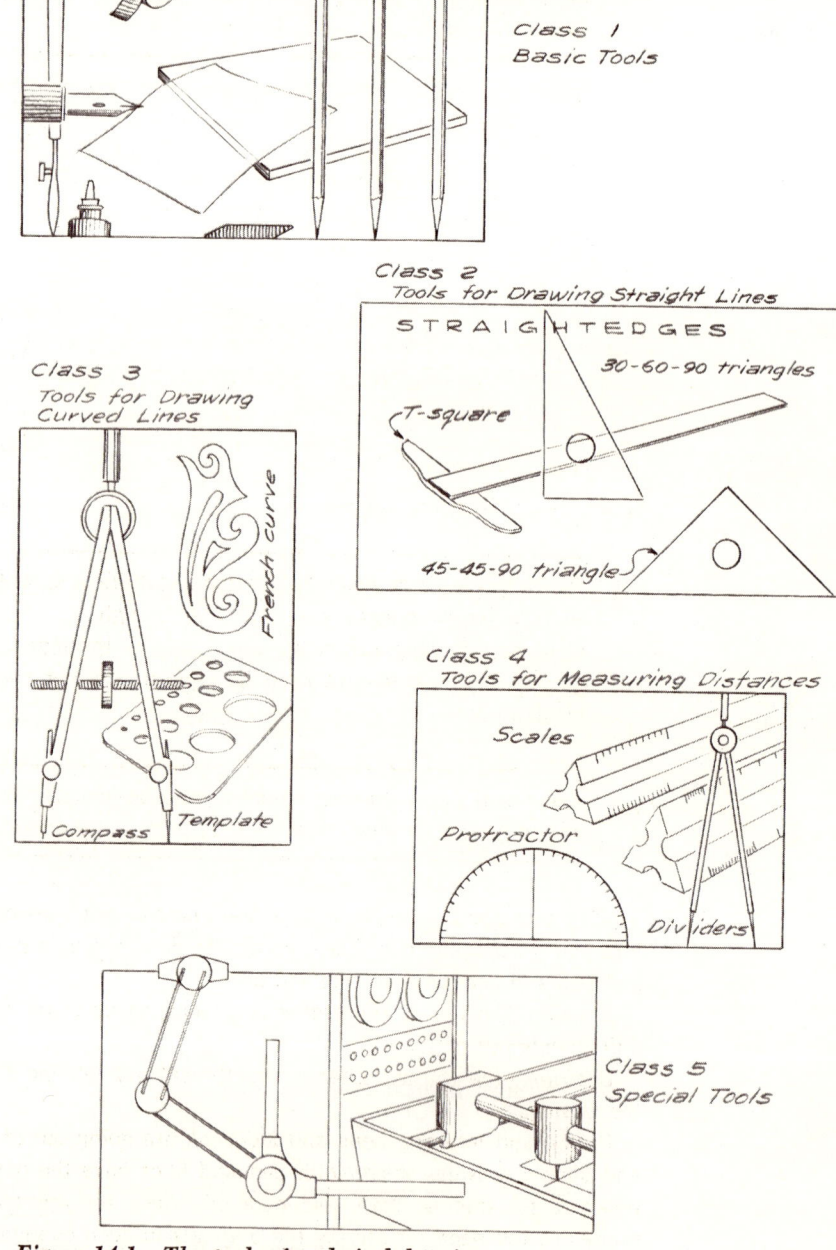

Figure 14.1 The tools of technical drawing.

DRAWING INSTRUMENTS AND THEIR USE 253

18; 7B; 9H

3 *Pencils.* Drawing pencils are available in _____ different grades of hardness. (See Figure 14.2A.) They range from very soft (grade _____) to very hard (grade _____).

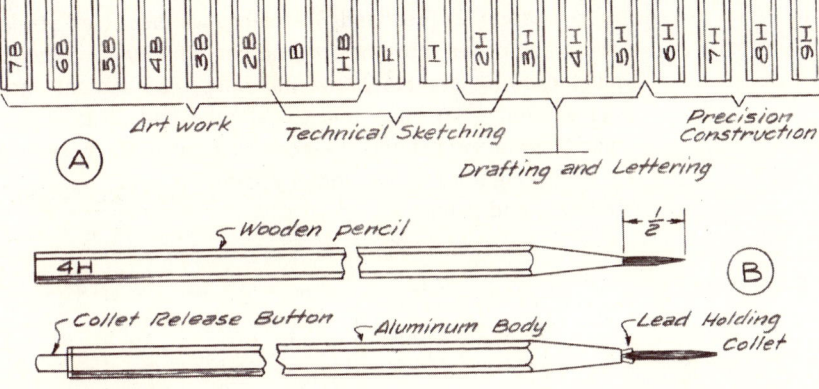

Figure 14.2 Drawing pencils.

B (through) 2H;
2H (through) 5H

4 Pencil grades recommended for technical drawing range from B through 9H. Sketching pencils are _____ through _____. For finished instrument drawing (drafting) use _____ through _____. The hard grades of 6H to 9H are used for precision layout and mathematic construction.

½; conical

5 The "lead" of a drawing pencil should be exposed about _____ inch and sanded to a long, thin _____ point. (See Figure 14.2B.)

With this point you can draw a fine line and get considerable mileage before you must resharpen. The best advice on choosing a pencil grade is to try them all and use the ones that you prefer.

Erasers. The use of the eraser is obvious to all of us who make mistakes and change our minds.

Drafting tape is the modern "thumbtack" for holding the paper firmly on the board.

Ruling and *lettering pens* and *india ink* are going out of style. New papers and plastic drawing sheets with penciled lines have the permanence formerly reserved for tracing cloth and india ink. Also, the *microfilming* of originals precludes the need for storing the originals in good condition for many years.

Class 2 instruments

T-square,
triangle

6 Class 2 instruments (Figure 14.1) are the tools for drawing straight lines. They are termed "straightedges" and are the _____ and the _____.

254 DRAFTING STANDARDS/UNIT 4

7 These tools are precision made for the purpose of having a straight edge against which you can guide a pencil for drawing a _____ _____.

straight line

In Chapter 2, Freehand Sketching, we discussed straight and curved lines. The types of straight lines or line combinations that concern us most in technical drawing are (1) horizontal lines, (2) vertical lines, (3) parallel lines, (4) perpendicular lines, and (5) angular lines. The *T-square* and *triangles* are instruments for guiding the construction of all these straight lines.

Study Figure 14.3. It shows a drawing board set up with paper, T-square, triangle, and pencil.

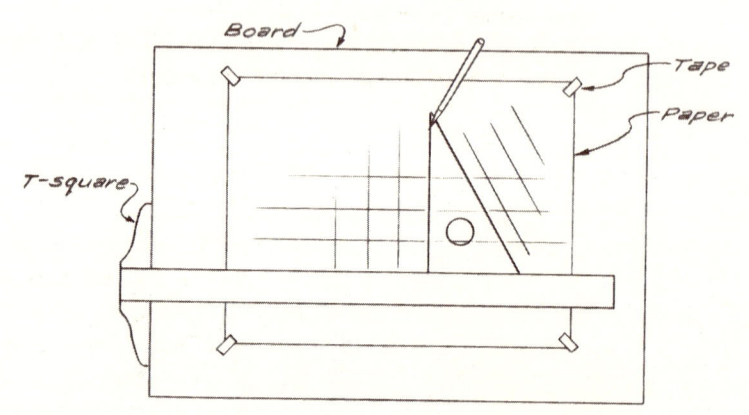

Figure 14.3 The principal use of the T-square and triangles.

8 What lines can be drawn with the T-square? _____

horizontal and parallel lines

9 What lines can be drawn with a triangle? _____

vertical, angular, and perpendicular (to horizontal lines)

10 The two standard triangles are the ____-____-____ and the ____-____-____.

45-45-90; 30-60-90

11 Two other angles can be easily drawn by using both triangles and the T-square. These angles are _____ degrees and _____ degrees. (See Figure 14.4.)

75 and 15

Thus, with these three instruments alone or in combination, you can draw lines at angles of 0, 15, 30, 45, 60, 75, and 90 degrees to the horizontal. Any angular lines in between these would have to be plotted with a protractor and then drawn using a triangle as a straightedge.

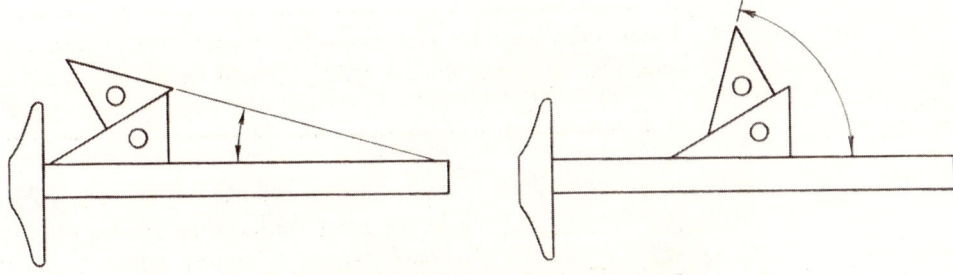

Figure 14.4 Combining triangles to obtain new angles.

Figure 14.5 shows a manipulation of two triangles or one triangle and the T-square that is very useful in technical drawing. A type of drawing problem that comes up in descriptive geometry is shown at 14.5A. Given a line AB, accurately construct a line P parallel to AB and two lines m and n perpendicular to both through A and B respectively.

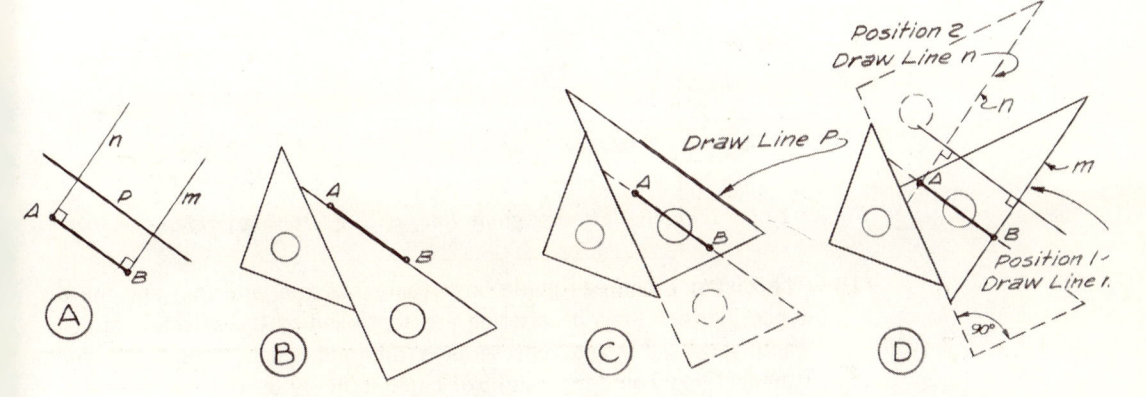

Figure 14.5 Manipulating triangles to draw a series of parallel and perpendicular lines.

parallel	12 First, position two triangles as shown at Figure 14.5B. The hypotenuse of a 30-60-90 triangle is positioned _____ to AB.
AB	13 Holding the 45-45-90 triangle firmly in place, slide the other triangle up as shown in 14.5C. Line P can now be drawn since the hypotenuse of the 30-60-90 triangle is still parallel to line _____.
90	14 Now, rotate the 30-60-90 triangle so the short leg is against the 45-45-90 triangle (Figure 14.5D). The hypotenuse has rotated through _____ degrees.
perpendicular	15 Draw line m through point B and slide the 30-60-90 triangle up and draw line n through point A. Lines m and n are parallel to each other and are _____ to lines AB and P.

256   DRAFTING STANDARDS/UNIT 4

The success of this construction rests in holding the stationary triangle firmly to the board. If it slips, it will change the angle. With practice, you will find this much-used technique very useful.

T-square

16   Freehand lettering should be drawn between horizontal guide lines. The easiest way to draw these is to estimate the height of the letter by eye and draw the lines lightly against the _____.

horizontal

17   There are, however, *lettering guides* on the market to aid in drawing these lines. Figure 14.6 shows two types. They have a series of holes through which you place a sharp pencil point. By dragging the device along the T-square you draw a(n) _____ line.

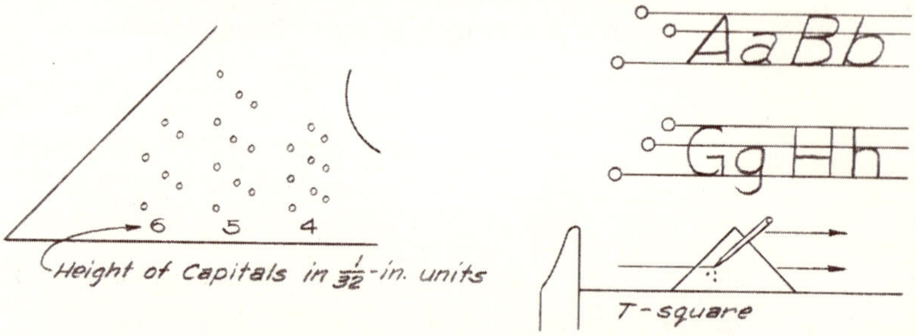

Figure 14.6   Drawing lettering guide lines with a lettering guide.

4/32 (1/8)

18   The distance between guide lines (height of the lettering) and the distance between lines of lettering are controlled by the selection of holes. Each series of holes represents a different height letter. The series numbers represent the height of capitals in 1/32 of an inch. Thus No. 4 guide lines give a letter height of _____ in.

Class 3 instruments

curved lines

19   The tools of class 3 are used for drawing _____ _____. (See Figure 14.1.)

circles;
circular

20   The most common tool is the *compass*. Compasses come in many sizes and styles but, regardless of differences, they are all used to draw _____ or _____ arcs.

For small-diameter circles, a *plastic radius guide* (or *circle template*) is a useful tool. It is especially useful in drawing rounded corners on objects.

A *french curve* is a curve-drawing device with a constantly changing radius. A smooth curve can be drawn through any set of plotted points by finding a match between the instrument and two or more adjacent points. A number of such matches are usually needed to complete a curve.

Class 4 instruments

The most important measuring tool of class 4 is the *scale*. The different scales useful in technical drawing were discussed and shown in Chapter 6, Proportion and Scale. The two most common types are shown in Figure 14.7A.

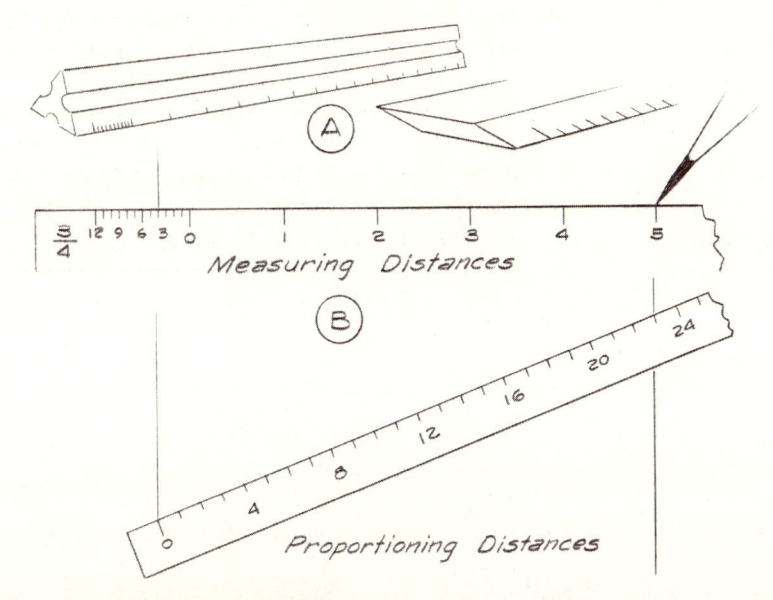

Figure 14.7 Two uses for a scale.

---

21 Figure 14.7B illustrates two uses of scales. The first is the measurement of distance. The distance shown being measured on an architect's scale is _____ to a scale of _____ = _____.

5'-4";
¾" = 1'-0"

---

22 The second use is the proportioning of distances. The 5'-4" distance is proportioned into _____ equal divisions.

23

The inspection of any scale will show it to be a precision device. To lay off a distance, always use a sharp pencil point to mark the distance. (See Figure 14.7.) Do not destroy the accuracy of the instrument by using a blunt point to mark the measure. Distances can be measured to the nearest 0.01 (one one-hundredth) in. with the engineer's 50-scale.

---

23 A scale is not a straightedge and should not be used as one. Measure distances with a _____ but draw straight lines with a(n) _____ or with a(n) _____.

scale; T-square, triangle

Distances may be transferred from a scale to a drawing by the use of *dividers.* Dividers are also useful in transferring a distance from one part of a drawing to another (see Figure 14.8A).

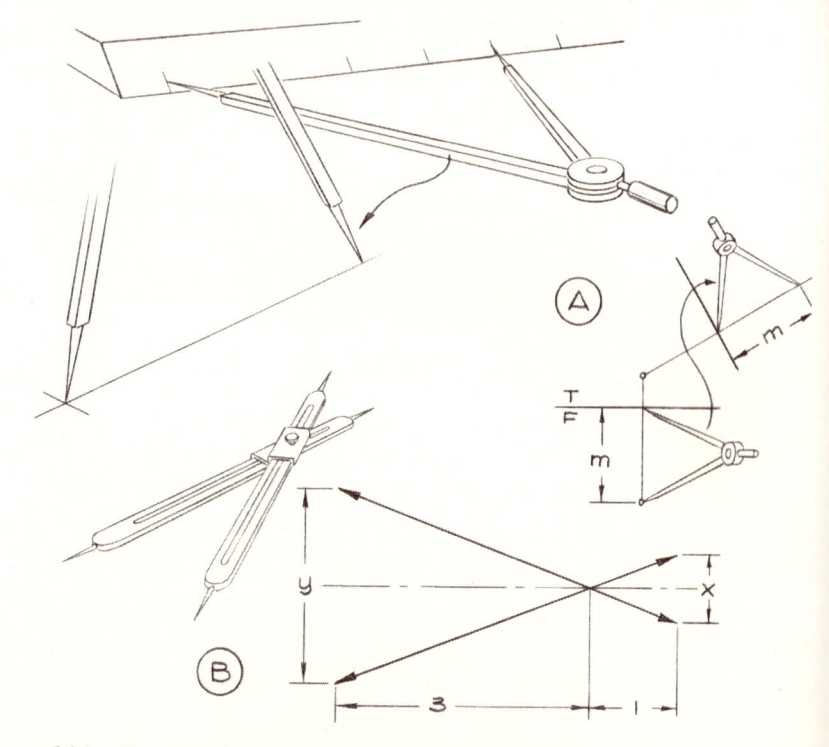

Figure 14.8   Uses for the dividers.

1:3

24   Figure 14.8B shows a set of *proportional dividers.* This instrument has needle points at both ends and a movable pivot. It is used for enlarging and reducing drawings. At the setting shown, the ratio is $x{:}y =$ _____._____ .

enlarge; reduce

25   Using this setting, you can either _____ a drawing to three times the original or _____ to one-third of the original. The arms of the instrument are calibrated to permit automatic setting to a desired ratio.

angles

26   Another measuring device essential in drafting is the *protractor.* It is used to measure _____ .

The most common type of protractor is the flat, semicircular clear plastic instrument. Because it is transparent, your construction is visible when you place it over your work. A 5- or 6-inch protractor is graduated in one-half degrees. This permits estimation of angles to the nearest one-quarter degree (15 minutes of arc). On a working drawing, the exact (to the minute) measurement of an angle is not important. The dimension that is put on the angle is important.

Class 5 instruments

T-square, triangle, scale, protractor

27 Class 5 instruments are special ones. One of the these is the *drafting machine*. (See Figure 14.9.) It is a device that is mounted on the drawing board and combines the functions of the _____, _____, _____, and _____.

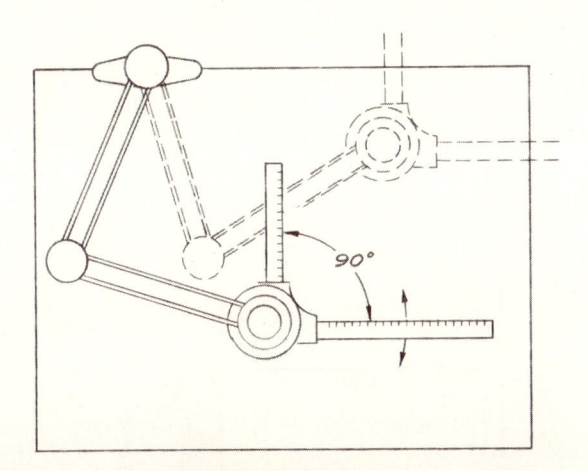

Figure 14.9 The drafting machine.

It does this efficiently and accurately. The drafting machine is ideal for production drafting because it is a great timesaver. However, it will not do any more than the T-square, triangles, etc.; it will just help you to do it faster.

Computer Drafting. Figure 14.10 is a look into the future of technical drawing. Information on a drawing is two-dimensional even though it depicts a three-dimensional object. Figure 14.10A shows an x-y plotter. A head unit carries a pencil or pen that maintains contact with the drawing surface.

The head can be driven by motors in both the x and y directions. Drawing information is programmed on a computer; that is, using the language of the computer, the xy-coordinate information for every point on every line is recorded on magnetic tape. When the computer is connected to the x-y plotter and the tape is played back, the signals drive the head in such a manner as to duplicate the original drawing.

Figure 14.10B is a schematic diagram of what drafting procedures may be in the future. At (1) the designer or drafstman makes a freehand sketch of an idea. At (2) he inserts the sketch into an electronic scanner which records the sketch information on magnetic tape (3) in xy-coordinate information. Whenever he wishes a copy, he plays the tape back and the x-y plotter produces a drawing (4) with all the niceties of a well-executed instrument drawing. As an alternative (5), he may wish just to look at the drawing. In this case he switches the tape output to a TV tube and sees the complete drawing as an image on the tube.

260 DRAFTING STANDARDS/UNIT 4

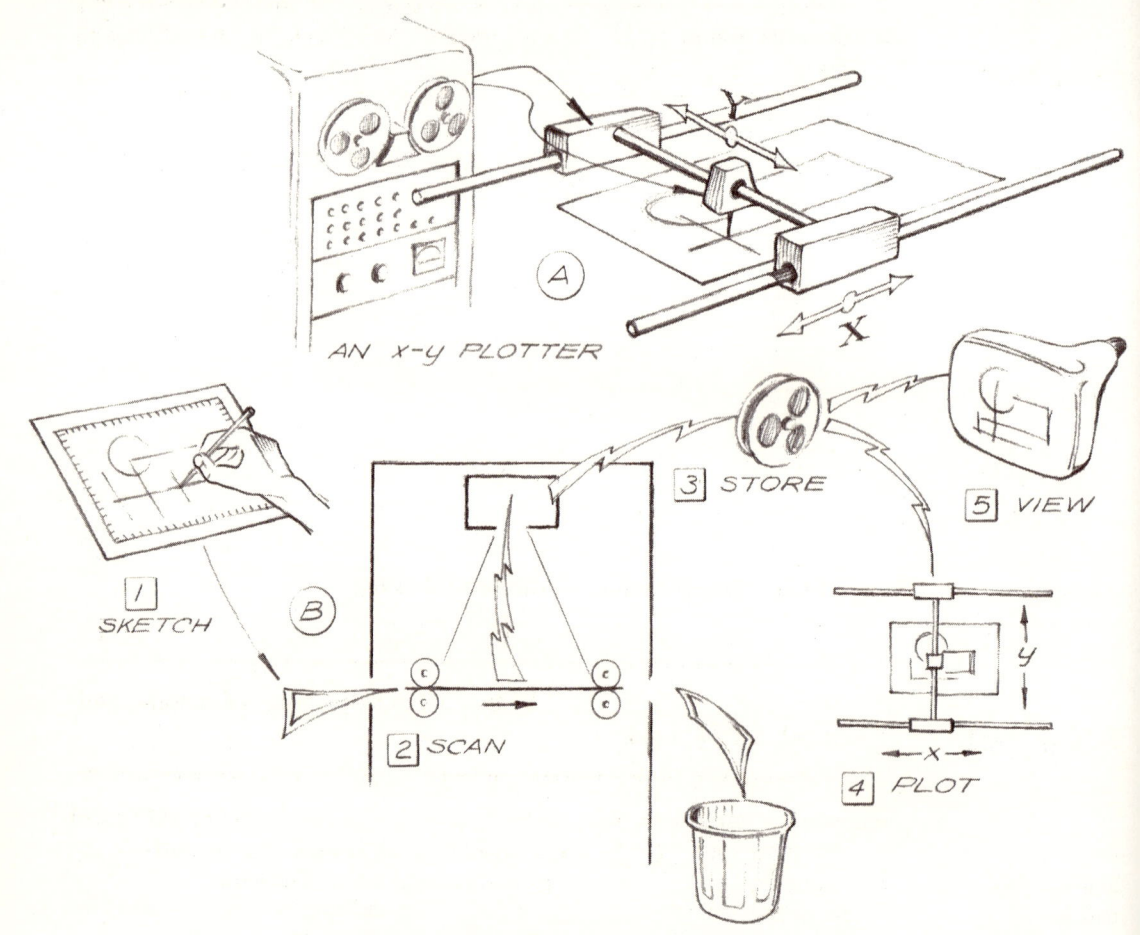

Figure 14.10 Is this the future of technical drawing?

	28 Machines of this sort are great labor-saving devices, but there will always be the need for the man to tell the machine what to draw. Such a system as is shown in 14.10*B* points up the need for people with an understanding of the principles of graphics and ability in _____
freehand sketching	_____. (See Step 1.)

This is the end of Part 1 of Chapter 14. Skip the rest of the chapter or proceed to the end if you wish to learn more detailed information about drawing instruments and their use. (See Introduction.)

Part 2

	29 Any drawing starts with _____ and _____. The drawing instruments are just a means to draw clear sharp lines that are straight, parallel, curved in the correct way, or any other configuration
paper (and) pencil	called for by the problem.

DRAWING INSTRUMENTS AND THEIR USE 261

drawing board

30 Consider Figure 14.11. To start, we wish to position the paper firmly on the _____ _____.

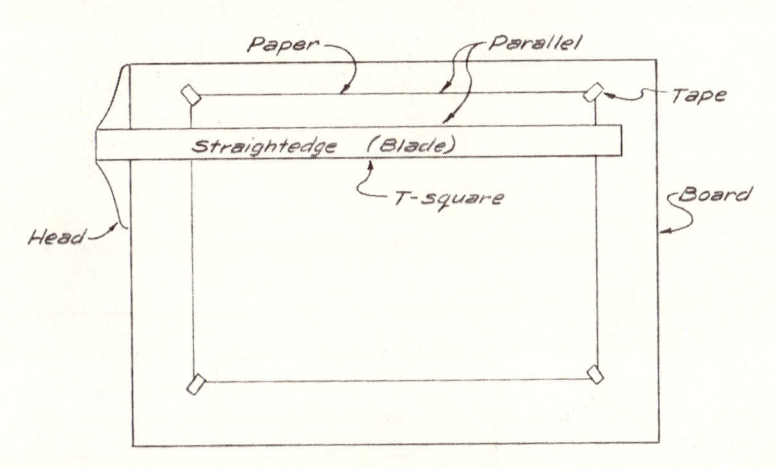

Figure 14.11 Setting up for an instrument drawing.

T-square

31 The _____ is the straightedge used for drawing horizontal, parallel lines.

drafting tape; parallel

32 Apply short pieces of _____ _____ at the corners of the paper to hold it firmly in position on the board. The top edge of the paper is _____ to the top edge of the T-square.

head

33 Never use the bottom edge of the T-square as a drawing edge. It may not be exactly parallel to the top edge.
The _____ of the T-square rides squarely along the side of the board.

parallel

34 Both the head and the side of the board must be straight. If not, the T-square will rock on the board and the horizontal lines will not be _____.

vertical; angular

35 The straightedge of the T-square is also a guide for the triangles. The triangles permit drawing _____ lines and _____ lines at 15, 30, 45, 60, and 75 degrees. (See Figure 14.12A.)

vertically

36 Figure 14.12B shows the correct method for drawing lines against a straightedge. The pencil should be held _____, not jammed into the corner of the T-square and the paper.

262 DRAFTING STANDARDS/UNIT 4

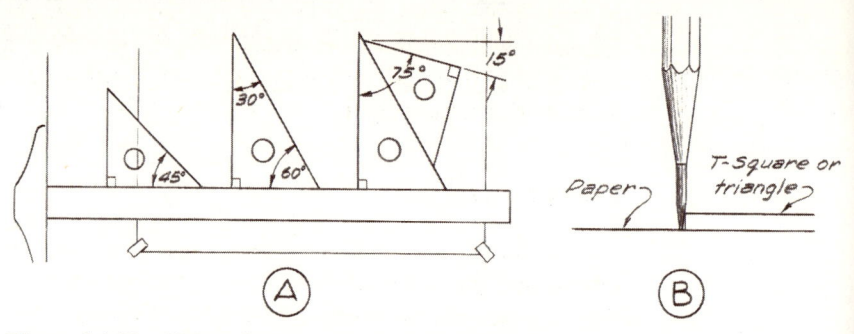

Figure 14.12 Using the T-square and triangles.

left; left

37 Figure 14.13 shows the correct way to work with triangles. Draw on the _____ side and have strong light coming from the _____ side so as not to cast a shadow on the line being drawn.

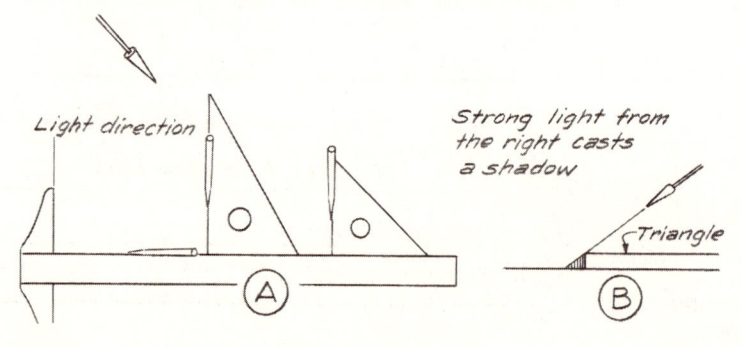

Figure 14.13 Draw on the left side of the triangles and on the top of the T-square.

head; triangle

38 Figure 14.14 shows the use of the left hand when drawing with T-square and triangle. The heel of the hand keeps the _____ of the T-square tight against the edge of the board while the fingers hold the _____ tight against the T-square straightedge.

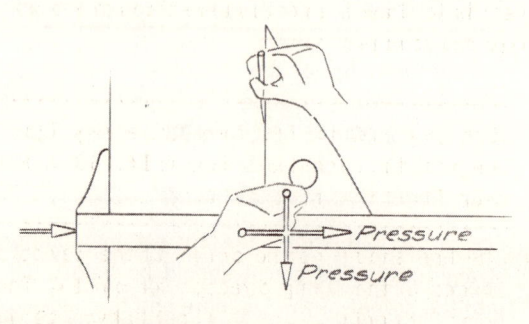

Figure 14.14 Hold the instruments firmly.

DRAWING INSTRUMENTS AND THEIR USE 263

These instructions apply to right-handed persons. Left-handed people can reverse the instructions or modify them to suit their habits. Remember to keep a long, thin conical point on the pencil. Accurate and neat work cannot be done with a blunt point.

The compass

The compass is a versatile tool. It is used for drawing full and partial circles, for transferring distances, and for constructions such as bisecting lines and angles. The compass lead is sharpened differently from the pencil lead. Figure 14.15 shows how to sharpen the lead.

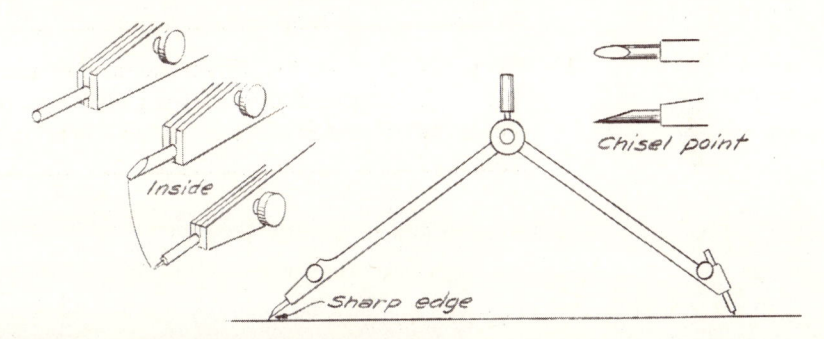

Figure 14.15 Sharpening the compass lead.

chisel

39 File or sand a long, flat _____-shaped point.

Keep the sharp end of the chisel point on the inside so that it will always be in contact with the paper, even for a circle of large radius. (See Figure 14.15.)

French curve

The French curve is used to draw irregular (noncircular) curves. As we stated earlier, this is a plastic instrument with many curves of constantly changing radii. Figure 14.16A shows some points forming an irregular curve. Our problem is to draw a smooth curve through the points, using a French curve to make it instrument perfect.

1, 2, and 3

40 It is good practice first to sketch a very light line through the points to see what the curve looks like. At 14.16B, a portion of a french curve has been found that matches points _____, _____, and _____.

41 The line should not be drawn all the way to point 3. If it is, it will be headed in the wrong direction for point 4. Thus the line is drawn from point 1 to point _____, about halfway between points 2 and 3.

A

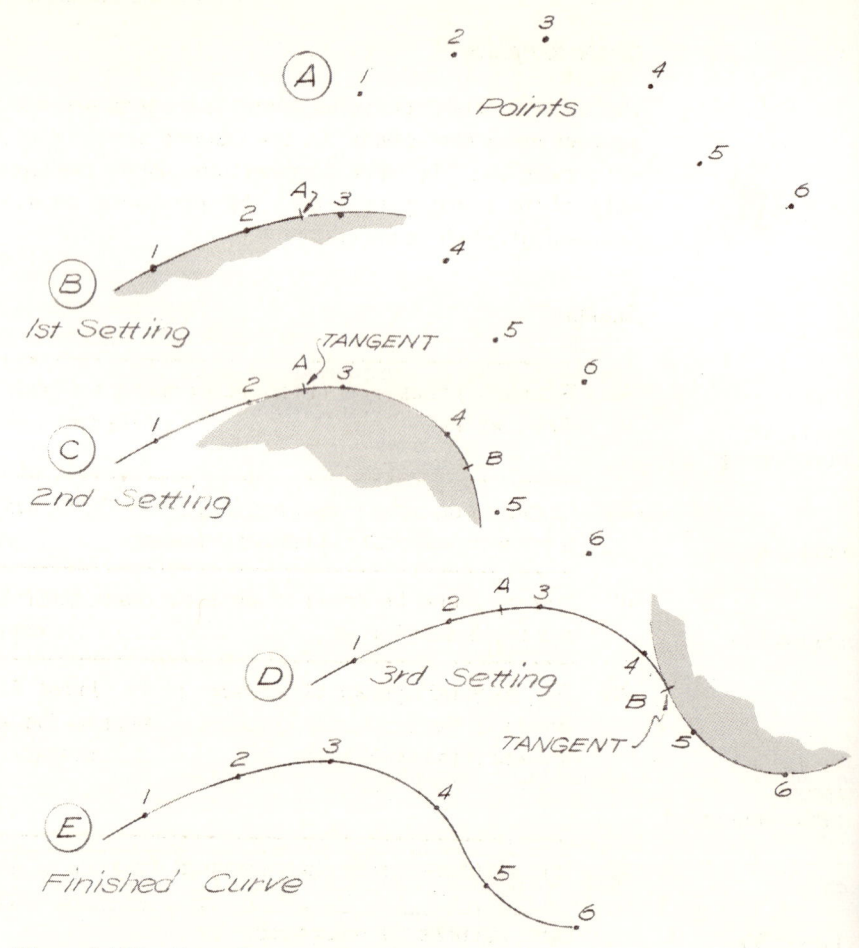

Figure 14.16 Procedure for using the French curve.

smooth	42 Figure 14.16C shows the next step. A new portion has been found that matches points A, 3, and 4. Note that the instrument is tangent to the previously drawn curve at point A. This is an important requirement if the results are to be a _____ curve. Inspection of the curve beyond point 4 shows an apparent point of inflection (change of direction) at B. The present setting seems to meet this point, so a line is drawn from A to B.
tangent; B	43 The final step is shown at Figure 14.16D. A position is found on the instrument that matches points B, 5, and 6 and is _____ to the previously drawn curve at ____.
tangent	44 The last portion of the curve is now drawn. Figure 14.16E shows the finished curve. It is often difficult to find good matches and smaller steps must be taken. On the other hand, sometimes you're lucky and find a portion of the instrument that will match every point. The important consideration is to make the beginning of a new step _____ to the curve already drawn.

Circle template

The circle template (or radius guide) is a useful tool and replaces the compass for many applications. Its use requires some skill, but a bit of practice will provide this. The circle diameters are slightly oversize to account for the width of the pencil point. A long, thin point must be used, however, or the resultant circle will be undersize.

Scales

measuring, proportioning

45 The many scales used in technical drawing are important for _____ distances and _____ distances.

proportion

46 Chapter 6 discussed scale and proportion. Of the two, _____ is the more important in technical drawing.

information

47 A drawing may be drawn to any scale (size), but if the dimensions are out of proportion, true _____ is not communicated.

inch-foot; metric (centimeter-meter)

48 The standard system of measure in the United States and in Great Britain is the _____-_____ system. The standard in most of the rest of the world is the _____ system.

$\frac{1}{64}$; $\frac{1}{32}$; $\frac{1}{16}$; $\frac{1}{8}$; $\frac{1}{4}$; $\frac{1}{2}$

49 The inch-foot system has standard fractions of an inch. These are _____, _____, _____, _____, _____, and _____. No other fractions are recognized.

decimal

50 If you wish to express a distance that lies between two of these fractions, you must change to a(n) _____ division of the inch.

architect's; mechanical, civil

51 Thus 1.0700 lies between 1$\frac{1}{16}$ (1.0625) and 1$\frac{5}{64}$ (1.0756).
There are three different scales used in technical work: the _____ scale, the _____ engineer's scale, and the _____ engineer's scale. (Refer to Figure 14.17.)

feet

52 The architect's scale changes _____ into fractions of an inch.

$^{20}\!/_4$ or 5

53 Thus, $\frac{1}{4}$″ = 1′-0″ is a typical architect's scale. A 20-foot building would appear as _____ inches on the drawing.

fractions

54 The mechanical engineer's scale changes inches into _____ of an inch.

$^{6}\!/_4$ or 1.5 ft. (18 in.)

55 $\frac{1}{4}$″ = 1″ (one-quarter scale) is a typical mechanical engineer's scale. A 6-ft-high object would be _____ on the drawing.

266 DRAFTING STANDARDS/UNIT 4

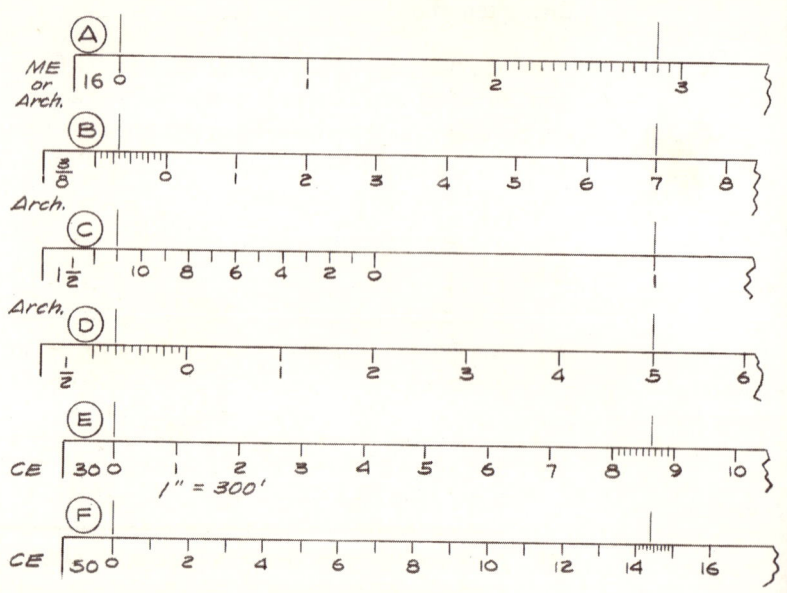

Figure 14.17 An exercise in measuring distances.

inches	**56** The civil engineer's scale divides _____ into decimal or multiple decimal (20, 30, 40, 50, 60) units.
(CE) 30-	**57** With the civil engineer's scale you can measure any scalar or vector quantity. Thus 1″ = 30 ft, 1″ = 300 lb, 1″ = 3000 ft per sec (fps) can all be measured on the same instrument. It is the CE _____-scale.
(A) 2⅞″ (B) 7′-8″ (C) 1′-11″ (D) 5′-9″ (E) 865 (F) 14.4 (1.44, 144)	**58** Figure 14.17 shows six scales measuring the distance between two vertical lines. What is the reading on each? (A) _____ ; (B) _____ ; (C) _____ ; (D) _____ ; (E) _____ ; (F) _____ .

The use of a scale for proportioning distances is a valuable technique to learn and use. Many times it is necessary to divide a line or distance into some uneven divisions. This can be simply done by laying an appropriate scale across the line and marking off the desired divisions.

3, 7, and 21 units	**59** Figure 14.18A shows a distance between two vertical lines and a scale laid across this distance. At the setting shown what unit subdivisions of the space can be easily made? _____

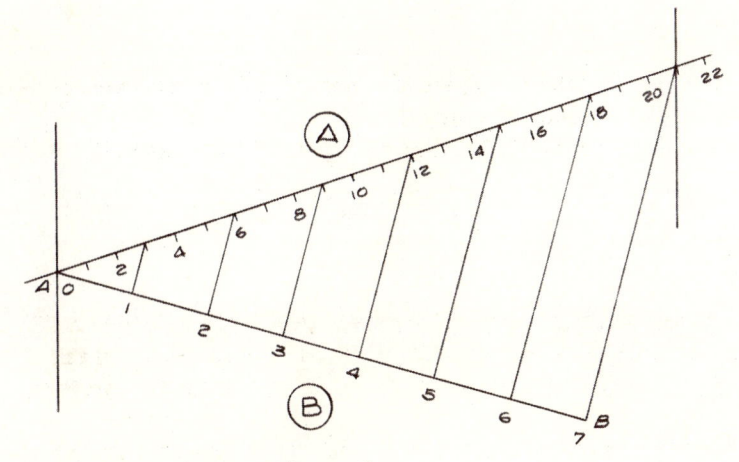

Figure 14.18 Proportioning with a scale.

---

60 Suppose our intent was to subdivide line *AB* into seven equal sections (see 14.18*B*). We could find these divisions by constructing six lines _____ to line *B*-21.

parallel

This has been but a brief account of drawing instruments and their use. It is just an introduction to the devices themselves. Skill in any art—including instrument drawing—comes through *practice*.

GRAPHICS
The Geometry of Technical Drawing

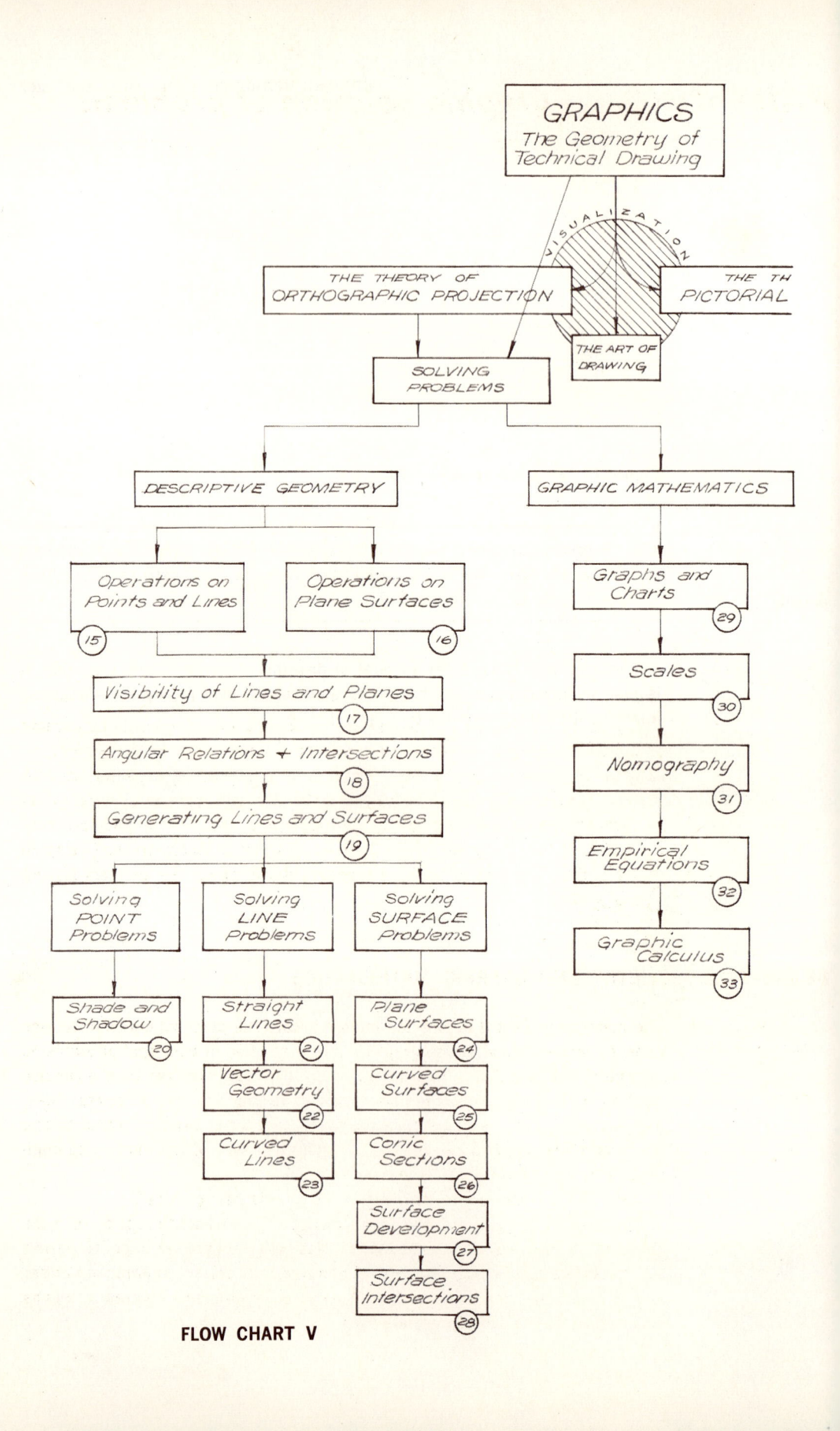

VISUALIZATION

THE THEORY OF ORTHOGRAPHIC PROJECTION

THE THEORY OF PICTORIAL

THE ART OF DRAWING

SOLVING PROBLEMS

DESCRIPTIVE GEOMETRY

- Operations on Points and Lines (15)
- Operations on Plane Surfaces (16)
- Visibility of Lines and Planes (17)
- Angular Relations + Intersections (18)
- Generating Lines and Surfaces (19)

- Solving POINT Problems
  - Shade and Shadow (20)

- Solving LINE Problems
  - Straight Lines (21)
  - Vector Geometry (22)
  - Curved Lines (23)

- Solving SURFACE Problems
  - Plane Surfaces (24)
  - Curved Surfaces (25)
  - Conic Sections (26)
  - Surface Development (27)
  - Surface Intersections (28)

GRAPHIC MATHEMATICS

- Graphs and Charts (29)
- Scales (30)
- Nomography (31)
- Empirical Equations (32)
- Graphic Calculus (33)

FLOW CHART V

PART TWO graphic solution of problems

INTRODUCTION

The Introduction to Part I presented two areas of the communications aspects of technical drawing—personal communications and organizational communications. In that part of the text, Chapters 1 to 14, we were concerned with the use of graphics in *picturing objects* either through *realistic pictorial forms* or through the *abstract but precise orthographic form*. These chapters covered all but two items on Flow Chart II: Descriptive Geometry and Graphic Mathematics. The rest of the text presents these topics, the *problem-solving* side of graphics.

DESCRIPTIVE GEOMETRY AND GRAPHIC MATHEMATICS

The material of Part II is directed mainly toward the personal communications aspect. Here we learn to use graphics in the solution of both spatial problems and numerical problems. In Unit 5, Descriptive Geometry, we have a precise method for solving space problems through further analysis of points, lines, and surfaces which is *descriptive geometry*. In Unit 6, Graphic Mathematics, we have a means of displaying and analyzing numerical information through graphic manipulation which is *graphic mathematics*.

Flow Chart V shows the detailed topics in this new material.

You, the reader, will not find this a completely detailed discussion of these subjects. It covers the important principles well enough, we hope, to permit you to understand orthographic projection and mathematic construction better and to use them confidently and skillfully to solve problems by graphic means where such means are appropriate.

DRAWINGS AND THE NEW DESIGN

For picturing objects the designer needs facility in translating mental images into three-dimensional reality. He first tries his ideas on paper to see if they will work. Drawings and sketches, although not positive proof of a successful design, can give valuable information as to the operational feasibility and dimensional details of the design. Perhaps a part of the physical arrangement of the new design involves the intersection of a cone and a plane. What is the exact line of intersection? If the two objects can be drawn in two orthographic views, the lines of intersection can be readily found by using the rules of descriptive geometry.

EVALUATING AND TESTING A NEW DESIGN

Once a design is put into the prototype or working-model stage, many tests must be run to compile numerical data. Most frequently the data are plotted on graph paper, and curves are drawn and analyzed. Here graphic methods of displaying data are quickly revealing. They show what new tests should be made and what data should be analyzed by computer techniques for a still more precise answer—or perhaps they show that the original design is successful without further study. Suppose an auto designer obtains test data for a new model showing the velocity attained from a zero start during a given time period (20 seconds). This data is shown graphically at (a) in the accompanying illustration. Through a knowledge of graphic calculus, he can quickly determine the distance traveled (b) and the acceleration pattern (c) over the same time period.

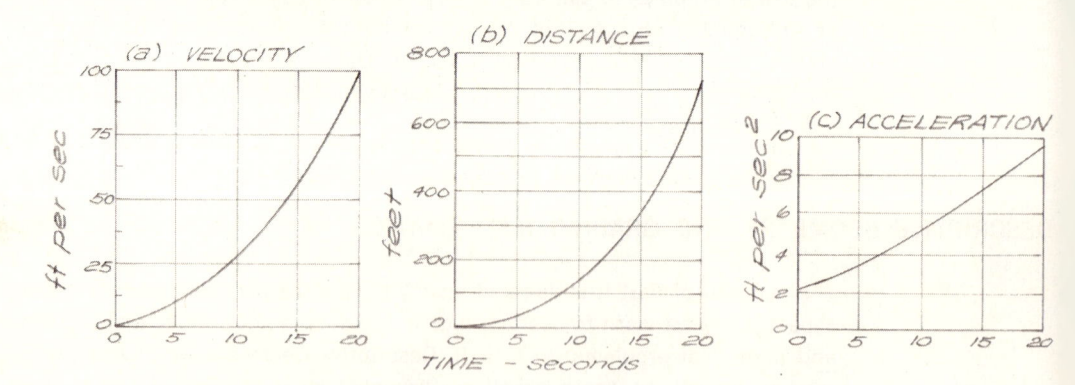

It is true that computers can do this more accurately. The important point here is that an ability to translate numerical information easily into a graphic display is important both for the quick, first-impression analysis of the data and for the formal presentation of final results in reports, papers, slides, and so forth.

UNIT 5 *descriptive geometry*

Descriptive geometry is the precise and unambiguous solution of space problems by orthographic projection techniques. Study of this subject shows you how to select and project auxiliary views to obtain useful information about an object positioned in space.

Chapters 15 through 18 present the basic operations of descriptive geometry. They show the use of the orthographic projection system for operating on points, lines, and plane surfaces in space. These four chapters in conjunction with the preceding fourteen chapters in Part I will enable you to gain precision in the orthographic projection techniques that you will find useful in describing objects graphically.

Chapters 19 through 28 relate the discussion of points, lines, and both plane and curved surfaces to more advanced ideas.

chapter 15   *graphic operations on points and lines*

In our study of descriptive geometry we will be working with abstract points, lines, planes, and curved surfaces. Any physical object can be considered to be made up of various combinations of these four concepts. We will begin the subject of descriptive geometry with a study of points in space.

FREEHAND SOLUTIONS FOR SKETCH EXERCISES

The sketch exercises in descriptive geometry should be executed by freehand sketching techniques even though the subjecet matter demands precise instrument construction. We are seeking understanding first and technique second. Proficiency in the approximate solution of problems by freehand techniques is a great timesaver. A quickly drawn freehand solution can provide the basis for either proceeding immediately to a rigorous computer analysis or abandoning that particular approach as being worthless.

The freehand techniques of importance are sketching a straight line parallel to a given line or given direction, erecting perpendiculars, and estimating distances to a known scale. These techniques were presented back in Chapter 2. A quick review of that material will be useful in solving the problems of this chapter.

GRAPHIC OPERATIONS ON POINTS AND LINES 273

POINTS

automobile	1	Do you remember the first drawing in this text? It was a dot-to-dot drawing. By connecting 55 numbered dots in sequence, you sketched a pictorial view of a(n) _____.
where	2	At that time we made the statement, "The main purpose of this text is to show you _____ to place the dots to draw any three-dimensional object on a two-dimensional surface."
8; 12; 6	3	Consider the cube that we have used extensively in the chapters on the theory of pictorial drawing (see Figure 15.1). It is made up of _____ points, _____ lines, and _____ planes.

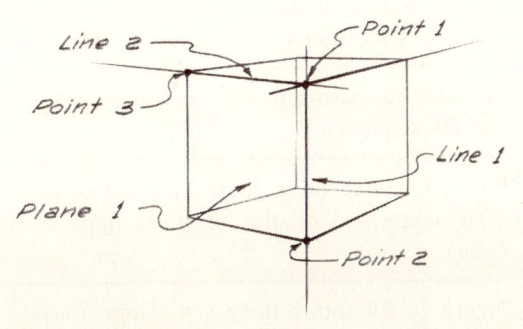

Figure 15.1 Points, lines, and planes combine to define a visual impression of a cube.

three (3)	4	A point is hard to define. It is a location in space and it is also nothing. You don't think of holding a point in your hand. In geometric terms, a point is the intersection of two lines. Point 1 in Figure 15.1 is actually a location in space where _____ lines come together.
line	5	If we create a second point (point 2 in Figure 15.1) we also create a(n) _____.
intersects	6	Adding point 3 creates a second line (line 2) which _____ line 1.
plane; three	7	Two intersecting lines (lines 1 and 2) define a(n) _____. In this example the intersecting lines were defined by _____ points.
triangular	8	Figure 15.2 shows the same chain of reasoning in a more abstract manner. The plane segment resulting from the placing of three points in space is _____ in shape.

274 DESCRIPTIVE GEOMETRY/UNIT 5

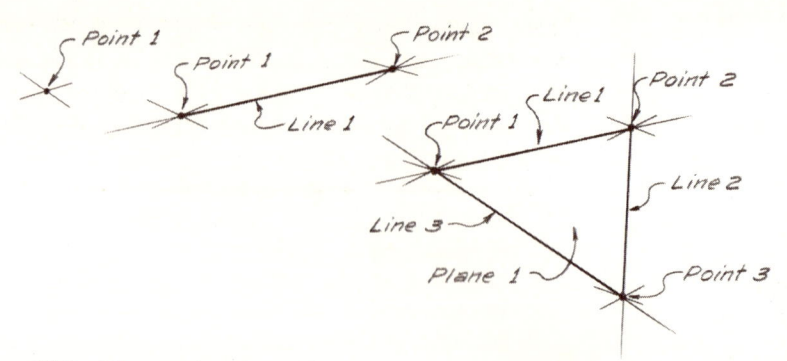

Figure 15.2 Three points in space define a plane surface.

8; cube	**9**	The cube of Figure 15.1 was built by first defining _____ points in space. Each of these points must have a definite relationship with all the others in order that the array of points define a(n) _____.
12	**10**	The second step is to join the points with _____ straight lines in a definite sequence.
planes	**11**	By their intersections, these lines define six _____ which, by their shape and relative positions, give us the visual impression of a cube.
½; ¾; 1	**12**	Figure 15.3A shows three orthographic views of a point (A). Point A is precisely placed within the frame of reference. A pictorial view of A is shown at 15.3B. Point A is: _____ in. to the left of the profile reference plane _____ in. below the top reference plane _____ in. behind the front reference plane

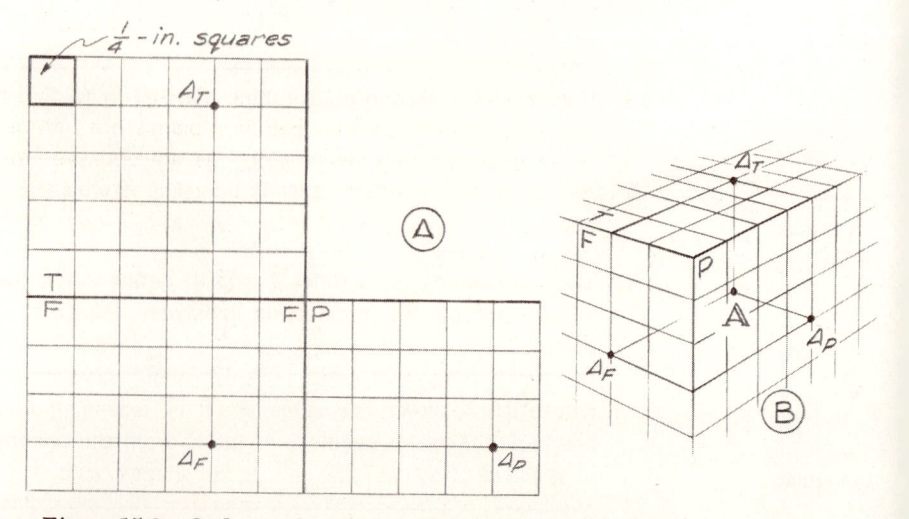

Figure 15.3 Orthographic views of point A, a point in space.

GRAPHIC OPERATIONS ON POINTS AND LINES 275

Directions in Space

These figures bring forth an important consideration that is essential to visualization and to the full understanding of the orthographic projection system. That consideration is the relationship of common, everyday directions such as left–right, up–down (above–below), and front–behind (forward–backward) to the frame of reference. We are dealing with objects in space and these are the words we use to describe directions in space.

13 The front view of any object shows the dimensions _____ and _____ of the object. The top view shows dimensions _____ and _____, and the profile view shows _____ and _____.

H and W;
W and D;
D and H

14 Relating the three dimensions of H, W, and D to such commonly used words as left, right, up, down, front, back, etc.:
H describes _____ and _____.
W describes _____ and _____.
D describes _____ and _____.

up (and) down;
left (and) right;
front (and) back (behind)

15 In describing the position of point A in Figure 15.3, we use the individual reference _____ of the space frame of reference as guides for establishing directions.

planes

16 Point A is _____ the top plane, _____ the front plane, and _____ of the profile plane.

below; behind;
left

17 It is important to remember that the three reference planes have a 90° relation with each other. This can be seen in the pictorial view of Figure 15.4A. When we look orthographically at any one of the reference planes we see two others as _____ views.

edge (straight line)

18 Figure 15.4B is the orthographic drawing of 15.4A. Here the lines of intersection between the reference planes are interpreted as the edge views depending on which plane we are considering. When looking at the front view, the F/T line is an edge view of the _____ reference plane and the F/P line is an edge view of the _____ reference plane.

top; profile

19 Similarly, when looking at the top view:
(1) T/F line is edge of _____ plane.
(2) T/P line is edge of _____ plane.
When looking at the profile view:
(3) P/T line is edge of _____ plane.
(4) P/F line is edge of _____ plane.

front; profile;
top; front

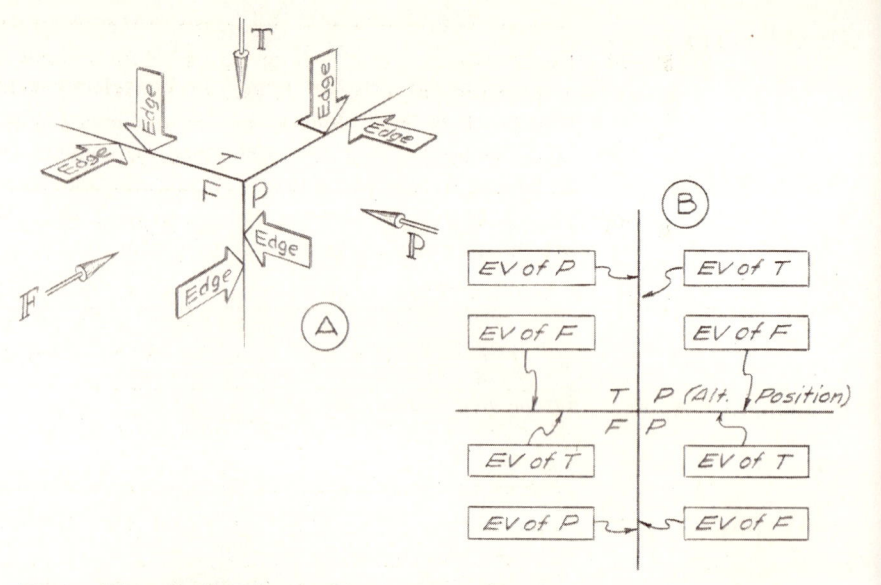

Figure 15.4 Each orthographic reference plane has a 90° relationship with any adjacent reference plane.

Reference Line Nomenclature

Understanding the role of the reference lines (lines of intersection between the reference planes) aids in mentally visualizing the pictorial equivalent of an orthographic drawing.

Before proceeding with a discussion of points, lines, and planes let's establish a system of nomenclature. We use the abbreviations, *T, F, P, R,* and *B,* to label the six principal reference planes. We may drop the subscripts denoting the left and right profile views unless both views are discussed and the subscripts are needed to distinguish between them.

edge	**20** The reference lines will be described by the symbol *T/F, F/T, F/P,* etc., where the first letter denotes the view we are considering and the second denotes the adjacent view which, in space, is actually a(n) _____ view.
view	**21** Points will be shown by a capital letter. Orthographic views of points will be a capital letter and a subscript such as A_T, A_F, M_P, X_B, etc. The subscript denotes the particular _____ of the point that we are describing.
subscript	**22** It is important to understand that point "A" without subscript appears only in a pictorial view. (See Figure 15.3*B*.) It is a particular point in space. An orthographic view is just one particular representation of the point and should carry a(n) _____ (Figure 15.3A).
2; 3	**23** Auxiliary views will be numbered consecutively starting with number 1. A notation of "the 2/3 line" would mean "we are looking at auxiliary view _____ and auxiliary view _____ appears as an edge view."

GRAPHIC OPERATIONS ON POINTS AND LINES 277

■ 24   Sketch 44A is a duplicate of Figure 15.3A showing a three-view orthographic drawing of point A. Add three views of point B to this sketch. B is 1½ in. to the left of P (profile reference plane), 1 in. below T and ¼ in. behind F. Label each view. Check your solution.

■ 25   The addition of point B creates a line AB in space. Sketch this line in three views. A pictorial view of line AB is given in the sketch.
Next, add three views of point C. C is 1 in. to the left of P, it is in plane T (0 in. below T), and is 1¼ in. behind F. Label all views and join the points with straight lines. Check your solution.

plane

26   The addition of point C creates a(n) _____, ABC. A pictorial view of ABC is given.

reference planes

27   The position of points A, B, and C were given as perpendicular distances from the three principal _____ _____.

28   We could have located points B and C from point A as long as our measurements were taken perpendicular to the three reference planes.
Point B is:
1 in. _____ of A; ¼ in. _____ A; and ¾ in. _____ of A.
Point C is:
¼ in. _____ A; ¾ in. _____ A; and ½ in. _____ of A.

point B:
left; below; (in) front
point C:
behind; above; left

Sketch 44A shows that the orthographic frame of reference becomes a precise tool for picturing an object in space.

Figure 15.5A shows the orthographic views of another point A. Suppose we wish to view point A in the direction of arrow 1, shown in T and F views. This is a specific direction. The front view shows that the line of sight is horizontal. The top view shows it to be from the right and slightly to the rear. This will be an auxiliary view. We know that the auxiliary reference plane needed to record view 1 must be perpendicular to our line of sight (orthographic viewing).

A requirement of the orthographic projection system dictates that the auxiliary reference plane must also be perpendicular to one of the given views. Since the line of sight is horizontal (parallel to the top plane), auxiliary plane 1 is hinged from the top plane (see Figure 15.5B). This is the only condition that will permit plane 1 to be perpendicular both to the top plane and to the chosen line of sight. Had we selected an oblique line of sight (not parallel to either top or front), we would have compounded the problem and necessitated a second auxiliary to find the desired view. (More of this later.)

278 DESCRIPTIVE GEOMETRY/UNIT 5

Figure 15.5 A demonstration of the basic measuring system in descriptive geometry.

auxiliary view 1	**29** Unfolding the frame of reference, we obtain, at Figure 15.5C, a three-view orthographic projection of point A consisting of the standard top and front views and _____ (your words).	

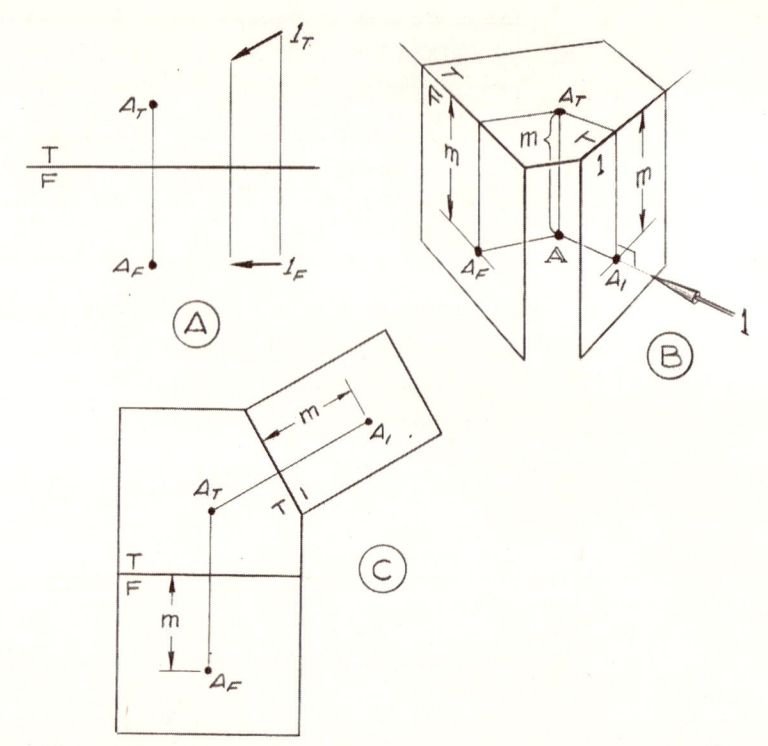

We had no need to take view 1 since it added no new information to the drawing. Figure 15.5 illustrates the *measuring system* used in orthographic projection and in the use of auxiliary views in descriptive geometry.

m	**30** In the pictorial of Figure 15.5B, we see that point A is a distance _____ below the top plane.
below	**31** This distance shows up in both view F and view 1 since, in these views, distances measured perpendicular to both the F/T line and the 1/T line, represent distances _____ the top plane. (*Choose one: above/below.*)
top; F/T; 1/T	**32** Thus, the distance *m* shows up as: (1) The true perpendicular distance of point A from the _____ plane. (2) The shortest distance (perpendicular) of A_F from the ____/____ line. (3) The shortest distance of A_1 from the ____/____ line.

GRAPHIC OPERATIONS ON POINTS AND LINES   279

90

33 The reference lines are actually folding lines between two adjacent reference planes and represent a(n) _____-degree relationship between the two planes.

equal

34 Thus, if we have any three adjacent planes, measurements made on the two "outside" planes in a perpendicular direction away from the reference lines will always be _____. (See Figure 15.6.)

This is important information. We have said that, given any two orthographic views, we can project as many auxiliary views as we wish. Thus, starting with two views (usually the top and front), the first auxiliary will be the third view of a three-view system and measurements are available to show distances away from the center view. Figure 15.6 shows this reasoning applied to a hypothetical situation involving auxiliary views 12, 13, and 14 of point B. Views 12 and 14 show that point B is a distance n away from reference plane 13.

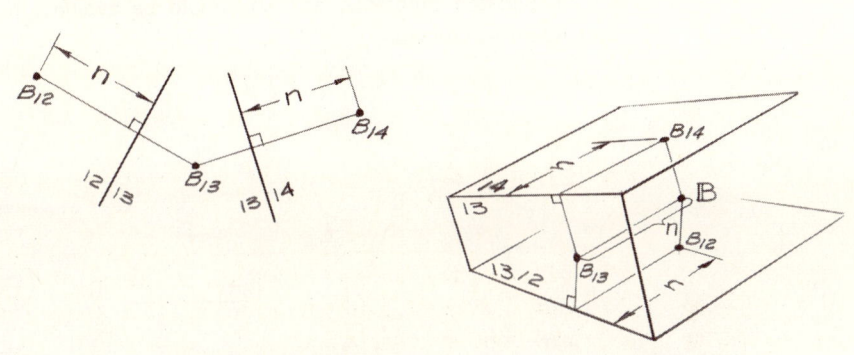

Figure 15.6 The three-view measuring system may be applied to any three adjacent orthographic views.

projection line

35 Figure 15.7 shows the orthographic solution of the simple problem of Figure 15.5 using standard techniques of projection and measurement. At 15.7A we sketch a(n) _____ _____ in the direction of the desired view (shown by arrow).

perpendicular

36 At Figure 15.7B we sketch the T/1 reference line a convenient distance from A_T and _____ to the projection line. This line separates the top reference plane and auxiliary reference plane 1.

m; F/T

37 Looking at the front view we can see that point A is a distance _____ below the top plane. This is shown by the perpendicular distance between A_F and the _____/_____ line.

38 Figure 15.7C is the final step in finding the auxiliary view of point A. The distance m is measured from the _____/_____ line along the projection line. The result is a third view of point A. It should be labeled

1/T; A_1

_____.

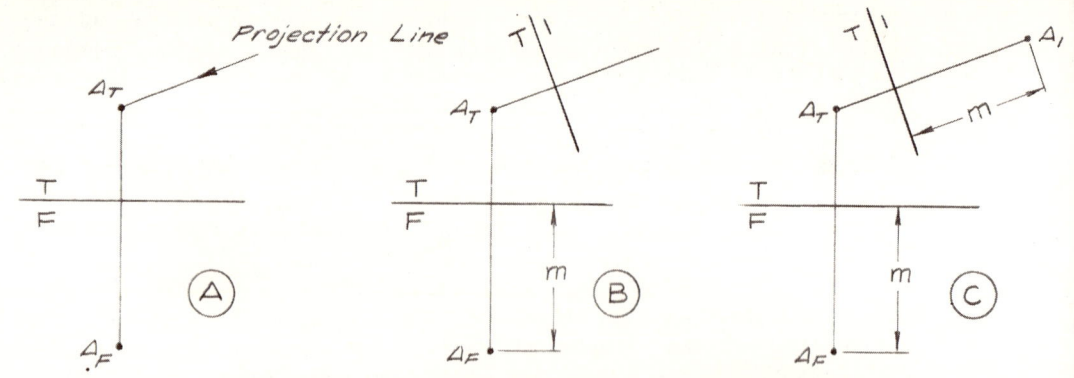

Figure 15.7 Constructing an auxiliary view of point A.

The placement of the T/1 reference line in Figure 15.7B was an arbitrary choice dictated solely by convenience. Figure 15.8 shows two other positions of the T/1 line. At 15.8A the T/1 line was placed at A_T. This can be confusing on complicated problems, and so should be avoided. Notice in the pictorial that A and A_1 coincide.

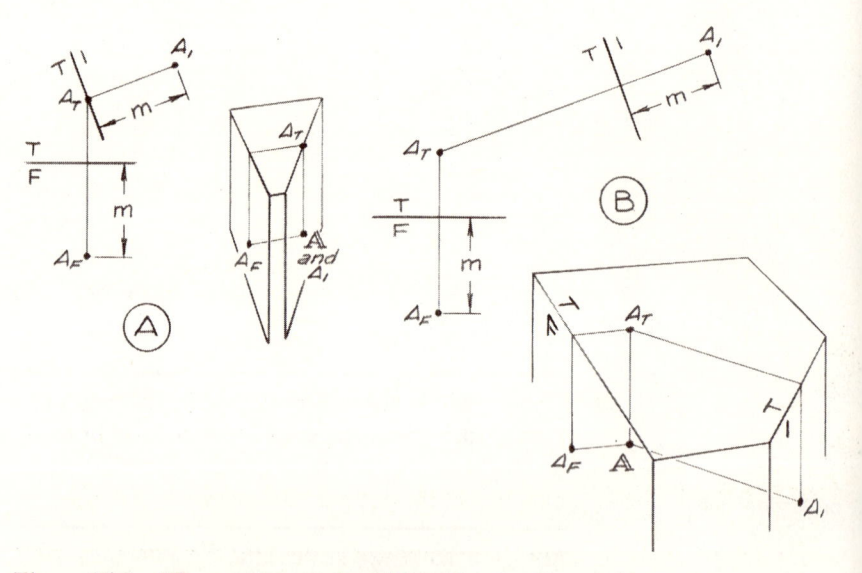

Figure 15.8 The position of the folding line for the auxiliary reference plane is arbitrary.

At 15.8B, the T/1 line has been moved far away from A_T. This wastes paper. A good choice is to position the the new reference line about ¼ to ½ in. from the object (A_T). Note that in both cases, none of the three views have changed except in their distances with respect to each other.

The actual transfer of the distance m in Figures 15.7 and 15.8 can be made in a number of ways. We are not interested usually in the actual distance in inches, centimeters, etc. Thus, as shown in Figure 15.9, we can use

GRAPHIC OPERATIONS ON POINTS AND LINES 281

dividers, any scale, the edge of a piece of paper, or—one of the best forms of practice for freehand sketching—estimate the distance. The use of dividers is the most accurate method.

39 We had no good reason for finding an auxiliary view of point A other than to illustrate the techniques and procedures. The creation of view 1 added no new _____ to what we already knew about point A, a particular point in space.

information

The foregoing discussion of single points is intended to show the projection and measuring techniques needed to gain any desired auxiliary view of a point. The assumption is that, if these techniques work for one point, they will work for two or more points together. Two points define a straight line. Adding a third creates two intersecting straight lines and thus a plane.

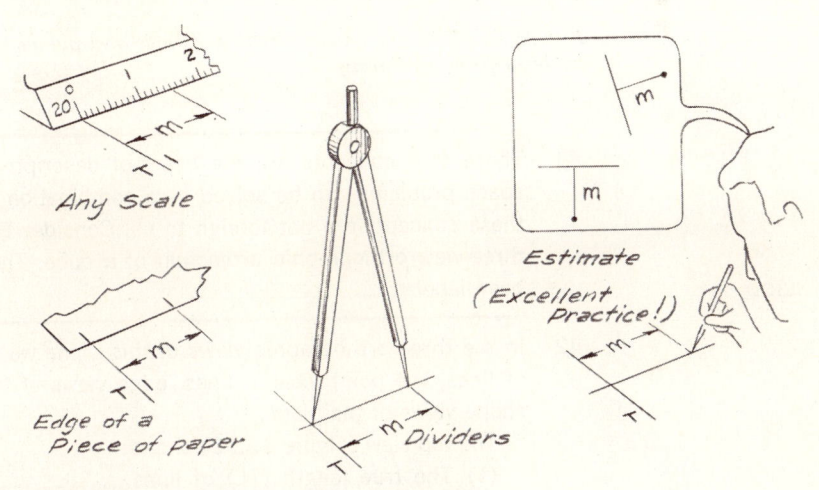

Figure 15.9 Methods for transferring distances.

LINES AND PLANES

40 Let's proceed to the discussion of lines and planes. Here we have good reason to take auxiliary views. Figure 15.10 shows four pieces of information about lines and planes that can be of great use to us. They are:
(1) The _____ _____ view of a line.
(2) The _____ view of a line.
(3) The _____ view of a plane.
(4) The _____ _____ view of a plane.

true-length *(TL)*;
point *(PV)*;
edge *(EV)*;
true-shape *(TS)*

282 DESCRIPTIVE GEOMETRY/UNIT 5

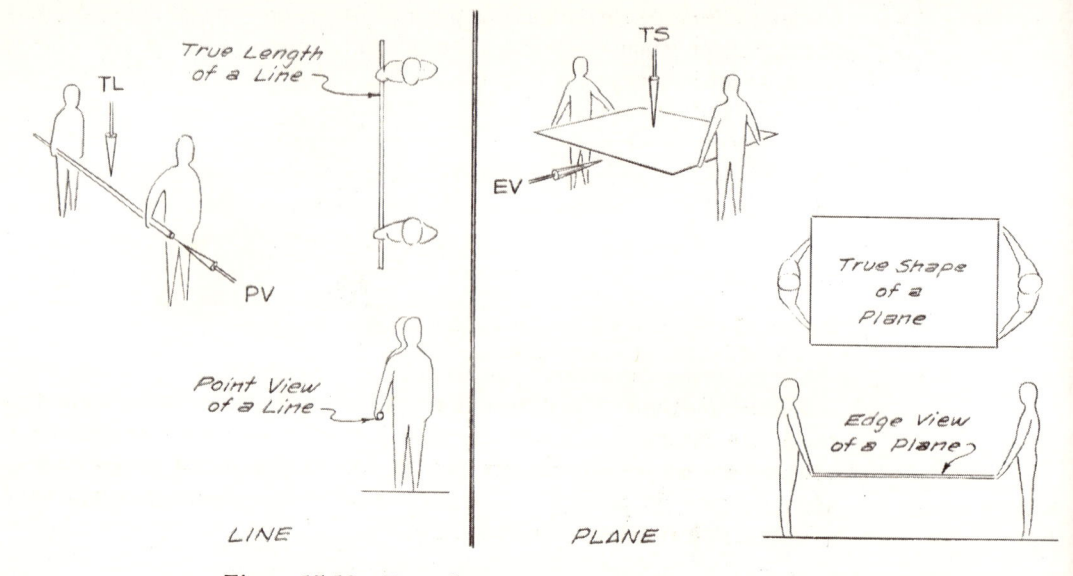

Figure 15.10 Four characteristics of lines and planes that are important in descriptive geometry.

ABCD	**41** These four concepts are the heart of descriptive geometry. Almost all space problems can be solved by a combination of these ideas. Actually these concepts are not foreign to us. Consider Figure 15.11. It shows a three-view orthographic projection of a cube. The square top plane has been labeled _____.
(TL): AB, BC, CD, and DA; (TS): ABCD	**42** In the three orthographic views of this plane we can see the true length of lines, the point view of lines, edge views of the plane, and the true-shape views of the plane. The top view (Figure 15.11) shows: (1) The true length *(TL)* of lines _____, _____, _____, and _____. (2) The true shape *(TS)* of plane _____.
(PV): AD and BC; (TL): AB and DC; (EV): ABCD	**43** The front view shows: (1) The point view *(PV)* of lines _____ and _____. (2) The true length *(TL)* of lines _____ and _____. (3) The edge view *(EV)* of plane _____.
(PV): AB and CD; (TL): BC and AD; (EV): ABCD	**44** The profile view shows: (1) The point view *(PV)* of lines _____ and _____. (2) The true length *(TL)* of lines _____ and _____. (3) The edge view *(EV)* of plane _____.

GRAPHIC OPERATIONS ON POINTS AND LINES 283

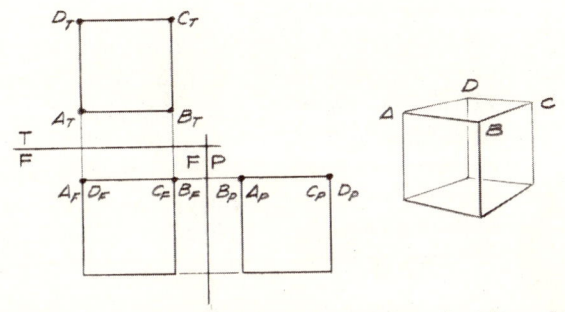

Figure 15.11 Any one face of a cube exhibits the TL and PV of lines and the EV and TS of the plane.

Thus, from the simple case of the cube in Figure 15.11, you can see that these four important characteristics of lines and planes are quite familiar to us. Our concern will be how to find this information in the general case of oblique lines and planes in space. The rest of this chapter analyzes lines. Chapter 16 does the same for plane surfaces.

ANALYSIS OF LINES

True Length of a Line

45 We will consider the true length of a line first. Hold your pencil up in front of your eyes. When do you see its true length? (See Figure 15.12.) When the pencil is _____ to your line of sight you see its *TL*.

perpendicular

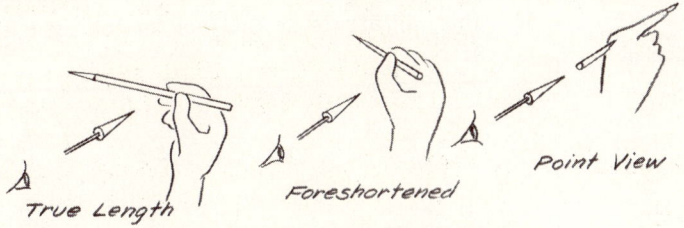

Figure 15.12 Visualizing the true length (TL) of a line.

46 In the orthographic system, our line of sight is *always* perpendicular to the reference planes. Thus if we position the pencil (or a line) in the frame of reference so that it is _____ to one of the reference planes, the projection on that plane will be a(n) _____ view of the line.

parallel;
true-length *(TL)*

47 In Figure 15.13, a pencil is positioned parallel to the front reference plane. Its projected image on the front plane is a *TL* view because our line of sight is perpendicular to the front plane and thus perpendicular to the _____.

pencil

284 DESCRIPTIVE GEOMETRY/UNIT 5

edge view	48	We know the pencil is parallel to the front plane because, when looking at the top view we see that the pencil is parallel to the T/F line. The T/F line represents a(n) _____ of the front reference plane.
inclined	49	The top view is not a true-length view because the pencil is _____ to the top plane resulting in a foreshortened view.

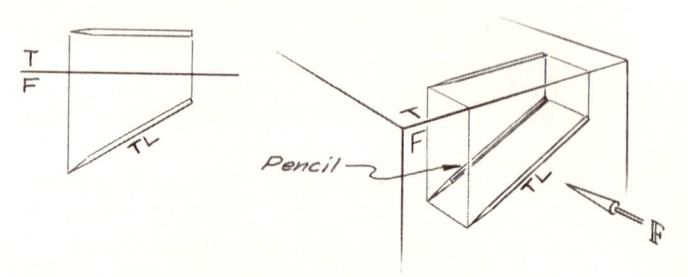

Figure 15.13 An orthographic true-length view.

T/F; F/T	50	Figure 15.14 shows the top and front views of an oblique line, AB. Neither view is a TL view because AB is not parallel to either of the two reference planes. This is evident because $A_T B_T$ is not parallel to the _____ line and $A_F B_F$ is not parallel to the _____ line.
parallel	51	We wish to find the TL of AB. We know that we must view it perpendicularly in order to see the TL. Therefore, let's create an auxiliary view that is _____ to AB.
perpendicular	52	This will assure us that our line of sight is _____ to AB.

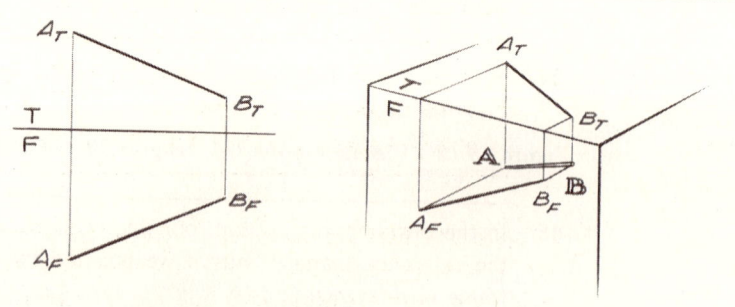

Figure 15.14 An oblique line—orthographic and perspective views.

Figure 15.15 shows, in steps, how we do this. First, at 15.15A, we sketch two projection lines from A_F and B_F perpendicular to $A_F B_F$. These represent our lines of sight in the orthographic viewing system.

perpendicular	53	Next, at Figure 15.15B, we sketch the new F/1 line a convenient distance from $A_F B_F$ and _____ to the projection lines of the first step.

GRAPHIC OPERATIONS ON POINTS AND LINES 285

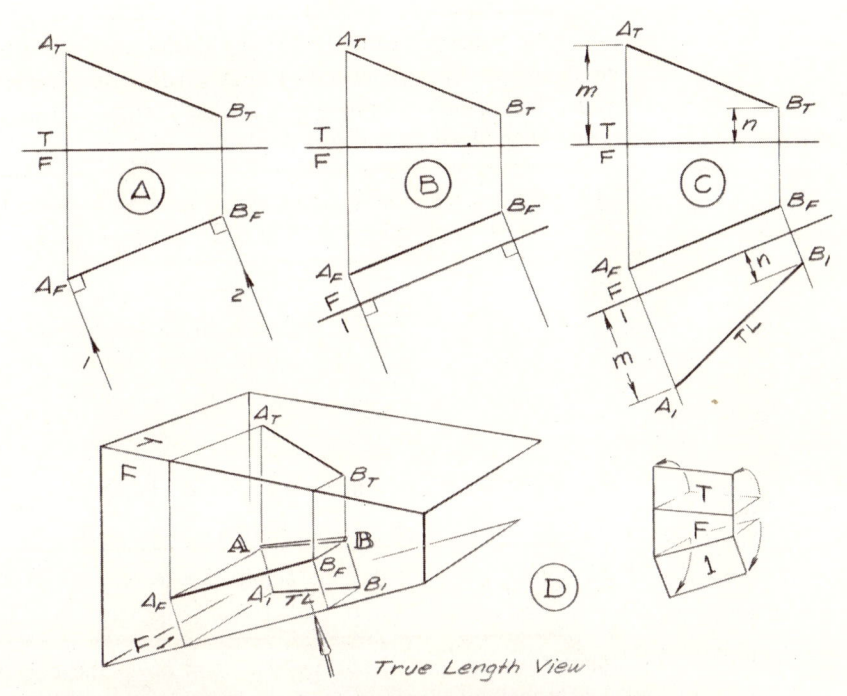

Figure 15.15 Finding the true length of the oblique line AB.

	54	It only remains now to measure distances along the projection lines to find points A_1 and B_1. Distances measured from the 1/F line into view 1 (front and back distances, that is) represent distances _____
behind		the front reference plane.
	55	These same distances are found in the top view measured from the T/F reference line. (Figure 15.15C.) Point A is _____ distance behind F and
m; n		point B is _____ distance behind F.
	56	Transfer these distances to view 1 along the appropriate projection lines and the problem is complete. (See Figure 15.15C.) A_1B_1 is a TL view
parallel		of line AB because auxiliary reference plane 1 is _____ to AB.

Figure 15.15D is a pictorial sketch of the completed problem. Auxiliary view 1 was projected from the front view. This was an arbitrary choice. It could have been projected from the top view with the same results, the true length of AB.

■ 57 Sketch 44B is a sketch exercise in finding the TL of a line. Analyze each problem and mark the TL if it exists in the two given views or find the TL by an auxiliary view. Use a sketching technique. To be as accurate as possible, be careful in (1) sketching parallel lines, (2) sketching perpendicular lines, and (3) estimating distances. Work quickly but deliberately. Check your solutions.

Point View of a Line

In Figure 15.12 you looked at a pencil to determine what view would show you a true length. Do the same now to analyze how we can see the pencil as a single point. Consult Figure 15.16.

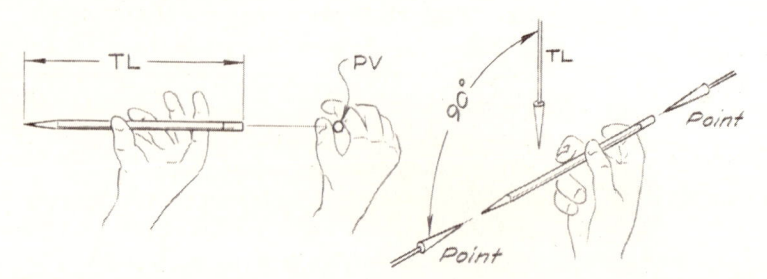

Figure 15.16 Visualizing the point view of a line.

---

58 If we hold a pencil perpendicular to our line of sight we see it in its _____.

true length

---

59 Turn the pencil so that you look down the end, that is, you see the pencil as a circle (or hexagon). The simplest movement required to change from a *TL* view to an end view is a 90° rotation. This fits the orthographic scheme well since any two adjacent views have a(n) _____-degree relationship in space.

90(-degree)

---

60 Figure 15.17A shows the *T* and *F* views of a line *AB*. The *T* view is a(n) _____ view because, as shown in the *F* view, both *A* and *B* are a distance *m* below the top plane and thus *AB* is _____ to *T*.

TL; parallel

Figure 15.17 Finding a point view (PV) of a line.

GRAPHIC OPERATIONS ON POINTS AND LINES 287

61 From our reasoning on how to see the point view of a pencil, we should see line AB as a point if we view in the direction of the two arrows of 15.17A.

Figure 15.17B shows the start of the procedure for creating an auxiliary view that will let us look in the direction of arrow 1. The first step is to extend the TL view by sketching a(n) _____ _____ in the desired direction.

projection line

62 Next we sketch the T/1 line perpendicular to the projection line a convenient distance away from B_T. (See Figure 15.17C.) This establishes the _____-degree relationship that we need.

90

63 Both A and B are projected along a single projection line. We must now find a measurement for each point to be laid off on the projection line from the 1/T line. This is actually the measurement for distances below the _____ plane.

top (T)

64 But, we already know that A and B are both a distance m below T, so we transfer this one measurement to view 1 and find A_1B_1 which is a(n) _____ _____ of line AB. (See Figure 15.17D.) Figure 15.17E is a perspective sketch of the solution.

point view (PV)

■ **65** This problem was simple because the TL of line AB was available in one of the two given views. Sketch 45A shows the T and F views of an oblique line CD.

PROBLEM: Find a point view (PV) of line CD. Following the instructions, sketch a solution to the problem.

STEP 1: We cannot find the PV of CD until we have a TL view. From either C_T and D_T or C_F and D_F, sketch projection lines perpendicular to CD.

STEP 2: Sketch the T/1 (or F/1) line perpendicular to the projectors and parallel to C_TD_T (or C_FD_F).

STEP 3: Transfer appropriate measurements to the new view to find C_1 and D_1. The line C_1D_1 is a TL of line CD. Check your solution to this point.

STEP 4: Sketch a projector extending the TL view of line CD in either direction. Establish a 1/2 line perpendicular to this projector.

STEP 5: Measurements made from the 2/1 line directed into view 2 are actually measurements of distance away from reference plane 1. Plane 1 was set up parallel to C_FD_F (or C_TD_T if you projected from the top view).

Remember that we are always working with a three-view system. The three views at this point are 2, 1, and either T or F. Taking measurements for C_2 and D_2 from the proper place should result in a single distance for both, and thus a single point, C_2D_2, which is a PV of line CD. Check your solution.

288 DESCRIPTIVE GEOMETRY/UNIT 5

66 In finding the *PV* of line *CD*, you constructed two auxiliary views. In both instances, the auxiliary view was created for a specific reason—the first to find the _____ of *CD* and the second to find the _____ of *CD*.

TL (true length);
PV (point view)

These were not just any old views. Their directions were specifically and geometrically defind. A perspective sketch of one possible solution of the problem of Sketch 45A is given. Visualizing auxiliary views is not easy. The sketch does a good job of illustrating the spatial arrangement of the four views. Some complicated problems need five or six auxiliary views for solution. It is very tedious (if not impossible) to sketch such a solution pictorially. The orthographic solution, however, is relatively simple if we follow the rules of projection and measurement that we have discussed here.

67 We can keep taking additional auxiliary views until our heads ache. This is a useless endeavor unless each view adds or clarifies _____ about the object in question.

information

68 It is important for you to understand that each orthographic view, whether principal or auxiliary, is a separate view of an object positioned in _____.

space

69 The object's position with respect to the orthographic frame of reference does not change. We change our _____ position.

viewing

We change our viewing position by (1) establishing a direction of view which we know will give us the information we are seeking, (2) passing a reference plane perpendicular to this direction, (3) projecting selected object points onto this plane according to a strict measuring system using the frame of reference as a base, and (4) finishing the view by connecting the points with appropriate straight lines.

■ **70** In Chapter 11 we described isometric projection as the orthographic view you get when you orient a cube so as to see a point view of a body diagonal. We now know how to find the point view of a line, so let's try this and see if it works.

Sketch 45B shows the *T* and *F* views of a cube. The corners have been lettered *A* through *H*. Subscripts have been omitted to keep the drawing simple. Following the instructions, find a view of the cube that shows a *PV* of the body diagonal *AG*.

First, find a *TL* view of body diagonal *AG*. A suggested direction for auxiliary view 1 is given by arrows. (This is an arbitrary choice and is only suggested so that the rest of the drawing will fit in the space provided.) *It is important to realize that all eight points must be projected into each new view.* All projection lines for view 1 must be parallel and must be perpendicular to the T/1 reference line. Check your solution this far.

You now have a view of the cube showing AG as a TL line. Following the procedures developed previously, project the cube into a second auxiliary view that will show body diagonal AG as a point. Check your solution.

The result is an isometric projection of a 1-in. cube. In a projected isometric, foreshortening is automatically accomplished, each of the space axes being 0.816 of its original length. Check your solution. If your sketching was reasonably accurate, each edge of the cube in view 2 should be about 0.8 times its original length (1 in.).

This is the end of Chapter 15. We will return to the subject of lines (both straight and curved) shortly, but first let's look at some basic projection techniques for analyzing plane surfaces by orthographic means in Chapter 16.

chapter 16 graphic operations on plane surfaces

We are looking at the four basic operations of descriptive geometry. These are the use of orthographic projection techniques to find:

(1) A true-length *(TL)* view of a line
(2) A point view *(PV)* of a line
(3) An edge view *(EV)* of a plane
(4) A true-shape *(TS)* view of a plane

The first two were developed in Chapter 15. The last two are the subject of this chapter. It will become evident in this chapter that finding an edge view and a true-shape view of a plane is quite dependent on knowing how to find a true length view and a point view of a line.

ANALYSIS OF PLANES

Edge View of a Plane

edge view *(EV)*

1 The third important consideration in solving space problems through descriptive geometry is finding the _____ of a plane.

straight line

2 An edge view of a plane surface is an orthographic view (principal or auxiliary) which displays the plane surface as a a(n) _____ _____.

GRAPHIC OPERATIONS ON PLANE SURFACES   291

3 Looking again at the orthographic views of a cube (Figure 16.1) we see that the top surface shows as an edge in the front and profile views. Actually other faces of the cube also show as edges in these views.

Considering only the top of the cube, what makes it appear as an edge in the front view? It is the fact that plane ABCD is _____ to the front reference plane.

perpendicular

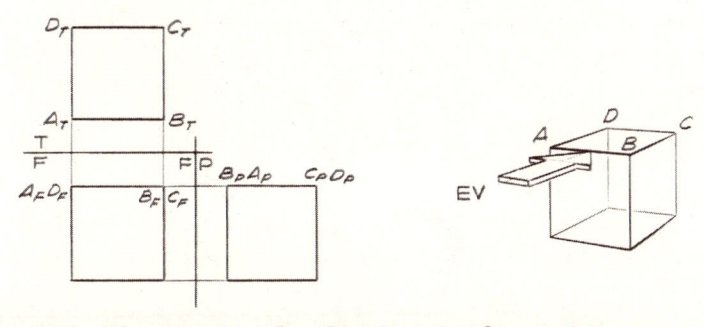

Figure 16.1   The concept of the edge view of a plane surface.

4 If perpendicularity is the criterion for finding the edge view of a plane, to find an edge view of an oblique plane such as MNO in Figure 16.2 we must find a(n) _____ reference plane that is perpendicular to plane MNO.

auxiliary

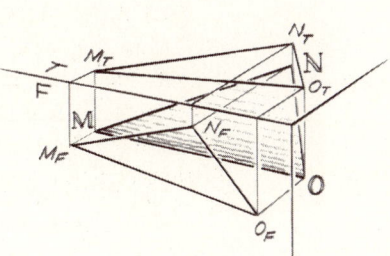

Figure 16.2   An auxiliary view perpendicular to plane MNO is not obvious.

5 Upon inspection of Figure 16.2 this doesn't seem easy. There are no obvious directions to assume for a new view that will assure a(n) _____ of MNO.

EV (edge view)

6 A further study of the cube in Figure 16.1 shows that lines BC and AD show as _____ in the front view.

points (PV)

292 DESCRIPTIVE GEOMETRY/UNIT 5

straight

7 Hold a triangle or any plane object so that you see a point view of one of the sides (see Figure 16.3). The plane of the triangle appears as an edge. No matter how you rotate the plane, always maintaining a point view of one side, you always see the plane as a(n) _____ line.

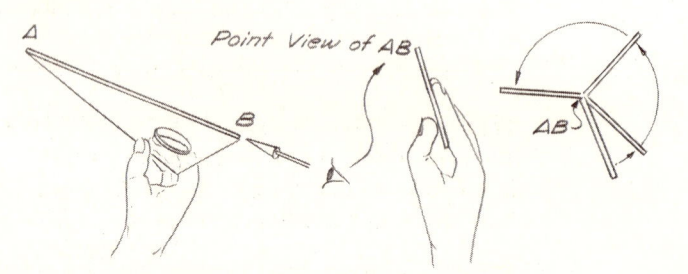

Figure 16.3 Visualizing the edge view of a plane surface.

point

8 From this we may write a rule: *To find an edge view of a plane, find a(n)* _____ *view of a line that lies in the plane.*

Let's test the rule with a simple example. Sketch 46A gives the T and F views of an oblique plane ABC.

PROBLEM: Find an edge view of ABC by finding a point view of a line that lies in the plane ABC.

TL (true-length)

9 Note that $A_T C_T$ is a(n) _____ line. ($A_F C_F$ is parallel to the F/T line.)

■ 10 On Sketch 46A make the necessary construction to find a point view of line AC. Remember that point B must also be projected into the new view. Check your solution.

EV (edge view)

11 The result is an edge view of plane ABC. Since view 1 gives a point view of line AC, points A and C coincide at $A_1 C_1$, and B_1 is a second point in view 1. With only two points defining what we know to be a triangular plane, it is obvious that the view is a straight line that represents ABC as a(n) _____.

Sketch 46B is a reproduction of the T and F views of oblique plane MNO from Figure 16.2. We can see that none of the lines MN, NO, or MO, are TL in the T and F views given. *Our problem:* Find an EV of plane MNO.

GRAPHIC OPERATIONS ON PLANE SURFACES   293

two (2)

12 From our experience with lines, we could find the *TL* of one of the three lines and then find a point view to satisfy the rule for finding an *EV* of a plane. This would take _____ auxiliary views.

■ **13** A shortcut is possible which will give us the desired edge view with only one auxiliary view. The trick is to create a *TL* line in either the *T* or *F* view that lies in the plane *MNO*. Find an *EV* of plane *MNO* (Sketch 46B) according to the following steps. Check your solution at each step.

STEP 1: In the *F* view (an arbitrary choice) sketch a line through point M_F, parallel to the *F/T* line, and intersecting line $N_F O_F$ at point X_F. This is a horizontal line (parallel to *T*). If we can find the *T* view of line *MX*, it will be *TL* and will lie in plane *MNO*.

STEP 2: Project X_F into *T* (projection line perpendicular to the *F/T* line) until you get an intersection with line $N_T O_T$. This intersection is point X_T.

STEP 3: Sketch line $M_T X_T$. This is a *TL* line. Project the line outward to the right. (We wish to find a point view of line *MX*.) Sketch a reference line (*T*/1) at a convenient position and perpendicular to this projection line. Project points N_T and O_T into view 1 with parallel projectors.

STEP 4: Transfer appropriate distances from the front view into view 1 for points *M*, *N*, *O*, and *X*. Line *MX* appears as a point. If your construction is correct and accurate, the three points will line up to form a straight line—an edge view of plane *MNO*.

NOTE ON ACCURACY: Do not expect exact solutions from freehand construction techniques. In this solution, the three points should be close enough to a straight line to satisfy you that, with precise instrument construction, the answer would be exact. If they are quite far off, check your construction against the answer sketches.

■ **14** Sketch 46C is an exercise in finding the edge view of a plane. Find edge views of the three planes given. Check your solutions.

Figure 16.4 shows a problem in which the edge view of a plane is needed for a solution. At 16.4A the *T* and *F* views of a plane *ABC* and a line *XY* are given. *XY* pierces *ABC*, but the piercing point (point of intersection) is not shown. Our problem then is to find the point of intersection between plane *ABC* and line *XY*.

It seems reasonable to believe that, if we had an auxiliary view that showed both line *XY* and an edge view of *ABC*, the point of intersection would be shown at the intersection of two straight lines. Figure 16.4B shows such a condition in auxiliary view 1.

294 DESCRIPTIVE GEOMETRY/UNIT 5

1/F and F/T

15 Once found in auxiliary view 1, the point of intersection, Z, is found in the T and F views by direct projection perpendicular to the _____ and the _____ reference lines. Since point Z is on line XY, it must show as being on $X_F Y_F$ and $X_T Y_T$. A perspective sketch of the problem is shown (16.4C).

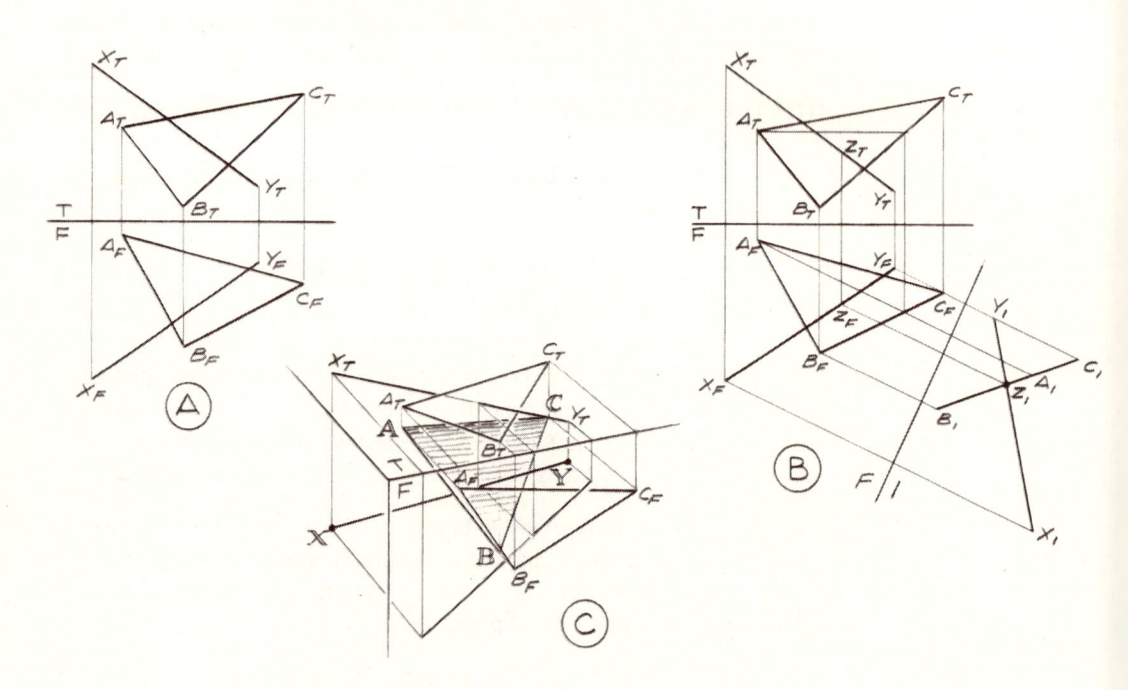

Figure 16.4 Using an edge view (EV) of a plane to find the point of intersection between a line and the plane.

Let's look now at how we use an edge view of a plane surface to find a true-shape view of the surface.

True Shape of a Plane

The fourth operation we must learn for solving space problems is to find a true-shape *(TS)* view of a plane.

ABCD

16 A true-shape view is one which shows the true aspect of all details that lie in the plane—the *TL* of all lines, the true angle between lines, etc. Consider the cube again (Figure 16.5A). The standard top view shows the true shape of plane _____.

GRAPHIC OPERATIONS ON PLANE SURFACES 295

17 Lines A_TB_T, B_TC_T, C_TD_T, and D_TA_T are all _____. The angles between these intersecting lines show, as is expected in a cube, as _____ degrees.

TL; 90

18 The reason plane ABCD is true shape in the top view is that it is _____ to the top reference plane.

parallel

19 This is the only condition that would make all four lines TL in one view. If plane ABCD is parallel to T, our line of sight is _____ to $A_TB_TC_TD_T$.

perpendicular

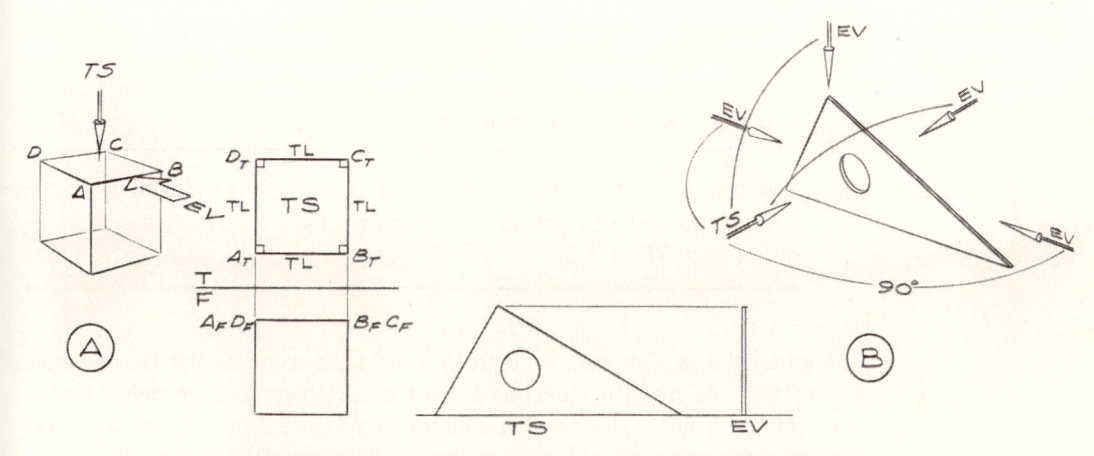

Figure 16.5 Visualizing the true shape of a plane surface.

20 The clear evidence in Figure 16.5A that ABCD is parallel to T can be seen in the front view. Here $A_FB_FC_FD_F$ is a(n) _____ that is parallel to the _____ reference line.

EV; F/T

We may now state a rule: *To find a true-shape view of a plane construct an auxiliary reference plane parallel to an edge view of the plane.*

21 This construction automatically puts our line of sight perpendicular to the plane. The sketches of the triangle in Figure 16.5B show that there is a(n) _____-_____ relationship between an edge view and a true-shape view of a plane.

90-degree

296 DESCRIPTIVE GEOMETRY/UNIT 5

We saw this same 90° relationship between a true-length view and a point view of a line.

■ **22** Sketch 47A is a simple example to test the rule for finding the true shape (TS) of a plane. Find the true shape of plane ABC by following the steps below.

Note that an edge view of ABC exists in the front reference plane. According to the rule, our line of sight must be perpendicular to this edge view and auxiliary view 1 must be parallel to ABC.

STEP 1: From points A_F, B_F, and C_F sketch projectors that are perpendicular to $A_F B_F C_F$. This establishes the direction of the new view. Check your solution.

STEP 2: Establish the 1/F line perpendicular to the projectors at a convenient distance from $A_F B_F C_F$. This line will be parallel to $A_F B_F C_F$. Transfer appropriate distances from view T to view 1 to find A_1, B_1, and C_1. Connecting these points with straight lines completes the TS view of plane ABC. Check your solution.

true lengths (TL)

23 You should be aware that what you did in solving the problem of Sketch 47A was to find an auxiliary view that gave the _____ of lines AB, BC, and CA.

Figure 16.6 is a repeat of Figure 5.6. It is the same as the last example except that the plane in question is part of an object and we didn't bother with reference lines. However, in constructing Figure 16.6 we had the same kind of measurement problem in laying off dimension D in the auxiliary view.

front–back

24 The dimension D relates _____ directions to the front plane. (*Choose one:* up–down/front–back/left–right.)

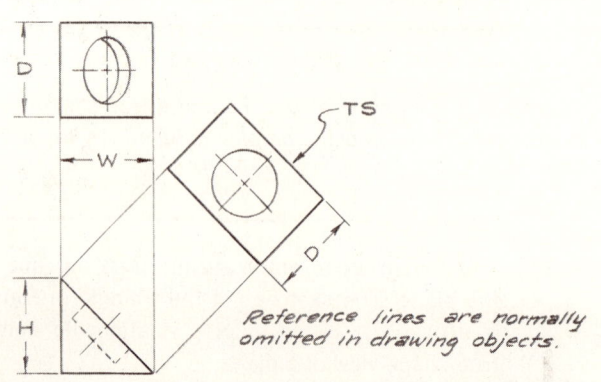

Figure 16.6 A true-shape (TS) auxiliary view of a sloping face on an object.

GRAPHIC OPERATIONS ON PLANE SURFACES 297

T/F; 1/F

25 We used the same concepts in finding the *TS* of *ABC* of Sketch 47A except that we made measurements from the front plane (using the ____/____ line and the ____/____ line).

It is standard procedure in picturing physical objects orthographically to omit reference lines. This means that the spacing between views is arbitrary and is usually dictated by the size of paper and the complexity of the drawing.

reference

26 In solving abstract problems of descriptive geometry we have the same arbitrary choice in positioning each new _____ line. This gives us control over the spacing of views.

27 Sketch 47B shows the *T* and *F* views of an oblique plane *PQR*.

PROBLEM: Find the *TS* of *PQR*. In thinking of how to proceed with this solution, you should reason as follows:

EV (edge view)

STEP 1: To find the *TS* of plane *PQR*, I first need a(n) _____ of the plane.

PV (point view)

28 **STEP 2:** To find an *EV* of *PQR*, I must first find a(n) _____ of a line that lies in plane *PQR*.

TL (true-length)

29 **STEP 3:** To find a *PV* of a line, I must first find a(n) _____ view of the line.

■ 30 **STEP 4:** The given views of plane *PQR* in Sketch 47B show no *TL* lines, therefore, I must create one.

Using sketching techniques, find the *TS* of plane *PQR* by following the above four steps in reasoning in *reverse order*. Check your solution.

■ 31 Sketch 48 shows the same type of problem except that the oblique plane is part of an object. A rectangular block has an oblique face made by slicing off a corner.

PROBLEM: Find the *TS* of the oblique plane. In solving problems of this type *always* look first for (1) *TL* lines, (2) *PV* of lines, (3) *EV* of planes, and (4) *TS* of planes. The two views of the oblique plane contains two *TL* lines.

Two orthographic views are given but not labeled. No reference line is given. By their positions, the two views could be front and right profile or left profile and front. Neither of these agree with the position of the block in the perspective sketch.

However, all of the given views are correct. By this time you should be able to visualize the object and turn it about in space at will. You are asked to find the *TS* of the oblique triangular plane. Label the corners of the plane in both views. It is not necessary to project the rest of the block into any auxiliary views.

Sketch a solution to the problem and check.

298 DESCRIPTIVE GEOMETRY/UNIT 5

SUMMARY

This completes the discussion of the orthographic operations on points, lines, and planes that are basic to descriptive geometry.

The material presented in frames 32 through 37 covers both Chapters 15 and 16.

Points

We have little use for studying points alone. However, the entire subject of descriptive geometry is built upon an ability to project points orthographically through any series of principal or auxiliary views.

2; 3

32 Any _____ points in space define the ends of a line segment and _____ or more points can define the corners of a plane segment.

Lines and Planes

The figures of Sketches 47 and 48 present a graphic summary of the four basic orthographic operations on lines and planes. The subject of these two sketch exercises is finding the true shape of a plane, the fourth basic operation.

(1) TL;
(2) PV;
(3) EV

33 To find the TS of a plane we must use the first three operations in the following order:
    (1) Find the _____ of a line that lies in the plane.
    (2) Find a(n) _____ of the line.
    (3) Find a(n) _____ of the plane.

parallel

34 We find a TL view of a line by projecting the line onto an auxiliary reference plane that is _____ to the line.

TL

35 We find a PV of a line by projecting the line onto an auxiliary reference plane that is perpendicular to a(n) _____ view of the line.

PV

36 We find an EV of a plane by finding an auxiliary view that shows a(n) _____ of a line that lies in the plane.

EV

37 Finally, we find a TS view of a plane by projecting all points in the plane onto an auxiliary reference plane that is parallel to a(n) _____ of the plane.

chapter 17   *visibility of lines and planes*

Before proceeding with our study of lines and planes in space, let's pause and consider the problems of visibility of orthographic projections.

---

pictorial

1   An orthographic projection is an exact and measurable drawing. It is not realistic. We draw objects in the _____ form if we wish realism.

Doing so sacrifices precision and measurability but increases our ability to visualize the object drawn. Since a pictorial drawing is realistic, there is usually no question as to the visibility of the object or objects portrayed. In the orthographic system there is often a question of visibility. Certain aspects of the object portrayed are not immediately evident.

VISIBILITY IN ORTHOGRAPHIC VIEWS

---

two

2   To translate an orthographic drawing into a recognizable mental image we must extract information from _____ or more views and combine it mentally.

---

visibility

3   Figure 17.1 is an example. Figure 17.1A shows the *T* and *F* views of two skew lines (nonintersecting and nonparallel). Which line is above the other? Which line is in front of the other? These are questions of _____ and are not readily answered.

299

300 DESCRIPTIVE GEOMETRY/UNIT 5

4 Figure 17.1B is a perspective sketch of the two lines with the two principal planes, T and F. Here visibility is easy and the two questions can be answered. Line _____ is above line _____ and line _____ is in front of line _____.

MN (above) AB;
AB (in front of) MN

5 It is true that this does not apply to the full lengths of lines AB and MN. The F view shows that point A is above points M and N and the T view shows that point N is in front of point B.

In trying to visualize the spatial relation of lines in orthographic views, we are interested in points of *apparent intersection* such as points _____ and _____ in 17.1A.

X (and) Y

6 To aid visualization from orthographic views, we must be aware of the commonly used space directions and their interpretations in each view. These directions are: up and _____; left and _____; and backward and _____.

down; right;
forward

■ **7** Sketch 49A shows these six directions by arrows in the T, F, and P views. Label each arrow with its direction. A perspective sketch is also shown. Check your solution.

8 Consider point X in the F view of Figure 17.1A. It is the apparent intersection of lines AB and MN. Sketch a projection line from point X perpendicular to the F/T line. In the T view, this projector represents a line of sight in the _____ direction.

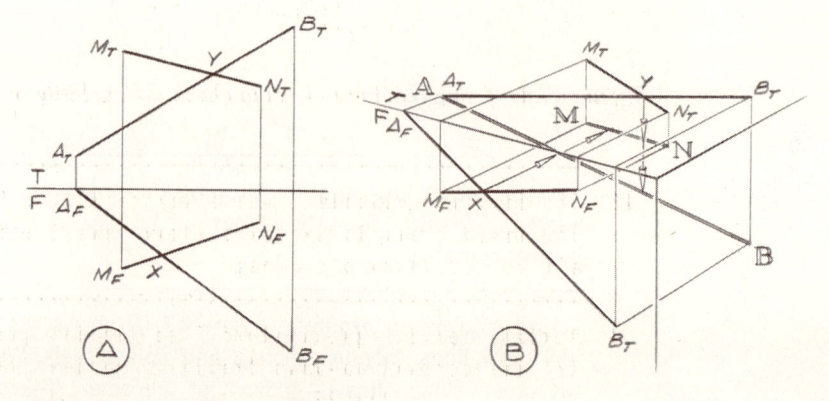

backward

Figure 17.1 Visibility of lines.

VISIBILITY OF LINES AND PLANES 301

$A_T B_T$; $M_T N_T$

9. The T view shows that in moving from front to back the projector meets line _____ before line _____.

MN (behind) AB

10. This means that, *at the horizontal level of point X,* line _____ is behind line _____.

MN; AB; below

11. A similar analysis can be made at Y, the apparent intersection of $M_T N_T$ and $A_T B_T$. Sketch a projector from Y perpendicular to the T/F line. This projector shows up–down directions in F. It meets _____ before _____ which proves that line AB is _____ MN at position Y.

intersecting

12. If the projectors from X and Y had coincided (if X was a direct projection of Y) lines AB and MN would be _____ lines.

The use of common space directions extends to first auxiliary views. After that (2nd, 3rd, etc.) it becomes difficult to imagine common directions. Figure 17.2 shows the T and F views of two skew lines PQ and RS. Their visibility can be analyzed in the same manner used for Figure 17.1.

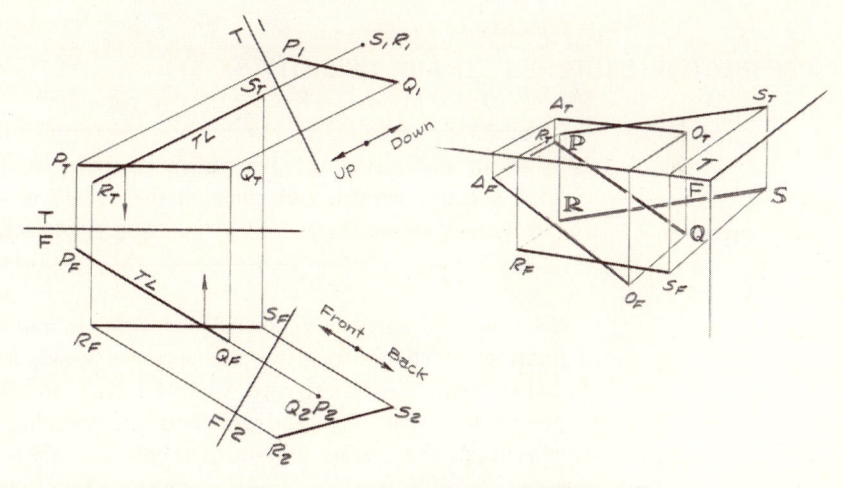

Figure 17.2 Using auxiliary views to check the visibility of lines.

PV

13. The lines are positioned in such a way that $R_T S_T$ is TL and $P_F Q_F$ is TL. This makes it easy to take two auxiliary views (1 and 2) that will show a(n) _____ of one of the lines.

above

14. Auxiliary view 1 is projected off T, so directions perpendicular to the 1/T line represent up–down directions. The view shows that line PQ is _____ line RS.

302 DESCRIPTIVE GEOMETRY/UNIT 5

front–back

15 Auxiliary view 2 is projected off F, so directions perpendicular to the 2/F line represent _____ – _____ directions. View 2 shows that PQ is in front of RS.

Analysis of 17.2 by the method of Figure 17.1 should bring the same conclusion about the relative positions of PQ and RS.

Do not be confused by the fact that the downward space direction goes upward on the drawing paper in view 1. These directions are determined by the T and F planes which are firmly established for every orthographic drawing once the T/F line is drawn.

■ **16** Sketch 49B shows the T and F views of a line (XY) intersecting a plane (ABC). The point of intersection (Z) has been found. Find the visibility of the two objects in each view. Make the visible parts of the line heavy and leave it thin where it disappears behind or below the plane. Use the method of Figure 17.1 and reduce the problem to two steps.
(1) Find the visibility of line XY and line AB.
(2) Find the visibility of line XY and line AC. Check your solution.

PERSPECTIVE SKETCHES TO AID VISIBILITY

orthographic

17 In our discussions of descriptive geometry so far we have presented perspective sketches of some of the problems even though all of the solutions are done by _____ projection.

We have used perspective sketches in order to improve the communication of information contained in the orthographic views. A pictorial drawing of a spatial concept is easier to read. Chapters 7, 8, and 9 presented the subject of perspective sketching. The problems in sketching orthographic views of points, lines, and planes are no different from those in sketching physical objects.

Consider Sketch 50A. It shows the T and F views of plane ABC. Line AC is parallel to the F plane and line BC is parallel to T.

VP_R; VP_L

18 A perspective sketch of the T and F planes is included in Sketch 50A. The two reference planes have been put in their true spatial positions. The T/F line (and any line parallel to it) will vanish to _____. The F plane has been bounded by vertical lines and the T plane by horizontal lines vanishing to _____. We did not have to do this—it merely adds to the realism. These lines do not exist in the orthographic views.

VISIBILITY OF LINES AND PLANES 303

■ 19 Note that points 1, 2, and 3, the intersection of the three projectors and the T/F line, have been placed on the T/F line in perspective. The distances between them are proportionally correct. Make a perspective sketch of plane ABC (Sketch 50A) by following the next five steps.

STEP 1: From points 1, 2, and 3 sketch projectors onto the T and F planes. On the T plane these projectors vanish to VP_L. On the F plane they are vertical.

STEP 2: Transfer points A_T, B_T, and C_T to lines 3, 2, and 1 respectively. Use measurements taken from the orthographic T/F line. Remember that $A_T C_T$ is parallel to the T/F line. In estimating measurements, remember that they must all be proportional and that foreshortening must be taken into account (Chapter 7). You now have the top view in perspective.

STEP 3: Do the same for the front view. Measure (by estimation) down from the F/T line on lines 1, 2, and 3 to find C_F, B_F, and A_F respectively. $B_F C_F$ is parallel to the F/T line.
  The orthographic projection of plane ABC has now been transferred to the perspective form. Check your solution to this point.

STEP 4: From A_T, sketch a vertical line downward. From A_F sketch a line vanishing back to VP_L. Point A (no subscript) is located at the intersection of these two lines. Find B and C in a similar manner.

STEP 5: Connect A, B, and C with straight lines. This is the perspective sketch of ABC. Darken the lines of the plane with bold, expressive line work. This will make ABC stand out from all the construction. If your construction is reasonably accurate, AC will appear parallel to $A_F C_F$ and BC will appear parallel to $B_T C_T$. (The final answer sketch shows the same problem sketched from another view point.) Check your solution.

20 Sketch 50B illustrates another method for making a quick perspective sketch of a space problem. The last example (plane ABC) is shown in the new form in the inset of Sketch 50B. Plane ABC has been sketched as if it were supported on a(n) _____ plane (M).

horizontal

This sketch makes the plane more visible, but does not relate it directly to the T and F planes.
  In Sketch 50B, a plane PQR is shown in T and F views along with a line XY that intersects the plane at point Z. A horizontal plane M (EV in F) has been passed through point R_F (so chosen because it is the lowest point of the five).

304 DESCRIPTIVE GEOMETRY/UNIT 5

■ **21** Plane M is shown in perspective in the sketch. Following the directions in the next few steps, erect plane PQR and line XY on plane M.

STEP 1: Transfer points 2, 3, 4, and 5 from the T/F line to the front edge of M (point 1 is given). Estimate distances between points, being careful to maintain relative proportions, and consider foreshortening.

STEP 2: We will now transfer the top-view information to plane M. From the five points sketch five projectors on plane M by directing them toward VP_L.

STEP 3: Transfer measurements between P_T, Q_T, R_T, X_T, and Y_T and the T/F line to the corresponding projectors on plane M. Again these measurements need not be exact, but should all be *proportional* especially as related to the distances between 1 and 2, 2 and 3, etc. Remember, this is a perspective sketch, so foreshortening must be taken into account (Chapter 7). The top-view information is now complete on plane M. Check your solution to this point.

STEP 4: Erect vertical lines at four points, P, Q, X, and Y. (R is in plane M.) Transfer vertical distances for each point. These are the distances between P_F, Q_F, X_F, and Y_F and the horizontal plane M in the front orthographic view. You have now found the actual space points, P, Q, R, X, and Y.

STEP 5: Connect appropriate points with straight lines. Estimate the position of point Z on plane PQR. Darken all visible lines. Check your solution.

In this last exercise you should recognize that, rather than plotting a correct top view in perspective form, you are dropping the top-view information down to a bottom or base plane. The sketch can be quickly completed by estimating heights from front-view information.

The foregoing discussion of perspective sketching was introduced to aid you in understanding the problems of descriptive geometry and to give you the confidence to use perspective sketching often. It is good practice to sketch these problems. You sharpen your ability to visualize as you meander through space with a pencil and a bit of perspective theory.

chapter 18 angular relations and intersections

We have been dealing with dimensional details of lines and planes in space. Let's look now at the angular relations between lines, between lines and planes, and between planes.

BEARING AND SLOPE OF A LINE

The position of a line in space can be defined by two angular quantities, bearing and slope. *Bearing* is the angle between the top view of a line and the north–south line. *Slope* is the true angle between a line and the horizontal (top) plane.

H, W, and D

1   Both of these concepts come to graphics from the civil engineering profession and other professions concerned with picturing the surface of the earth through maps. A map is an interesting graphic construction in that it pictures the three principal space dimensions, ____, ____, and ____ in one view—and yet is not a pictorial drawing.

Figure 18.1 shows a section of a contour map. The one view is the top or *plan* view. Dimensions W and D are given by the principal dimensions of the map. Height dimensions are given by contour lines—lines of constant horizontal *elevation* (height). Each line carries a number which represents feet above mean sea level.

305

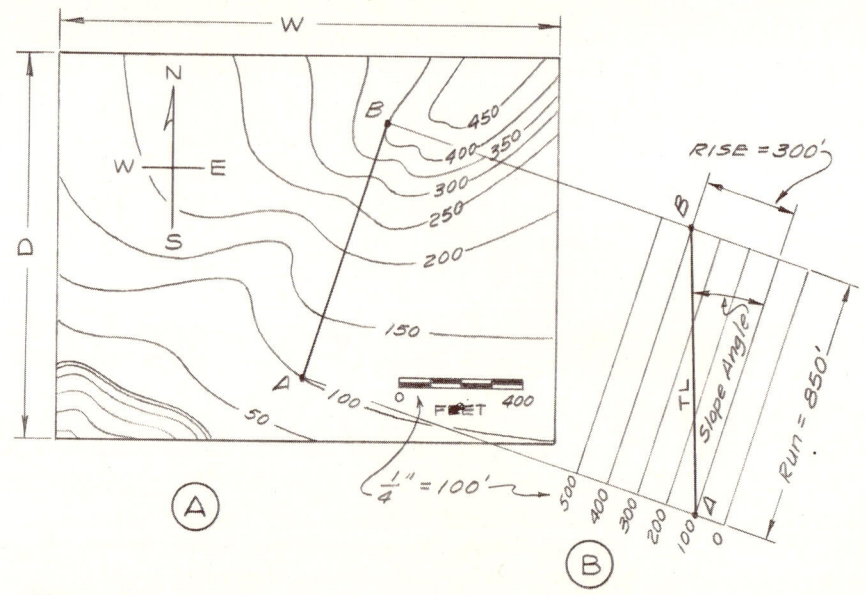

Figure 18.1 A contour map shows H, W, and D in one view.

top	2	As shown in Figure 18.1, the north–south line is a vertical line with north at the _____. This is the usual case, but by no means a standard.
100; 400	3	Consider line AB. It could be the centerline of a proposed section of highway. It starts at _____ ft above sea level and ends at _____ ft.
300	4	This means that, in going from A to B, the line rises _____ ft.
is not	5	Thus we know that line AB as shown on the map _____ a TL line because it slopes upward from A to B or, conversely, slopes downward from B to A. (Choose one: is/is not.)
north–south (N–S)	6	AB is merely the top view of a sloping line (according to the system of notation we use, it should be labeled $A_T B_T$). The angle that AB makes with the _____–_____ line is about 19°.
top (T)	7	This angle is called the *bearing* of the line AB. The bearing of a line is the acute angle the _____ view of the line makes with the north–south line. The bearing of AB is written: N 19° E.
S 19° W	8	Line AB is directed in a NE direction. If we consider the line as going from B to A, it has a different bearing. The bearing of line BA is _____.
100	9	Line BA is a SW line. The bearing angle is the same. At Figure 18.1B we have taken a vertical section into the earth along line AB. A zero point (sea level) is selected and a vertical scale with contour planes for every _____ feet of elevation is drawn.

ANGULAR RELATIONS AND INTERSECTIONS 307

Adding line AB to this section, we see that this is a TL of AB. We have created a TL view without the front view from which we normally take measurements. Actually, the vertical scale and the horizontal contour planes gave us the same information that we could have received from a standard front view of AB.

---

angle

10 The elevation view also shows the *slope* of line AB. The slope of a line is the true _____ the line makes with the horizontal (top reference plane).

We can measure this angle with a protractor if the vertical and horizontal scales are equal. Otherwise, we can calculate the angle from the fact that slope (S) is the angle whose tangent (arctan S) = rise/run.

300; 850

11 Line AB rises _____ ft in a _____-ft run (horizontal distance). (See Figure 18.1B.)

−19.44°; −35.3%

12 Thus the tangent of the slope angle (S) = $300/850$ = 0.353 or S = 19.44°. Slope is also expressed as a percentage which is called the *grade* of the line. The grade of line AB = rise/run × 100 = $300/850$ × 100 = 35.3%. If a line slopes upward, its slope and grade are positive (+). If a line slopes downward, they are negative (−).
Considering line BA, slope = _____ and grade = _____.

---

top

13 Figure 18.2 shows the road centerline, AB, of Figure 18.1 drawn in standard orthographic form. The rise, run, and bearing can be determined from this drawing. The true slope of line AB we know is 19.44°.
True slope of a line is defined as the true angle that a line makes with the _____ reference plane.

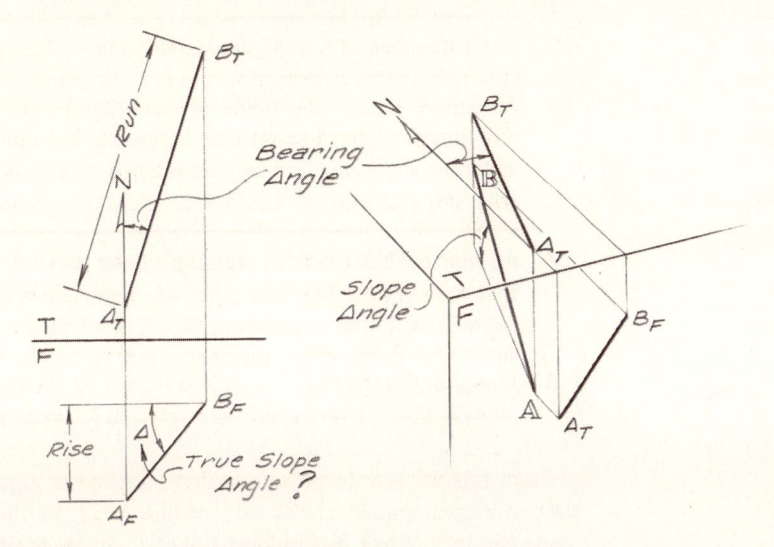

Figure 18.2 Line AB of Figure 18.1 in standard orthographic form.

308 DESCRIPTIVE GEOMETRY/UNIT 5

At first glance, we might look at the front view of AB in Figure 18.2 and say that the angle $A_F B_F$ makes with the F/T line (angle A) is the true slope. *This is not true* however. $A_F B_F$ is a foreshortened view of AB, so true slope cannot show in the front view. What, then, is needed to find the true angle between a line and a plane?

TRUE ANGLE BETWEEN A LINE AND A PLANE

Specific Cases

TL; true angles

14 Consider Figure 18.3A. A 45-45-90 triangle is shown positioned vertically on a horizontal plane. We see a true shape of the triangle. All lines are _____ lines and all angles are _____ _____.

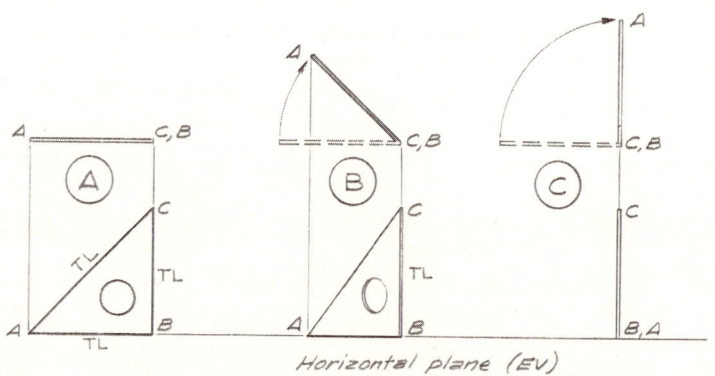

Figure 18.3 The concept of true angle between a line and a plane.

45

15 The true slope of line AC (angle BAC) is _____ degrees.

16 At Figure 18.3B, the plane of the triangle has been turned away from our line of sight keeping line BC vertical. Angle BAC no longer appears to be 45°; it is something greater. However, we know intuitively that the true slope of line AC has not changed and is still _____ degrees.

45

17 At Figure 18.3C, the triangle has been rotated about BC still farther so that we see an edge view. Line AC makes an apparent angle of _____ degrees with the horizontal. Again we know this is not true. It is the same triangle and is still positioned vertically so the _____ of line AC is 45°.

90; slope

From this we see that the true slope showed in Figure 18.3A only. In 18.3A we had a combination of (1) a TL of line AC, (2) an edge view of a horizontal plane, and (3) the true slope of line AC. All three of these did not exist to-

ANGULAR RELATIONS AND INTERSECTIONS 309

gether in Figure 18.3B and C. True slope is defined as the true angle between a line and a horizontal plane.

TL; EV

18 Let's write a general rule from these considerations: To find the true angle between a line and a plane find a single view that shows a(n) _____ of the line and a(n) _____ of the plane.

EV; TL

19 Going back to the question of Figure 18.2, "Is angle A the slope of line AB?", we can see now why the answer is no. The F/T line represents a(n) _____ of the T plane, but $A_F B_F$ is not a(n) _____ line.

top; 1/T

20 Any view, principal or auxiliary, projected off the top reference plane shows an EV of T. This edge view is the (any view)/T reference line. Figure 18.4A shows the construction required to find the slope S of line AB. A TL view of AB was found in view 1 projected off the _____ plane. To satisfy the rule, the necessary edge view is the ___/___ line.

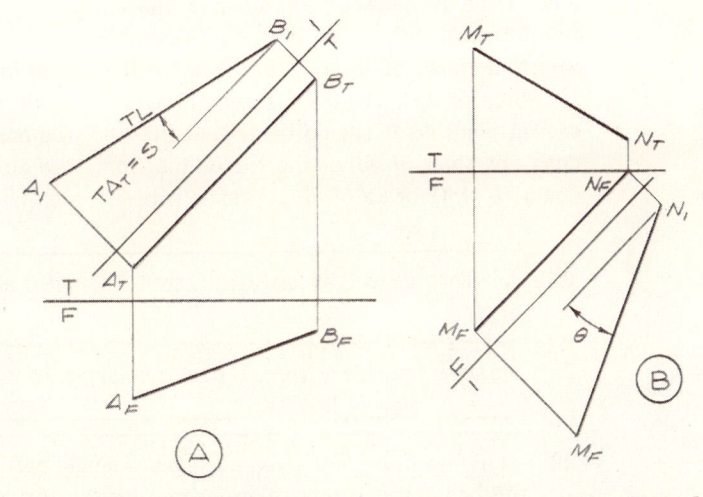

Figure 18.4 True angle (TA) between a line and one of the principal reference planes.

$\theta = TA_F$

21 We will use the abbreviation TA to indicate the *true angle* between lines and planes. TA with a subscript means a true angle with a specific line or plane. Thus:
   TA_T = true angle with top plane = S (slope)
   TA_F = true angle with front plane
   TA_1 = true angle with auxiliary view 1
   TA_{ABC} = true angle with plane ABC, etc.
Figure 18.4B shows the T and F views of an oblique line, MN. View 1 is a TL view of MN projected off the F view. What is angle θ? (Check one or more: S _____, TA_T _____, TA_F _____, or TA_1 _____?

310 DESCRIPTIVE GEOMETRY/UNIT 5

EV; F	22	Angle θ is the angle that M_1N_1, a TL line, makes with the 1/F line. The 1/F line represents a(n) _____ of plane _____.
top	23	To find TA_T (slope) of line MN, we would have to project the TL auxiliary view off the _____ view of line MN.
TL; EV	24	Using the general rule, we can find the true angle between any line and any plane by finding a single view that shows a(n) _____ view of the line and a(n) _____ of the plane.
■	25	Sketch 51A is an exercise on bearing, grade, slope, and true angle between a line and one of the principal reference planes. Sketch the construction necessary to find the quantities asked for. Estimate angles by considering them a subdivision of 90°. Check your solutions.

General Case

We have been considering the TA between a line and specific planes, the principal orthographic reference planes. Let's look now at the problem of finding the TA between any oblique line and any oblique plane. The general rule for doing this is: *To find the TA between a line and a plane, find a single view that shows both the TL of the line and an EV of the plane.*

Sketch 51B illustrates the general case. A plane ABC and a line XY intersect at point W. It should be evident that the true angle between ABC and XY does not show in either the top or the front view since neither of the conditions, TL of XY or EV of ABC, is satisfied.

EV	26	Auxiliary view 1 shows a(n) _____ of ABC and another oblique view of XY.
TL	27	However, auxiliary view 1 does not show TA because X_1Y_1 is not a(n) _____ view of line XY.
TS; EV	28	Thus, we must find a way to *select* a single particular edge view of plane ABC that shows line XY in its true length. But how do we do this? Consider Figure 18.5A. It shows a disk on a vertical shaft. The top view is a(n) _____ view of the disk and the front is a(n) _____.
TS	29	Suppose we look horizontally at the disk in directions 1, 2, 3, and 4 (random choices). Figure 18.5B shows what we get if we project auxiliary views 1, 2, 3, and 4. In each case we obtain *another edge view*. From this we can reason that any view projected off of a view showing the _____ of a plane will yield an EV.
TS	30	Thus, if we wish to select a particular edge view of a plane, we must first find a(n) _____ view of the plane.
2	31	On Sketch 51B, a TS view of plane ABC is given (auxiliary view _____). Line XY is projected into this view.

ANGULAR RELATIONS AND INTERSECTIONS 311

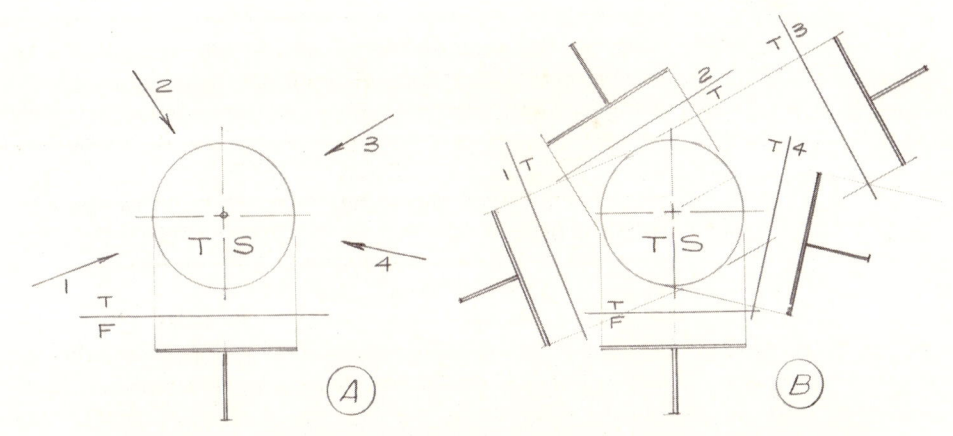

Figure 18.5 Condition necessary for selecting one particular edge view of a plane.

■ 32   Knowing that any view projected from a view showing the true shape of a plane will yield an edge view of the plane, you should see the final solution to our problem. (See Sketch 51B.)

Sketch auxiliary view 3 which shows both an edge view of plane ABC and a TL of line XY, and thus the true angle between ABC and XY (label it TA). Check your solution.

PERPENDICULARITY

33   When are two intersecting lines in space perpendicular? Sketch 52A shows the T and F views of two intersecting lines AB and AC. Point A is the point of intersection. Are the lines truly perpendicular? _____yes; _____no; _____I don't know.

no

34   You now respond (*Check one*):
  (1) _____Well, oh yeah? Angle $C_T A_T B_T$ is shown as 90°, so they must be perpendicular.
  (2) _____Prove it!
  (3) _____OK if you say so.

prove it!

35   We can't be sure until we prove it one way or the other. The two intersecting lines actually form a plane, ABC. We know that the TS view of a plane shows all lines in their _____ and all angles as _____.

TL; TA (true angles)

■ 36   Therefore, if we find the TS of plane ABC we will be able to measure angle BAC to determine if it is 90°. In Sketch 52A, make the construction necessary to find the TS of plane ABC. Note that $B_T C_T$ is parallel to the T/F line, therefore $B_F C_F$ is a TL line. Check your solution. If you projected the TS correctly, you have proved whether angle BAC is truly 90°.

312 DESCRIPTIVE GEOMETRY/UNIT 5

intersecting

37 From the last example, we can see that two _____ lines are not necessarily perpendicular even though one view shows them as being perpendicular. What, then, are the conditions for perpendicularity?

TS

38 Consider Figure 18.6. It uses the 45-45-90 triangle again to study perpendicularity of intersecting lines. At 18.6A we see a _____ view of the triangle.

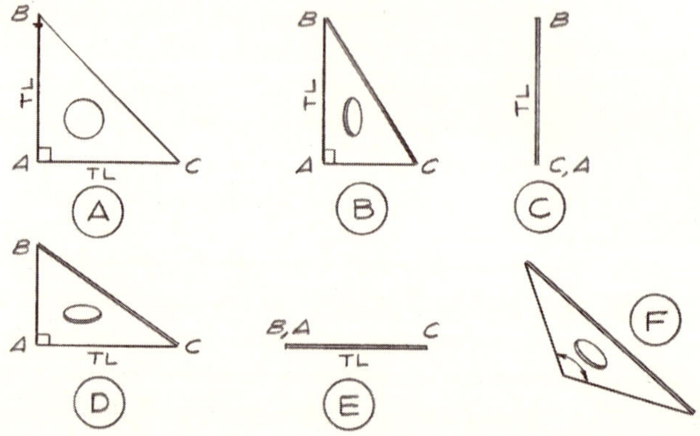

Figure 18.6 When do perpendicular intersecting lines show their 90° relation?

90

39 All lines (AB, BC, and AC) are TL. AB and AC are perpendicular intersecting lines and the view shows it. (Angle BAC is 90°.)
   At 18.6B the triangle has been rotated about AB (vertical) so that line AC is no longer TL but line AB is still TL. Angle BAC is _____ degrees.

   At 18.6C we see an EV of the triangle. AB is still TL and AC is a PV. Angle BAC is still 90°. (With AB as a TL and AC as a point, there is no other interpretation.)

90

40 Figure 18.6D and E show the same changes except that line AC is kept in its TL and line AB changes from TL (at A) to a point (E). Angle BAC still shows as _____ degrees.

TL

41 Figure 18.6F shows the triangle in an oblique position. None of the three edges (lines) shows as a(n) _____ line and none of the angles (including the 90° angle) shows its true value.

TL (true length)

42 From this analysis, we can say that: *Two perpendicular, intersecting lines in space will show as being perpendicular (90°) in any view in which one or both of the lines are _____*.

ANGULAR RELATIONS AND INTERSECTIONS   313

yes

43 Sketch 52B is an exercise on perpendicularity. At 52B(1) the T and F views of two intersecting lines, UV and WX, are shown. Are these perpendicular lines? Yes _____. No _____.

■ **44** They are perpendicular because $U_T V_T$ is a TL line and angle $W_T X_T V_T$ is 90 degrees.

52B(2) is the same problem we solved in Sketch 52A. We used two auxiliary views to prove that AB is not perpendicular to AC. Prove this fact again using one auxiliary view and the statement on perpendicularity.

52B(3) gives the T and F views of line MN and point A. Find line AB in both views. Point B is on MN and AB is perpendicular to MN. Check your solutions.

ANGULAR RELATION BETWEEN PLANES

TL; T (top)

45 The *slope* of a line is found by creating a view that shows the _____ of the line and an EV of the _____ plane.

horizontal (top) plane

46 We often need to know the *slope of a plane*. By definition, the slope of a plane is the true angle between the plane and the _____ _____.

Let's analyze what is needed to find the true angle between any two planes. We draw segments of planes (triangles, rectangles, etc.), but a plane should be considered infinite in size—extendable in all directions within the plane.

straight

47 Thus, if two planes are not parallel, they will meet somewhere and their intersection will be a(n) _____ line.

48 Figure 18.7A shows a horizontal rectangular plane, ABCD, and an oblique triangular plane, XYZ. They are not parallel.

If we wish to find the intersection of these two, we can extend lines _____ and _____ to find their points of intersection with ABCD. (See 18.7B.)

XY (and) XZ

line (of) intersection

49 These two points (V and W in 18.7B) define the _____ of _____ between ABCD and XYZ.

314   DESCRIPTIVE GEOMETRY/UNIT 5

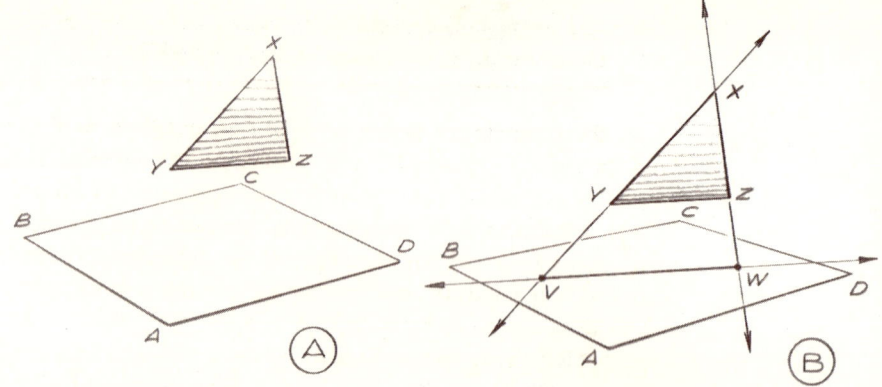

Figure 18.7 Any plane segment may be extended indefinitely.

	50	Hold two triangles (or any stiff plane surfaces) so that they intersect in a straight line. (See Figure 18.8A) What view of these planes will show the true angle between them? It is the view that shows the line of intersection as a(n) _____ .
point (PV)		
EV	51	We know that if we see the PV of a line that lies in a plane, we are also seeing a(n) _____ of the plane.
both	52	Thus, if we see the line of intersection between two planes as a point, we will see the edge view (EV) of _____ planes. (See Figure 18.8B.)
EV	53	Thus we may write a rule: *To find the true angle (TA) between two intersecting planes, find a view which shows the _____ of both planes.*

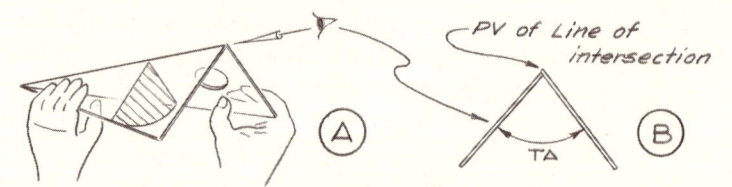

Figure 18.8 Visualizing the TA between planes.

TA (true angle)	54	Consider the figure in Sketch 53A. It shows the T and F views of a plane, ABC. We wish to find the slope of the plane. Remember that slope (S) is labeled TA_T which is the _____ with the top plane.
EV	55	To find the slope we must find a view in which both planes appear as edges. The reference line between any auxiliary view and the T plane represents the _____ of the top reference plane when looking at the auxiliary view.
■	56	Thus we must find an EV of ABC in an auxiliary view projected from T. Sketch the necessary construction to find this edge view of ABC. Label the slope of plane ABC with an "S". Check your solution.

ANGULAR RELATIONS AND INTERSECTIONS   315

1/T

57 The angle between the edge view $A_1B_1C_1$ and the _____/_____ line is the slope, S, of ABC. Slope is also marked TA_T.

In the example of Sketch 53A, we did not have to find the line of intersection between plane ABC and the top reference plane. This is true because we could get an EV of plane ABC in a first auxiliary view projected off the top plane. Since the 1/T reference line represents an EV of T when looking at view 1, we had met the requirement for finding the TA between two planes, namely, find a single view that shows an edge view of both planes.

What about the general case of two oblique planes in space? The figure of Sketch 53B shows two intersecting planes, PQR and QRS.

PROBLEM: Find the TA between PQR and QRS.

QR

58 We know that we must find a single view which shows an EV of both planes. To do this we must find a PV of the line of intersection. The line of intersection between planes PQR and QRS is _____.

■ **59** In Sketch 53B, make the construction necessary to find the TA between planes PQR and QRS. Two auxiliary views are required.

STEP 1: Find a TL view of QR, the line of intersection. (This step is given.)

STEP 2: Find a PV of QR.

Be sure to carry points P and S into auxiliary view 2. Label the TA. Check your solution.

INTERSECTIONS BETWEEN LINES AND PLANES AND BETWEEN PLANES

is not

60 Consider Figure 18.9A. It shows two oblique planes, ABC and NOP. Their line of intersection _____ shown in either the T or F views. (*Choose one:* is/is not.)

an EV of plane NOP

61 If we wish to find the TA between planes ABC and NOP we must first find their line of intersection. Figure 18.9B shows the auxiliary-view method for doing this. View 1 was selected to show _____ _____ (your words).

X (and) Y

62 In view 1, the points of intersection between line AC and BC with plane NOP are evident. These are points _____ and _____.

line of intersection

63 Since points X and Y are *common to both planes*, line XY is the _____ _____ (your words).

PV

64 If we wished to find the TA between planes ABC and NOP, we would have to find a _____ of line XY.

The construction of Figure 18.9 required an auxiliary view. We can find the line of intersection between two planes without the need for auxiliary views. We will use *cutting planes* to do this.

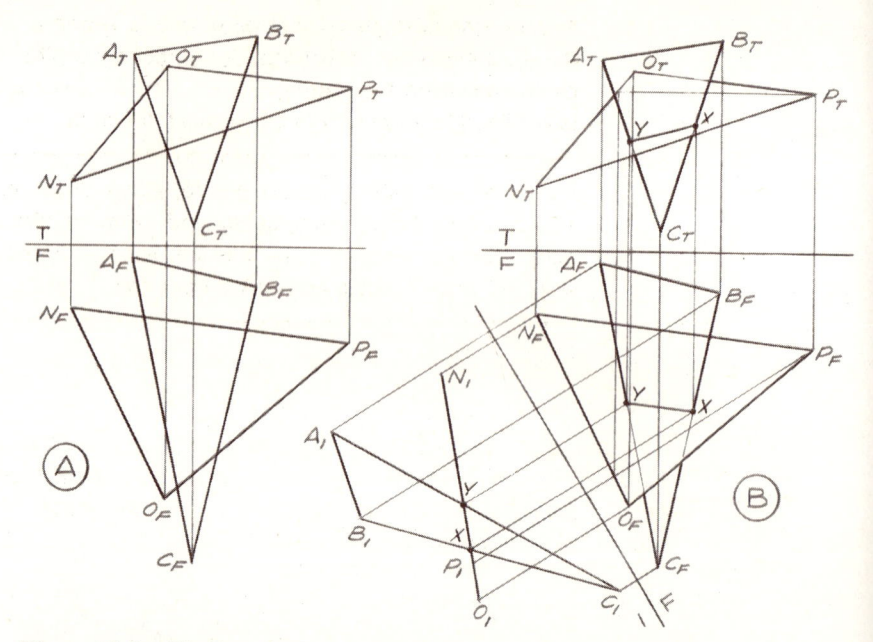

Figure 18.9 Finding the line of intersection between two planes—auxiliary view method.

PLANES AS CUTTING DEVICES

65   Whenever you slice a salami, you are creating a cutting plane (see Figure 18.10A). The motion of the knife, if accurate, creates a plane which sections the salami. The shape of the section is a(n) _____ or a(n) _____ depending upon the angle between the axis of the salami and the imaginary cutting plane (plane of the knife blade).

circle, ellipse

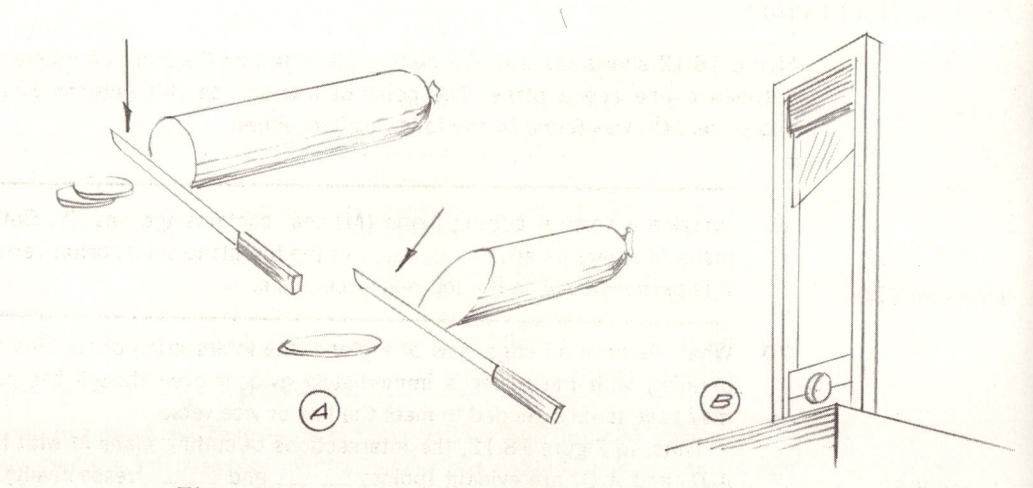

Figure 18.10 Concept of a cutting plane.

ANGULAR RELATIONS AND INTERSECTIONS 317

plane

66 A more dramatic planar cutting device is shown in Figure 18.10B. It is the guillotine—an instrument for inflicting capital punishment by decapitation. Here the cutting _____ is the blade itself, and it is confined to move only in the vertical direction.

top

67 We have used cutting planes before, either drawn or imaginary. Figure 18.11 shows both orthographic and pictorial views of a mixer-valve body. The top view is a full section which pictures the object as if it were cut in two pieces horizontally and the _____ half removed.

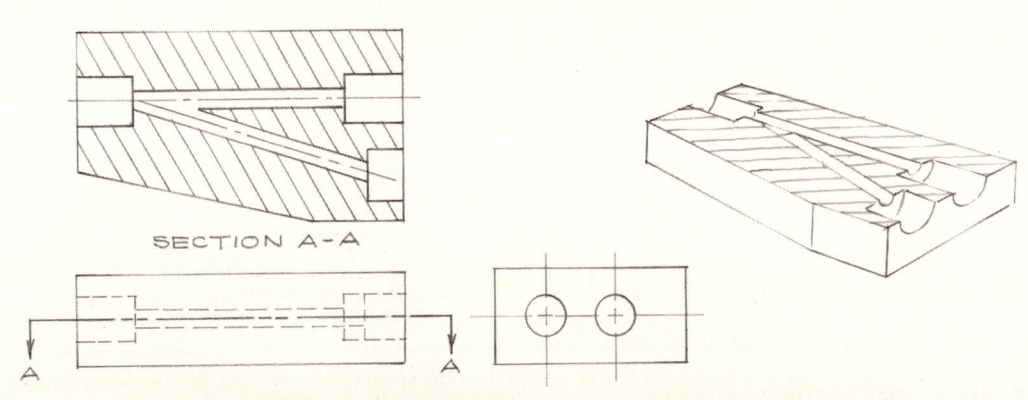

Figure 18.11 The sectioned view—an important use of the cutting plane.

edge

68 The front view shows a(n) _____ view of the cutting plane. This tells us how the object was sectioned.

CUTTING PLANE METHOD

Point of Intersection

Figure 18.12 shows the use of a cutting plane to find the point of intersection between a line and a plane. The point of intersection (W) between line XY and plane ABC was found by the following procedure.

edge view (EV)

69 Imagine a vertical cutting plane (M) that contains the line XY. Cutting plane M shows as a(n) _____ in the top plane since, being vertical, it is perpendicular to the top reference plane.

U_T and V_T

70 When we have an edge view of a plane, the intersection of any line (not parallel) with the plane is immediately evident even though the plane may have to be extended to meet the line or vice versa.
   Thus, in Figure 18.12, the intersections of cutting plane M with lines $A_T B_T$ and $A_T C_T$ are evident (points _____ and _____ respectively).

318 DESCRIPTIVE GEOMETRY/UNIT 5

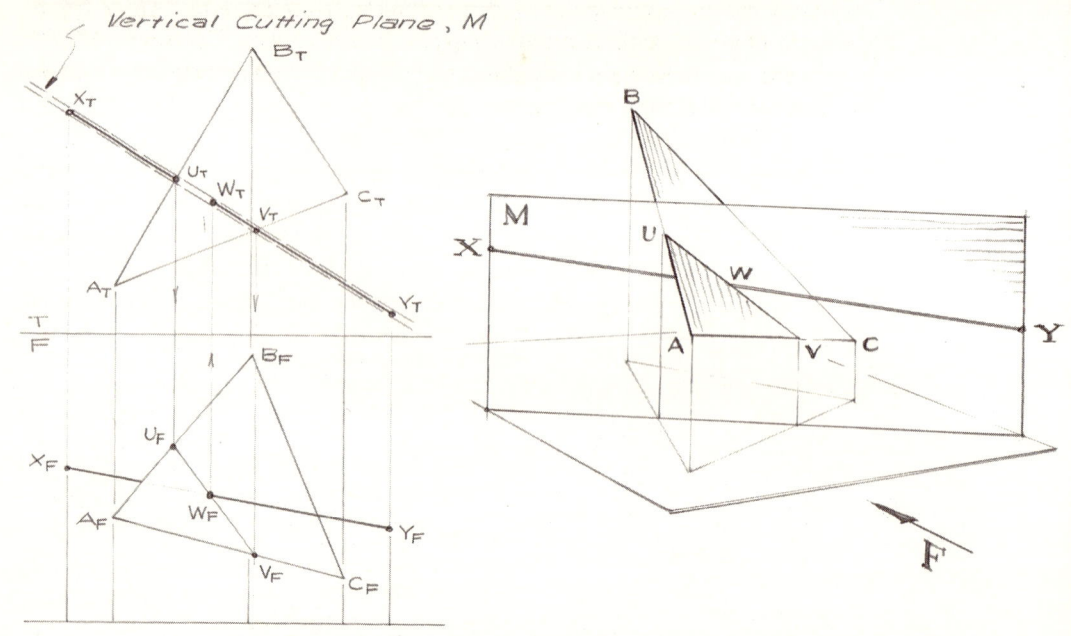

Figure 18.12 Intersection of a line and a plane—cutting-plane method.

projecting	71	Having points U and V in one view (top), we can find a second view of both by _____ U_T to $A_F B_F$ and V_T to $A_F C_F$.
intersection	72	Two points define a straight line, therefore $U_F V_F$ can only be the line of _____ between planes M and ABC.
(plane) ABC, (plane) M; (line) XY	73	We know (by construction) that line XY lies in plane M, therefore, the point W_F where $X_F Y_F$ intersects $U_F V_F$ must be common to three things: (1) plane _____; (2) plane _____; and (3) line _____.
$X_T Y_T$	74	W_F is, therefore, the desired point of intersection between line XY and plane ABC. The top view of W is found by projecting W_F to line _____.
imagined	75	We solved this problem by drawing four lines and no auxiliary views. We never had to draw the cutting plane; we merely _____ it.
perpendicular	76	Figure 18.13 shows the same problem solved by making the cutting plane _____ to the front reference plane.

The same result (point W) was achieved. It is often advantageous where there is a choice, to imagine a *vertical cutting plane* as in Figure 18.12 because it is easier to *visualize*.

The use of a cutting plane is a powerful aid in solving many types of space problems. You will see more examples in the next few chapters. In many instances the cutting plane may be imagined as an edge in one of the given views. In others, the cutting plane must be constructed as a triangle. In this latter case, we know we must define the triangle in at least two views before we have created a cutting plane.

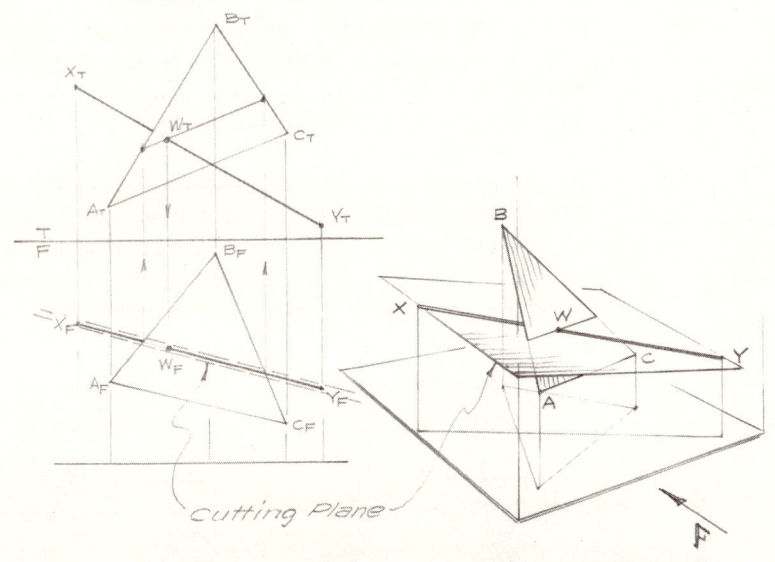

Figure 18.13   The cutting plane can be perpendicular to the front plane.

■ 77   Sketch 54A shows a plane *DEF* and a line *RS*. Line *RS* intersects plane *DEF* but the point of intersection is not shown. Find the point of intersection (*T*) without using an auxiliary view. Show the visibility of the line and plane in both views. Check your solution.

Line of Intersection

Figure 18.7 showed pictorially how we can find the line of intersection between two oblique planes by extending two side of one plane until we find point intersections with the second. These two points define a line, the line of intersection between two planes. In Sketch 54A we found the line of intersection between a cutting plane and *DEF* (the given plane) as a step in finding the intersection of a line and a plane.

■ 78   In Sketch 54B, find the line of intersection between the two oblique planes *ABC* and *DEF*. The procedure is exactly the same as that used in solving the line and plane intersection problem of Sketch 54A. Two cutting planes will be used since you must find the *points of intersection* between two lines on one of the planes (for example, lines *DE* and *DF* of plane *DEF*) and the other plane (*ABC* in this instance). Label these two points *X* and *Y*. Points *X* and *Y* are common to both plane *ABC* and plane *DEF*. Therefore, line *XY* is the desired *line of intersection*.

The complete solution is shown pictorially. Complete the orthographic solution, show visibility of the two planes, and check the results.

AUXILIARY VIEWS OF PHYSICAL OBJECTS

We have worked with auxiliary views in the last few chapters by considering abstract points, lines, and planes in space. Let's take a brief look at the uses of the techniques developed in these chapters for more practical purposes, the complete orthographic description of physical objects.

Many physical objects have faces or other features that are inclined (not parallel) to any of the six principal views when the object is placed in a normal position within the frame of reference.

inclined

79 Consider Figure 18.14. No matter how we establish the object with respect to the T, F, and P reference planes, one face will always be _____ to these planes.

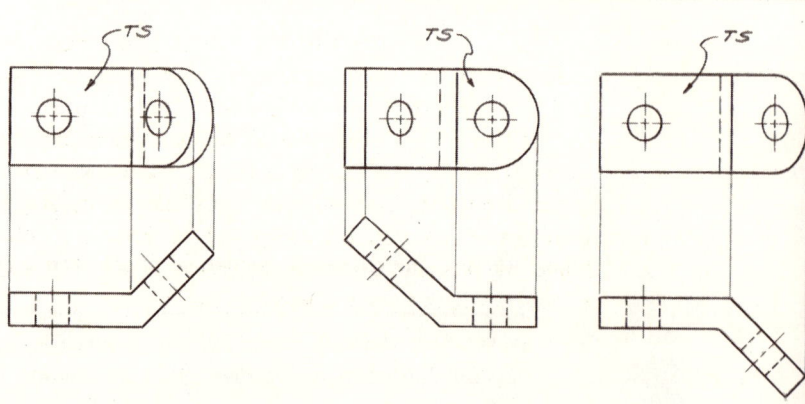

Figure 18.14—Reorienting the object will not produce a view showing the true shape of both faces.

auxiliary

80 To show a true shape of the inclined face we must use a(n) _____ view.

angular (bent)

81 Figure 18.15 shows a three-view orthographic projection that completely defines the object. Note that the F view is the only complete view and shows the true _____ nature of the object.

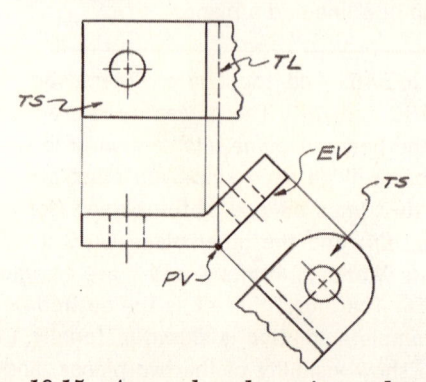

Figure 18.15 A complete three-view orthographic projection.

ANGULAR RELATIONS AND INTERSECTIONS 321

ellipses

82 The top view and auxiliary view 1 show the foreshortened face broken off to avoid the necessity of drawing _____ for the hole and the rounded end.

EV (edge view)

83 The problems of making the drawing of Figure 18.15 are identical to the problems of finding the TS of a plane. In fact, they are somewhat simpler because a(n) _____ of the inclined plane is available in one of the given views.

■ **84** Sketch 55 is an exercise in using auxiliary views for finding the TS of sloping faces on objects. Complete the views given, using any necessary auxiliary views to find the information requested. Refer to Figure 18.15. Check your solution.

This is the end of Chapter 18. It presented some methods for solving space problems involving angular relations and intersections between lines and planes and between two planes.

In addition to a discussion of line-and-plane and plane-and-plane problems we have seen how a cutting plane (or planes) can be useful in solving certain space problems. You will see more use of cutting planes in the following chapters. In visualizing these problems, keep in mind the image of the cutting plane as a knife blade or guillotine blade that slices through the three-dimensional problem and exposes the interior.

chapter 19   generating lines and surfaces

Chapters 15 through 18 presented the graphic operations on points, lines, and planes which are the foundation for descriptive geometry. These operations permit us to solve space problems using the orthographic projection systems. The techniques are standard and involve three main steps:

    (1) *Visualize* the known information.

    (2) *Reason* out the sequence of orthographic views needed to find the unknown information.

    (3) *Project and measure* according to a standardized, right-angle projection and measuring system.

Chapters 19 through 28 show examples of how these techniques are used to solve some useful space problems.

CONCEPT OF LINE AND SURFACE GENERATION

Lines and surfaces can be viewed as being generated by a moving point (for lines) or a moving line (for both plane and curved surfaces). This latter concept is particularly important for the orthographic description of certain types of curved surfaces.

Line Generation

generates

1   A line—whether straight or curved—can be said to be *generated* by a point moving in space. A point that moves in a specific manner in space _____ a line.

segment

2   We have said that a straight line (or at least a segment of a straight line) is defined by two points. In Figure 19.1A, AB is a(n) _____ of a straight line.

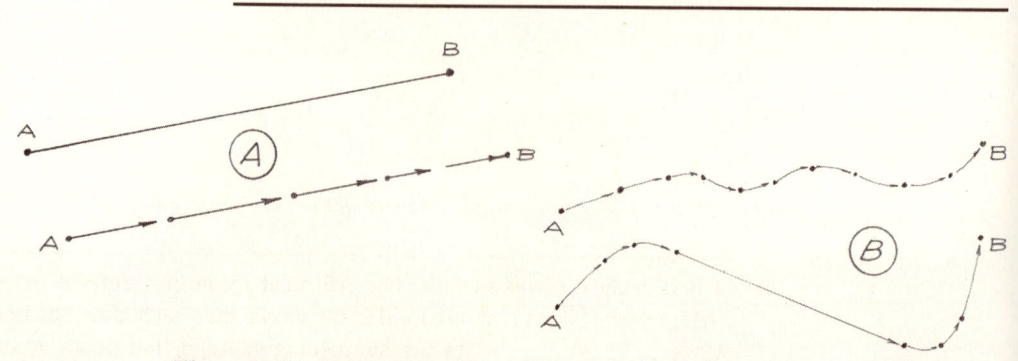

Figure 19.1   A moving point generates a line—either straight or curved.

GENERATING LINES AND SURFACES 323

generating

3 Another way to look at the line segment AB is to consider it as the locus of a number of positions of a *point*. As shown in Figure 19.1A, a point starts at A and moves to B along the shortest possible path, thus _____ the straight-line segment AB.

moving

4 If the point takes any other path than the shortest one, another line is generated, which is made up of either curved or straight segments or both. Figure 19.1B shows two non-straight lines generated by a(n) _____ point.

moves

5 A description of line AB is thus dependent upon the manner in which the point _____ in traveling from A to B.

This is a simple concept and not of particular significance since most line segments can be described and recognized as *lines* and not as paths of a moving point. However, let us consider what happens when a *line* moves in a predetermined fashion.

Surface Generation

surface

6 When a line moves in space, a *surface* is generated, the nature of the surface being a function of the type of line and the manner in which it moves. A moving line generates a(n) _____ in space.

A'B'

7 Consider Figure 19.2A. AA'B'B is a plane surface generated by line AB moving to a new position, _____.

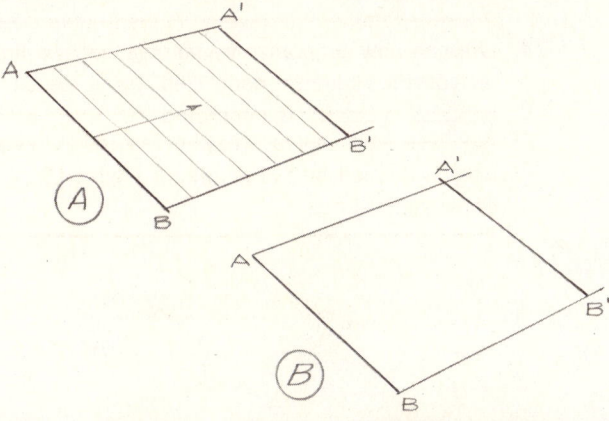

Figure 19.2 *A moving straight line can generate a plane surface.*

parallel;
AA' (and) BB'

8 If it is truly a *plane surface*, line AB must move in a definite manner. The pictorial sketch of Figure 19.2A shows that each new position of AB is _____ to the succeeding one and that points A and B constantly touch two straight-line segments, _____ and _____.

324 DESCRIPTIVE GEOMETRY/UNIT 5

moving

9   As with a line, it is easier for us to "see" the plane AA'B'B as four intersecting lines (AA', A'B', B'B, and BA) rather than as a *path* of a(n) _____ line AB. (Figure 19.2B.)

curved

10 In Figure 19.3A line AB moves in space in such a way that each new position is parallel to the succeeding position but, in this instance, the lines directing the motion are _____.

generating (moving)

11 The resulting curved surface is the general case of the *cylinder*. Figure 19.3B shows the same surface but without the successive positions of the _____ straight line, AB.

Note that the surface of 19.3A looks more like a curved surface than that of 19.3B. The inclusion of the different positions of the generating line aids the visual interpretation of the surface.

Nomenclature

generatrix; directrix

12 Figure 19.3A shows the nomenclature used to describe a surface generated by a moving line. The generating (moving) line is called the _____. The directing (stationary) line is called the _____.

elements

13 The different positions of the generatrix are called _____ of the surface.

surface

14 We can now generalize by stating that if a line, the generatrix, moves in a specific manner controlled by a second line, the directrix, a(n) _____ is generated.

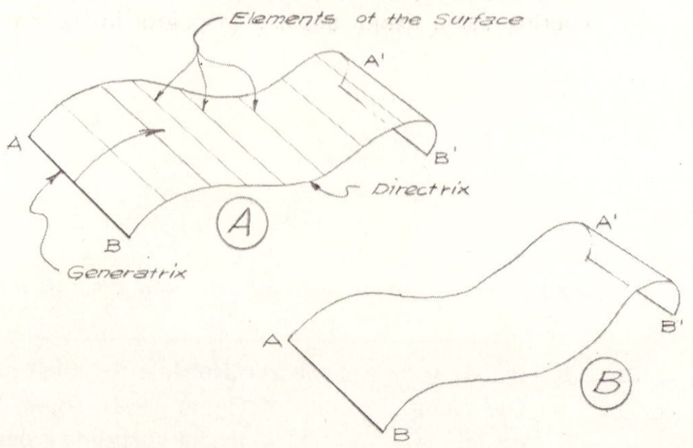

Figure 19.3   A moving straight line can generate a curved surface.

15 The Table of Figure 19.4 is a simple classification of surfaces based upon the nature of the generatrix and the directrix. This table is by no means complete, but it does show how surfaces are _____ by a moving line.

generated

GENERATRIX	DIRECTRIX	RESULTING SURFACE	
Straight	Straight	Plane 	
Straight	Curved	Single Curved 	
Curved	Straight	Single Curved 	
Curved	Curved	Double Curved 	

Figure 19.4 Classification of surfaces generated by a moving line.

The idea of a straight-line generatrix which generates a curved surface will be discussed in more detail in Chapter 25. Each successive position of the moving generatrix is a straight-line element on the surface. We already know how to project points and straight lines in the orthographic system, and so, using the concept of many elements making a curved surface, we will be able to project the curved surface by merely projecting points and lines.

For the moment, let's continue the discussion of points and lines by considering some useful aspects of points in the next chapter.

chapter 20   points—projecting shade and shadow

POINTS IN SPACE

The first figure (1.1) in this text was a dot-to-dot drawing of the type found in children's books. By joining points (dots) with straight lines in a predetermined sequence, a perspective sketch of an automobile resulted. The inclusion of this exercise was intended to show that *any* object can be considered to be made up of a series of points and that the problem of learning to draw any object is essentially the problem of learning *where to place the points*.

A single point is of no particular concern. We do know, however, that two points define a straight line; three points define two intersecting straight lines or a plane surface; and many points can define a very complicated curved or planar, three-dimensional object as long as we know how the points are to be connected with straight or curved lines. In effect then a point is most useful in graphics in defining the ends of a straight line or successive positions on a curved line.

PROJECTED SHADE AND SHADOW

There is one interesting application of the technique of plotting points. That is the orthographic plotting of shade and shadow. This subject is presented here because it offers the following advantages:

    (1) It is an excellent exercise in spatial visualization.
    (2) It provides good practice in plotting points in the orthographic system.
    (3) It aids pictorial sketching by showing how objects cast shadows.

POINTS—PROJECTING SHADE AND SHADOW   327

shadow

1. Figure 20.1 shows a sundial—a clock that shows the time of day by casting a(n) _____ of a line on a calibrated time scale.

2. We consider rays from the sun to be _____. Thus, as the sun moves across the sky a sharp image (shadow) of the line will move across a horizontal plane.

parallel

Parallel Rays from the Sun

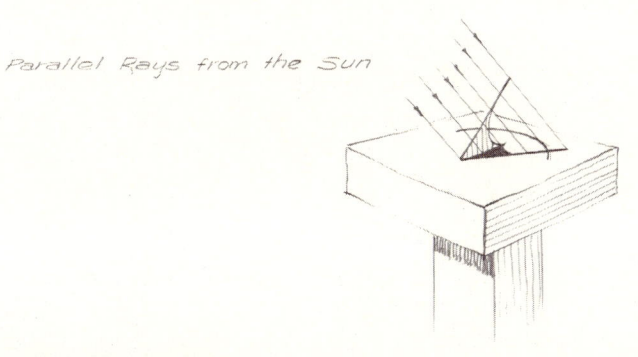

Figure 20.1 The sundial.

vertically

3. The simplest sundial is shown in Figure 20.2A. It consists of a rod stuck _____ into the ground.

direction

4. At any given time of day, the sun's position is known and thus the _____ (AB) of the parallel rays can be known as related to the orthographic frame of reference.

Figure 20.2B shows the orthographic views of the rod and its shadow. The orthographic views of arrow AB, the direction of parallel light rays, are also shown.

front

5. For a simple object such as a straight, vertical rod, only *one* point must be plotted to find the shadow on the ground plane. This is the top of the rod (point X).
   The shadow of point X is point X'. The position of point X' on the ground is found first in the _____ view.

edge

6. This is necessary, since, in the front view, the ground plane shows as a(n) _____ view.

X'_F

7. Thus a ray projected through X_F parallel to $A_F B_F$ intersects the ground plane at X'_F. The shadow does not show at all in the front view, but shows completely in the top view. A ray drawn parallel to $A_T B_T$ through point X_T can be extended indefinitely without knowing where the *intersection with the ground occurs.* This information can be found in the top view by projecting point _____ to an intersection with the ray through X_T.

328 DESCRIPTIVE GEOMETRY/UNIT 5

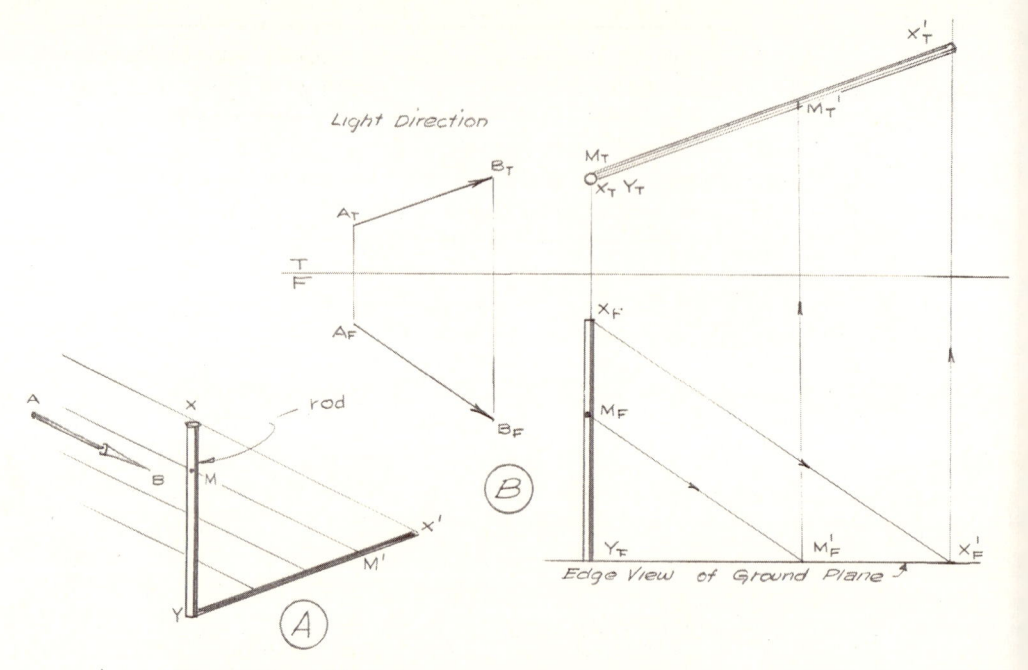

Figure 20.2   The geometry of projected shadows.

shadow	8	Line $X'_T Y_T$ is the _____ of rod XY on the ground plane.
$X'_T Y_T$	9	No other point on XY need be found. This can be shown by considering any point such as M on the rod XY. If the above procedure is followed, the "shadow" of point M will be found to lie on line _____, which is the shadow of rod XY on the ground plane.

In other words, point M'_F is found in the front view by finding where a ray through M_F intersects the grounnd plane (M'_F). A ray through M_T coincides with the ray through X_T (top of the rod) therefore M'_T (the top view of the shadow of point M) lies on the existing shadow, $X'_T Y_T$.

two (2)	10	In the example of Figure 20.2, the direction of the light rays (line AB) was selected at random. *Any direction can be chosen.* The only requirement is that a line representing the direction must be drawn in at least _____ adjacent orthographic views.
	11	Figure 20.3 shows five different sun positions and the effect of these positions on the shadow cast by a vertical rod. (1) At dawn or sunset the sun's rays are essentially horizontal and thus cast an infinitely long shadow. (2) A low sun casts a(n) _____ shadow. (3) A middle sun casts a shadow equal in length to the height of the rod. (4) A high sun casts a(n) _____ shadow.
long; short; no		(5) A vertical sun (high noon on the equator) casts _____ shadow.

POINTS—PROJECTING SHADE AND SHADOW 329

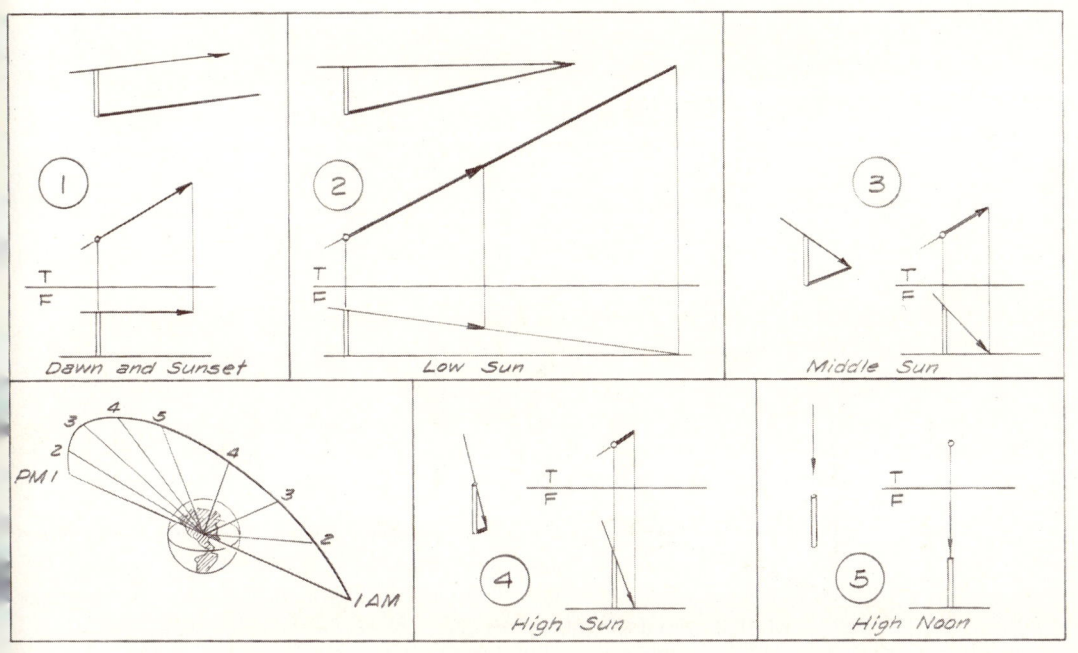

Figure 20.3 The effect of the sun's position on a cast shadow.

length

12 Architects and others who use the techniques of orthographic projection of shade and shadow have a complete range of light directions from which to choose. Indeed, they can show time of day by the _____ of the shadow cast by the object pictured.

Standard Direction of Light

In technical drawing, there is no reason to show time of day in a drawing (we are not trying to show a mood). Architects, illustrators, and industrial designers have standardized on one particular direction to show shade and shadow. This is near the middle sun position, shining over the *left shoulder* as one views the object.

45

13 The standard direction coincides with a body-diagonal of a cube oriented as shown in Figure 20.4. In the orthographic drawing, the arrow representing this *standard direction* makes an angle of _____ degrees with the T/F, F/T, F/P, and P/F reference lines.

45-45-90

14 Because of this, projecting shade and shadow in the orthographic system is made easy since the standard _____ drawing triangle may be used to project rays in any one of the six principal views.

DESCRIPTIVE GEOMETRY/UNIT 5

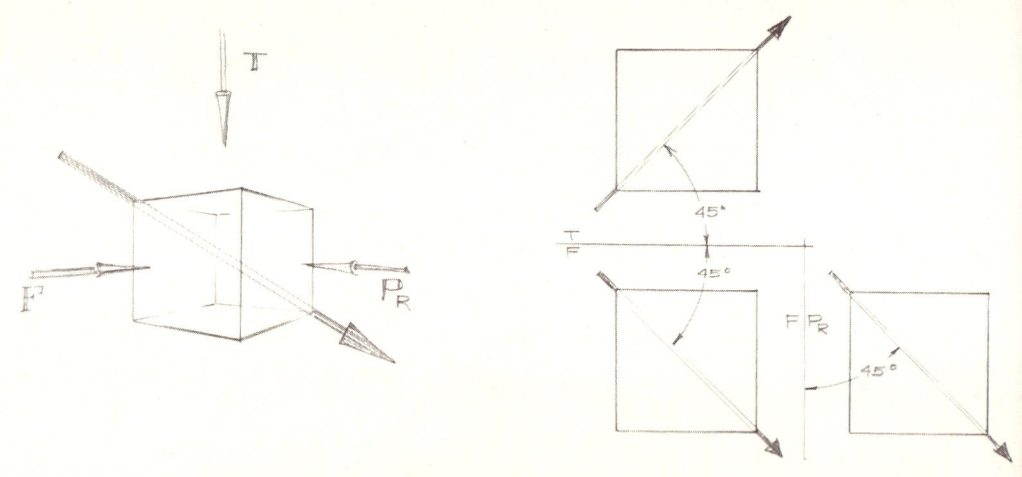

Figure 20.4 A standard light direction for projecting shade and shadow.

---

15 Architects use this standard light-ray direction so that the shadows cast by recesses in buildings will appear orthographically the same height or width as the *depth* of the recess.

Figure 20.5 shows this. The height of the shadow cast by the top edge of the doorway and the width of the shadow cast by the left vertical edge are both equal to D, the _____ of the doorway recess.

depth

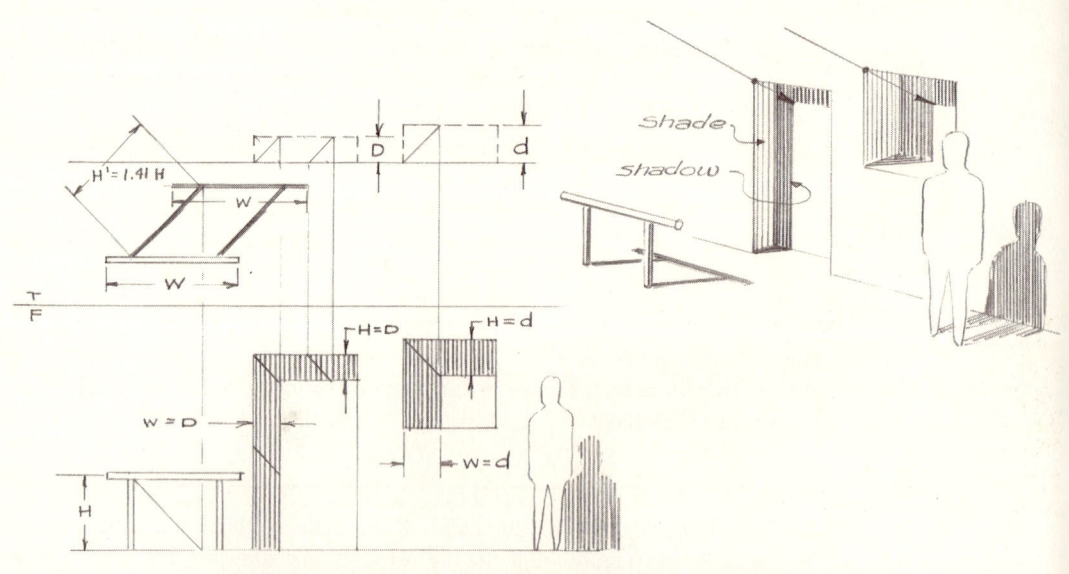

Figure 20.5 The standard light direction gives a shadow whose height and width equal the depth of the recess producing the shadow.

---

16 The same is true of the recessed window and the top of the hitching post. The dimension, W, of the ground shadow of the horizontal bar on the hitching post is _____ to the actual width of the bar.

equal

POINTS—PROJECTING SHADE AND SHADOW   331

larger

17 Note, however, that the shadow dimension H', representing the height H, of the hitching post is _____ than the actual dimension. (*Choose one:* smaller/larger.)

It will actually be $\sqrt{2} = 1.41$ times the true height since the true angle between the standard light direction and the horizontal plane (ground) is 35°16'.

18 Before proceeding, we should define the difference between *shade* and *shadow*. Consider Figure 20.6. It shows a pictorial sketch of a square, vertical post with shade and shadow indicated.

opposite (away from)

Shade is the absence of light on the sides of a single object, that is, the sides _____ (your words) a single light source.

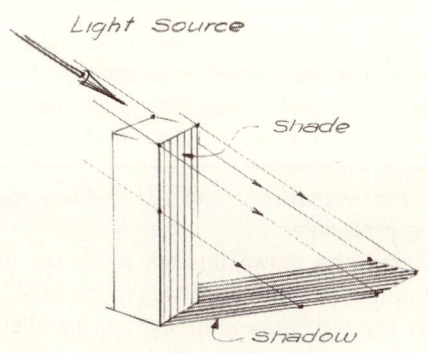

Figure 20.6   Shade versus shadow.

19 *Shadow* is the absence of light on one object due to the intervention of a second object in the light beam. When we walk in the sun, our body casts a(n) _____ on the ground but the side of the body away from the sun is in _____.

shadow; shade

20 By these definitions, we do not "sit in the shade of the old apple tree" as the old song tells us, but we sit in the _____ of the old apple tree.

shadow

21 Since we are waxing romantic about this subject, let's look to the moon for a good example. Figure 20.7A shows three phases of the moon. We see only part of the moon although we know it is a full sphere. The part we cannot see against the night sky is in _____.

shade

22 Figure 20.7B shows the geometry existing during a full eclipse of the moon. Earth has intervened between the Sun and the Moon, and, therefore, the Moon is in Earth's _____.

shadow

332 DESCRIPTIVE GEOMETRY/UNIT 5

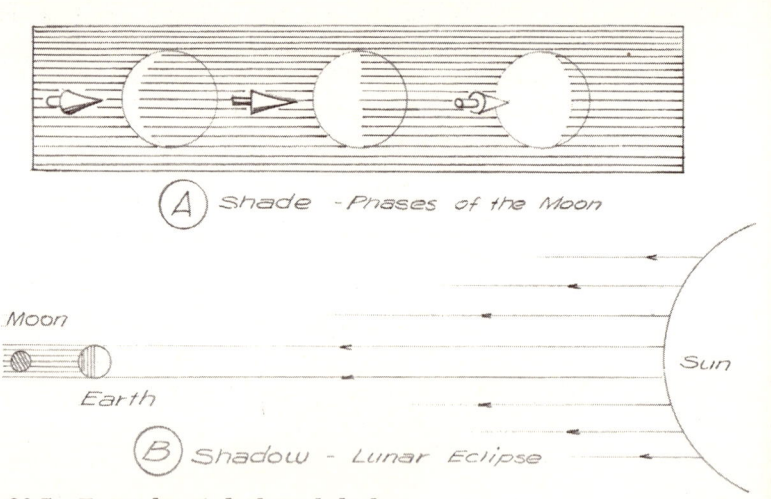

Figure 20.7 Examples of shade and shadow.

ABCD ($A_TB_TC_TD_T$)

23 Let's now go back to the square post and find out how the shadow on the ground can be plotted accurately in the orthographic system. The figure in Sketch 56A shows the standardized light direction (L) and the post in top and front views. The top of the post is labeled _____.

24 The ground shadow of point A has been found in both views by the following procedures:
(1) A ray was drawn through A_F in the direction L_F until an intersection with the ground plane was found at A'_F.
(2) A ray was drawn through A_T parallel to L_T and extended (no intersection is obvious).
(3) Point A'_F was projected orthographically into the top view until it intersected the ray through A_T. This gives the shadow of point A on the _____ in both views.

ground

■ 25 Complete the shadow according to the following procedures (Sketch 56A):

Sketch the construction necessary to find the shadows (B', C', and D') of points B, C, and D in both views. Point C doesn't actually have a shadow on the ground, but include it in the construction anyway.

You now have the shadow of the square post top on the ground. Note that it is identical in size and shape to the actual top. The two vertical edges BE and DF will control the rest of the shadow.

Complete the ground shadow of the post by drawing straight lines from E_T to B'_T and from F_T to D'_T.

Complete the problem by shading in the area bounded by $F_TD'_TA'_TB'_TE_TA_T$. Use simple, bold, line shading that is parallel to L_T. Check your solution.

26 The shadow of point E_T is at E_T and the shadow of point F_T is at F_T. (True? _____. False _____.)

True

POINTS—PROJECTING SHADE AND SHADOW   333

This is true because both points E and F lie on the ground and therefore cast no shadow. Any point on $B_T E_T$ will cast a shadow on line $E_T B'_T$. Try it! Select any point X that lies on line BE. Show it in two views and find its shadow by the procedure used for A', B', C', and D'. The answer to Sketch 56A shows a typical point X.

$F_T D'_T$

27 The same would be true for *any* point on line DF. Its shadow would fall on line _____.

---

28 This orthographic solution provided the information necessary to make and shade the pictorial sketch of Figure 20.6. Note that none of the _____ (dark) side of the post shows in the orthographic views of Sketch 56A.

shaded

---

29 The solution was obtained by _____ a series of points in two orthographic views and then connecting the points in correct _____ with straight lines to achieve the desired solution.

plotting (projecting); order (sequence)

This is the procedure of descriptive geometry no matter how complicated the problems become: *Plot points in at least two adjacent views and then connect the points in proper sequence.*

---

30 It was stated earlier that the ground shadow of the top of the square post was a same-size square. Figure 20.8 shows that *the shadow of a plane surface cast upon a surface parallel to the plane wil be an exact duplicate of the original plane surface.* This is true only if we assume _____ light rays.

parallel

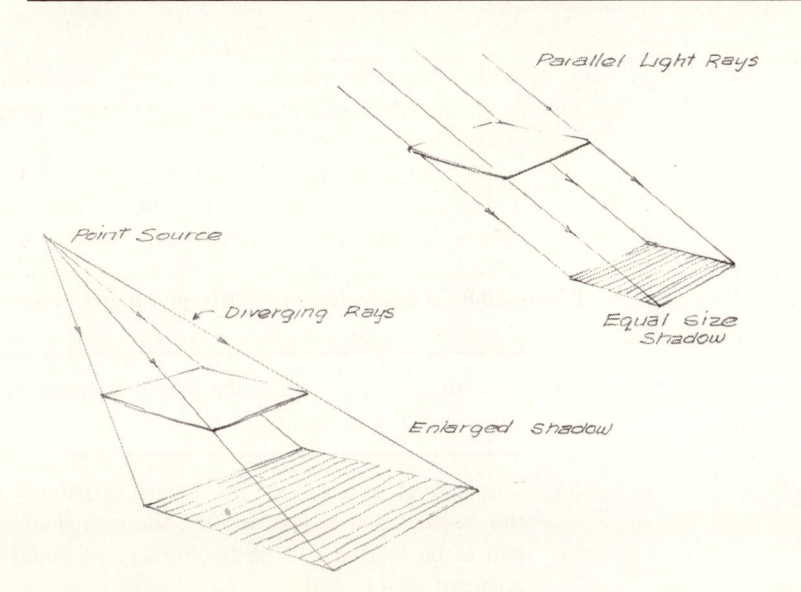

Figure 20.8   Parallel rays give equal-size shadows on parallel surfaces.

334 DESCRIPTIVE GEOMETRY/UNIT 5

31 This is an important concept and understanding it helps to visualize in advance what the shadow of an object will be.

Sketch 56B shows a circular disk mounted horizontally on a thin vertical rod. Find the shadow of the disk on the ground.

NOTE: Remember the statement of the last frame. You should know how the shadow of the disk will appear and this should suggest an easy way of finding it. The answer sketch shows the simplest construction needed to solve this problem. Check your solution.

32 If projecting a shadow on a parallel surface gives an equal image, what happens when the surface is not parallel? Consider Figure 20.9. It is the same as the example of Sketch 56B except that a(n) _____ wall has been placed behind the object.

vertical

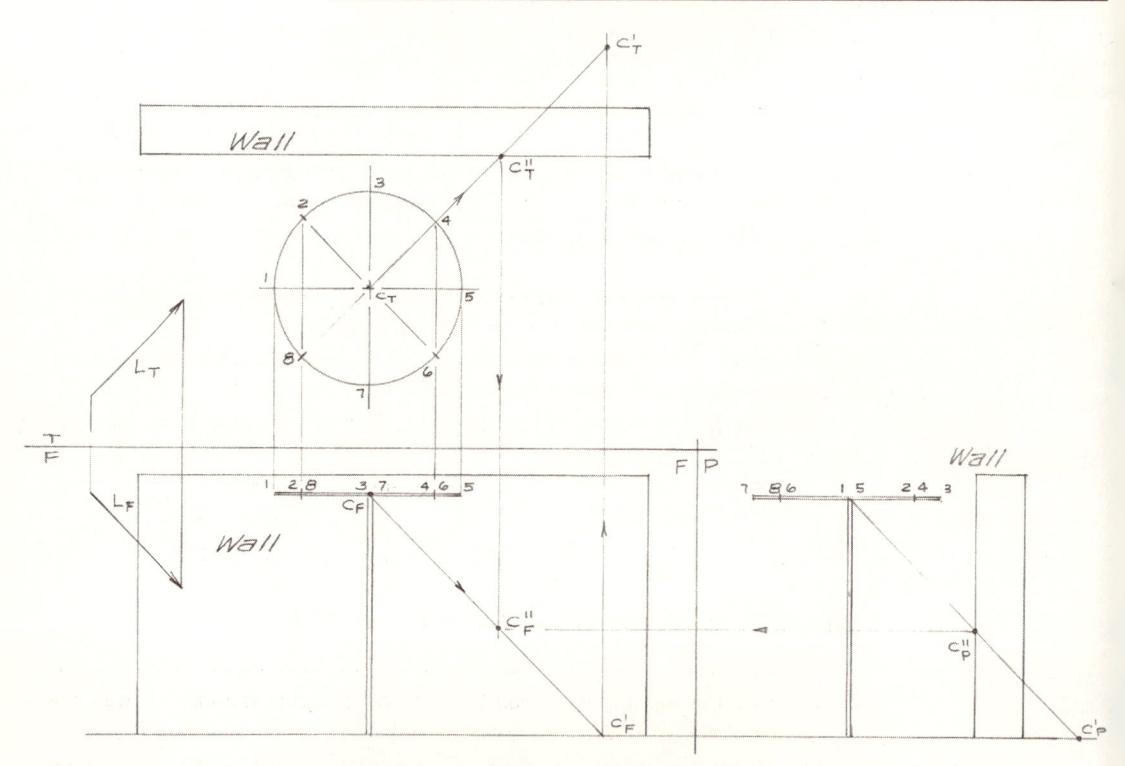

Figure 20.9 The shadow of a horizontal plane on a vertical surface.

33 Construction based upon the previous examples shows that the ground shadow of the center, C, of the disk lies _____ the wall.

behind

34 This cannot be true, so the shadow of point C must lie on the wall. In the construction of Figure 20.9, the wall shadow, C'', of the disk center had to be plotted by first finding _____ and then projecting into the adjacent view to find _____.

C''_T; C''_F

POINTS—PROJECTING SHADE AND SHADOW   335

points

35   However, the plane of the wall is not parallel to the plane of the disk. Therefore, we can assume that the shadow on the wall will not be a same-size circle. Since we cannot draw a circle using C''_F as a center, we must divide the circular outline of the disk into a number of _____ and plot each one separately to find the shadow of the disk on the wall.

ellipse

36   The circular view (top) of the disk is divided into eight equal segments each providing a numbered point on the circumference. Figure 20.10 is a complete solution of this problem. The shadow of the circular disk on the wall is a(n) _____.

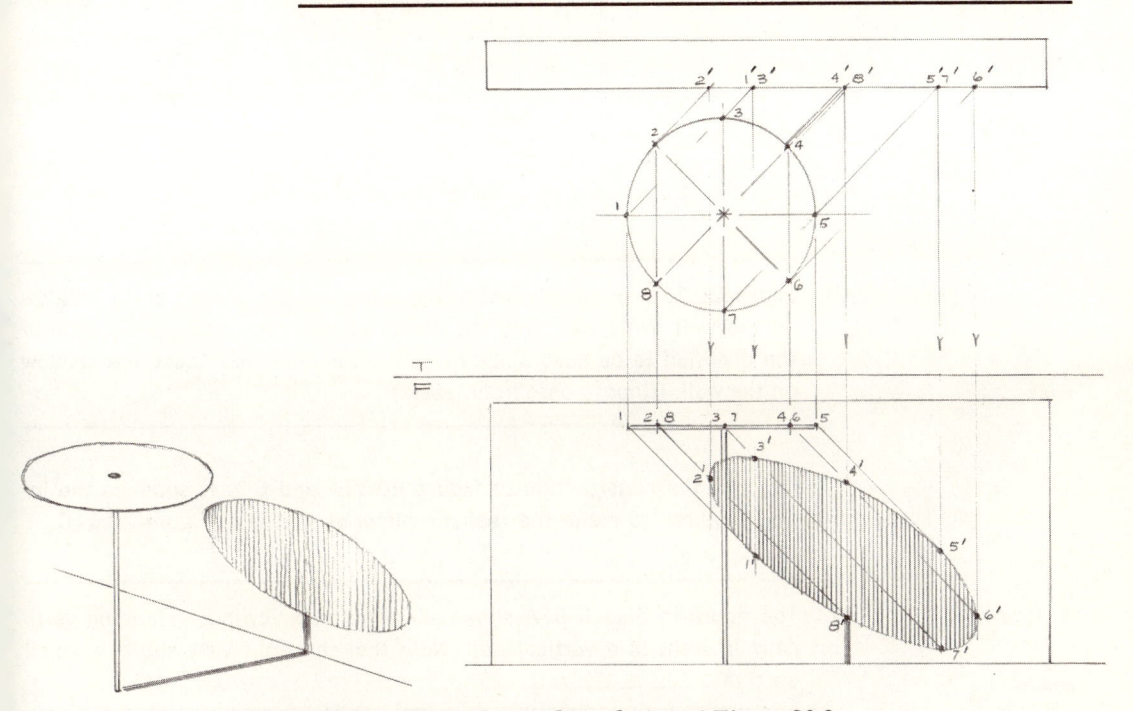

Figure 20.10   The orthographic solution of Figure 20.9.

smooth curve

37   Each of the eight points was plotted in the same manner as was the center of the disk, C. Once plotted, a(n) _____ _____ was drawn through points 1', 2', 3', 4', 5', 6', 7', and 8' to produce the elliptical shadow.

circular;
elliptical

38   Figure 20.11 is the same problem except that the wall has been moved back from the disk. The resulting shadow will lie partly on the ground and partly on the wall. The part on the ground will be _____ and the part on the wall will be _____.

behind

39   At 20.11A the circular part on the ground is determined by finding the ground shadow of the center C, even though it lies _____ the wall.

336 DESCRIPTIVE GEOMETRY/UNIT 5

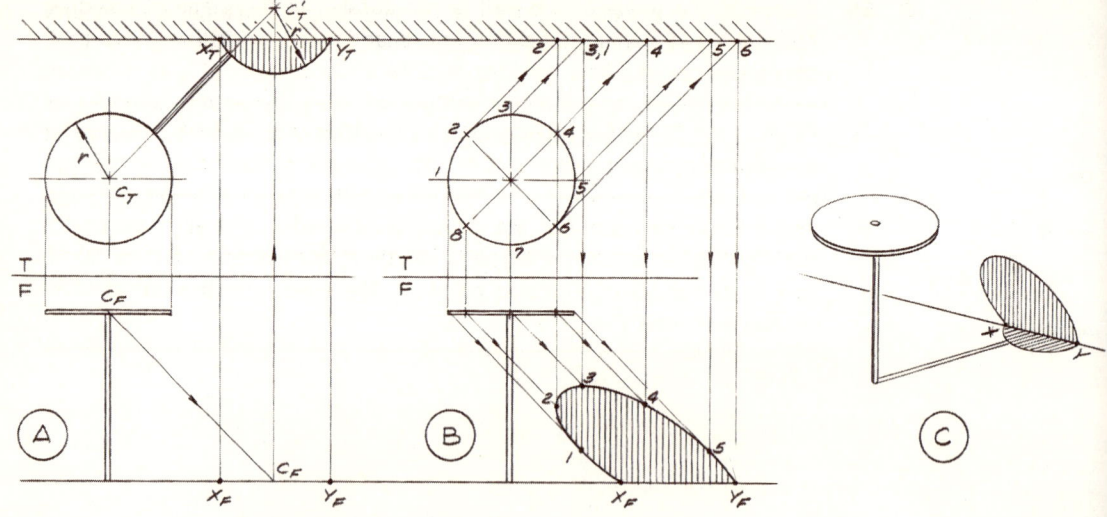

Figure 20.11 A shadow may lie on two or more surfaces.

back	**40**	Figure 20.11B shows the rest of the solution—the portion of the shadow on the wall. Note that only the points on the _____ side of the disk had to be used since these are the ones that cause the shadow on the wall. (*Choose one: front/back.*)

Again, it is the construction of Figure 20.11A and B that supplies the information required to make the realistic pictorial sketch of Figure 20.11C.

shade	**41**	The figure in Sketch 57A shows a right-circular cylinder standing vertically in front of a vertical wall. Note that the front view shows a small strip of _____.
2 (and) 6	**42**	Shade is defined as the absence of light on the parts of the object which are opposite a single source of light. The top view shows rays of light striking the cylinder. Points _____ and _____ define the limit between the light side of the cylinder and the dark side.
6	**43**	Point _____ was used to determine how much of the dark side would show in the front view.
	■ **44**	Complete the views of Sketch 57A by finding the shadow of the cylinder on the wall and on the ground. Check your solution.
	■ **45**	Objects can cast shadows on themselves. Consider the object of Sketch 57B shown in front and profile views. It is a wall plaque with a rectangular niche and a triangular shelf. Sketch the construction required to find the shadow on the object. Check your solution.

POINTS—PROJECTING SHADE AND SHADOW 337

■ 46 Sketch 57C is another part that combines circular and sloping members. Find the shadow of the object. Point A' on the shadow has been located. Check your solution.

47 Figure 20.12A shows a compound cylindrical part—a part made up of two different-sized cylinders. Let's consider the shadow cast by the upper cylinder on the lower one. Will this shadow be cast by the top edge or the bottom edge of the large diameter cylinder? _____.

Bottom

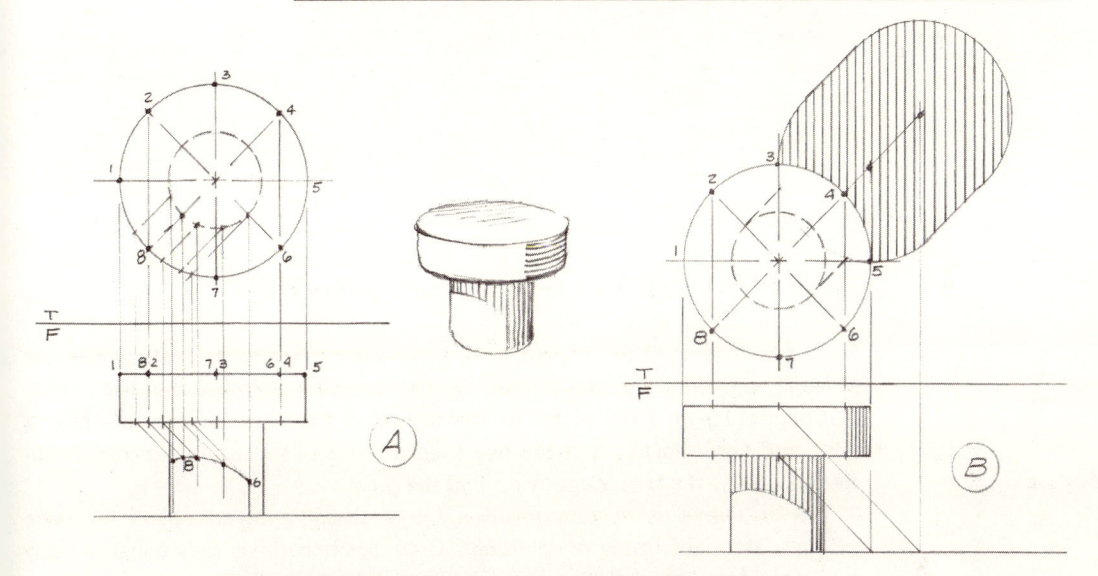

Figure 20.12 Object casting a shadow upon itself.

A few points on the bottom edge on each side of point 8 will be sufficient to find the shadow cast by the upper cylinder on the lower cylinder. The successful solution of shade and shadow problems involves some preliminary visualization to determine what the shadow should look like. This visualizing gives the clue as to where to plot the points to achieve the desired results.

Figure 20.12B shows the complete solution to the problem including the shadow on the ground plane. Note that the shadow on the ground is caused by (1) points 2, 3, 4, 5, and 6 on the top surface of the large diameter cylinder; (2) points 6, 7, 8, 1, and 2 on the bottom surface; (3) the vertical lines at 2 and 6 joining the top and bottom surfaces; and (4) the vertical sides of the small-diameter cylinder.

From the foregoing discussion, you can see the value of considering three-dimensional physical objects as being made up of a series of points. In the orthographic projection system, it is easy to project the points into any desired view and then join them with straight or curved lines to find the desired information.

This brief discussion of projected shade and shadow is intended to sharpen your ability to visualize in the orthographic system. The knowledge of how shadows are cast and how the shaded side of an object can be determined is a powerful aid in making realistic shaded perspective sketches. You need not have the training of an artist to make your sketches show exactly what you intend.

chapter 21   *straight lines*

From the consideration of points, we extend our thinking to straight lines or, at least, segments of straight lines—the shortest distance between two points. In Chapter 15 we have examined the orthographic characteristics of straight lines and know how to find the two special views of most interest in descriptive geometry, the true-length view and the point view.

The orthographic system involves the positioning of an object in space within a standard frame of reference. Once positioned, we may construct any auxiliary views needed to supply further information about the object. A projected auxiliary view is merely the view we would see if we changed our position with respect to the object. We move around in space until we see the object in a desired manner (true length, point view, etc.).

There is one useful operation that involves moving the object within the frame of reference. This operation is finding the true length of a line by the rotation method.

TRUE LENGTH OF A LINE BY ROTATION

Often we are interested in finding only the *magnitude* of the true length of a line as opposed to finding a *true-length view*. This occurs in vector geometry (Chapter 22) and in the development of polyhedrons and certain simple curved surfaces (Chapter 27).

parallel

1   Figure 21.1 shows the top and front views of line *AB* and an auxiliary view which shows the true length of *AB* because the auxiliary reference plane was set up _____ to line *AB*.

magnitude

2   Using the method of *rotation* to find the true length of a line eliminates the need for an auxiliary view. As stated above, this method is useful when we wish to find just the _____ of the true length.

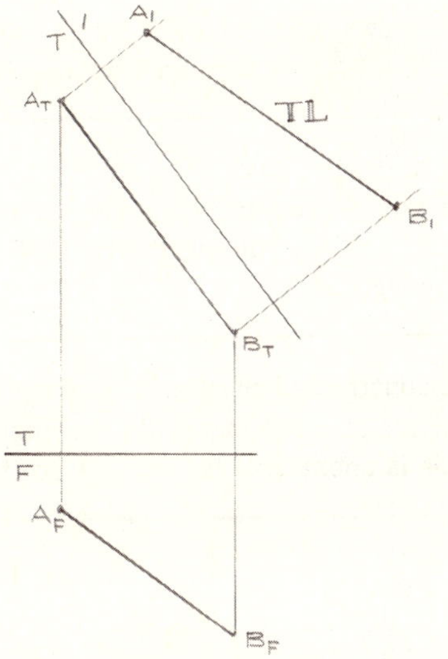

Figure 21.1 Review: TL of a line by the auxiliary-view method.

---

3 Figure 21.1 shows a *true-length view* which includes the magnitude.
The criterion for finding the true length of line is that the line must be ________ onto a reference plane that is parallel to the line.

projected

---

4 The method of rotation involves rotating the line in such a manner that it is parallel to one of the two given views, usually the ________ view or the ________ view.

top, front

---

5 In the orthographic system, an operation performed in one view must be performed in ________ views.

all

Moving the line within the frame of reference must be done in a deliberate manner or an entirely new line will result and the *TL* information will not necessarily be correct.

If you analyze the different ways of changing the position of a line within the frame of reference, you find that you can rotate the line about a specific axis and not change its relative position with respect to one of the principal reference planes.

---

6 Figure 21.2*A* shows orthographic and pictorial views of line *AB*. An *axis of rotation (XY)* has been included. The axis intersects line *AB* at point *A*. It shows as a point in the top view and as a *TL* line in the front view.

This means that the axis has been constructed ________ to the top reference plane and to plane *M* which is parallel to the top plane.

perpendicular

---

340 DESCRIPTIVE GEOMETRY/UNIT 5

cone

7 Figure 21.2B shows what happens when line AB is rotated about axis XY. A(n) _____ is generated by the moving line.

constant

8 Since this is a right-circular cone and the base (plane M) is perpendicular to the axis of rotation, the angle θ, that line AB makes with plane M is _____ during the full 360° of rotation. (Choose one: constant/changing.)

rotate

9 Plane M was constructed parallel to the top reference plane, therefore if we _____ AB about XY, the angle between AB and the top reference plane is unchanged for any new position of AB.

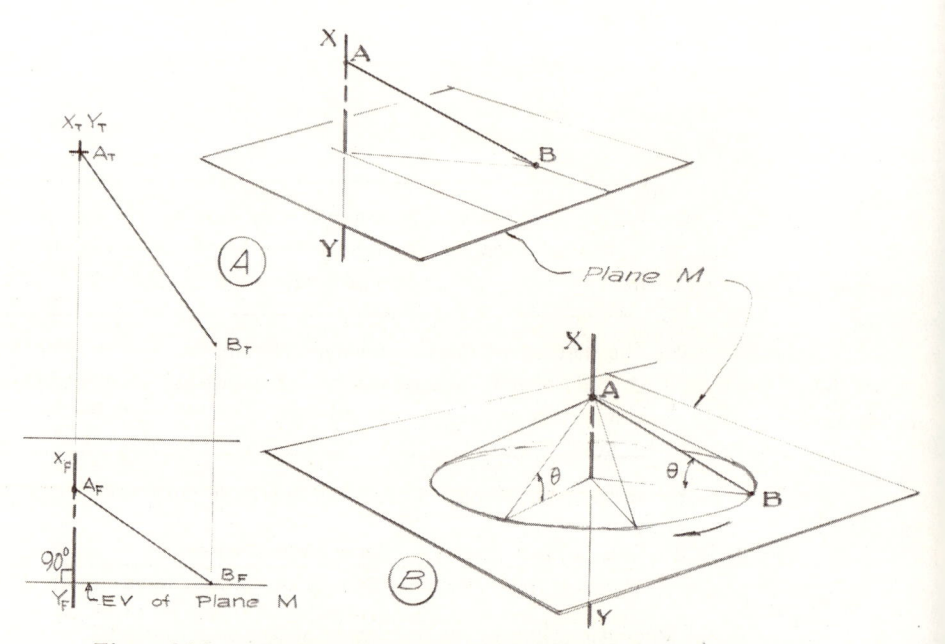

Figure 21.2 Rotating a line while still keeping its true angle with one of the principal planes constant.

circle

10 Figure 21.3 shows the orthographic views of the cone generated by rotating AB about XY. The top view is a circular view of the cone with the vertex at A_T and point B_T describing a plane _____.

front

11 The front view of the plane circle is an edge view perpendicular to axis XY. The entire _____ view appears as a typical triangular contour.

True

12 The two extreme contour positions of AB, $A_F B'_F$ and $A_F B''_F$, are both TL views of AB. True _____? False _____?

STRAIGHT LINES 341

13 The top view shows that these two positions, $A_T B'_T$ and $A_T B''_T$, occur
when the rotation is stopped while line AB is _____ to the
T/F line (and thus to the front reference plane).

parallel

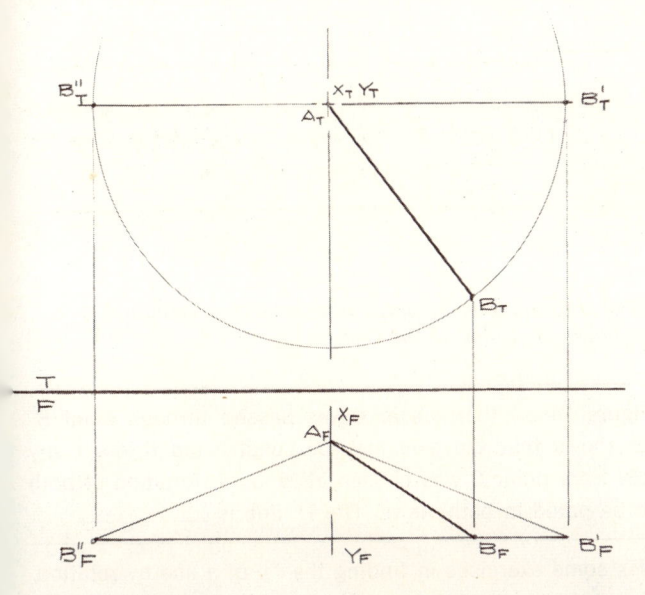

Figure 21.3 Rotating line AB about axis XY generates a right-circular cone.

Figure 21.4 Simplest construction for finding the TL of a line by rotation.

14 This is the criterion for finding the *TL* of a line.
In practice, the full cone need not be drawn. Figure 21.4 shows the
minimum construction needed to find the *TL* of *AB* by rotation. The *TL*
line is line _____.

$A_F B'_F$

15 It was implied earlier that the line must not be moved in any way that
will change its *relative position* with respect to one of the principal
reference planes. By making the axis *XY* _____ to one of
the two given planes (top), the true angle between line *AB* and the top
reference plane was not changed by rotation.

perpendicular

16 The problem could have been solved by setting axis *XY* perpendicular
to the front reference plane rather than to the top. Figure 21.5A shows
this construction. The *TL* line is _____.

$A'_T B_T$

342 DESCRIPTIVE GEOMETRY/UNIT 5

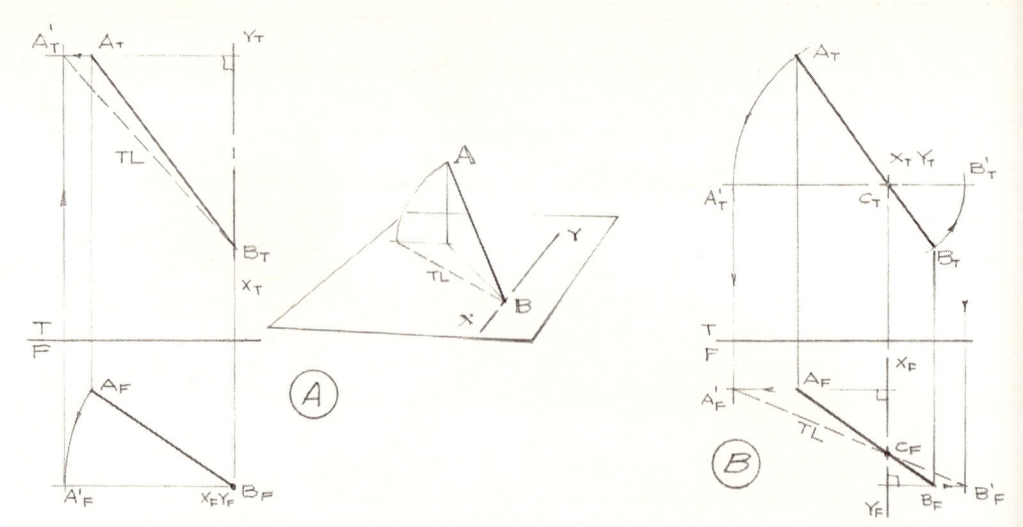

Figure 21.5 The axis of rotation may be perpendicular to any reference plane and may pass through any point on the line.

A'$_F$B'$_F$

17 Note also in Figure 21.5A that the axis was passed through point B.
    Figure 21.5B shows that the axis may be constructed through any point on line AB. Here point C was chosen at random. Rotation of both A and B had to be noted in both views. The *TL* line is _____.

■ 18 Sketch 58A gives some exercises in finding the *TL* of a line by rotation. Try these and check your solutions with the answer sketches. The answers show alternate solutions for each problem.

    The examples in 21.5 show that putting the axis through either end of the line results in a simpler solution.

chapter 22 vector geometry

Before leaving the study of straight lines for that of curved lines, let's look at an interesting and important use for straight lines—the graphic solution of *vector* problems. All of society, and especially our technical society, is concerned with numerical quantities. These quantities fall into two categories.
  (1) Scalar quantities
  (2) Vector quantities
A scalar quantity shows a *magnitude* to an appropriate scale. Temperature is a good example. We talk of 72° Fahrenheit which is a numerical description of room temperature according to an established scale; 22.2° centigrade describes the same condition to a different scale. Other scalar quantities that describe just a magnitude are hours, dollars, angular degrees, inches, etc.

On the other hand, certain quantities imply a *direction* and a *line of action* (position) as well as a magnitude. Force, weight, velocity, and acceleration are examples and are called *vector quantities*.

Vector analysis is most useful in the field of *mechanics* which treats of the action of forces on bodies. *Statics* is the part of mechanics concerned with the action of forces in producing rest or equilibrium. *Dynamics* relates to the action of forces in producing motion.

This chapter is concerned with some simple techniques of analyzing vector problems by graphic means. Any vector problem can be solved precisely by analytical means—especially now, with the ready availability of electronic computers. Some problems lend themselves to graphic solutions for, at least, a first-try approximate answer. Familiarity with the techniques of vector geometry adds another tool to the scientist's or engineer's capability for making quick but accurate value judgments.

---

force

1 A vector expresses the magnitude, direction, and line of action (or position) of certain numerical quantities. Figure 22.1A shows a boy pulling a sled. He is exerting a(n) _____ on the sled through the rope in an effort to cause the sled to move over the ground.

344 DESCRIPTIVE GEOMETRY/UNIT 5

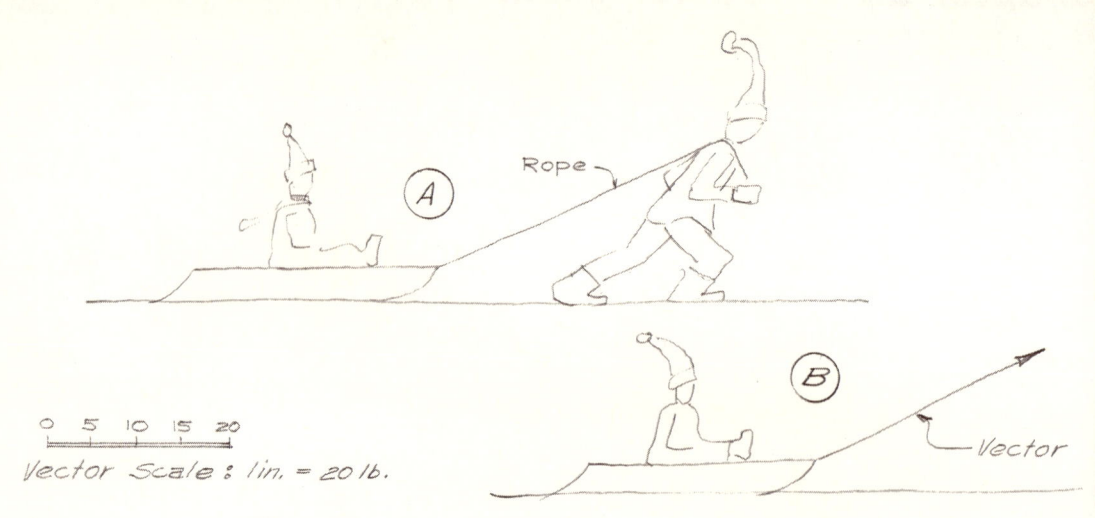

Figure 22.1 The graphic concept of a vector quantity.

arrow
(arrowhead)

2   The concept of the taut rope helps us to visualize a graphic image of the situation. We can remove the boy (22.1B) and let a straight line, a vector, replace the rope. The length of the line can represent the magnitude of the force, the direction is shown by the _____, and the position is determined by the point on the sled where the rope was fastened.

Though the picture now does not show how the sled is being pulled, it does give us a graphic vector upon which we may operate to learn more about the forces exerted on the sled.

We are concerned with the action of forces in producing rest or equilibrium. This is the field of statics. Dynamics is a field that treats of the action of forces in producing motion.

rest;
equilibrium

3   The sled is at _____ until the instant it starts to move (*statics*). While it is accelerating, its motion is dynamic. If it reaches a *constant velocity* (zero acceleration) it is said to be in _____ (static).

concurrent;
coplanar

4   The table of Figure 22.2 classifies vectors into four groups. Two or more vectors that act through a single point on a body are called _____ vectors. Two or more vectors that act in a single plane are _____ vectors.

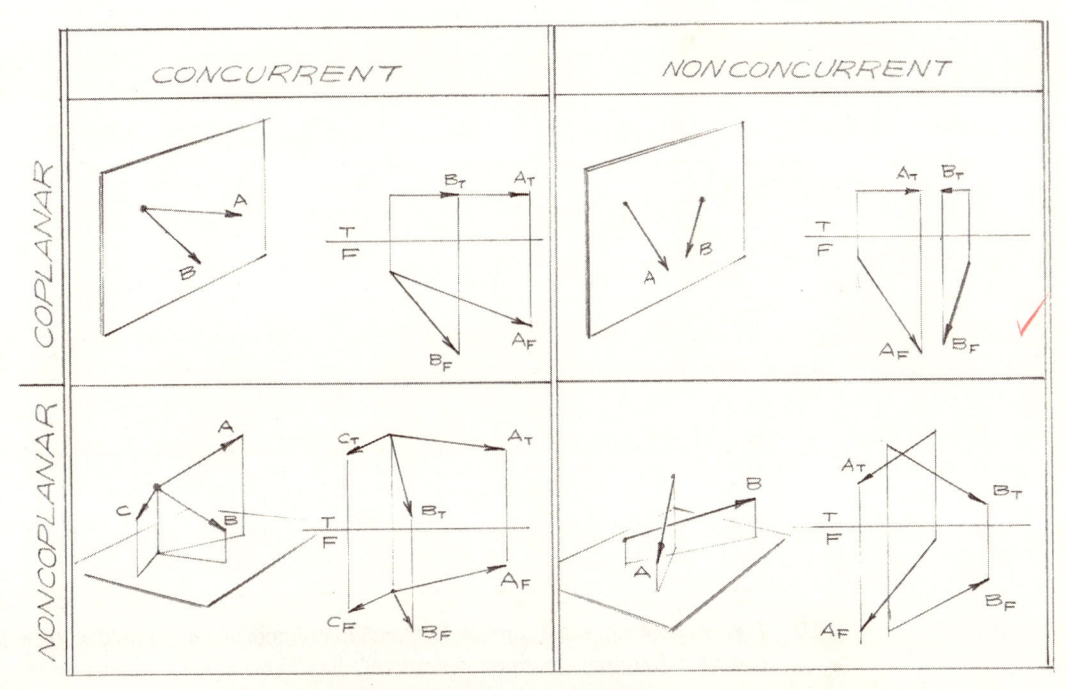

VECTOR GEOMETRY   345

Figure 22.2   The classification of vectors.

Of course, the opposite is true in the case of *noncurrent* and *noncoplanar* vectors respectively as the names imply.

CONCURRENT COPLANAR VECTORS

magnitudes

5 In Figure 22.3A, vectors AB and AC are needed to keep the body upon which they act at rest. The lengths of AB and AC are drawn to a vector scale and show the _____ of the vectors.

directions;
A

6 Their _____ are shown by the arrowheads and their positions by the fact that they both act through point _____.

rest
(equilibrium)

7 We can *add* these two vectors graphically to find their *resultant*—the one force that would replace AB and AC without altering the state of _____ of the body.

diagonal

8 Figure 22.3B shows vector addition. Two lines have been constructed parallel to AB and AC forming a parallelogram. The _____ of the parallelogram is the vector sum, or resultant, of vectors AB and AC.

346 DESCRIPTIVE GEOMETRY/UNIT 5

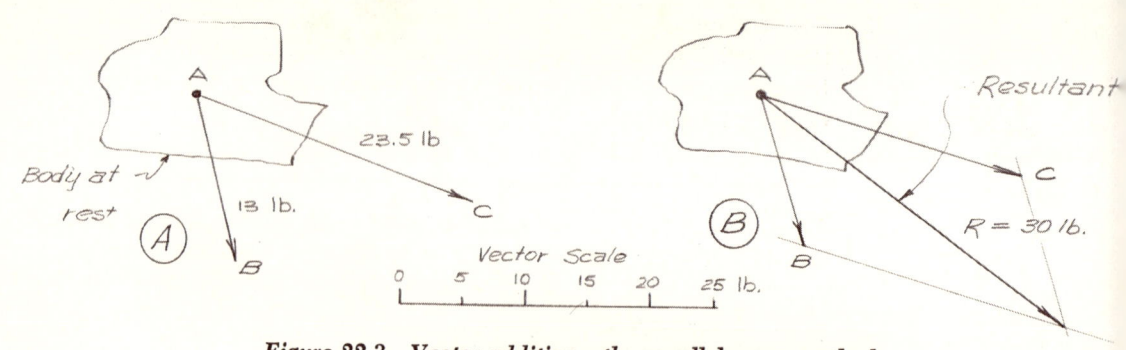

Figure 22.3 Vector addition—the parallelogram method.

scale

9 The resultant is shown in magnitude, direction, and position. Its exact magnitude can be found by measuring line AD to the correct vector _____.

10 Figure 22.4 shows a similar but easier method of vector addition. It is called the triangle method. Vector AC has been moved and its tail end connected to the arrow end of AB. Drawing a line from point A to the arrow end of the new position of AC closes the triangle and yields the

resultant _____.

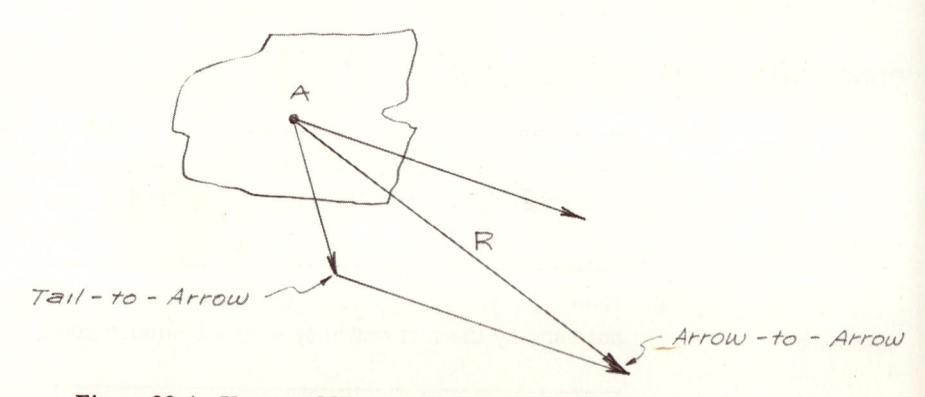

Figure 22.4 Vector addition—the triangle method.

11 Geometrically the triangle method is identical to the parallelogram method. It is an easier method when more than two vectors are involved. Consider Figure 22.5A.

Four forces, A, B, C, and D act on a body through point O. The resultant R has been found using the triangle method. Starting with force C, the others were added, tail to arrow, in the sequence C, ____, ____,

B, A, D   ____.

VECTOR GEOMETRY   347

arrow end
(arrowhead)

12   The resultant (R) was found by closing the *vector polygon* with a line from point O to the arrow-end of force D. The direction of R is definite. The _____ of R must meet the arrow end of the last vector in the addition.

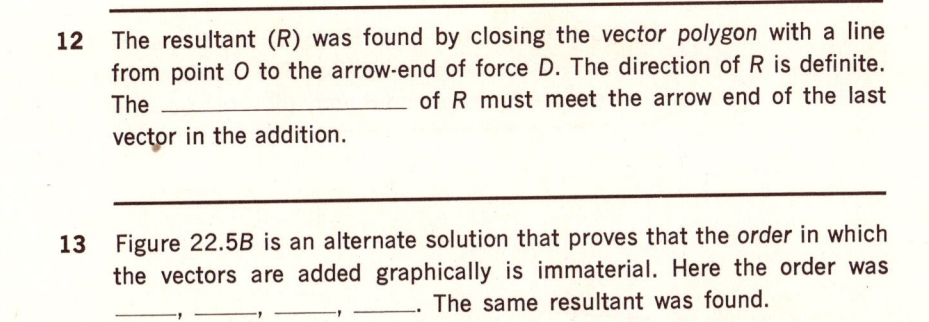

A, C, D, B

13   Figure 22.5B is an alternate solution that proves that the *order* in which the vectors are added graphically is immaterial. Here the order was ____, ____, ____, ____. The same resultant was found.

The problem could have been solved using the parallelogram method. The procedure is: Find the resultant R' of any two of the given vectors (A and B for instance) using the parallelogram method. Combine R' with C and find R''. Combine R'' with D and find the final resultant, R. Sketch this solution on one of the figures of 22.5. You should arrive at the same answer for the resultant, R.

motion

14   In the last two examples, construction was made to *close* the vector polygon. This is a necessary condition for considering the body to be *at rest* or *in equilibrium*. The vector sum of the forces must be zero.

If the vector polygon is open after the resultant has been drawn, we can deduce, by inference, that the body is in _____.

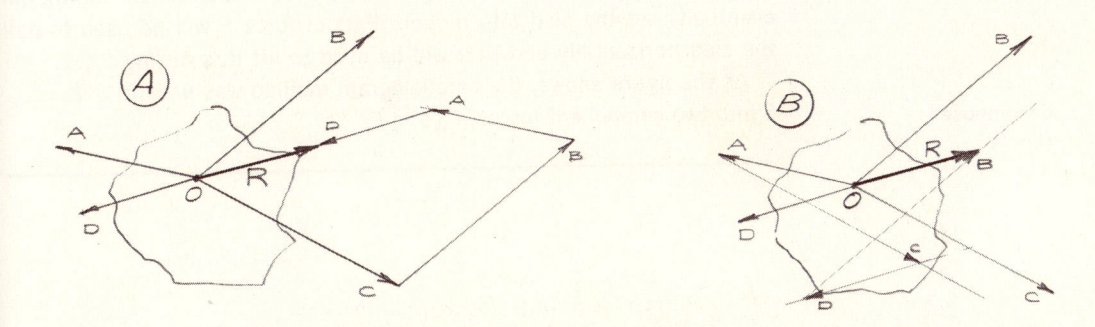

Figure 22.5   Addition of four concurrent vectors by the triangle method.

If two or more forces act on a body to *put it into motion,* finding the resultant will merely show the one force that would provide the same motion.

It is often desirable to know what force is required to keep such a body at rest. This force will be the *reaction*, a force that is equal in magnitude and position (line of action) but *opposite* in direction to the resultant of the original force.

348 DESCRIPTIVE GEOMETRY/UNIT 5

motion

15 Consider Figure 22.6. Forces A and B operate through point O and tend to set the body in motion. The resultant R is the one force that can replace forces A and B and still set the body in the same _____.

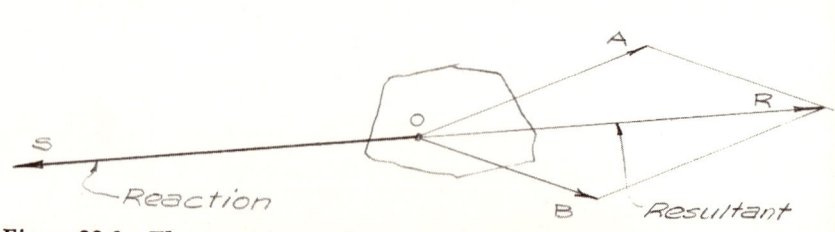

Figure 22.6 The reaction exactly counteracts the resultant.

reaction; rest

16 The force S is the _____ and will counteract A and B (or R) and keep the body at _____.

■ **17** Sketches 58B and C are exercises in resolving concurrent, coplanar forces. At 58B, find the resultant R of the five forces, A, B, C, D, and E.
At 58C, three forces are given in *direction only*. Their values are given along with a vector scale. These forces, F, G, and H tend to set the body in motion. Find the reaction to F, G, and H. Check your solutions.

Components

decompose
(divide)

18 Just as we can find the sum of two vectors, so can we *decompose a single vector into two components*. Consider Figure 22.7, the sled problem. We know that exerting a force F in the direction shown will eventually set the sled into motion. Part of force F will be used to pull the sled horizontally and part will be used to lift it vertically.
As the figure shows, the parallelogram method was used to _____ F into two component forces, F_V and F_H.

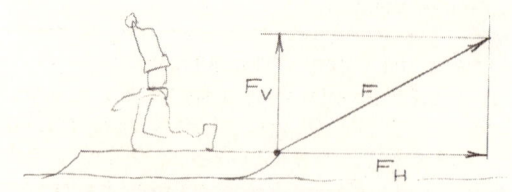

Figure 22.7 Decomposing a single vector into component vectors.

components

19 The directions of F_V and F_H were determined in advance since we wished to know how much of force F was used to lift the sled and how much was used to slide it along the ground. F_H and F_V are called _____ of F.

VECTOR GEOMETRY 349

■ 20   Sketch 59A shows a single force M acting on a body. It is desired to know the magnitude, direction, and position of forces A and B. Their directions are given at a and b. Sketch a solution. Check it.

NONCONCURRENT COPLANAR VECTORS

21   Figure 22.8A shows two forces, A and B, acting on a body at rest. A acts through point O and B through point M. These are nonconcurrent, coplanar forces. This means that the two forces _____ acting in a single plane and _____ acting through a single point. (Choose one: are/are not.)

are; are not

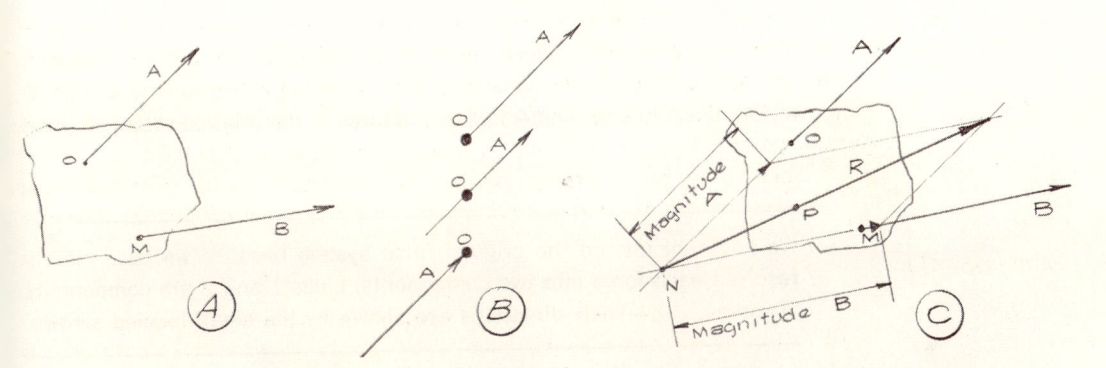

Figure 22.8   *Nonconcurrent vectors and the transmissibility of vectors.*

22   It is not readily apparent how we may find the resultant of A and B. Figure 22.8B shows an operation that can be performed on any vector *without altering the effect of the vector on the body.* In three successive sketches, vector A has been slid along its _____ of _____ through point O.

line (of) action

23   This is the *theory of transmissibility* of vectors and is valid as long as the vector maintains its original line of action. Figure 22.8C shows the solution of the original problem using the theory of transmissibility.

Vectors A and B have been slid along their respective lines of action to an intersection at point N thus converting the problem to a _____ force system.

concurrent

■ 24   Find the resultant of the coplanar, nonconcurrent forces A, B, and C acting on the bar of Sketch 59B.

HINT: Find the resultant of any two of the forces and then match it up with the third force to find the final resultant in magnitude, direction, and position. Check your solution.

The String Polygon

Refer to Sketch 59C. The *space diagram* shows a problem that, at first glance, looks as if it is the same as that in Sketch 59B. However, forces A, B, and C are so close to being parallel that they would have to be slid too far to find an intersection. There is another method for finding the magnitude, direction, and position of the resultant of forces A, B, and C. It involves the use of a *string polygon*.

25 The *vector polygon* (Sketch 59C) has been drawn using the triangle method. This gives the resultant R in _____ and _____ but not in _____.

magnitude,
direction;
position

Position is not found because we do not know where R acts on the bar.
A new vector polygon was constructed by selecting *any convenient point O* and drawing lines 1, 2, 3, and 4 to the junctures of the original vector polygon as shown.

26 We have not altered the original force system because we have merely resolved each force into two components. Lines 1 and 2 are components of force _____. Their directions are shown by the small labeled arrows.

C

27 Similarly, lines 2 and 3 are components of force _____.

B

28 Note that the direction c is opposite to direction b on component 2. This means that component 2 is canceled out because it consists of equal and opposite forces.
Lines 3 and 4 are components of A and component _____ cancels out.

3

29 This leaves only components 1 and 4 which form a closed triangle with _____ and thus cause no change in the system.

R

■ 30 Going back to the space diagram (59C) we can construct a *string polygon* to find the line of action of R, which is the only thing missing in our solution.
Sketch a solution according to the following procedures:

STEP 1: Choose *any point* on vector A of the space diagram. Through this point sketch:
(a) A line parallel to component 3 until you get an intersection with vector B. Component 3 is the common canceled component between A and B. You may extend B in either direction if necessary.
(b) Another line parallel to component 4 (the common component between A and R). Make it about 2 in. long to the right.

VECTOR GEOMETRY 351

STEP 2: From the intersection of B and 3, sketch a line parallel to component 2 until you get an intersection with C. 2 is the common canceled component between B and C.

STEP 3: From the intersection of C and 2, sketch a line parallel to component 1 (common between C and R).

Where lines 1 and 4 intersect there occurs a new point (call it M) which closes the string polygon. Point M is a point on the line of action of R.

STEP 4: Transfer R in direction and magnitude from the vector polygon, placing it through point M, and the problem is complete. R is now known in magnitude, direction, and position. Check your solution.

Parallel Forces

The string-polygon technique is most useful in solving parallel-force problems for, with parallel forces, it is never possible to slide any two along their lines of action to an intersection.

■ 31   Sketch 60A shows a horizontal beam acted upon by three vertical forces, A, B, and C. B is the weight of the beam acting downward at the center.

Using the techniques of the last example, find the magnitude, direction, and position of the resultant force R. The procedures are:

STEP 1: Construct the vector polygon which gives R in direction and magnitude (Given).

STEP 2: Select a convenient point O and add canceled components to each force on the vector polygon.

STEP 3: Transfer the components to the space diagram in proper sequence, completing a closed string polygon which locates a point on the line of action of R.
Check your solution.

32   In a parallel-force, beam problem such as Sketch 60A, it is often desirable to know the magnitudes of the two _____, R_1 and R_2 at the ends of the beam (See Figure 22.9). A roller support is used at one end to assure that the reaction at that point is vertical.

reactions

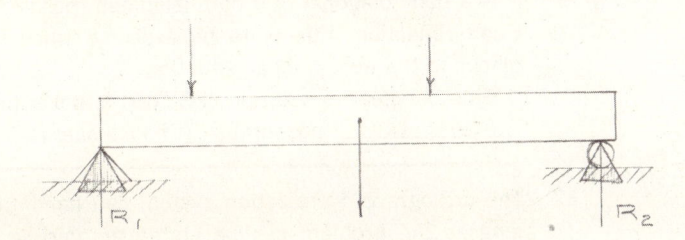

Figure 22.9   Forces R_1 and R_2 keep the beam at rest.

352   DESCRIPTIVE GEOMETRY/UNIT 5

■ 33   The directions and positions of the two reactions are known. Only their magnitudes are missing.

Using the final solution of Sketch 60A, find the magnitudes of R_1 and R_2 by following the next few steps:

STEP 1: Taking the two component strings of the resultant R (the ones that gave a point M on the line of action of R), extend them *in the opposite direction* from point M until you obtain intersections with the lines of action of R_1 and R_2 (extended if necessary). The left-hand string (No. 1 in the answer sketch) should intersect R_1; the right-hand string (No. 4) should intersect R_2.

STEP 2: Draw a line between these two intersections. This line is actually a common, canceled component between R_1 and R_2.

STEP 3: Transfer this line to the vector polygon. It must pass through point O and be parallel to the line found in Step 2. The intersection of this new line with R on the force polygon divides R into R_1 and R_2. R_1 is the upper portion; R_2 is the lower.

Check your solution.

CONCURRENT NONCOPLANAR VECTORS

do not

34   Figure 22.10A is a pictorial sketch of three concurrent, noncoplanar vectors, A, B, and C (space vectors). We have been concerned only with coplanar vectors up to this point. Noncoplanar vectors are vectors which _____ act in a single plane. (*Choose one: do/do not.*)

Do the parallelogram and triangle methods of the coplanar form apply to the noncoplanar situation? A study of the pictorial sketch of Figure 22.10B will show that they do.

R_1

35   First, the parallelogram law was applied to forces A and C (since two intersecting lines create a plane) yielding the resultant _____.

36   Then R_1 was resolved with force B to give the final resultant, R. Actually, R is a body diagonal of a parallelepiped (polyhedron with parallel sides). Upon reflection, this is to be expected when we switch from the coplanar to the noncoplanar situation.

This perspective solution (22.10B) is not satisfying, however, since it _____ measurable. (*Choose one: is/is not.*)

is not

37   The orthographic projection system can handle this type of problem as long as the problem can be stated graphically in at least _____.

two views

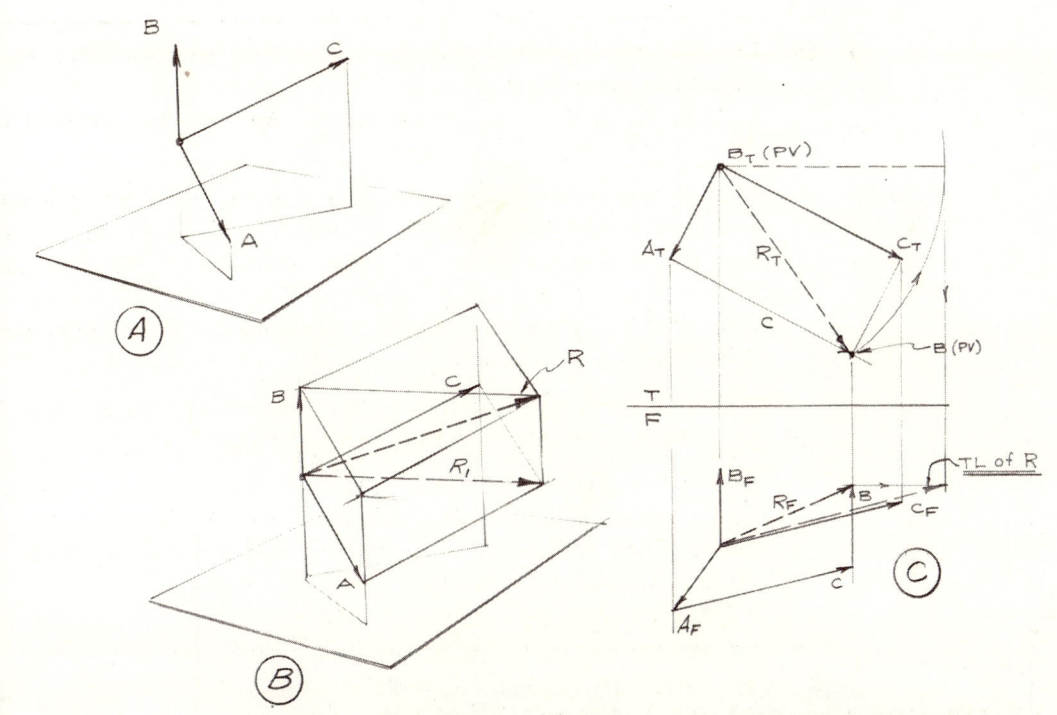

Figure 22.10 The resultant of three concurrent, noncoplanar forces is the body diagonal of a parallelopiped.

---

38 Figure 22.10C shows the complete orthographic solution of the problem. The triangle method was used to sum the three forces in two views. Remember that any construction made in one view must be duplicated in the second view or else you are violating the rules of orthographic projection.

rotation

The true length of R was found by the method of _____.

■ **39** Sketch 60B shows four concurrent, noncoplanar forces, P, Q, T, and S acting through point O. Find the magnitude *(TL)*, direction, and position of the resultant R.

In both views, add the vectors tail to arrow in any order. Find the resultant by closing the polyhedron in both views. Find the TL of R. Check your solution.

NOTE: The top and front views of R must project between views.

A useful problem in vector geometry is finding the forces in the legs of simple space frames (3-dimensional structures) when only the applied load and the physical configuration (space diagram) of the frame is known. Consider Figure 22.11. It shows a wall-mounted truss (space frame) with a known load W hanging from point O.

In this type of problem, one must assume that the forces in each member act along the centerlines of the members. This is essentially true as long as we consider the members rigid.

354 DESCRIPTIVE GEOMETRY/UNIT 5

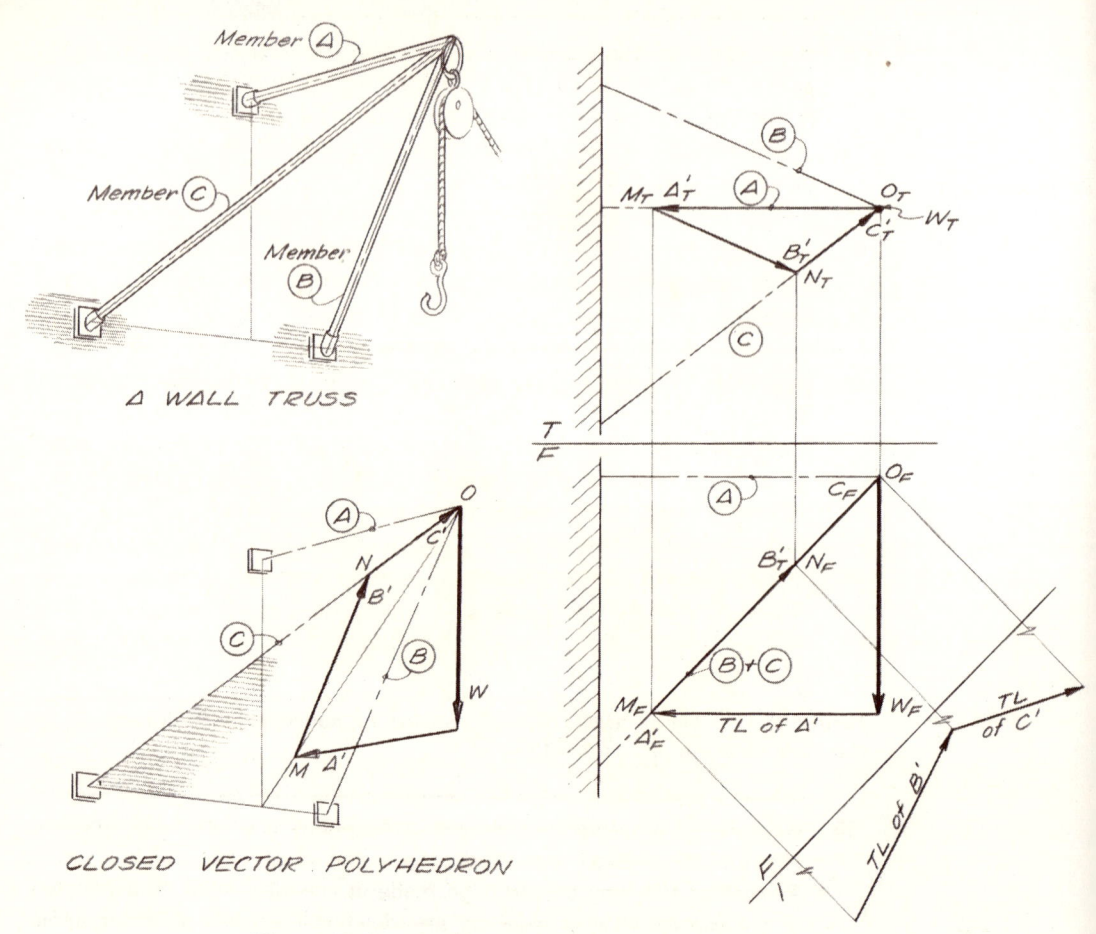

Figure 22.11 Finding three unknown forces in a space frame.

lines (of) action	**40** The vector problem thus becomes: What are the forces in members A, B, and C given the load W and the _____ of _____ of the forces in A, B, C?
TL (true length)	**41** A solution depends upon our ability to create a *force polyhedron* in two views. This would give two views of each force from which we could find their magnitudes by finding the _____ of each.

With only the one force W known, we must start the polyhedron by drawing from the arrow end of W. But where to go? If we choose the direction of C, we get nowhere because we do not know where to terminate that component.

front	**42** Note that the triangle formed by legs B, C, and the wall appears as an edge in the _____ view.

VECTOR GEOMETRY 355

43 It is logical to assume that a vector drawn from the arrow end of W_F parallel to A_F should stop on the plane formed by B and C. The reason is that this reduces the problem to a coplanar situation.

Accordingly, line A'_F was drawn parallel to A_F to the _____ of the plane giving point M_F in the plane.

EV (edge view)

44 The same step was taken to find M_T. The line from the arrow end of W_T drawn parallel to A_T happens to coincide with A_T so M_T was actually found by _____ from M_F to line A_T.

projecting

45 Now, from M_T (in the plane of B and C) a vector was drawn parallel to B_T to an intersection, N_T, with C_T. N_F was found by projection. This intersection is a valid choice since we know the polyhedron must _____, and the only way to get back to point O in a direction parallel to member C is directly along member C.

close

46 All that remains is to find the true length of A, B, and C. Vector A is TL in the front view. The TL of B and C can be found in a single auxiliary view set up _____ to the edge view of the plane of B and C (front view).

parallel

47 Success in solving this problem was dependent upon having a(n) _____ view of one of the three planes formed by the concurrent members A, B, and C.

edge

The solution of the wall-truss problem was made easy because an edge view of one of the planes formed by the three concurrent legs was available. Had it not been available, a single auxiliary view could have been drawn that provided a view of the truss in which one of the planes showed as an edge. The solution could then have proceeded as before.

Figure 22.12 is a solved example of the solution of a general case of a concurrent, noncoplanar force system. The space frame is a tripod with legs of unequal length. A known force W is shown true length in the front view. The problem was to find the forces acting in the three legs, OD, OE, and OF.

48 An edge view of one of the three planes created by the legs of the tripod _____ show in the top and front views. (Choose one: does/does not.)

does not

49 Using standard projection techniques, an edge view of plane _____ was found in auxiliary view 1.

ODF

50 The solution of the problem then proceeded exactly as in the wall-truss example of Figure 22.11.

The magnitudes of the forces in the three legs was determined by finding a(n) _____ of each of the three vectors.

TL

356 DESCRIPTIVE GEOMETRY/UNIT 5

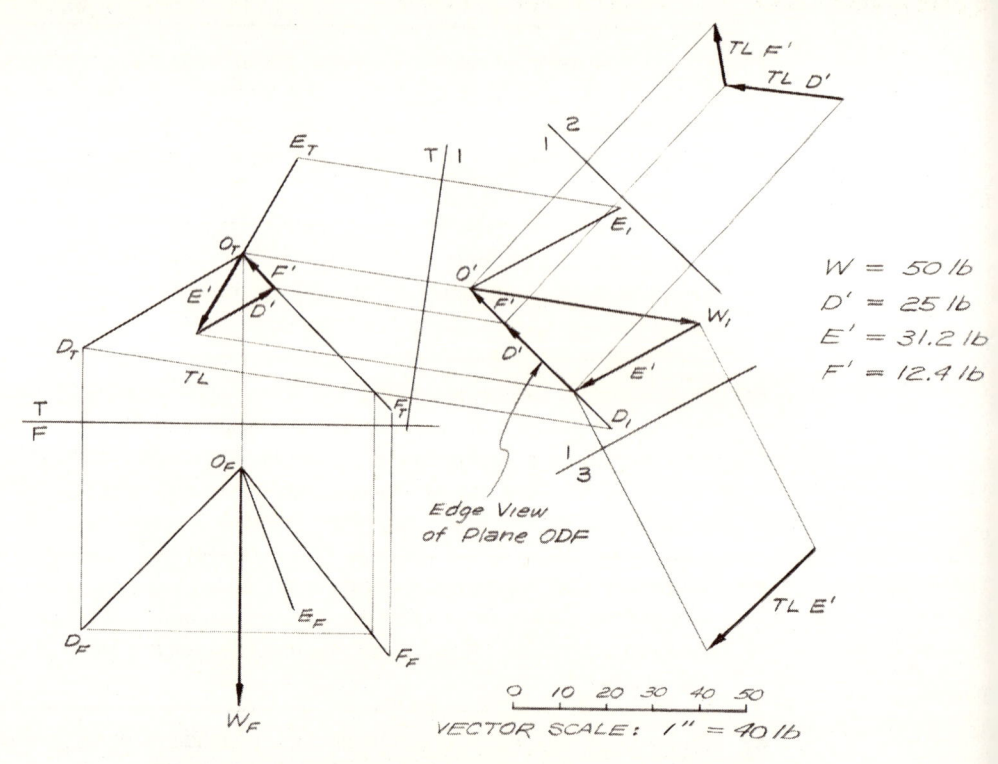

Figure 22.12 Finding the forces in the legs of a loaded tripod.

W = 50 lb
D' = 25 lb
E' = 31.2 lb
F' = 12.4 lb

VECTOR SCALE: 1" = 40 lb

vector | **51** The actual numerical magnitude can be determined by measuring each TL vector to an appropriate _____ scale.

This is the end of Chapter 22. We have discussed mechanical force alone. You should understand that the techniques developed can be applied to any vector *quantity* whether force, weight, velocity, acceleration, electric current, magnetic force, etc.

chapter 23   curved lines

We have finished with the analysis of straight lines, but have by no means finished with their use in solving the problems of descriptive geometry. Let's look briefly at curved lines. There are two classes of curved lines.
(1) Single curved lines—lines that lie in a single plane. They are plane curves.
(2) Double curved lines—lines that do not lie in a single plane. These are often called space curves.

SINGLE CURVED LINES

There are many single curved lines that are important to the world of science and industry. Most of the important ones give a graphic representation of a mathematic expression. This chapter provides a brief introduction to a few of these lines.

generated	1	Lines (straight or curved) can be considered as being _____ by a moving point.
plane	2	When considering a single curved line, we picture the point as moving about in a single _____.
mathematic	3	In the technical world there are many lines of importance whose movement follows a(n) _____ law.

357

358 DESCRIPTIVE GEOMETRY/UNIT 5

Single curved lines may be classified into two main groups: (1) open curved lines and (2) closed curved lines. Figure 23.1 shows general examples of these two classes.

The most common curved line is the *circle*. It is one of a family of useful curves called the *conic sections* or, *conics*. The others are the ellipse, parabola, and hyperbola. Their name comes from the fact that all of them can be produced by cutting a right-circular cone with a plane. More of this in Chapter 26.

Simple Closed Curves

Open Curves

Closed Curves

Figure 23.1 *The classification of single curved lines.*

---

radius; origin

4 Figure 23.2 shows the circle. By definition, a circle is the locus of all points in a plane that are equidistant from a fixed point in the plane. The fixed point is the *center* of the circle.

In Cartesian (rectilinear) coordinates the equation for a circle is $x^2 + y^2 = r^2$ where r is the _____ of a circle with its center at the _____.

(semi)major;
(semi)minor

5 Figure 23.3 shows the ellipse. An ellipse is the locus of a point the sum of whose distance from two fixed points is constant. The two fixed points are called the *foci*.

The equation of an ellipse in Cartesian coordinates is $x^2/a^2 + y^2/b^2 = 1$ where a is the semi_____ axis and b is the semi_____ axis.

■ **6** We know the ellipse as being an oblique view of a plane circle. Sketch 61A shows a plane circle in two principal views. An auxiliary reference plane has been established (the F/1 line). The true-shape view has been divided into eight points. Sketch the view of the circle found in auxiliary view 1 by projecting points 1 through 8. Check your solution.

CURVED LINES 359

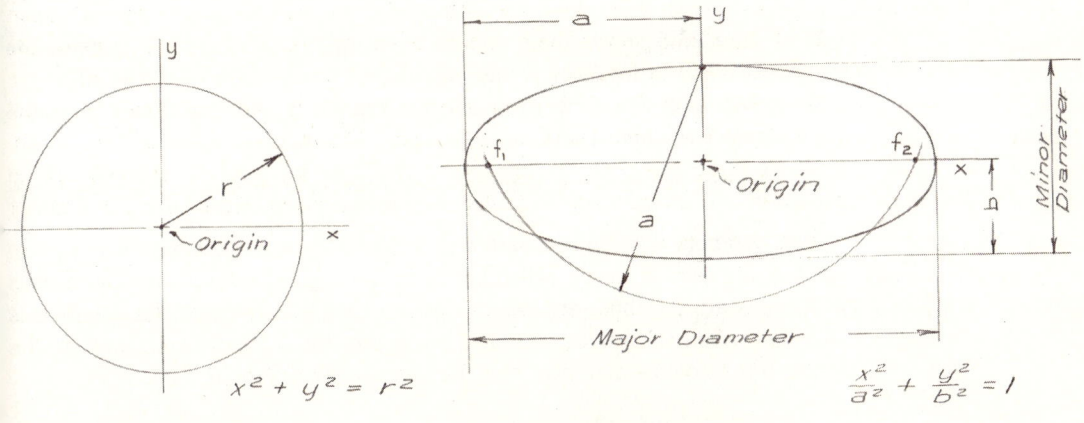

Figure 23.2 The circle.

$$x^2 + y^2 = r^2$$

Figure 23.3 The ellipse.

$$\frac{x^2}{a^2} + \frac{y^2}{b^2} = 1$$

■ 7 Figure 23.4 shows the simplest way to plot any ellipse when you know the major and minor diameters. It is the *trammel method*.

Construct a trammel by marking on the edge of a piece of paper one-half the major axis, a. Label the marks 1 and 2. From one end (mark 1) of this space measure one-half the minor axis, b. Label the new mark 3.

Lay this on the axes so that point 2 touches the minor axis and point 3 touches the major axis. As long as this relation is maintained, mark 1 will be a point on the ellipse.

Sketch 61B gives the major and minor axes of an ellipse. Make a trammel and plot the ellipse. Plot enough points in each quadrant to permit you to sketch a smooth curve through them. No solution is given.

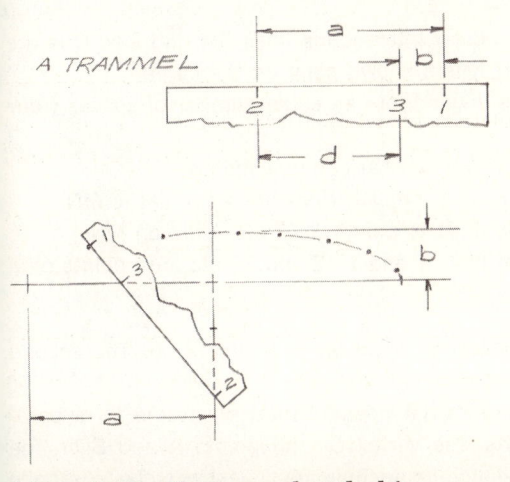

Figure 23.4 The trammel method for constructing an ellipse.

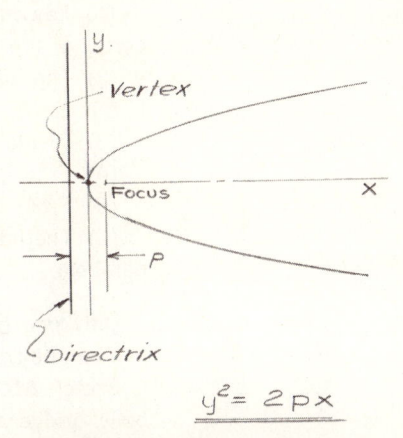

Figure 23.5 The parabola.

$$y^2 = 2px$$

360 DESCRIPTIVE GEOMETRY/UNIT 5

open

8 Figure 23.5 shows the parabola. It is a(n) _____ curved line which is the locus of points equidistant from a given fixed point and a given fixed line. The fixed point is the *focus* and the fixed line is the *directrix*. (Choose one: open/closed.)

The parabola is useful in nature and design. Lamp reflectors have a parabolic cross section as do reflecting mirrors on giant telescopes. The arches of bridges are parabolic for reasons of economy and strength. Its equation is $y^2 = 2px$ when the vertex is at the origin and the x axis passes through the focus; p is twice the distance from the focus to the origin.

■ **9** In graphics, most commonly, when there is need to draw a parabola, the axis, the vertex, and one point on the curve are known. Figure 23.6 illustrates the simplest method for constructing a parabola under these conditions.

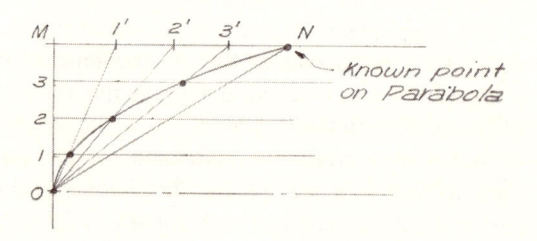

Figure 23.6 Constructing a parabola given the vertex and one point on the curve.

(1) Lay off two perpendicular intersecting lines OM and MN. O is the vertex of the parabola and N is a known point on the curve.
(2) The distance OM is divided into an equal number of spaces (four in this case).
(3) Divide MN into the *same number* of equal spaces.
(4) From points 1, 2, 3, etc., on OM, draw lines parallel to MN.
(5) From point O, draw lines to points 1′, 2′, 3′, etc., on MN.
(6) The intersections of lines 1 and 1′, 2 and 2′, etc., are points on a parabola.

Only one half of the parabola is shown in Figure 23.6. The second half can be constructed in the same manner.

Sketch 61C shows a rectangular opening that is a frame for a parabolic bridge arch. The axis, the vertex O, and points A and B on the curve are given. Using sketching techniques, construct the parabolic arch of the bridge. No solution is given.

CURVED LINES **361**

The hyperbola is illustrated in Figure 23.7. It is defined as the locus of points the difference of whose distance from two fixed points is constant. The fixed points are the foci. The hyperbola is a double figure with two open branches.

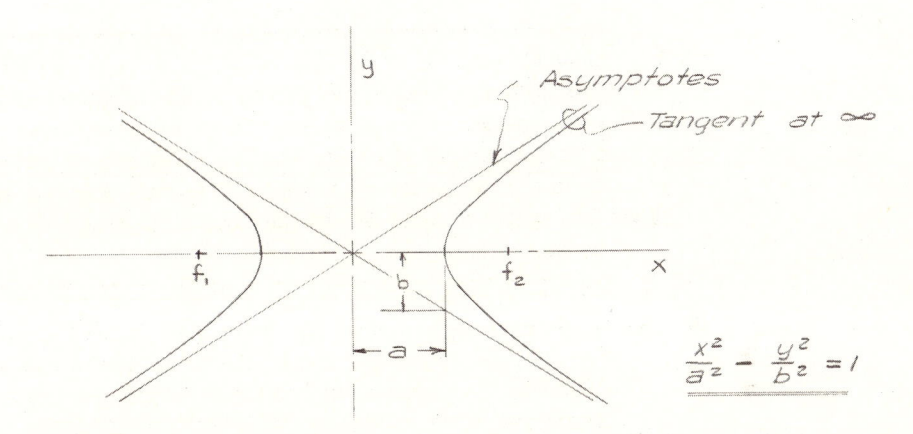

Figure 23.7   The hyperbola.

The equation of a hyperbola with its center at the origin and the x axis passing through the foci is $x^2/a^2 - y^2/b^2 = 1$.

One other important single curved line is the sinusoid or sine wave. Many phenomena of nature obey the fluctuations of the simple sine wave. Mechanisms being driven with *simple harmonic motion* are following a sine curve.

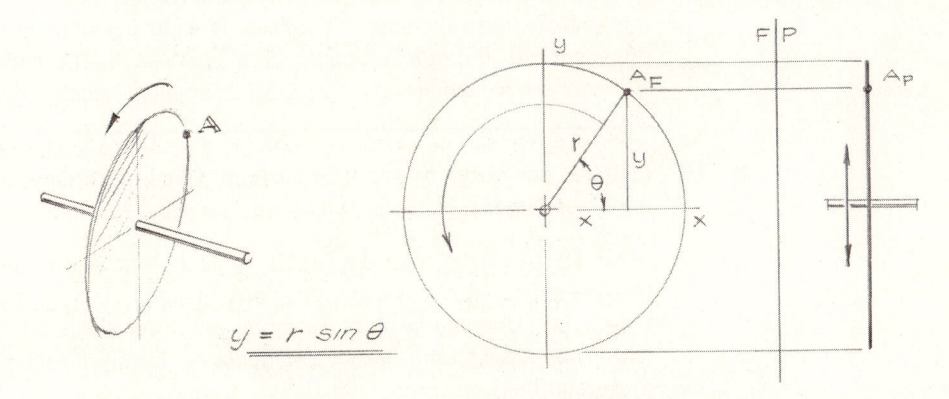

Figure 23.8   The basis for the sinusoid (simple harmonic motion).

10 Consider Figure 23.8. A wheel is rotating on an axle. A point A is marked on the rim of the wheel. If the wheel is rotated at a constant angular velocity, point A will move in a(n) _____ path at a constant linear speed.

circular

362 DESCRIPTIVE GEOMETRY/UNIT 5

TS (true-shape); edge

11 Two principal orthographic views of the wheel are also shown. The front view is a(n) _____ view and the profile view is a(n) _____ view.

diameter

12 In the true-shape view we see the circular path of point A. However, in the edge view, point A merely oscillates up and down between two limits. The maximum travel of this linear motion is equal to the _____ of the wheel.

variable

13 The speed of point A_P will be _____. The speed of A_F is constant. (*Choose one:* constant/variable.)

constant

14 Point A_P is observing simple harmonic motion as long as the angular velocity of the wheel is _____.

The front view shows the *xy* coordinates of point A at one particular instant in its travel. From this we can write the equation $y = r \sin \theta$ which expresses the *y* distance of A at any point on the circle in terms of the radius *r* and the sine of angle θ. This is the equation for a sine curve.

constant

15 Sketch 61D shows how we can plot this vertical *y* distance against time for one complete revolution of the wheel. The full circle is first divided into eight equal segments. This gives us eight positions of point A that represent the duration of equal time intervals in the complete circle since we are assuming _____ angular velocity.

■ **16** Plot the sine curve on the time scale provided, according to the following procedures:

STEP 1: Divide the total time scale (t_0 to t_f) into eight equal divisions to correspond with the eight positions of point A in its travel around the circle. (This can be done easily by eye. First divide the full scale in half. Then bisect each half to get quarters. Finally bisect each quarter to get eighths.)

STEP 2: Sketch vertical lines through these time markers. Consider the starting point ($t = 0$) to be position 1. The *y* coordinate of point A at this position is zero.

STEP 3: Plot $y = 0$ at t_0. After one-eighth of the full travel, the point is at position 2. Because the time axis coincides with the *x* axis, project a horizontal line across to the second time line. This transfers the *y* distance of position 2 directly to the time curve.

STEP 4: Repeat this procedure for the rest of the points, sketch a smooth curve through the points, and the construction is complete. Check your solution.

This has been a discussion of a few of the more important single curved lines with emphasis on how to construct them graphically without resorting to actual solutions of their equations. There are others such as cycloids, epicycloids, spirals, and involutes that have direct application in some phases of technical work.

DOUBLE CURVED LINES

Double curved lines—also called space curves—are lines that do not lie in a single plane. In contrast to plane curves, there are not many space curves that are important in the scientific and technical world.

However, one space curve, the *helix,* is the most used and useful curved line. It is the basis for the screw thread, the screw conveyor, and many other devices that are used to translate rotary motion into linear motion.

Figures 23.9 and 23.10 show the cylindrical and the conical helix. Since the helix is a line, it can be considered to be generated by a moving point. One way to visualize the generation of a helix is shown at 23.9A. Suppose a pencil is positioned at the bottom of a right-circular cylinder. Imagine what would happen if the pencil were moved vertically at uniform velocity *in the same time* that the cylinder rotated one revolution. The pencil would draw a helix on the surface of the cylinder.

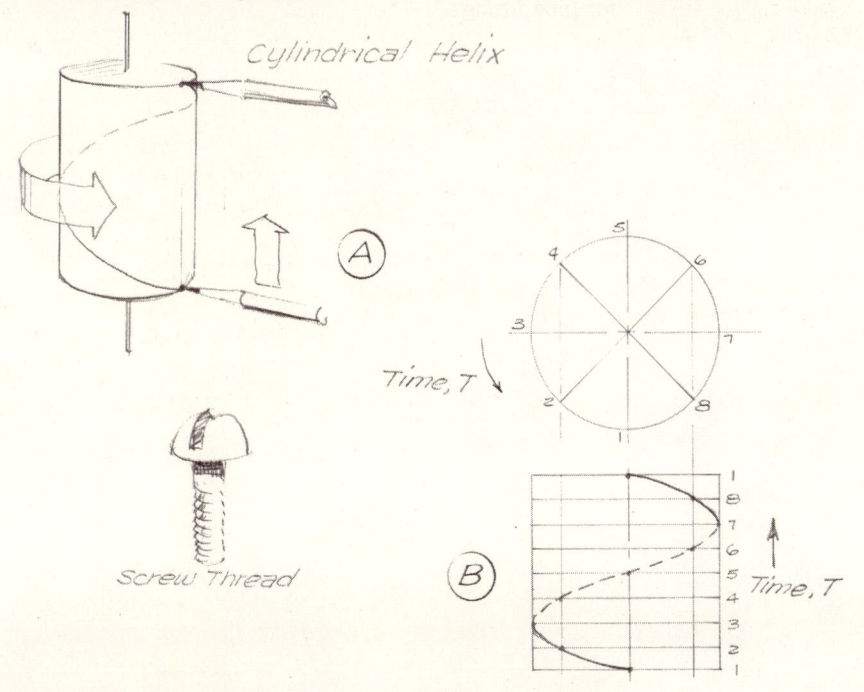

Figure 23.9 The cylindrical helix—a double curved line.

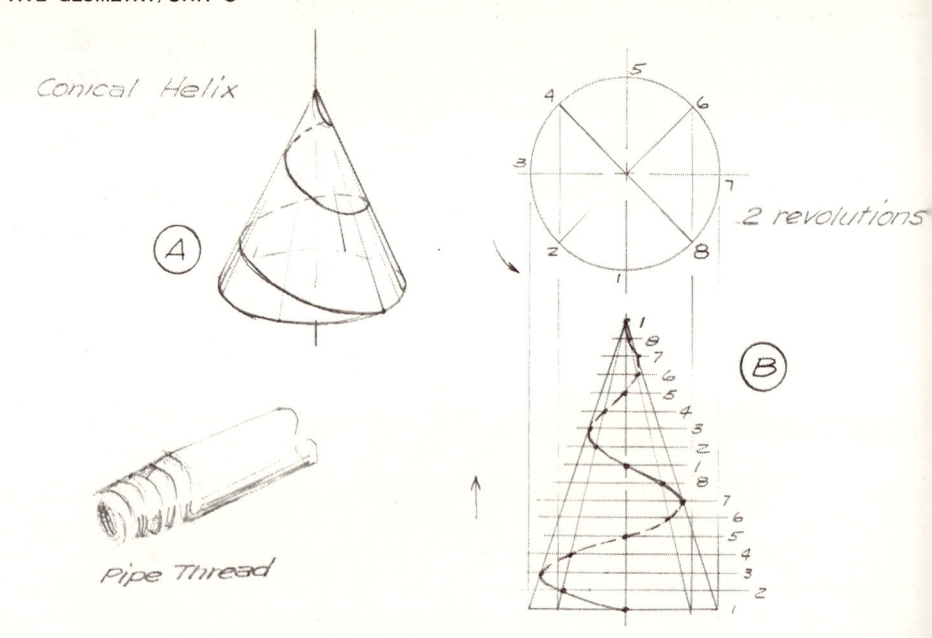

Figure 23.10   The conical helix.

Figure 23.9B shows this concept in orthographic form. This figure is identical to the solution of Sketch 61D, which showed the generation of a sine curve. The orthographic profile of a helix is a sine curve.

Figure 23.10A and B shows similar pictorial and orthographic views of a conical helix. Its principal use is the pipe thread where the conical shape permits two parts to be jammed together to provide a strong, leak-proof seal for pipe fittings.

chapter 24   *plane surfaces*

We have already seen one important use of plane surfaces in graphics. In Chapter 18 we presented planes as cutting devices. A plane can be used to cut through any three-dimensional-space problem and expose the interior detail for inspection. Another important use of plane surfaces in graphics is as the boundaries of solids.

PLANES AS BOUNDARIES OF SOLIDS

When a surface forms a boundary completely enclosing a portion of space of three dimensions, it becomes a *solid*. The word "solid" is used here in the mathematic sense. A *mathematic solid* is considered hollow—a container rather than the thing contained.

Any solid bounded by plane surfaces is a *polyhedron*. Figure 24.1 shows some of the polyhedrons that are of most interest to mathematicians, scientists, and engineers.

This chapter begins a study of solid figures. We will start with simple polyhedrons and, in later chapters, extend into the familiar curved solids: the cylinder, cone, sphere, etc.

Two areas relating to solids are of technical interest:
   (1) The *development* of the surfaces of solids.
   (2) The *intersection* of two or more solids.

This chapter explores the development and intersection of polyhedrons.

366 DESCRIPTIVE GEOMETRY/UNIT 5

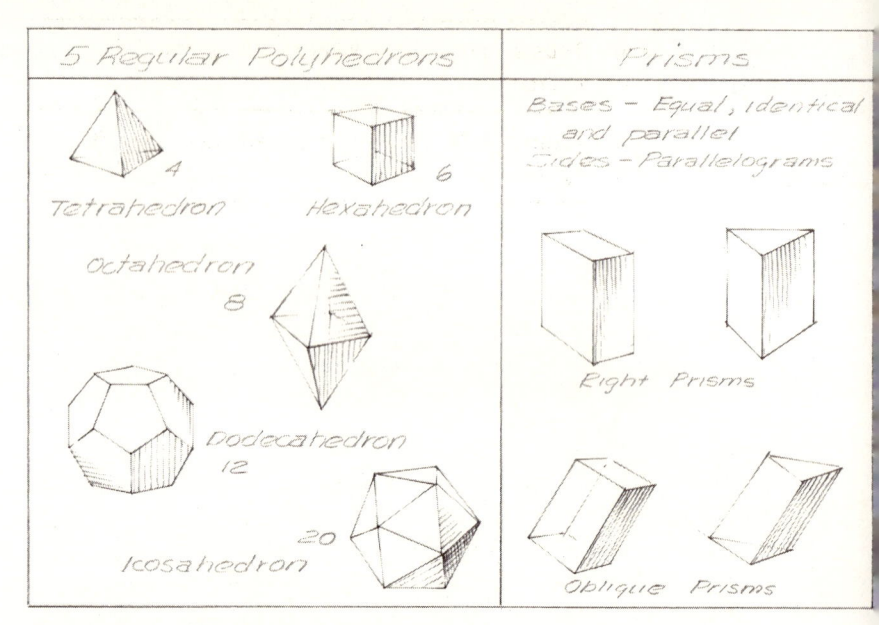

Figure 24.1 Classification of polyhedrons.

POLYHEDRONS

plane

1 We have seen that any three points define a(n) _____ in space.

plane

2 This is true because three points define two intersecting straight-line segments, which in turn define a(n) _____ surface, such as triangular plane 123 in Figure 24.2A.

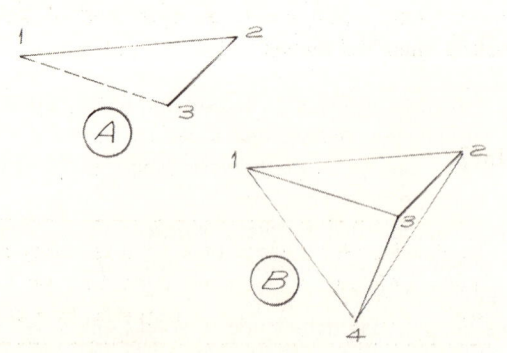

Figure 24.2 Creating planes with points in space.

three

3 The addition of a fourth point not in plane 123 creates _____ additional plane surfaces. (Figure 24.2B)

PLANE SURFACES 367

tetra(hedron)	4 Thus a four-sided polyhedron is formed. This is called a(n) _____-hedron.
hexahedron	5 Figure 24.3A shows that the addition of a fifth point creates a six-sided figure, a(n) _____.

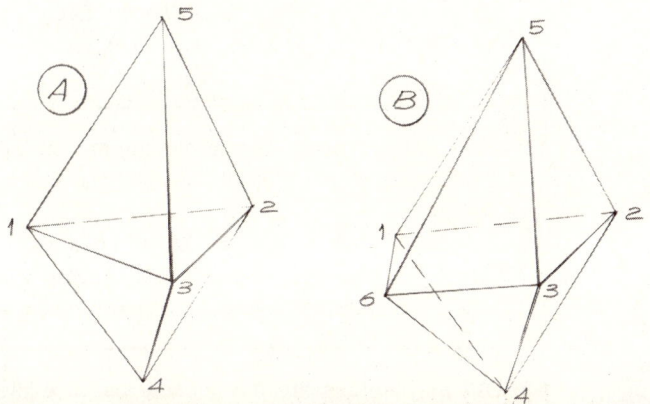

Figure 24.3 A polyhedron is a composite of points, lines, and planes.

8 (points); 12 (lines); 6 (planes)	6 The most common hexahedron is the *cube*. We have seen before that a cube is made up of _____ points, _____ lines, or _____ planes.

The addition of a sixth point (Figure 24.3B) creates an *octahedron*, an eight-sided solid.

The foregoing emphasizes the fact that any polyhedron is merely an array of points, lines, or planes. To operate on polyhedrons in the orthographic projection system requires only an extension of the techniques developed for projecting lines and planes.

SURFACE DEVELOPMENT

development	7 To *develop* the surface of a solid is to lay the surface out on a plane. If you unfolded a cardboard cube and laid it flat, you would be making a(n) _____ of the surface of the cube.
■	8 Sketch 62A shows a sketch of a ¾-in. cube and a ¼-in. grid. Sketch the *development* of the cube. Check your solution.

Chances are that you developed the cube intuitively, knowing that it consisted of six, ¾-in. squares joined together in any manner.

368 DESCRIPTIVE GEOMETRY/UNIT 5

cube

9 Surface development is most useful for manufacturing solid objects from flat sheet material. If you cut out the developed surface of Sketch 62A, you could make a ¾-in. paper _____ by folding along five lines.

TL (true-length);
TA (true angle)

10 Examination of the developed surface reveals that all six faces of the cube are shown in their *true shapes*. Each line is shown as a(n) _____ line and each angle is shown as a(n) _____.

¾; 90

11 This is the criterion for developing any surface that can be developed—find the true length of all lines and the true angles made by intersecting lines.

For the cube of Sketch 62A, all lines are _____ in. long and all angles are _____ degrees. Knowing this enabled you to *develop* the surface of the cube intuitively without any orthographic construction.

plane

12 All polyhedrons are developable because they are made up of four or more _____ surfaces.

TL; TA

13 Figure 24.4 shows an oblique triangular pyramid (tetrahedron). To develop the surface, we must find the _____ of all lines and the _____ between intersecting lines.

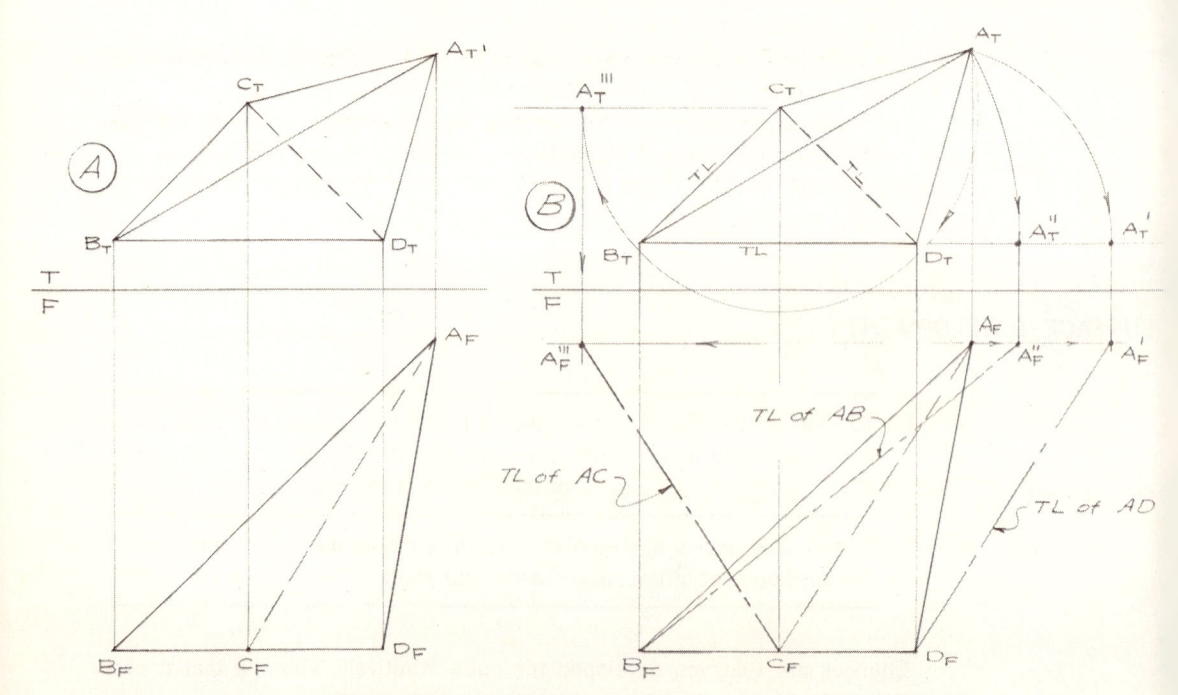

Figure 24.4 Preparing an oblique triangular pyramid for the development of its surface.

PLANE SURFACES 369

The pyramid is made up of four triangles. To construct any triangle, we must know one of the following:
(1) The *TL* of all three sides (SSS)
(2) The *TL* of two sides and the size of one angle (SAS)
(3) The size of two angles and the *TL* of one side (ASA)

BD, DC, and CB

14 The simplest method is to find the *TL* of all three sides. The triangle can be constructed using only a compass and straightedge.
Looking again at Figure 24.4A, what lines on the pyramid are *TL* in the given views? _____

15 We must find the true lengths of the other three, *AB*, *AC*, and *AD*. We are interested only in the *magnitude* of the true lengths of these lines and not in finding a *TL* view. Chapter 21 presented a handy method for finding the *magnitude* of the *TL* of a line. It is the method of _____ and does not require any auxiliary views.

rotation

16 Figure 24.4B shows the construction required to find the true lengths of *AB*, *AC*, and *AD* by the method of rotation.
We do not need to know the true size of any angles of a triangle as long as we know the true lengths of *all* lines. Thus, we are now ready to lay out the *development* of the _____ of pyramid *ABCD*.

surface

17 Figure 24.5A shows triangle *ABC* developed into its true shape. The procedure was as follows:
(1) The *TL* of *BC* was laid off on any straight line.
(2) Setting the compass to a radius equal to the *TL* of *AB* and using *B* as a center, arc *a* was drawn.
(3) Setting the compass to a radius equal to the *TL* of *AC* and using *C* as a center, arc *a'* was drawn to intersect arc *a*.
The intersection of arcs *a* and *a'* locates point _____ and the true shape of the triangle is complete.

A

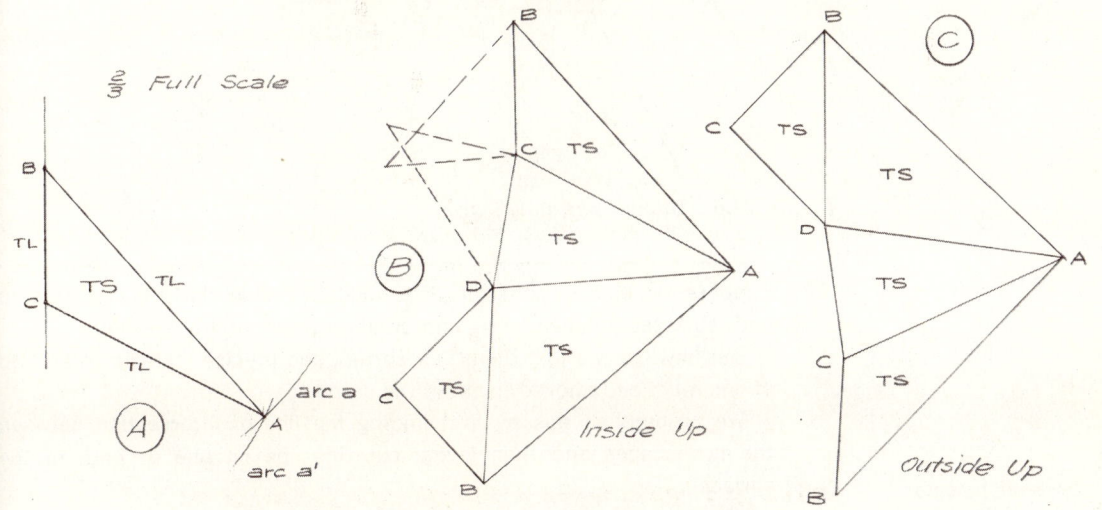

Figure 24.5 The development of the surface of the pyramid of Figure 24.4.

370 DESCRIPTIVE GEOMETRY/UNIT 5

development

18 Side ACD adjoins triangle ABC with the common side AC already laid out. Repeating the construction yields the TS of ACD; and then ADB; and finally the base triangle, BCD (See 24.5B). This is the complete _____ of the surface of the oblique triangular pyramid.

The order in which the triangles are laid out is not important. The base, BCD, can be placed at any one of three positions.

It is conventional to develop a surface *inside-up*. In manufacturing shapes from sheet metal, the layout is made on the inside, or *forming side*.

The pyramid development at 24.5B is developed inside-up. Figure 24.5C shows the same development presented *outside-up*.

The *procedure* used to develop the pyramid of Figures 24.4 and 24.5 is standard for laying out the true shape of any developable surface. The steps are:

(1) Describe the entire surface by a series of straight lines which form triangles.

(2) Find the true lengths of all the lines.

(3) Lay them out in sequence.

INTERSECTION OF POLYHEDRONS

prism; pyramid

19 Figure 24.6 shows a *composite* solid made up of a rectangular _____ and a triangular _____.

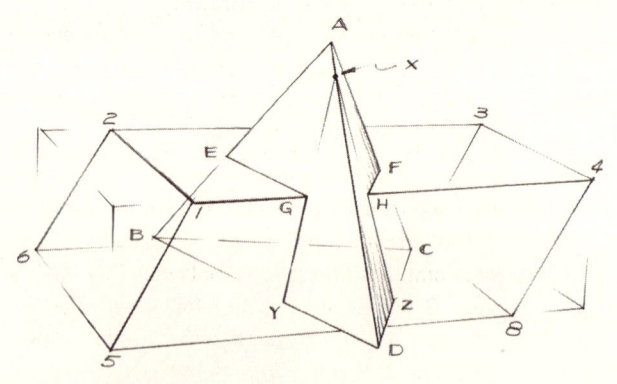

Figure 24.6 Intersecting polyhedrons.

developments

20 Suppose we wish to construct this solid. We know that we can develop both surfaces independently and create a prism and a pyramid.

But how can we join them in a correct and precise manner? A portion of one must be removed to admit the second.

We accomplish this by first finding the *line of intersection* between the two shapes and then laying out this line on one or both of the surface _____.

PLANE SURFACES 371

21 To find the line of intersection between the prism and the pyramid, we must first draw the composite shape in two orthographic views. This is done in Sketch 62B. The line of intersection is missing.

The corners of the prism have been labeled 1 through 8 and the pyramid corners labeled ABCD (subscripts have been omitted).

An examination of Figure 24.6 and Sketch 62B shows that the intersection on the prism will occur on faces _____ and _____. On the pyramid, the intersection occurs on faces _____, _____, _____, and _____.

1234 and 1458; ABC, ABD, ACD, (and) BCD

---

22 Thus the problem of finding the _____ (your words) between the two shapes becomes one of finding the intersections between planes—a technique which we have already discussed (Chapter 18).

line of intersection

■ **23** First let's find the intersection between plane ABC and the top of the prism (1234). Plane 1234 shows as an edge in the front view. It can be seen in this view where lines AB and AC intersect plane 1234. (See Sketch 62B.)

Sketch these two points in both views and label them E and F. Line EF is the intersection of planes ABC and 1234. Check your solution for this step.

Next, we will find the line of intersection between planes ABD and 1234. Using the same technique to find the point of intersection between line AD and 1234, we find that AD does not intersect 1234.

This means that we must find the point of intersection between line 1–4 and plane ABD. Plane 1485 shows as an edge in the top view. Imagine a *vertical* cutting plane through 1485. Sketch the front view of the points of intersection between *lines AD and BD* and the *cutting plane*. Label these points X and Y respectively. Line XY is the line of intersection between plane ABD and the cutting plane.

The intersection of line XY and line 1–4 (in the front view) yields the point of intersection between line 1–4 and plane ABD. Label this point G and project to the top view. Line GE is the line of intersection between planes ABD and 1234. Check your solution for this step.

Repeat this procedure to find:

(1) Line XZ (intersection between a cutting plane through 1485 and plane ACD)

(2) Point H (intersection between line 1–4 and plane ACD)

(3) Line HF (intersection between planes ACD and 1234)

The intersection between the pyramid and plane 1234 is now complete. Actually, the intersection between the pyramid and plane 1485 is also complete—line YG is the line of intersection between 1485 and ABD; line ZH between 1485 and ADC; and line YZ between 1485 and BCD. Check the complete solution.

372 DESCRIPTIVE GEOMETRY/UNIT 5

points

24 Thus the complete intersection is line *EFHZYGE*. The problem was reduced to finding the intersections between *specific lines* and *specific planes* which yield _____ which are common to both solids.

line of
intersection

25 Connecting these points with straight lines in both views yields the _____ (your words) between the two solids.

Figure 24.7 shows the development of both solids with the intersecting portions removed. These parts could be formed out of any sheet material and put together to yield the composite form shown in Figure 24.6.

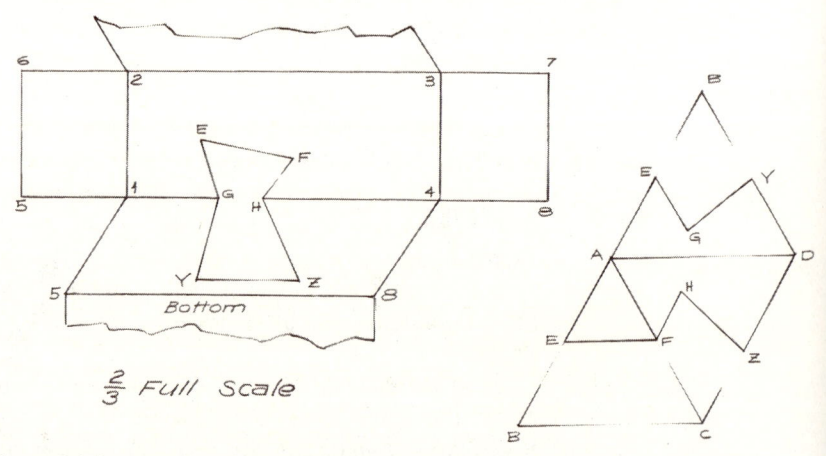

Figure 24.7   Developments of the prism and pyramid of Figure 24.6.

Success in solving this type of problem depends upon being able to *visualize* the physical arrangement of the two objects. A pictorial sketch is a great aid in solving the problem. Make a sketch first and approximate the line of intersection. This picture shows you which lines and planes to concentrate on in the orthographic solution.

■ **26** Sketch 63 shows both the orthographic and pictorial views of a composite shape consisting of a triangular pyramid intersecting a triangular prism. Using the procedures of the last problem, sketch a solution by finding the line of intersection between the two shapes. Check your solution.

This is the end of Chapter 24. The subject of developments and intersections will be taken up again in Chapter 27, where we will deal with cones and cylinders. Except for some differences in our approach to these problems, the basic operations are the same as you have seen here with polyhedrons.

chapter 25 curved surfaces

We leave the subject of plane surfaces for that of curved surfaces. Figure 25.1 shows a classification of curved surfaces. The two main classes of technical interest are the *ruled surfaces* and the *double curved surfaces*.

Ruled surfaces are surfaces generated by a moving straight line, while double curved surfaces are generated by a moving curved line. This chapter will be concerned with ruled surfaces and, in particular, with *single curved surfaces* as opposed to *warped surfaces*.

Single curved surfaces are ruled surfaces that are developable. Warped surfaces are ruled surfaces that are not developable.

SINGLE CURVED SURFACES

Chapter 19 discussed briefly the generation of surfaces by moving lines. Any line moving in space generates a surface. The line is called the *generatrix*. The manner in which it moves is controlled by another line (or plane) called the *directrix*. Any particular position of the generatrix is called an *element* of the surface.

If a surface is generated by a straight or curved line which has a circular directrix causing it to rotate about an axis, it is called a *surface of revolution*. Figure 25.1 shows that the most commonly used curved surfaces are surfaces of revolution.

Chapter 25 will show how simple single curved surfaces are described orthographically and will then present some problems involving the intersection of lines and surfaces. It ends with a discussion of how to describe a curved surface in a precise manner for orthographic projection.

374 DESCRIPTIVE GEOMETRY/UNIT 5

RULED SURFACES		DOUBLE CURVED SURFACES
Single Curved (Developable)	Warped (Not Developable)	Sphere
Cone		Ellipsoids — Oblate, Prolate
Cylinder	Hyperbolic Paraboloid	Paraboloid, Hyperboloid — All surfaces of Revolution

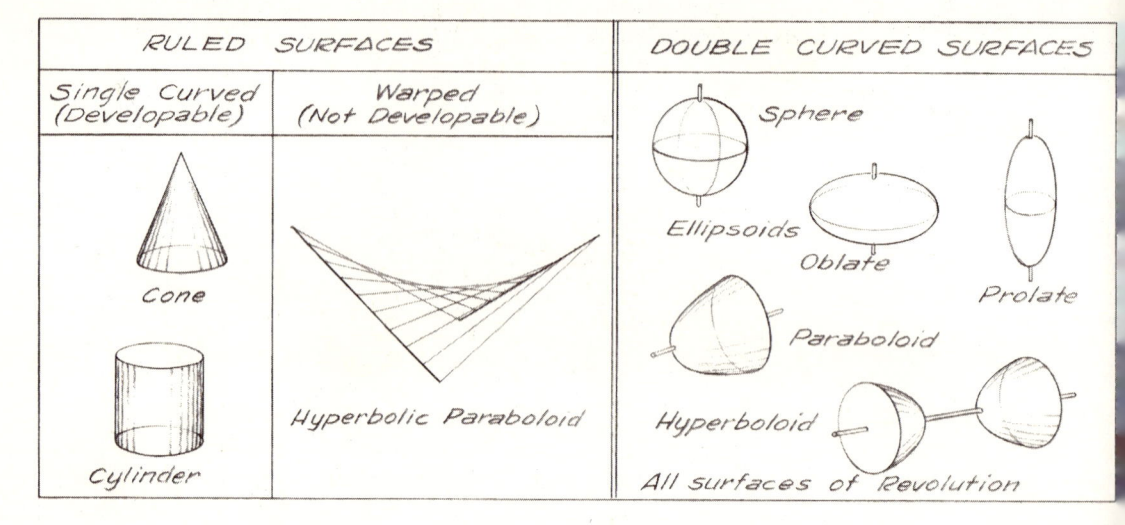

Figure 25.1 Classification of curved surfaces.

Summary and Pretest

In a few places in the following material you will find a summary figure and a pretest. The summary figure summarizes graphically some or all of the material of the chapter. The pretest presents a few simple problems for you to solve. Study the summary figure and work the problems. Answers are given. Check your answers. If you worked the problems correctly, skip to the next chapter or to an advanced frame as directed.

The purpose of the pretest and the summary is to let you decide whether you need to go through all of the material of the chapter.

Figure 25.2 presents a graphic summary of part of the material of Chapter 25. After studying these drawings, solve the following three problems to determine whether you should proceed through the next 50 frames or skip them.

PRETEST A: (See Sketch 64A.) Point P lies on the back side of the cone shown orthographically. Find the top view of P using an element on the surface of the cone that contains P. Compare your solution with the answer sketch.

PRETEST B: (See Sketch 64B.) Point R (shown in the front view) lies on the front side of the cone. Find the top view of R using an element on the surface that contains R. Compare with the answer sketch.

PRETEST C: (See Sketch 64C.) The axis (XY) of a right-circular cone is parallel to the top plane and inclined to the front plane. Complete the top view of the cone and sketch the view that shows the axis as a point and the base as a circle. Compare with the answer sketch.

If your solutions of Pretests A, B, and C are correct, proceed directly to the material following frame 50. *Otherwise*, proceed through the following material.

CURVED SURFACES 375

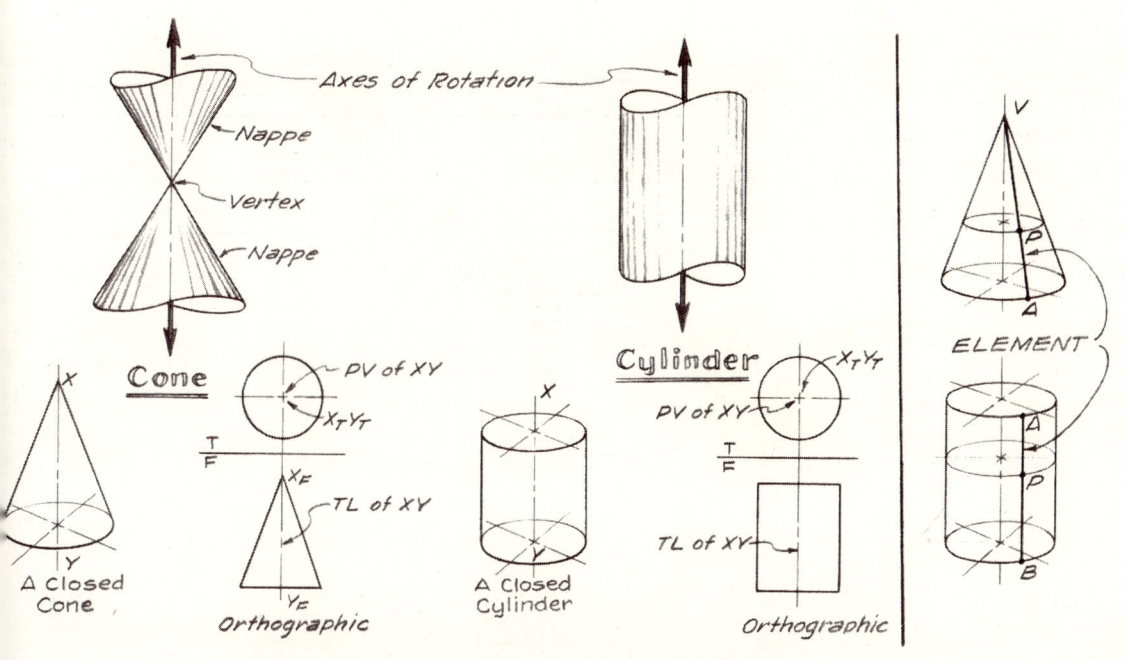

Figure 25.2 Summary sketch: Cones and cylinders.

CONES AND CYLINDERS

generated

1. The circular cone and the circular cylinder are the two most common surfaces of revolution. They are _____ by rotating a straight line about an axis.

intersect

2. To generate a circular cone, the straight-line generatrix must _____ the axis. (See Figure 25.3A.)

parallel

3. To generate a circular cylinder, the straight-line generatrix and the axis must be _____. (See Figure 25.3B.)

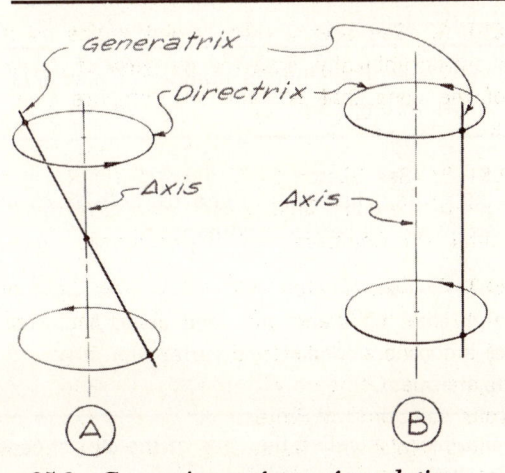

Figure 25.3 Generating surfaces of revolution.

Cones

surface (of) revolution

4 Figure 25.4A shows a line AB rotating about an axis XY. A(n) _____ of _____ called a circular cone is generated. The point of intersection between AB and XY is called the vertex of the cone.

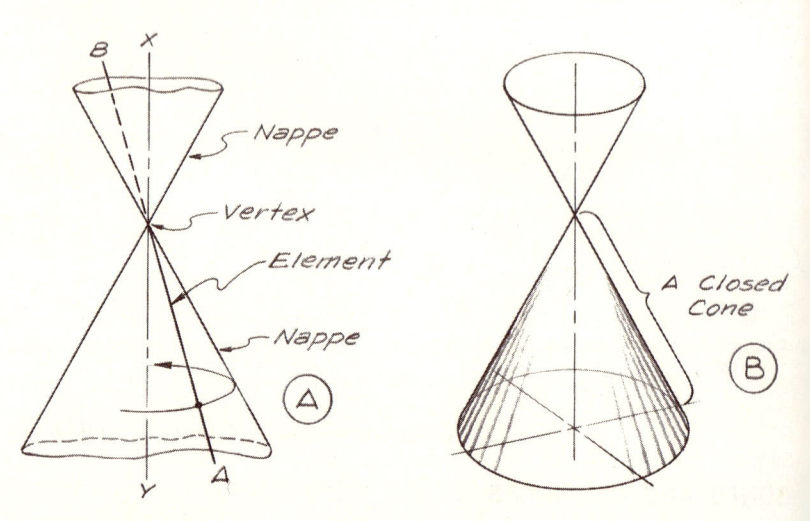

Figure 25.4 The cone.

5 This is the general case of the right-circular cone (hereinafter referred to simply as *cone* since we will not be discussing many other types). The point of intersection of the axis and the generating line is the

vertex

_____.

6 As line AB generates a cone by rotating about axis XY, each successive position of AB defines an element on the surface of the cone. The surface of the cone is defined by an infinite number of straight lines called

elements

_____.

vertex

7 Every one of an infinite number of elements on the surface of a cone passes through the _____.

The generating line can extend on both sides of the vertex. Therefore, a cone has two equal parts called nappes.

plane

8 Figure 25.4B shows a closed cone. A closed cone is the specific case of a cone bounded by the vertex and a(n) _____ perpendicular to the axis.

circular

9 This plane is called the base of the closed cone. The base is _____ in shape.

CURVED SURFACES 377

This concept points up the important fact that the intersection of a surface of revolution and any plane perpendicular to the axis of revolution is a circle. These circular intersections are typical of all surfaces of revolution. We can arbitrarily establish cutting planes perpendicular to the axis of surfaces of revolution to help us solve problems related to the surface.

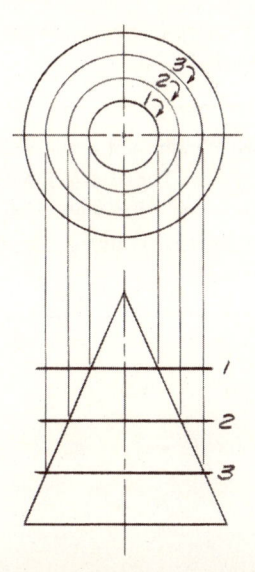

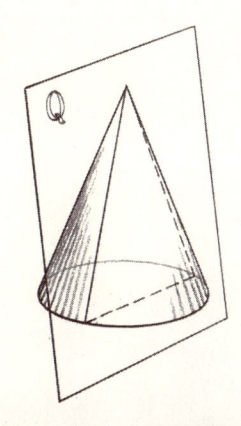

Figure 25.5 Cutting planes perpendicular to the axis of a cone describe circles on the surface.

Figure 25.6 A cone cut by an oblique plane.

---

edge

10 In Figure 25.5, the _____ views of three cutting planes (1, 2, and 3) are seen in the view showing the true length of the axis.

point

11 The cutting planes have a circular intersection with the surface in the view of the cone that shows the axis as a(n) _____.

■ 12 In Sketch 64D, two views of a cone are shown along with a cutting plane M, perpendicular to the axis. Sketch the top view of the line of intersection between plane M and the cone. Check your solution.

vertex

13 Other cutting planes are useful in studying conical surfaces. In Figure 25.6, plane Q is an oblique cutting plane which passes through the _____ of the cone.

elements

14 Two _____ are created by the intersection of plane Q and the surface of the cone.

vertex

15 These two elements are described by drawing straight lines between the _____ and the two points of intersection that plane Q makes with the circular base of the cone.

■ 16 Sketch 64E shows a cone cut by a plane Q (EV in F). Sketch the top view of the two elements cut by Q (use projection techniques). Label the front one AV and the rear one BV. Check your solution.

■ 17 Sketch 64F shows the top view of three elements, VC, VD, and VE, on a vertical cone. Sketch the front views of these elements and label. Check your solution.

■ 18 We now have two convenient ways to describe graphically a point P on the surface of a cone. These are shown pictorially in Figure 25.7.
Sketch 65A shows the top view of a point P on the surface of a cone. Sketch the front view of P using a cutting plane similar to Q of Figure 25.7 that contains P.

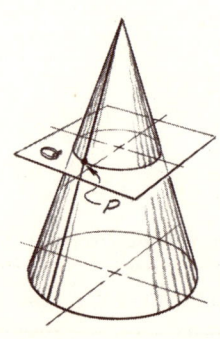

Figure 25.7 Two ways to describe a point on the surface of a cone.

■ 19 Sketch 65B shows the front view of a point P on the surface of a cone. Sketch the top view of P using an element that contains P.

NOTE: P is on the *back* side of the cone.

Sketch 65C shows the front view of a point P on the surface of a cone. P is on the front side of the cone. We cannot find the top view of point P by using the technique of Sketch 65B. (Using an element containing point P.)

■ 20 We know that point P lies on element VA. We can use the principle of rotation to find the top view, P_T, easily. Rotate element VA to a position parallel to the front plane (the TL position). Show the new rotated positions of A_F and P_F. Label A'_F and P'_F. Next project P'_F to element $V_T A'_T$. Having this top view of point P, labeled P'_T, we can reverse the rotation, putting element VA back in its original position, carrying point P with it. Perform this operation in the top view. Check your solution.

Cylinders

infinity	**21** A right-circular cylinder can be considered to be a special case of the right-circular cone with its vertex at infinity. (See Figure 25.8A.) Thus, we can say that elements on the surface of a cylinder intersect at _____.

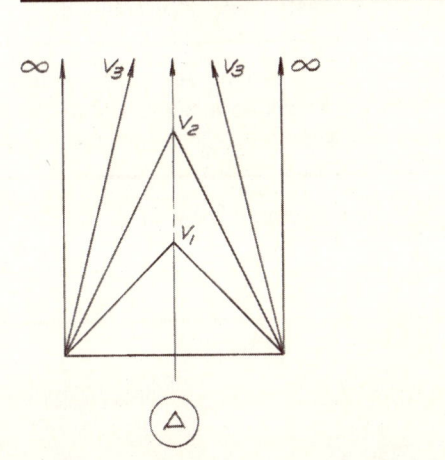

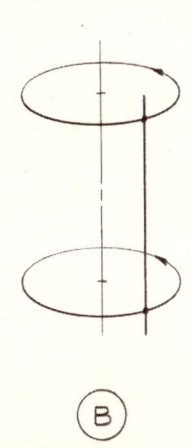

Figure 25.8 The cylinder is a special case of the cone whose vertex is at infinity.

parallel	**22** Straight lines intersecting at infinity are said to be _____.
elements	**23** The axis of a right-circular cylinder (hereinafter called simply cylinder) is parallel to all _____ of the cylinder.
parallel	**24** A cylinder is generated by rotating one line about another. The two lines must be _____. (See Figure 25.8B.)
elements	**25** Successive positions of the generating line as it rotates about the axis are called _____ of the cylinder.
	■ **26** There are an infinite number of elements on a cylinder. In Sketch 65D, AB is one of these elements shown in the front view. Sketch the top view of AB. Check your solution.
points; top	**27** In Sketch 65D, both the axis XY and the element AB appeared as _____ in the top view, meaning that these lines are perpendicular to the _____ reference plane.
TL (true length)	**28** The axis XY and element AB are perpendicular to the top projection plane, therefore, they must be parallel to the front plane and thus show as _____ in the front view.

380 DESCRIPTIVE GEOMETRY/UNIT 5

points; front

29 If a cylinder is oriented in space so that its axis is perpendicular to the top plane, all of its elements show as _____ in the top view and as true-length lines in the _____ view.

■ 30 The top and front views of a vertical cylinder are shown in Sketch 65E. Project the cylinder into auxiliary views 1, 2, and 3. (Use freehand sketching techniques.) Check your solution.

TL (true-length)

31 From Sketch 65E it can be seen that, if any orthographic view of a cylinder shows the axis as a point, the axis and all of the elements of the cylinder will show as _____ lines in any adjacent view.

elements

32 Figure 25.9 is a pictorial view of a cylinder and a cutting plane Q, parallel to the axis. Two _____ are defined by the intersection of Q and the surface of the cylinder.

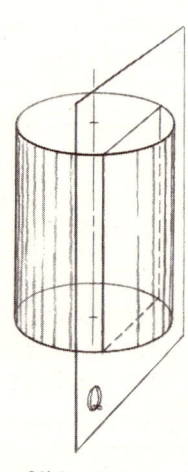

Figure 25.9   Intersection of a cylinder and a cutting plane parallel to the axis.

■ 33 Sketch 65F shows the two principal orthographic views of a cylinder and an edge view of a cutting plane Q, in the front view. Sketch the top view of the two elements cut by Q. Label the front element AB and the rear element CD in both views. Check your solution.

surface (of) revolution

34 Since a cylinder is generated by rotating a straight line about an axis parallel to the line, it is called a(n) _____ of _____.

circles

35 As is true with all surfaces of revolution, the intersections of the surface with cutting planes perpendicular to the axis of revolution are _____.

CURVED SURFACES 381

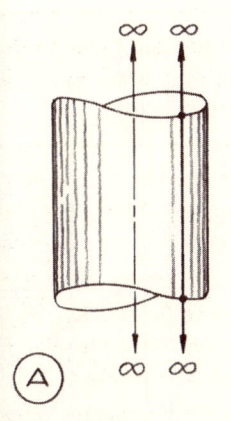

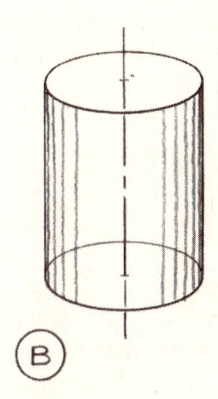

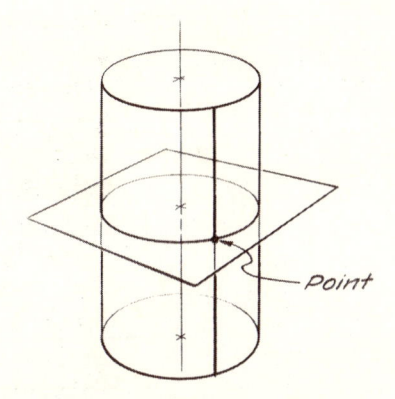

Figure 25.10 The general case of the cylinder (A); the closed cylinder (B).

Figure 25.11 Two ways to describe a point on the surface of a cylinder.

36 Figure 25.10A shows the general case of a cylinder in which both the axis and the generating line extend to infinity. For convenience in solving descriptive geometry problems, we use closed cylinders—cylinders bounded by two planes _____ to the axis (Figure 25.10B).

perpendicular

■ 37 Sketch 66A shows the front view of a cylinder cut by three cutting planes, P, Q, and R, perpendicular to the axis. Complete the top view of the cylinder showing the intersections of the cylinder and P, Q, and R.

38 To describe a point on the surface of a cylinder (see Figure 25.11) we can either find a(n) _____ on which the point lies or a(n) _____ made by a cutting plane which is perpendicular to the axis and passes through the point.

element; circle

39 Figure 25.12A shows a cylinder cut by an oblique cutting plane Q, shown as a(n) _____ in the front view.

edge

40 Figure 25.12B is a pictorial sketch of the cylinder and the cutting plane Q. The intersection of the oblique plane Q and the cylinder is a(n) _____.

ellipse

41 Figure 25.12C shows a general cylinder cut by four planes (Q_1, Q_2, Q_3, and Q_4) in such a way that the angle θ between Q and the axis varies from 90° to 0°. The intersections between these planes and the surface of the cylinder are:

Q_1 _____ Q_3 _____
Q_2 _____ Q_4 _____

circle; ellipse;
ellipse; two
elements

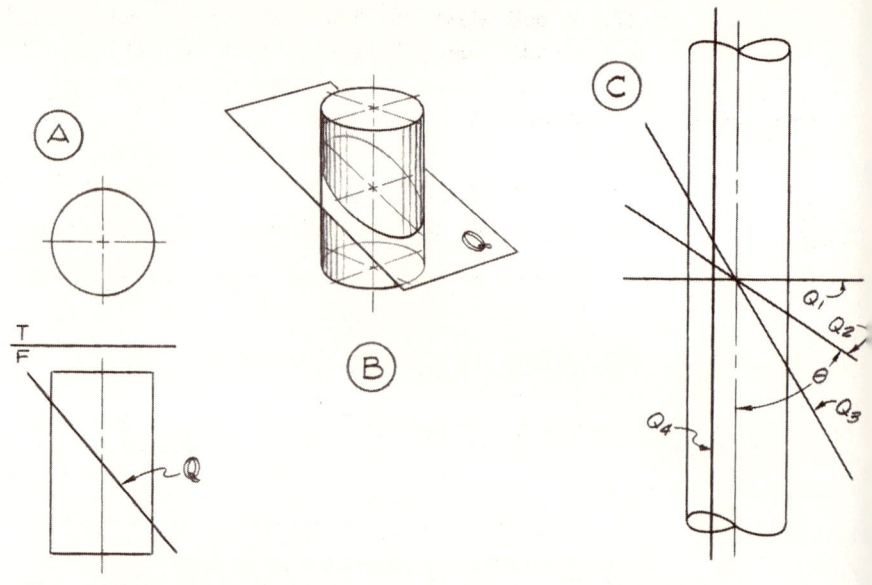

Figure 25.12   Intersections between a cylinder and oblique planes.

increases	**42** An ellipse is usually described graphically by determining its major and minor diameters. From the previous step, it can be said that as angle θ decreases, the ratio of major diameter to minor diameter _____. (*Choose one:* increases/decreases.)
equal; diameter	**43** A circle is a special case of an ellipse in which the major and minor diameters are _____. The *minor diameters* of all ellipses cut from any one cylinder are constant and are equal to the _____ of the cylinder.
ellipses; TL (true-length)	**44** Sketch 66B shows a cylinder whose axis is parallel to the front reference plane and inclined to the top reference plane. The circular ends of the cylinder appear as _____ in the top view. The axis is a(n) _____ line in the front view.
	■ **45** To find the circular view of the cylinder in 66B, we must first find a point view of the *axis*. Sketch the circular view in its correct orthographic position. Check your solution.
top; front	**46** Figure 25.13A shows the principal orthographic views of a cylinder with an oblique axis. The axis is inclined to both the _____ and _____ projection planes.

CURVED SURFACES

TL (true-length); point

47 In both views, the limiting end planes appear as ellipses. To find the circular view of the cylinder we must find two auxiliary views, first a(n) _____ view of the axis and then a(n) _____ view of the axis.

infinity

48 The cylinder is a special case of a cone with its vertex at _____.

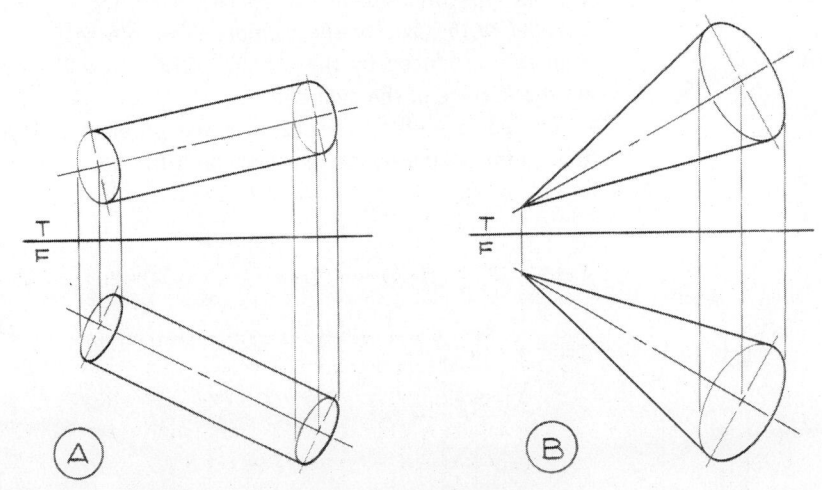

Figure 25.13 Cylinders and cones with oblique axes.

■ **49** Views of cones with inclined and oblique axes can be analyzed in the same manner as cylinders. Sketch 66C shows a cone with its axis parallel to the top reference plane and inclined to the front reference plane. Label the true-length (*TL*) view of the axis and sketch the circular view of the cone. Check your solution.

50 Figure 25.13B is a sketch of the top and front views of a cone having an oblique axis. The base of the cone appears as an ellipse in both views. To find the circular view of the cone, we must find two auxiliary views, a(n) _____ view of the axis and then a(n) _____ view of the axis.

TL (true-length); point

These first 50 frames cover the material presented as summary and pretest at the beginning of Chapter 25. Go back to Pretests (A, B, and C, and Sketch 64A, B, and C) and test yourself to see if you understand basic operations for locating points on the surface of a cone.

INTERSECTION OF A LINE WITH A CONE OR A CYLINDER

Problems occur in the technical world that involve finding the points of intersection between lines and cylinders, cones, or (for that matter) any surface of revolution.

Summary and Pretest

Figure 25.14 shows pictorial sketches of a line intersecting a cylinder at (A) and a cone at (B). These sketches suggest a method of orthographic solution to this type of problem. At 25.14A a cutting plane *contains line XY and is parallel to the axis of the cylinder*. The intersections of line XY with the two elements produced by the cutting plane define the two points of intersection on the surface of the cylinder.

The same method was used in the problem of Figure 25.14B except that the cutting plane passes through both the line and the *vertex* of the cone.

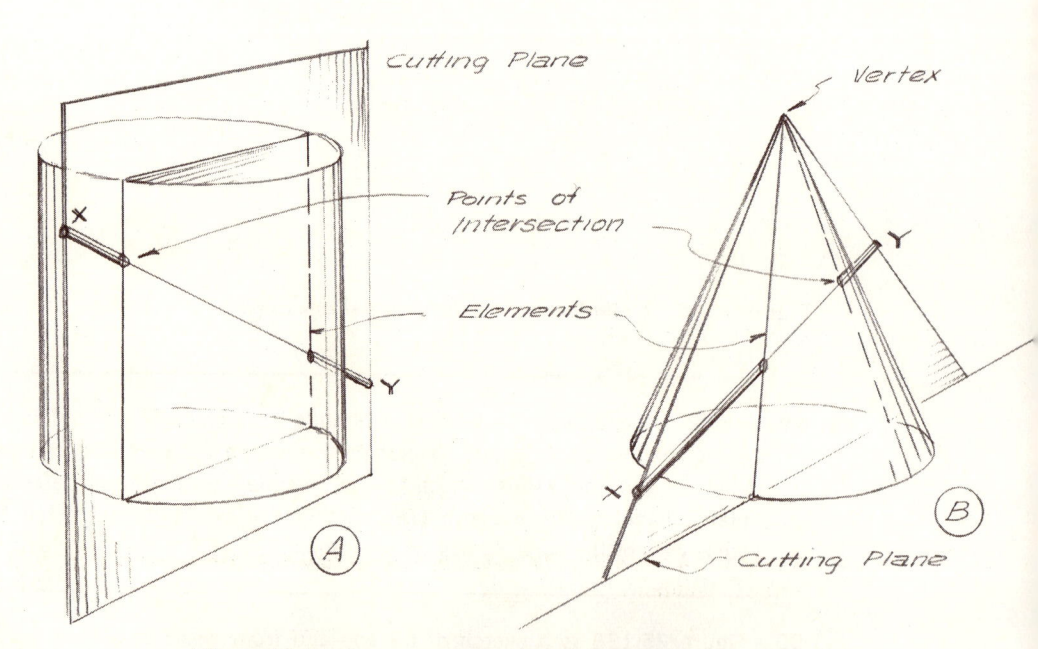

Figure 25.14 Summary sketch: The intersection of a line with a cylinder and with a cone.

PRETESTS D AND E: Sketch 66D shows the principal orthographic views of an oblique cylinder and a line XY. Sketch 66E shows a cone and another line XY. Find the points of intersection between the solid and the line for each. Compare your answer with the answer sketch. If your solution agrees with the answer, proceed to the discussion of surface description following frame 60. If you did not solve the problems correctly or if you do not understand the solution, continue through the rest of Chapter 25.

NOTE: In both problems, the cutting planes *will not* appear as edges in either view.

Intersection of Line and Cylinder

elements

51 By passing a cutting plane through a cylinder parallel to the axis of the cylinder, we obtain two _____ as an intersection.

52 The solution of the line-cylinder-intersection problem of Sketch 66D is dependent upon being able to pass a cutting plane:
 (1) Parallel to the axis of the cylinder
 (2) Containing line XY

Figure 25.15 shows a simpler problem involving a vertical cylinder. Here, passing (or imagining) the cutting plane is easy since the axis shows as a point in the top view. Therefore, any plane parallel to the axis will appear as a(n) _____ in that view.

EV (edge)

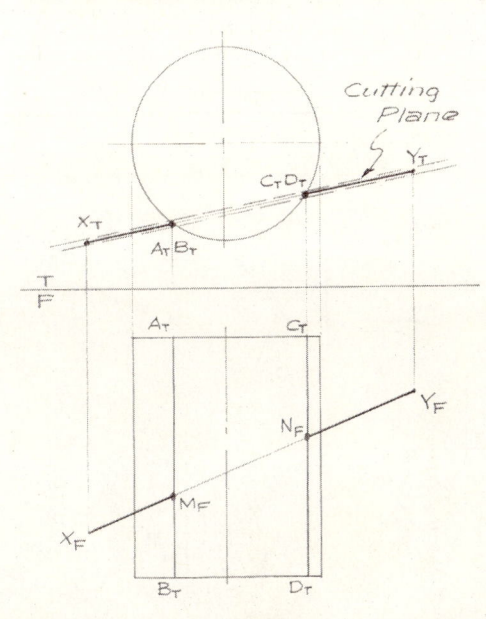

Figure 25.15 Finding the points of intersection between a line and a vertical cylinder.

53 The cutting plane appears as an edge in the top view and cuts two elements, AB and CD, on the surface of the cylinder. It also contains line XY. Therefore, points _____ and _____ are common to (1) line XY, (2) the cylindrical surface, and (3) the cutting plane and are the desired points of intersection.

M (and) N

The cylinder of Sketch 66D is an oblique cylinder and therefore does not show the axis as a point in either principal view. An auxiliary view could be taken to show the axis as a point. The problem could then be solved in the same manner as that of Figure 25.15.

386 DESCRIPTIVE GEOMETRY/UNIT 5

There is a simpler method by which we can solve the problem of Sketch 66D without resorting to an auxiliary view. The fact that we cannot draw the desired cutting plane as an edge in one of the given views does not deter us from using the cutting-plane method as long as we are careful to:

(1) Construct a plane that is useful to us.
(2) Make the same construction in two views.

---

parallel; contain

54 To be useful, the plane we are looking for must be _____ to the axis of the cylinder and _____ line XY.

55 The solution involves constructing a plane parallel to a given line (the axis), so let us first look at how this can be done.

A plane is parallel to a line if any line in the plane is parallel to the given line.

parallel

This idea is illustrated in Figure 25.16. At 25.16A, line AB is parallel to line CD. Planes 1, 2, and 3 all contain line CD, therefore, all three planes are _____ to line AB.

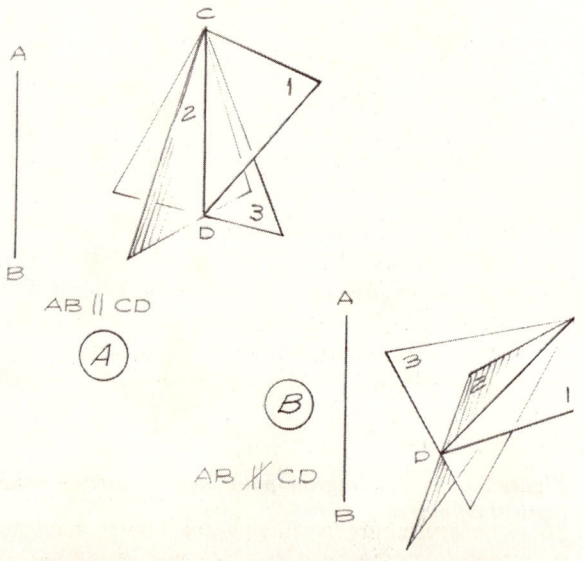

Figure 25.16 A plane is parallel to a line if any line in the plane is parallel to the given line.

CD

56 In Figure 25.16B, planes 1, 2, and 3 all contain line CD but they are not parallel to line AB because _____ is not parallel to AB.

■ **57** Sketch a solution to the problem of Sketch 66D by following the procedures of the next few steps:

STEP 1: The cutting plane that we wish to construct will cut two elements from the cylinder. To find the positions of these two elements, one side of the cutting plane must pass through one of the circular end surfaces of the cylinder. The top of the cylinder shows as an edge in the

CURVED SURFACES 387

front view. Therefore, first sketch a horizontal line through this edge view. This is one side of the cutting plane and we are assured that it lies in the plane of the top of the cylinder. Check your solution.

STEP 2: Next, extend line XY to the left from X_F to an intersection with this line (point M_F). From point Y_F sketch a line parallel to the axis ($A_F B_F$) upward to an intersection (point N_F) with the horizontal line through the top of the cylinder. The front view of the cutting plane is now complete.

Plane $M_F N_F Y_F$ is the desired cutting plane since:
(1) It contains line XY.
(2) It is parallel to axis AB (line $Y_F N_F$ is parallel to $A_F B_F$).
Check the solution.

STEP 3: We must now construct the top view of cutting plane MNY. First project points M_F and N_F into the top view. Extend $X_T Y_T$ to the left to an intersection with the M_F projection line. This locates point M_T.

From Y_T, sketch a line parallel to axis $A_T B_T$ to an intersection with N_F projection line. This locates point N_T.

The top view of triangle MNY can now be drawn by sketching line $M_T N_T$.

STEP 4: Note that line $M_T N_T$ crosses the circular view of the top of the cylinder—cutting the circle in two places. These two points (R_T and S_T) are the locations of the two elements which are the intersection of cutting plane MNY and the cylinder.

Sketch the two elements (parallel to axis AB) in both views. Where line $X_T Y_T$ intersects the two elements are the two points of intersection between XY and the surface of the cylinder.

Find the front view of these points by direct projection or by projecting points R and S to the edge view of the top of the cylinder and find the two elements. Check the complete solution.

Intersection of Line and Cone

vertex	58	The procedure for finding the intersection between a line and a cone (Sketch 66E) is the same as that of the line and cylinder problem with one notable exception: The cutting plane will not be parallel to the axis AB, but must pass through the _____ of the cone.
elements	59	This is true because only planes which pass through the vertex of a cone cut straight-line _____ on the surface.
■	60	Sketch 66E shows a right-vertical cone (axis AB) and an oblique line (XY). Find the points of intersection between XY and the surface of the cone by the following steps: **STEP 1:** Start the problem by sketching a line through the base of the cone (edge view in F). This line will eventually locate the two elements produced by the cutting plane.

Next, from the vertex (A_F) sketch two lines, one passing through X_F and the other through Y_F to intersections with the horizontal line through the base. (Label these points M_F and N_F respectively). Check your construction.

STEP 2: The triangular cutting plane AMN is now complete in the front view and satisfies the requirements that (1) the plane must contain line XY and (2) it must pass through the vertex.

Construct the top view of AMN. Project M_F and N_F into T. Sketch lines from A_T (vertex) through X_T and Y_T to intersections with the M and N projection lines yielding M_T and N_T respectively. Draw line $M_T N_T$.

$M_T N_T$ intersects the circular base of the cone at two points. These points are the ends of the two elements cut by plane AMN. The intersection of line XY and these two elements yields the two desired points on the surface of the cone.

The front view of the two points of intersection can be found by direct projection to $X_F Y_F$. Check the complete solution.

It is interesting to note that if the top view of the line through the base had not crossed the circular base, there would be no intersection between XY and the cone. Also, if $M_T N_T$ ended up *tangent* to the circular base, this would prove that line XY (and plane AMN) was tangent to the cone. (See Figure 25.17).

The same reasoning applies to the intersection of a line and a cylinder.

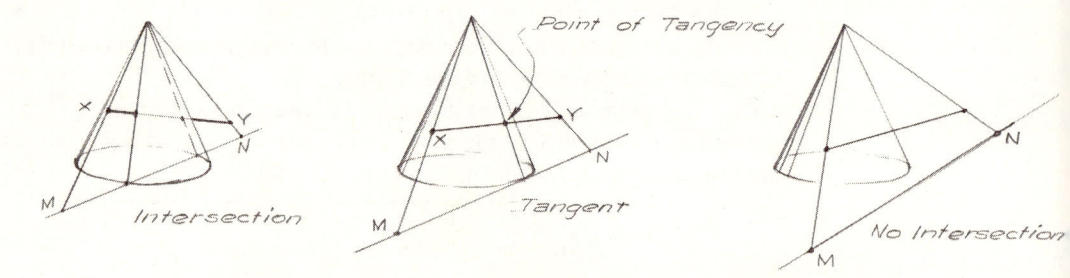

Figure 25.17 Line MN shows how and/or if line XY intersects the cone.

SURFACE DESCRIPTION

To project a curved surface in the orthographic system, the surface must be described in precise graphic terms.

Summary and Pretest

Figure 25.18 summarizes in pictorial terms the accepted manner of describing cones and cylinders graphically for orthographic projection.

PRETEST F: Sketch 67A gives the principal views of a cone and a cylinder. Sketch eight equally spaced elements on the top and front views of both the cylinder and the cone. Number the elements consecutively in both views.

Refer to answer sketches for the solutions.

If your solutions are CORRECT, proceed directly to Chapter 26.

CURVED SURFACES 389

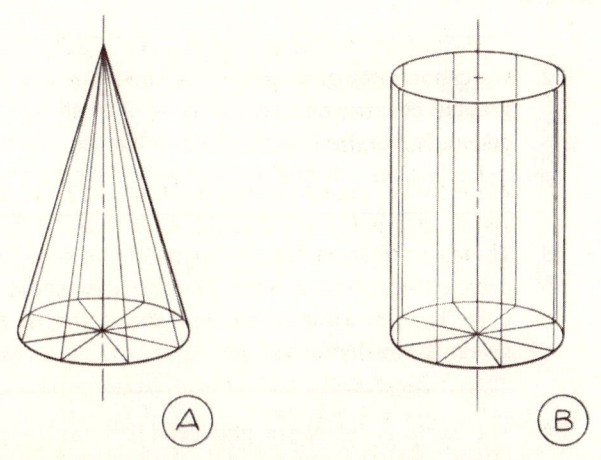

Figure 25.18 Summary sketch: Precise surface description of a cone and a cylinder.

If your solutions are INCORRECT, proceed through the remainder of Chapter 25.

---

61 The object shown in the two principal orthographic views in Figure 25.19A is unmistakably recognized as a(n) _____.

cylinder

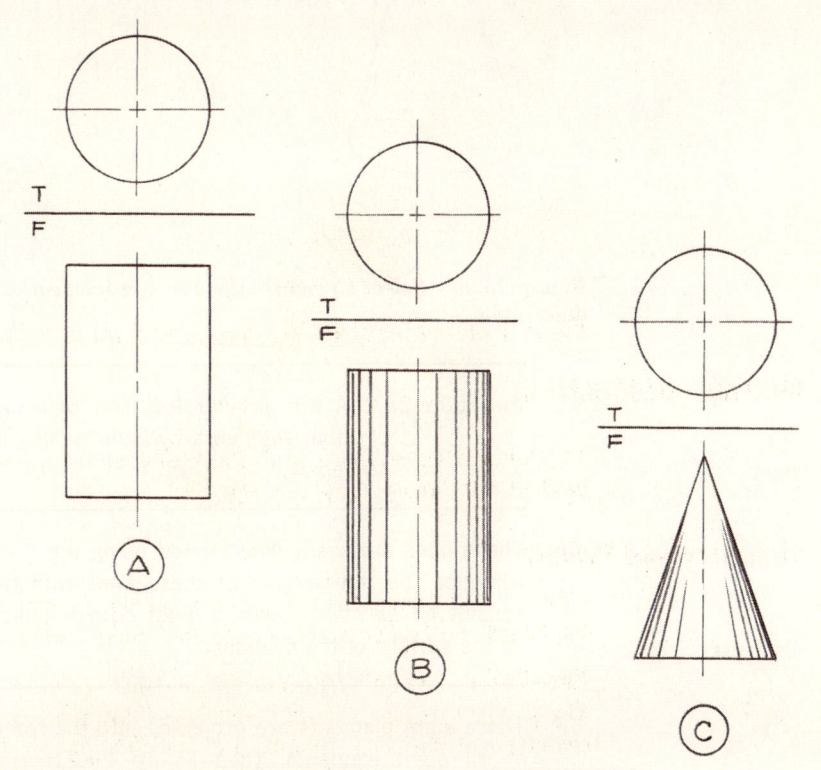

Figure 25.19 Shading gives realism to a curved surface but not precision.

390 DESCRIPTIVE GEOMETRY/UNIT 5

shading

62 We cannot, however, "see" the surface of the cylinder let alone perform graphic construction on it. Figure 25.19B has had _____ added to the front view. This enhances our ability to see the surface but still does not give us a precise description of the surface.

precise

63 The same analysis can be made of the cone. Figure 25.19C shows the principal views of a cone. The top view shows the diameter of the circular base and the front view shows the height of the cone. Shading gives us a realistic, but not _____, description of the surface.

select

64 The surfaces of cones and cylinders can be precisely defined by creating *select* elements on the surface. (Refer to Figure 25.20.) To define the surfaces of cones and cylinders precisely, we must draw _____ elements on the surfaces.

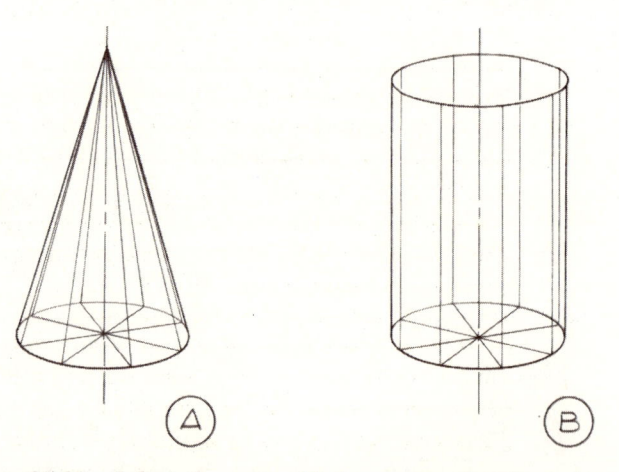

Figure 25.20 Select elements added to a curved surface give a precise description.

eight

65 In Figure 25.21A the circular top view of a cylinder was divided into _____ equal segments by constructing lines through the center of the circle 45° apart.

elements

66 These lines are easily constructed using the T-square and the 45-45-90 triangle. The intersection of these lines with the circular view can be considered as *point* views of eight equally spaced _____ on the surface of the cylinder.

numbered

67 These eight elements are projected into the front view yielding *TL* views of all eight elements. (See Figure 25.21B.) They are consecutively _____ for identification purposes.

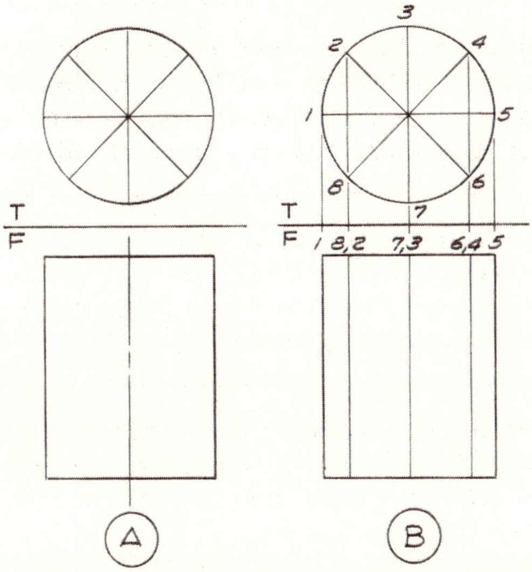

CURVED SURFACES 391

Figure 25.21 Orthographic construction of select elements on a cylinder.

These eight, equally spaced, numbered elements precisely define the cylindrical surface. We could now project the surface into any orthographic view by projecting just the eight elements.

---

68 The two principal views of a cone are shown in Figure 25.22A. The circular top view has been divided into _____ equal segments by constructing lines through the center of the circle 30° apart.

12

---

69 This is easily done using a T-square and a 30-60-90 triangle. The intersection of these lines and the circular top view of the cone *cannot* be considered as point views of elements on the surface of the cone, but merely as _____ on the base of the cone. These points have been projected to the base. (See Figure 25.22B.)

points

---

70 In Figure 25.22C the front view of the 12 _____ has been completed by joining the 12 points on the base with the _____ of the cone. The elements have been consecutively numbered in both views for identification purposes.

elements; vertex

---

71 In Figure 25.22C, the original circle divisions were laid out to be symmetrical with the front reference plane.
   Five elements (No. ___, ___, ___, ___, and ___) lie on the back of the cone.
   Five elements (No. ___, ___, ___, ___, and ___) lie on the front.
   Two elements (No. ___ and ___) lie on the sides (contour elements).

Back—2, 3, 4, 5, 6;
Front—8, 9, 10, 11
12; Side—1, 7

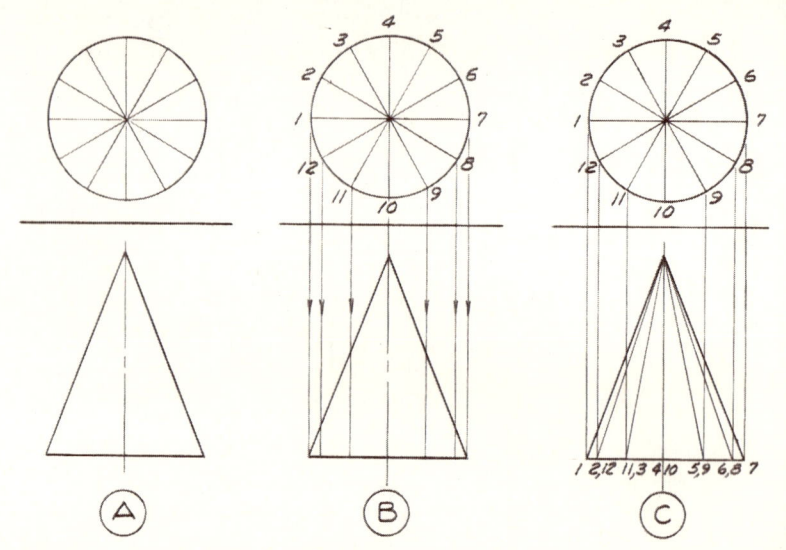

Figure 25.22   Orthographic construction of select elements on a cone.

Symmetrical arrangement of the elements saves drawing time. The number of elements chosen is a function of the accuracy desired. A large number of elements gives a better description of the surface and thus yields greater accuracy. Normally, 8 or 12 are used because of the ease of construction with standard instruments.

■ 72   Sketch 67A (pretest F) shows the principal views of a cone and a cylinder. Construct eight equally spaced elements on each, showing both front and top views. Number the elements in both views.

NOTE: The starting point and direction of numbering is immaterial, but the numbering must be consistent in both views.

chapter 26 *the conic sections*

INTERSECTION OF A PLANE AND A SURFACE OF REVOLUTION

The material of Chapter 23 presented a discussion of a family of single curved lines called the conic sections or conics. They are the *circle, ellipse, parabola,* and *hyperbola*. The conic sections are plane curves formed by the intersection of cutting planes with a right-circular cone.

This chapter discusses how the four curves are produced by cutting planes and then shows how the curves can be plotted by finding the true shape of the line of intersection between a cutting plane and a cone (surface of revolution).

Summary and Pretest

Figure 26.1 summarizes pictorially the relationships of the four cutting planes to the cone axis in producing the four conic sections.

angle θ = angle between the cutting plane and the cone axis
angle α = the half angle of the cone

PRETEST: Referring to Figure 26.1, state the *value* or *range of values* of angle θ for the four conic sections.

Circle _____   Ellipse _____
Parabola _____   Hyperbola _____

Check your answer with the answer to frame 18.

If you were correct on all four conics: Proceed with the material following frame 19.

If you were wrong on the ellipse: Turn to frame 7.

If you were wrong on the parabola and hyperbola: Turn to frame 11.

Otherwise, proceed through frames 1 to 19.

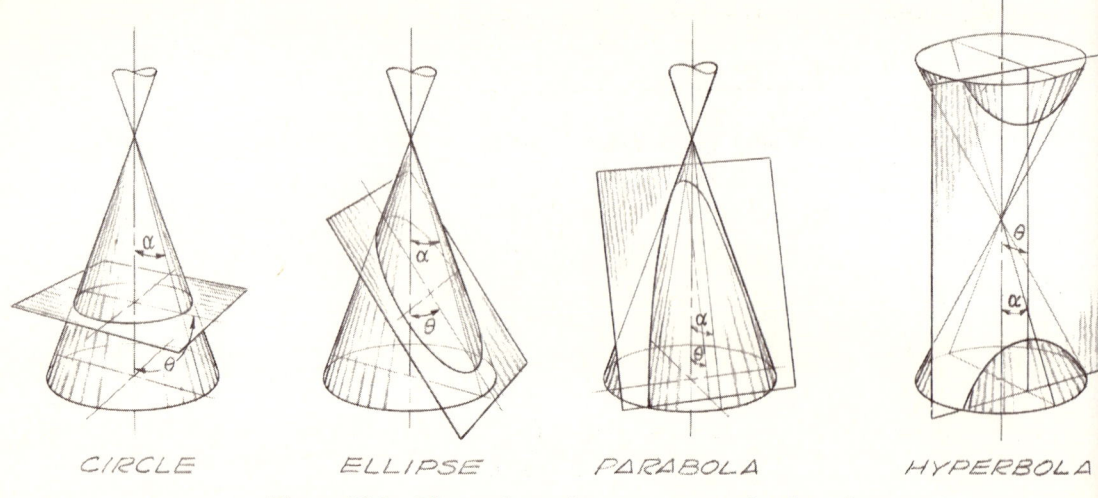

Figure 26.1   The conic sections—summary sketch and pretest.

CIRCLE ELLIPSE PARABOLA HYPERBOLA

cutting

1. The conic sections—circle, ellipse, parabola, and hyperbola—are so called because they can be described by finding the line of intersection of a cone with four specific _____ planes.

circle, ellipse, parabola, hyperbola

2. The particular cutting plane chosen determines which of the conic sections (often called conics) is described, the _____, the _____, the _____, or the _____.

The selection of the cutting plane used to describe any one of the four conics depends upon the angle between the plane and the axis of the cone.

elements; constant

3. We will call this angle θ. (See Figure 26.2.) Of interest also is the angle α (alpha), the true angle between all _____ of the cone and the axis. Angle α is _____ for any one particular cone.

Circle

perpendicular; 90

4. Refer to Figure 26.3. We have seen previously that a circle is described by the intersection of a cone with a plane _____ to the axis of the cone. Angle θ equals _____ degrees.

THE CONIC SECTIONS 395

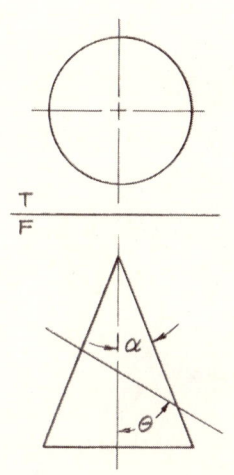

Figure 26.2 Graphic definition of angles alpha (α) and theta (θ).

90; intersection	5 Thus, to find the first conic, the circle, we must draw a cutting plane that makes an angle of _____ degrees with the axis of a cone and find the line of _____ between the plane and the cone.
■	6 Sketch 67B shows the top and front views of a cone. By sketching, establish a cutting plane that will yield a circular intersection with the cone. Show the line of intersection in both views. Check your solution.

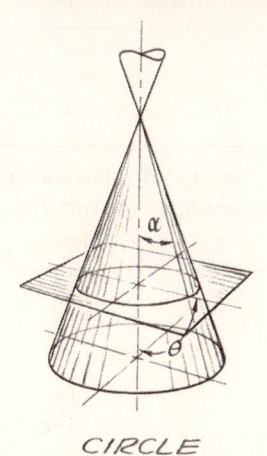

Figure 26.3 The circle.

Ellipse

ellipse	7 As soon as angle θ becomes less than 90°, the section cut by the plane changes from a circle to a(n) _____. (See Figure 26.4.)
ncreases	8 As angle θ continues to decrease, the ellipse becomes longer, i.e., the ratio of major diameter to minor diameter _____ (Figure 26.5A). (*Choose one:* increases/decreases.)

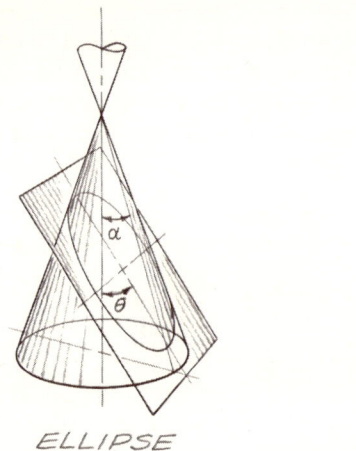

396 DESCRIPTIVE GEOMETRY/UNIT 5

Figure 26.4 The ellipse. 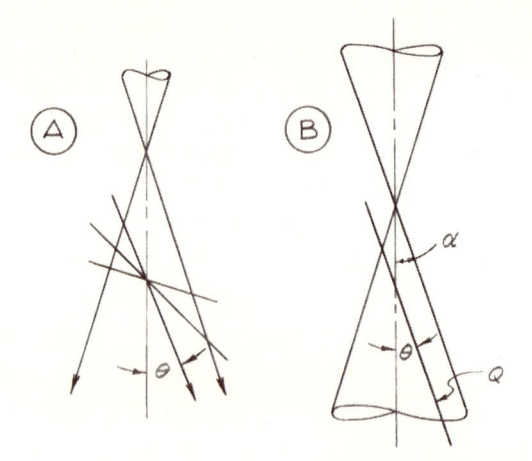   *Figure 26.5 The range of angle θ for ellipse-producing cutting planes.*

α (alpha)

9   A limit of ellipse-producing cutting planes is reached when the plane no longer cuts through all of the elements of the cone. Figure 26.5B shows this limit—plane Q is parallel to the contour element of the cone and θ = _____.

■ 10   Sketch 67C shows two principal views of a cone cut by a plane Q. The line of intersection will be an ellipse. Make an approximate sketch of the top view of the line of intersection. Check your solution.

Parabola

α; parallel

11   Consider Figure 26.6. The line of intersection between a cutting plane and a cone is a parabola when θ = _____. The plane Q is _____ to the element VA.

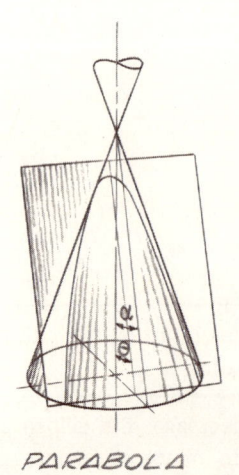

Figure 26.6 The parabola.

THE CONIC SECTIONS 397

■ **12** Sketch 67D shows T and F views of a cone and a cutting plane Q. Since plane Q is parallel to the contour element VA, it does not cut all of the elements of the cone. Thus the line of intersection cannot be an ellipse but is a parabola. Make an approximate sketch of the top view of the line of intersection.

NOTE: Three points can be found easily. Check your solution.

Hyperbola

13 Continued decrease of angle θ produces the hyperbola as the _____ of _____ between a cutting plane and a cone. See Figure 26.7A.

line (of) intersection

14 In Figure 26.7B, plane Q cuts a hyperbola on the cone. Angle θ is now _____ than angle α. The plane does not cut all of the elements of the lower nappe of the cone but now cuts the upper nappe as well as the lower. (*Choose one: less/greater.*)

less

15 A cutting plane that cuts both nappes of a cone (but does not contain the vertex) has a(n) _____ as its line of intersection.

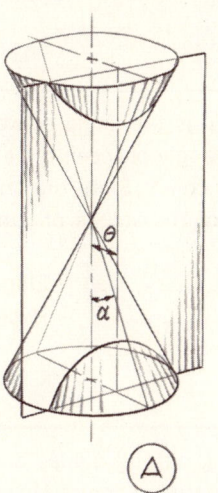

hyperbola

Figure 26.7   The hyperbola.

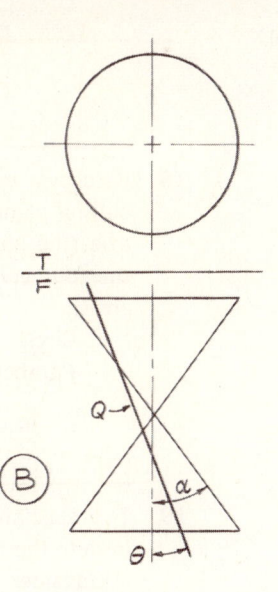

■ **16** The criterion for producing a hyperbola by passing a cutting plane through a cone is that the angle θ be less than the angle α. On Sketch 67E construct an approximate top view of the hyperbola cut by plane Q. Three points on each nappe can be readily obtained. Check your construction.

398 DESCRIPTIVE GEOMETRY/UNIT 5

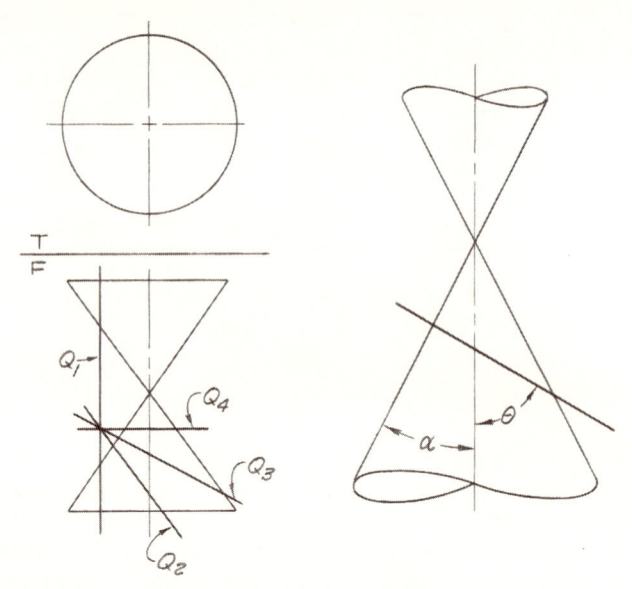

Figure 26.8   Conic-producing cutting planes.

17   Figure 26.8 shows a cone cut by four planes. The line of intersection of each of these planes with the cone is:

Q_1 _____ Q_3 _____
Q_2 _____ Q_4 _____

(Q_1) hyperbola
(Q_2) parabola
(Q_3) ellipse
(Q_4) circle

18   As shown in Figure 26.8, θ is the angle between the cone axis and the cutting plane and α is the angle between the elements of a cone and the axis (the half-angle of the cone). State the criterion for finding each of the four conics in terms of the values or range of values of angles θ and α.

Circle _____   Ellipse _____
Parabola _____   Hyperbola _____

Circle: $\theta = 90°$
Ellipse: $90° > \theta > \alpha$
Parabola: $\theta = \alpha$
Hyperbola: $\theta < \alpha$

19   Two special cases involving cutting planes and cones that do not conform to the foregoing discussion of conics should be noted.
      Consider Figure 26.9. Planes P_1 and P_2 contain the vertex and plane R contains the axis. The intersection of the cone and these planes are:

P_1 _____
P_2 _____
R _____

(P_1) a point
(P_2) two elements (2 straight lines)
(R) two elements

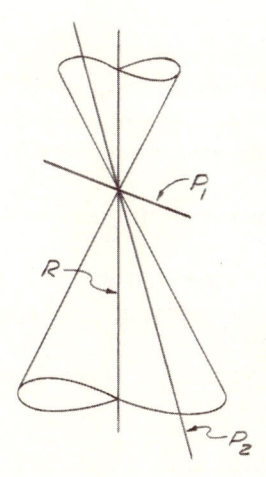

Figure 26.9   Other intersections between a cone and cutting planes.

TRUE SHAPE OF THE LINE OF INTERSECTION

The material of the remainder of Chapter 26 presents the method of finding the true shape of the line of intersection between surfaces of revolution (cones and cylinders in this instance) and any cutting plane.

Summary and Pretest

Figure 26.10 is a pictorial sketch of a vertical cone (axis *XY*) cut by an oblique cutting plane *ABC*. The line of intersection between the cone and the plane is one of the conic sections.

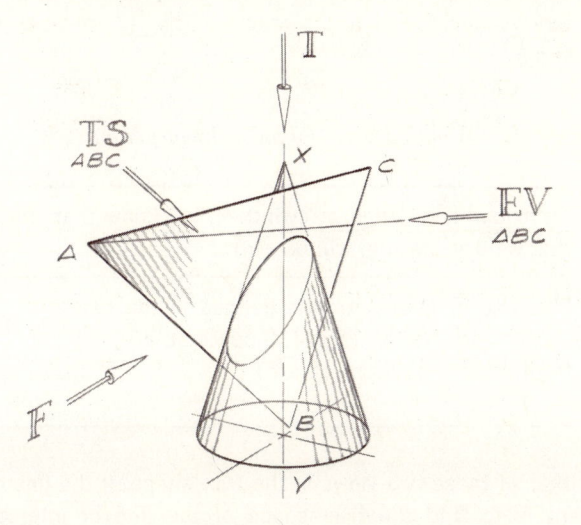

Figure 26.10   Intersection of a plane and a cone—the general case.

400 DESCRIPTIVE GEOMETRY/UNIT 5

PRETEST: Do one of the following:

(1) Using freehand sketching techniques on Sketch 68A, find the approximate true shape of the line of intersection between the cone and plane ABC.

(2) If you are *unable* to complete (1) above, proceed directly to frame 20. If you did item (1) above, compare your answer with the answer sketch (for 68A) and proceed as follows:

(a) If your answer agrees completely with the answer sketch, proceed to Chapter 27.

(b) If you obtained the correct views, but do not agree on the details of the true shape (the eight select points that define the intersection), proceed through frames 28 to 37.

(c) If your answer *does not* agree with the answer sketch, proceed through all the rest of the material of Chapter 26.

---

20 Consider Figure 26.11A. A right-circular cylinder is cut by an inclined plane. We see that the cutting plant ABCD shows as a(n) _____ in the front view. The top view of the cylinder shows the axis XY as a(n) _____ .

edge; point

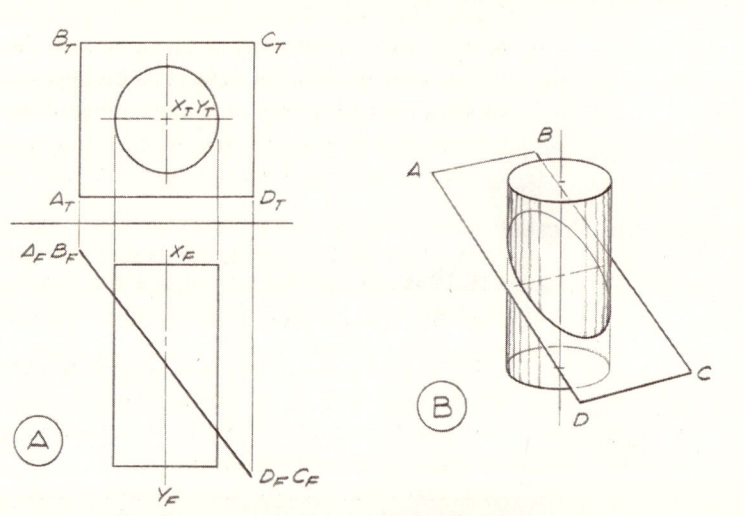

Figure 26.11 A cylinder cut by an inclined plane.

---

21 From our knowledge of cylinders, we know that the line of intersection that we are seeking, will be a(n) _____ .

ellipse

22 This line of intersection is defined in the top and front views of Figure 26.11A. In the top view it appears as a(n) _____ and in the front view it appears as a(n) _____ _____ .

circle;
straight line

Neither of these two views is the true shape of the line of intersection. Our problem is to find the true shape of the line of intersection between the cylinder and the plane.

THE CONIC SECTIONS 401

	23	Figure 26.11B shows a pictorial sketch of the cylinder and cutting plane. The ellipse (line of intersection) lies wholly on both the _____ and the _____ of the cylinder.
plane; surface		
	24	If we can define the ellipse on the two available views of plane ABCD, we can show the true shape of the ellipse by creating an auxiliary view which shows the _____ of plane ABCD.
TS (true shape)		
	25	Figure 26.11A shows the contours of the cylinder but does not precisely describe the surface. We know that we can precisely describe the surface of a cylinder or a cone by drawing select _____ on the surface.
elements		
■	26	Sketch 68B shows a cylinder and a cutting plane similar to Figure 26.11A and B. Find the true shape of the line of intersection between plane ABCD and the cylinder. The following steps show the procedure:

STEP 1: Sketch eight equally spaced elements on the two views of the cylinder. Number them consecutively in both views.

NOTE: The direction of numbering and the starting position are immaterial as long as the numbers agree on both views.

The surface of the cylinder is now precisely defined by eight elements. The line of intersection lies wholly on both the plane and the surface of the cylinder. Therefore, the ellipse cut by plane ABCD will be defined by the intersection of the eight elements and plane ABCD.

STEP 2: We are now ready to establish the auxiliary view that will give us the true shape of plane ABCD and thus the true shape of the line of intersection. Sketch in the reference line for auxiliary view 1 which will give us the true shape of ABCD. Label the reference line F/1.

STEP 3: By sketching and estimating distances (use your dividers if you wish) project the eight points of intersection between the elements of the cylinder and plane ABCD into auxiliary view 1. Number the points in view 1. Check your solution.

	27	These eight points can now be joined with a smooth curve which is an ellipse, the true shape of the _____ of _____ between the cylinder and plane ABCD.
line (of) intersection		
	28	Sketch 68A (the Pretest sketch) shows a vertical cone (axis XY) cut by an oblique plane ABC. It differs from the previous example (68B) in that the plane does not appear as a(n) _____ in either one of the given views.
edge		
	29	Without an edge view of plane ABC we cannot find the points of intersection of select _____ on the cone nor can we find the true shape view of the _____ ABC.
elements; plane		

edge; cone

point

point

TL (true length)

top

30 Therefore, we must first draw a single auxiliary view that shows two things:
(1) The plane as a(n) _____.
(2) Another view of the _____.

31 We find the edge view of a plane by finding a(n) _____ view of a line that lies in the plane.

32 We have a choice of projecting the edge view of ABC from either the top or the front view. However, since we must also project the cone into the auxiliary view, we must choose the view that is easiest to draw and shows the cone to advantage.

The top view of the cone in the figure (Sketch 68A) shows the axis XY as a(n) _____.

33 This means that XY is perpendicular to the top plane. *Any* auxiliary view projected off the top view will show axis XY in its _____.

34 On the other hand, *any* auxiliary view projected off the front view will show a foreshortened view of the axis XY. Therefore, it is to our advantage to find the edge view of ABC in an auxiliary view projected off the _____ plane.

■ **35** Find the true shape of the line of intersection between plane ABC and the cone of Sketch 68A. The following steps show the procedure:

STEP 1: Sketch a line in plane ABC that will show as a true length in the top view. Draw in the reference line for an auxiliary view that will show this line as a point and thus give an edge view of ABC.

Now project plane ABC and the cone into auxiliary view 1. (Use sketching techniques.) Check your solution at this point.

STEP 2: Sketch eight equally spaced elements on the surface of the cone in views T and 1. Number these elements in both views. For convenience arrange two of these elements to be contour elements in view 1. Check this step with the answers.

Sketch in a reference line 1/2 for an auxiliary view that will show plane ABC in its true shape.

The intersection that we are seeking is defined in view 1 by the intersection of the eight select elements and the edge view of plane ABC. *This is the only view that we have of the eight points so far.*

STEP 3: We cannot project these eight points into view 2 until we have two views of the eight points. Since the points lie on the elements, we can project each point back from view 1 to view T, placing each on its correct numbered element. Perform this projection.

REMEMBER: To find the top view of the points on the center elements (points 1 and 5 on the answer sketch) you must use *rotation*. Check this construction.

THE CONIC SECTIONS 403

STEP 4: We now have two views of the eight points that represent the intersection between the cone and plane ABC. Project the eight points in view 2. Number the eight points found.

Joining these eight points with a smooth curve yields the true shape of the ellipse that is the line of intersection between the cone and plane ABC.

■ 36 In a problem of this type it is often desirable to show the front view of the intersection. On Sketch 68A we now have enough information to find this front view. Using standard projection techniques, find the front view of the intersection. Show visibility. The answer to step 4 of frame 35 shows this construction.

■ 37 Sketch 68C presents a problem of finding the true shape of a parabola. The figure shows two principal views of a cone.

STEP 1: Pass a cutting plane whose line of intersection will be a parabola. Sketch eight equally spaced elements on the surface of the cone (top and front views). Number the elements.

The front view shows the intersection of some of the eight elements with the cutting plane. Project these points to the top view and sketch the top view of the parabola. Check your solution.

STEP 2: We now have two views of the intersection. Complete the problem by finding the true shape of the line of intersection between the cutting plane and the cone. The intersection is a parabola. Check your final solution.

Note in the answer sketch that two extra elements (9 and 10) were added between 3 and 4 and 6 and 7 respectively. This was done to give a more even distribution of points on the final true-shape curve.

This is the end of Chapter 26. No example of finding the true shape of a hyperbola has been given. The method for solving this would be identical to those shown. The basic ingredients for a solution are:
(1) An edge view of the cutting plane showing the intersection of a number of select elements of the solid with the plane.
(2) A second view of these points of intersection—usually the view showing the axis of the solid as a point.

chapter 27 development of curved surfaces

Figure 25.1 gave a classification of curved surfaces. Ruled surfaces (surfaces generated by a moving straight line) are divided into two groups:
  (1) Developable or single curved surfaces
  (2) Undevelopable or warped surfaces

The cylinder and the cone are the best examples of single curved surfaces since they can be developed.

The purpose of developing a surface is to permit construction of a solid (in the mathematical sense) from flat material such as paper, sheet metal, plastic, etc. Cylinders and cones can be cut out of sheet material and rolled into their solid forms. Figure 27.1 shows this pictorially.

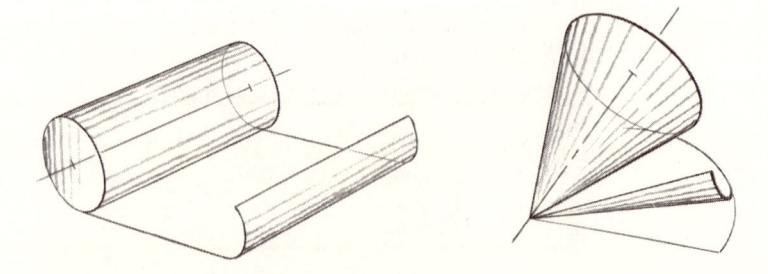

Figure 27.1 The concept of rolling (or unrolling) a cylinder and cone.

Chapter 27 shows how to develop full cylinders and cones and also portions of these solids as cut by planes or other solids. It ends with an example of the development of a *transition piece*—a solid whose surface is a combination of planes, cylinders, and cones.

CYLINDERS

DEVELOPMENT OF CURVED SURFACES 405

rectangle

1 Figure 27.2 shows a right-circular cylinder of altitude H and radius r. The development of the surface is simple, being merely a(n) _____ of dimensions $H \times 2\pi r$.

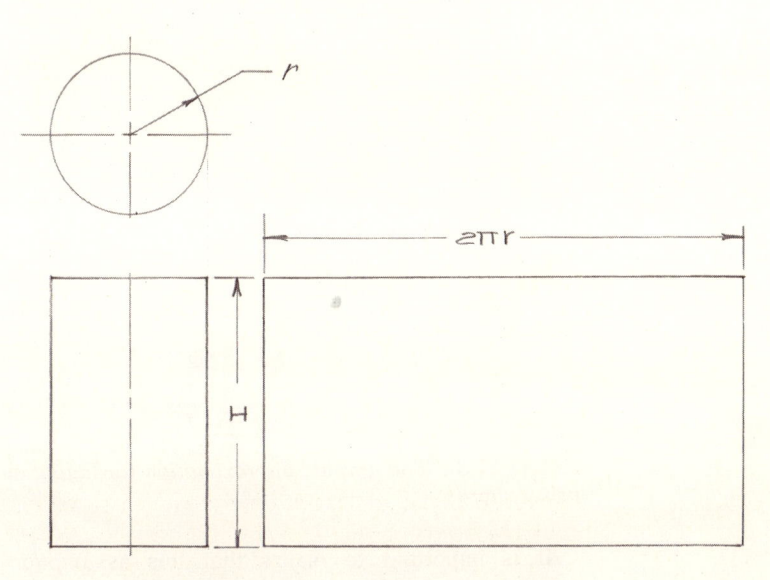

Figure 27.2   The development of a right-circular cylinder.

To lay out the development, H can be transferred directly from the orthographic views and the circumference ($2\pi r$) can be calculated and scaled.

A completely graphic solution, however, would be one in which all distances were transferred directly to the development from the orthographic views. To transfer the circumference involves dividing the circular view of the cylinder into small equally sized segments and then transferring these with dividers one at a time.

circumference

2 Figure 27.3 shows a 1½-in. diameter circle divided into eight equal-sized segments which divide the _____ into eight equal lengths.

arc

3 Setting dividers to any one of these lengths (1-2, 3-4, 8-1) and then stepping off this distance eight times along a straight line will transfer the total circumference to the line *approximately*.

This is a *graphic approximation*. It is not accurate since we are assuming that the length of the chord (See 27.3B) is equal to the length of the circular _____.

406 DESCRIPTIVE GEOMETRY/UNIT 5

4   By actual calculation, each arc in Figure 27.3 is 0.590 in. long while the chord is 0.575 in. long. In laying out the circumference graphically, we multiply the error eight times so that the graphic method produces a line 0.120 in. _____ than that obtained by the calculated method. This is a 2.5 percent error. (*Choose one: longer/shorter.*)

shorter

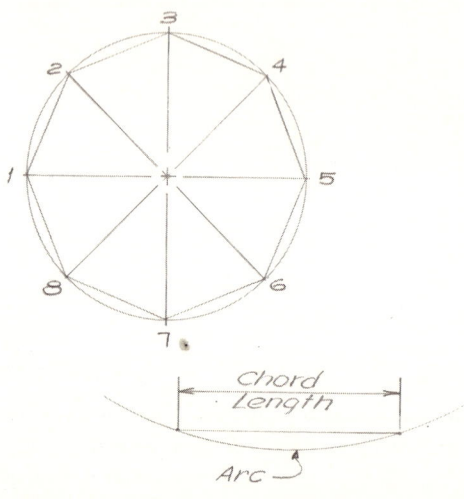

Figure 27.3   The graphic approximation for laying out a circumference using chords.

It is important to realize that this *assumption* approaches *reality* the smaller the divisions of the circle become. If we divided the circle into 120 divisions, the error would be reduced to 0.17 percent. However, the complexity of the graphic method increases with the number of divisions taken. For most practical purposes, a circle is divided into 8 or 12 parts since this can be done quickly and accurately with standard instruments.

CONES

5   Figure 27.4A shows a right-circular cone of altitude H, slant height R, and base radius r.
   The development of any right-circular cone is a pie-shaped wedge whose radius is _____ and whose circular length is _____. (See Figure 27.4B.)

R; $2\pi r$

6   A flat disk can be visualized as a cone with zero altitude (H = 0). The other extreme is the cylinder, which can be visualized as a cone whose vertex is at _____. (H = _____)

infinity; ∞

7   To develop a right-circular cone we first draw an arc of radius R and one radial line. We then divide the base circle into equal-sized divisions and lay off (graphically) a distance that approximates the base circumference, _____, along the arc and draw a second radial line.

$2\pi r$

DEVELOPMENT OF CURVED SURFACES   407

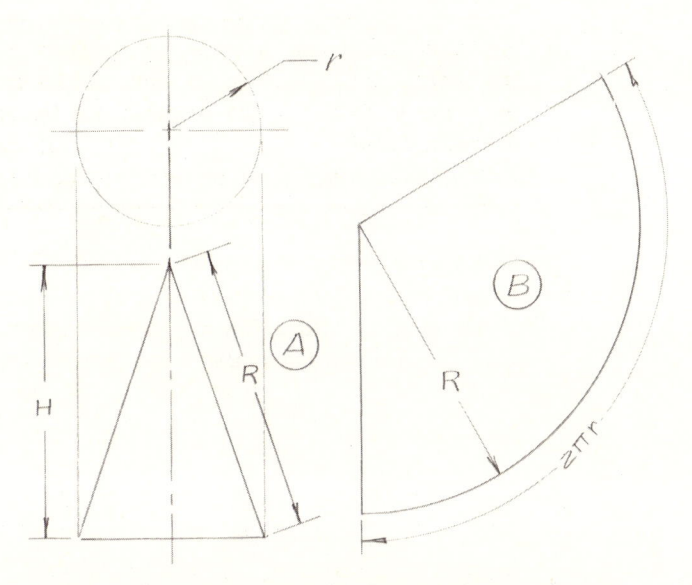

Figure 27.4   The concept of the developed surface of a cone.

assumption	8	By this method, we are making a(n) _____ that, for each division of the base circle, the arc length equals the chord length.
$2\pi R$; $2\pi r$	9	For greater accuracy, we can use a calculated method. Consider Figure 27.5. It shows a developed cone (base radius *r* and slant height *R*) included in a full circle of radius *R*. If angle A is known, we can measure this angle with a fair degree of accuracy using a protractor. Angle A will then automatically cut a portion of the full circle circumference, _____, into a segment equal in length to _____.

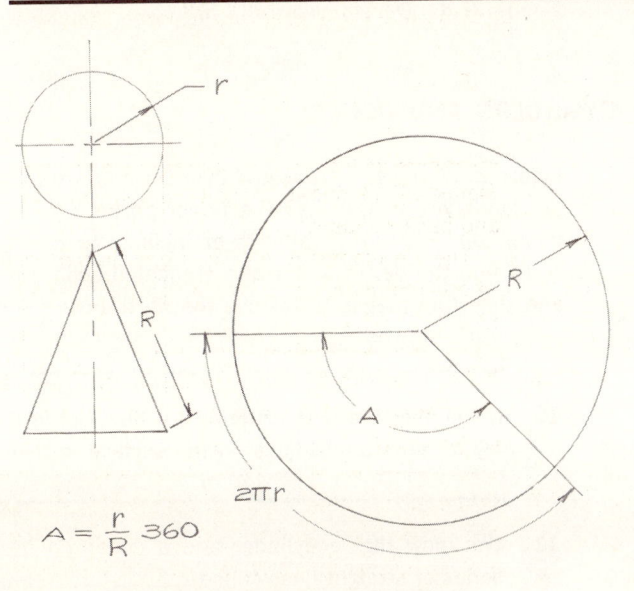

$$A = \frac{r}{R} 360$$

Figure 27.5   Calculating the developed angle A.

408 DESCRIPTIVE GEOMETRY/UNIT 5

An examination of this figure shows that we can write an equality of ratios, involving angle A and the known parameters, R and r, which relates the development to the full circle. This is

$$\frac{A}{2\pi r} = \frac{360°}{2\pi R}$$

Canceling out the constant term 2π, we find that

$$A = \frac{r}{R} \times 360$$

Both R and r are known—or can be measured from orthographic views. The development produced by calculating and measuring angle A will be more accurate than that constructed by the graphic method.

Figure 27.6 shows an example with calculations.

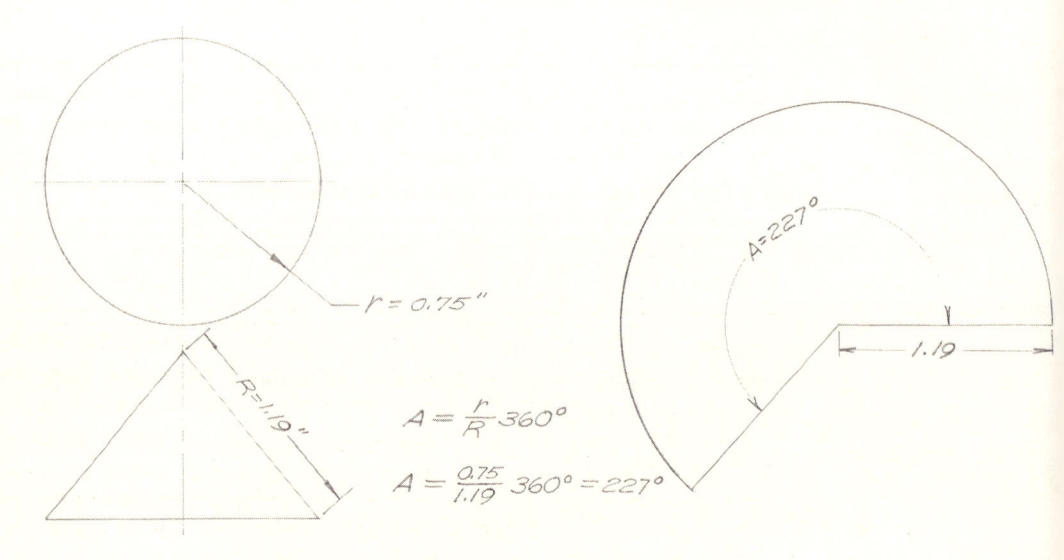

Figure 27.6   Development of a full cone.

TRUNCATED CYLINDERS AND CONES

Figure 27.7 shows a truncated cylinder and a truncated cone. These are regular solids which have had the upper portions removed. The top surfaces are planes and may be parallel to or inclined to the bases.

To find the development of a truncated solid, we can develop the full solid and then find means to remove the part above the cutting plane.

TL (true lengths)

10 Remember that the criterion for developing the surface of a solid is to lay off all straight lines on the surface in their _____.

---

elements

11 We know that a cylinder and a cone can be described precisely by a series of straight lines called _____.

DEVELOPMENT OF CURVED SURFACES   409

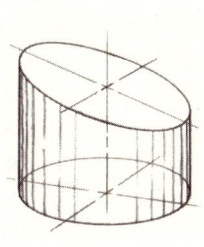

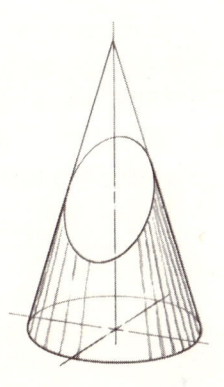

Figure 27.7   Truncation.

points; TL	12	Sketch 69A shows a truncated cylinder cut by a plane M. The plane shows as an edge in the front view. Eight equally spaced elements of the surface are given. The elements appear as _____ in the top view and as _____ lines in the front view.
eight	13	The full-cylinder development is also given—a rectangle H × 2πr. The distance 2πr has been divided into _____ equally spaced divisions.
5; shortest	14	This can be done accurately by proportioning with a scale or by bisecting with a compass. (First bisect 2πr, then bisect the halves, then the quarters.) The elements on the development have been numbered starting and ending with No. _____ the _____ element. (*Choose one:* longest/shortest.)
■	15	The numbering is arranged so that the development will be *inside-up*. With the cylinder and its full development laid out as shown, it is only necessary to put the true length of each element on the correct line on the development. This can be done by direct projection since the front view of the cylinder shows the true lengths of all elements between the base and plane M. On Sketch 69A project each element to the correct line on the development. Check your solution.
curve	16	This yields eight points through which a smooth _____ can be drawn and the development of the truncated cylinder is complete.
TL	17	Sketch 69B shows a cone cut by a sloping plane M, shown as an edge in the front view. The development of the full cone is given. **PROBLEM:** Develop the portion of the cone that lies between the base and cutting plane M. As with any development problem, we must find a way to lay out the _____ of all straight lines on the surface in their proper sequence.
elements	18	The only straight lines on a cone are _____ of the surface of the cone.

19 Eight equally spaced elements are shown on the orthographic views of the cone. These can be transferred to the development by dividing the circular length ($2\pi r$) into eight equal divisions. Each point is then joined by a straight line to the vertex. The TL of every element is equal to the slant height R _____ (your words).

■ **20** This is done most accurately with a compass by first bisecting the total arc, then bisecting the halves, and finally, bisecting the quarters.

STEP 1: Divide the arc of Sketch 69B into eight equal spaces. For this purpose you can bisect by eye—it is good practice. Draw the full-length elements.

STEP 2: Next, number the elements on the development so that it will be *inside-up* and cut on the *shortest element*. Remember that the finished development will be identical whether inside-up or outside-up; only the numbering shows the difference. Check your solution.

The developed surface has now been *triangulated*—divided into eight narrow triangles. This is not completely true since the narrow end is an arc rather than a straight line (chord).

However, had we used the graphic method for laying out $2\pi r$ rather than the calculated method, the sectors would be triangles since we would have made the assumption that the length of the chord (1-2, 2-3, ...8-1 in the top view) equals the length of the arc.

The problem now is to find the true lengths of the portions of the eight elements that lie between the base and plane M. Two elements, 1 and 5, are true length in the orthographic views. These are the contour elements of the cone and are always TL in a view showing the profile of the cone.

STEP 3: Consider elements 2 and 8. They coincide in the front view. If they are both rotated into position 1, they become TL. Rotate the point of intersection of elements 2 and 8 with plane M into position 1. Check your solution.

STEP 4: Find the true lengths of 3 and 7 and 4 and 6. Transfer each TL portion to the corresponding element on the development (measuring from the circular arc which represents the base of the cone).

Join the points obtained with a smooth curve and the development is complete. Check the complete solution.

21 Development by *triangulation* is a basic method. We first saw it in Chapter 24 (development of polyhedrons), and we will see it again in this chapter with the development of transition pieces.

Triangulation is a concept based upon the fact that, if _____ three (3) sides of a triangle are known, the triangle can be constructed.

point; TL

22 The orthographic views required to complete the development of a cylinder or a cone are:
(1) A circular view showing the axis as a _____
(2) A view showing the profile of the solid (_____ view of the axis) and an edge view of the cutting plane

If you are not given one or both of these views, they must be obtained. Figure 27.8 shows the same problem used in Chapter 26—a cone is cut by an oblique plane ABC. All that is missing for a development is the edge view of the cutting plane. Auxiliary view 1 was taken to provide the edge view and, at the same time, give another view of the cone showing its contour. Enough information is now available to develop the cone.

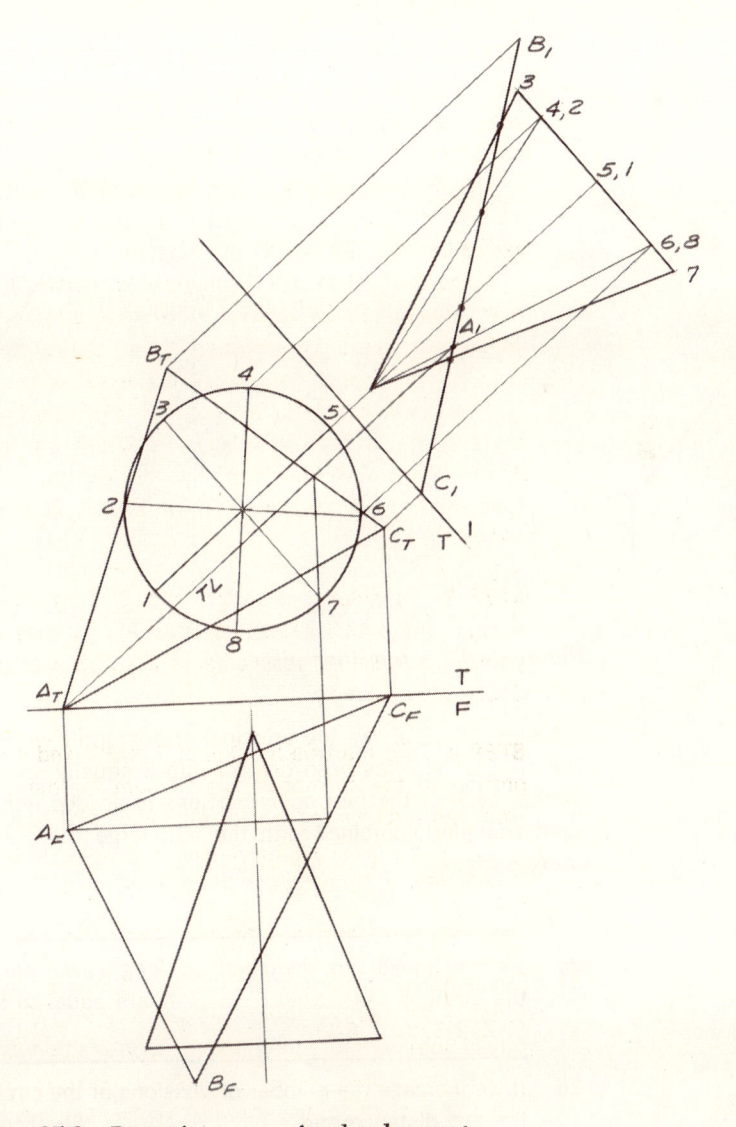

Figure 27.8 Preparing a cone for development.

412 DESCRIPTIVE GEOMETRY/UNIT 5

TRANSITION PIECES

Figure 27.9A illustrates a situation that occurs in many ways in technical work. A cylinder M must be transformed from its circular cross section to a square cross section as at N. This can be accomplished by designing a transition piece—a multiple-form piece that is developable.

transition piece	23 At Figure 27.9B, a connecting surface has been sketched between M and N. It is a composite of four triangular planes and four sections of a cone. It is called a(n) _____ _____.
developable	24 The conical portions are parts of oblique cones. (Figure 27.9C) Since planes and cones are developable, the entire transition piece is _____.

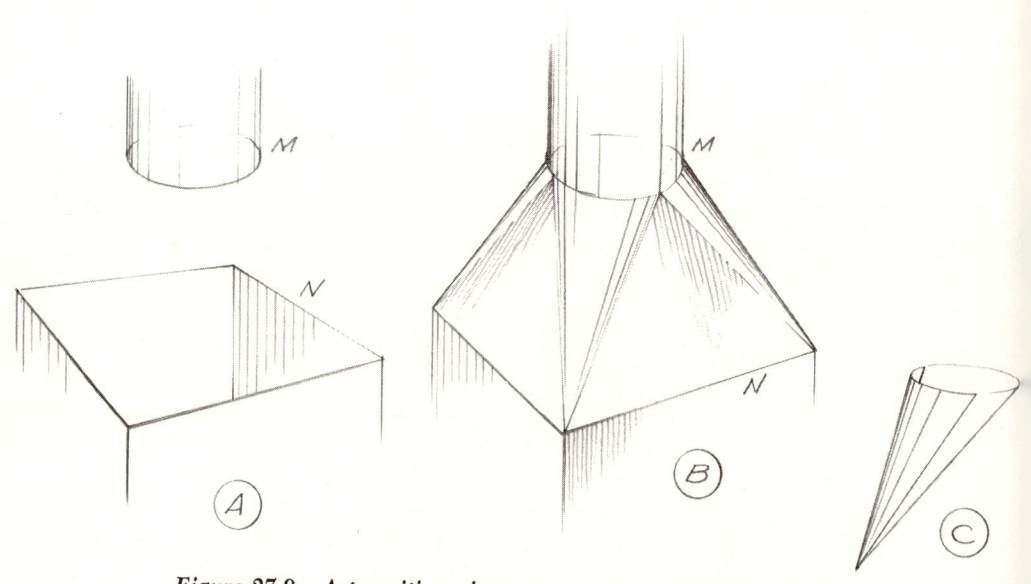

Figure 27.9 A transition piece.

Figure 27.10 shows the principal orthographic views of the transition piece. The circular top has been divided into 8 equally spaced points. This is done so that each of the four conic portions is divided into two "triangles." These eight triangles combined with the four large triangular planes comprise the entire surface.

chords; arcs	25 We can make this deduction as long as we are willing to assume that the straight-line _____ are equal to the circular _____ (1-2, 2-3, . . . , 8-1).
shorten; increase	26 If we increase the number of divisions of the circle, we _____ the arc distance and _____ the accuracy of the approximation. (*Choose one:* lengthen/shorten; increase/decrease.)

DEVELOPMENT OF CURVED SURFACES

rotation

27 The development can proceed in exactly the same manner as that used for polyhedrons in Chapter 24. The surface is made up of 12 connecting triangles. Start with any one triangle. Find and lay out the *TL* of three sides. The true lengths can be found most easily by the method of _____. (See Figure 27.10.) Finally, lay out all the rest in succession.

Figure 27.10   The piece triangulated and ready for development.

28 Figure 27.11 shows the development of *one-quarter* of the piece bounded by lines 1-*X* and 7-*Y*. The full transition piece is _____ about both centerlines and therefore only one-quarter need be developed. The remaining three-quarters can be duplicated as shown in Figure 27.11B.

symmetrical

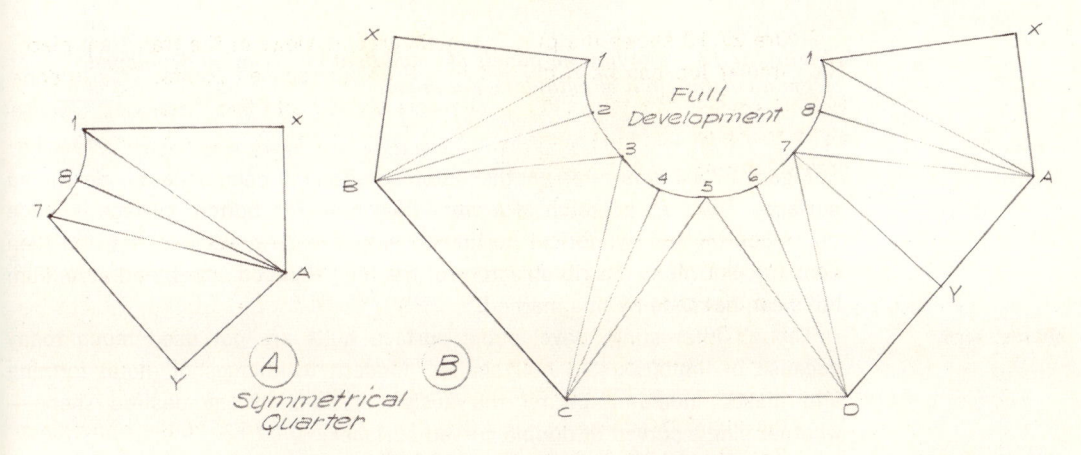

Figure 27.11   The development of a transition piece.

■ **29** Sketch 70A shows the top and front views of a hopper used in moving granulated material (wheat, sugar, salt, etc.). The top is a rectangle which is transformed into a circular cross section. Sketch 70B shows an air-conditioning duct which has a transition piece to transform a *horizontal* cylindrical pipe into a *vertical* rectangular duct.

Sketch the lines necessary to triangulate these two pieces in preparation for developing the surfaces.

Check your solutions with the answer sketches. These sketches also include the development of a symmetrical part of each.

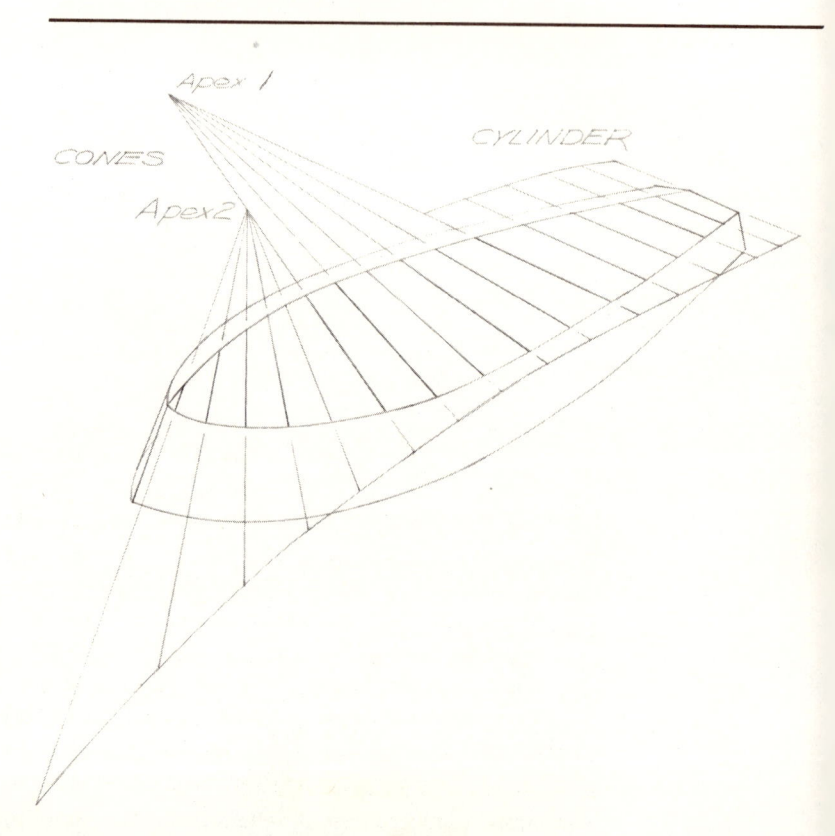

Figure 27.12 The bottom panels of small-boat hulls can be developed from a combination of cylinders and cones.

Figure 27.12 illustrates another interesting use of composite, single curved surfaces. It shows a sketch of a small-boat hull. The bottom surface is made up of conical and cylindrical portions. These can be developed flat and then bent to conform to the rib structure of the hull. Plywood, steel, and aluminum hulls can be made by this method.

Though interesting, developable surface hulls are not used much today because of limitations in hull shapes. Modern techniques of metal forming and plastic moulding permit the design of hulls of any desired shape—whether single curved or double curved surfaces.

chapter 28 intersection of curved surfaces

We have defined a mathematical solid as a container rather than the thing contained. If we talk about the intersection of a cone and a cylinder (two familiar solids), we are thus talking about the line of intersection of two curved surfaces. This line of intersection is a double curved line.

In Chapter 24 we considered the intersection of polyhedrons—solids bounded by plane surfaces. You will recall that we used the cutting-plane method to find the point of intersection between a single line on one of the surfaces and one or more of the planes of the second surface. Upon finding a series of these points, we then connected them in an appropriate sequence with straight lines to describe the orthographic views of the line of intersection between the two polyhedrons.

This identical process can be used to describe the line of intersection between two surfaces of revolution.

straight

1 We have been working in the last few chapters with single-ruled surfaces, curved surfaces generated by a moving _____ line.

element

2 Any particular position of the moving straight line is called a(n) _____ of the curved surface.

cutting

3 One way to define one or more selected elements on the surface is to pass _____ planes through the surface.

415

416 DESCRIPTIVE GEOMETRY/UNIT 5

elements	4 Figure 28.1 shows a cylinder intersecting a cone. A single cutting plane (Q) has been passed through the two surfaces. This one plane defines two _____ on *each* of the surfaces.
4 (four)	5 These four straight-line elements all lie in plane Q. They intersect at _____ points.
cylinder, cone, plane Q	6 These four points are points on the line of intersection between the cylinder and the cone. They represent four points common to: (1) _____; (2) _____; (3) _____
pass through the vertex	7 To cut straight-line elements on a cone, a cutting plane must _____ _____ (your words)
be parallel to the axis	8 To cut straight-line elements on a cylinder, a cutting plane must _____ (your words)

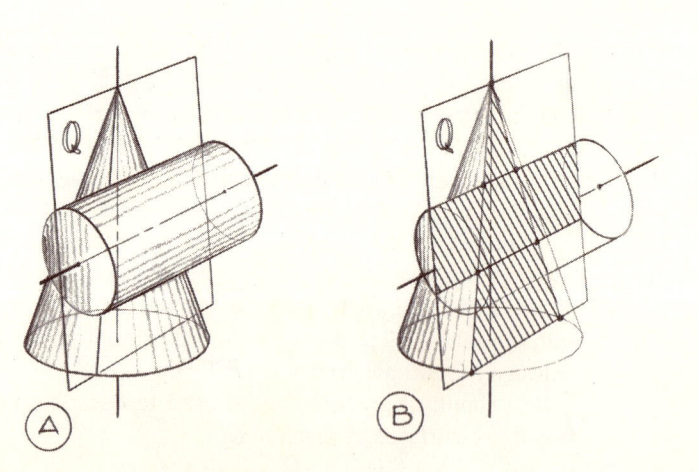

Figure 28.1 Using a cutting plane to locate four points on the line of intersection between a cone and a cylinder.

Cutting plane Q in Figure 28.1 was positioned so that it passed through the vertex of the cone and was parallel to the axis of the cylinder. The result was two elements described on each surface which, in turn, described four points on the line of intersection. The full intersection is found by constructing more cutting planes to yield more points.

INTERSECTION OF CURVED SURFACES 417

TWO CYLINDERS

Consider Figure 28.2. It shows top, front, and profile views of two intersecting cylinders.

---

9 The axis of one cylinder is a horizontal line and the axis of the other is a vertical line.
   Axis _____ is horizontal
   Axis _____ is vertical

UV (horizontal)
XY (vertical)

---

10 Four cutting planes, A, B, C, and D, are drawn. They are vertical planes and appear as follows in the three principal views:
   (1) Top: _____.
   (2) Front: _____.
   (3) Profile: _____.

(1): EV (edge view)
(2): TS (true-shape view)
(3): EV

---

11 All four planes are parallel to axes UV and XY. Therefore each plane *should* cut _____ elements on each cylinder.

2 (two)

---

12 However, note planes A (profile view) and D (top view). Each one is _____ to one of the cylinders and therefore will describe only one element on the surface.

tangent

---

13 How many *points* on the line of intersection will each of the four cutting planes provide?
   Plane A: _____ points.
   Plane B: _____ points.
   Plane C: _____ points.
   Plane D: _____ points.

(A): 2
(B): 4
(C): 4
(D): 2

Planes A and D are limiting planes. They describe the limit of the line of intersection. Planes B and C are general planes. General planes will always appear between the two limiting planes. They may be placed anywhere within the range of the intersection. They are usually evenly spaced as in Figure 28.2. The criterion for positioning cutting planes is to give well-distributed points on the line of intersection so that they may be joined with an accurate curved line.

418 DESCRIPTIVE GEOMETRY/UNIT 5

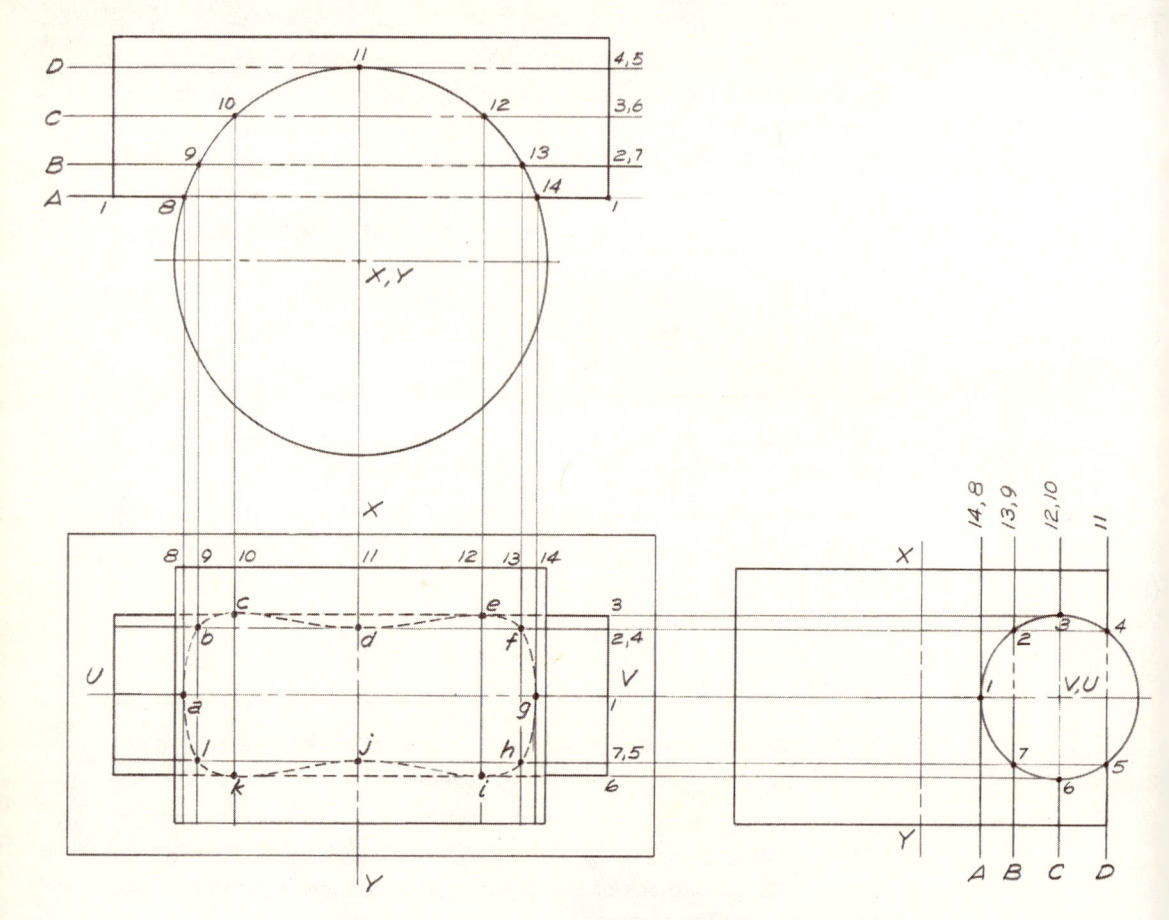

Figure 28.2   Two intersecting cylinders.

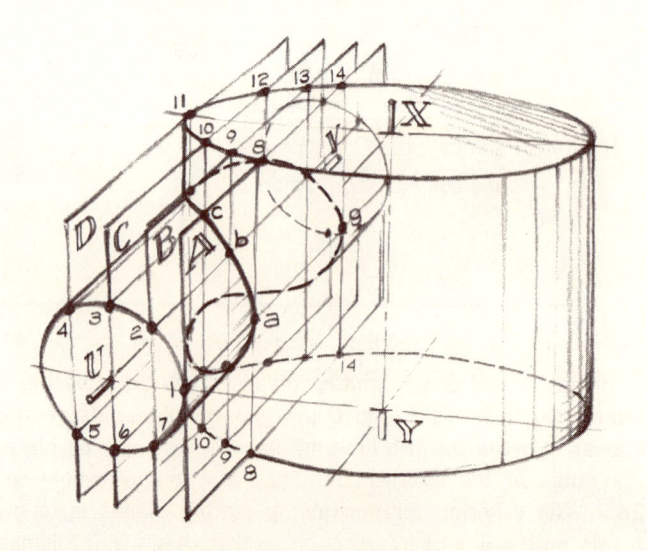

Figure 28.3   Pictorial drawing of the problem based on 28.2.

INTERSECTION OF CURVED SURFACES 419

14 The complete line of intersection between the two cylinders is defined by 12 points, a through l (front view).

Elements are numbered 1-1 through 7-7 on the small-diameter cylinder (axis UV) and 8-8 through 14-14 on the large-diameter cylinder (axis XY). Complete the following table by matching numbered elements on the two cylinders with lettered points on the intersection.

      Cylinder
    UV   XY

(a) _____ _____
(b) _____ _____
(c) _____ _____
(d) _____ _____
(e) _____ _____
(f) _____ _____
(g) _____ _____
(h) _____ _____
(i) _____ _____
(j) _____ _____
(k) _____ _____
(l) _____ _____

	UV	XY
(a):	1–1	8–8
(b):	2–2	9–9
(c):	3–3	10–10
(d):	4–4	11–11
(e):	3–3	12–12
(f):	2–2	13–13
(g):	1–1	14–14
(h):	7–7	13–13
(i):	6–6	12–12
(j):	5–5	11–11
(k):	6–6	10–10
(l):	7–7	9–9

15 Points common to both cylinders were found by matching _____ created by a single cutting plane.

elements

16 The actual points are the _____ of elements which lie in a single plane.

intersections

Figure 28.3 is a pictorial drawing of the complete problem.

CONE AND A CYLINDER

Figure 28.4 shows top and front views of an oblique cone and an oblique cylinder. Figure 28.5 is a pictorial drawing of the problem and solution.

PROBLEM: Find the line of intersection. Both objects are elliptical in cross section. This can be deduced from the fact that the top view of the horizontal bases are circular. Both axes are oblique lines and therefore sections taken perpendicular to the axes cannot be circular but must be elliptical.

The solution of the problem will differ from that of Figures 28.2 and 28.3 in that we will not be able to use the edge views of selected cutting planes.

420 DESCRIPTIVE GEOMETRY/UNIT 5

17 From the pictorial example of Figure 28.1, we know that we can find points on the line of intersection by establishing a series of cutting planes which: (your words)
(1) _____ vertex of the cone.
(1) pass through;   (2) _____ axis of the cylinder.
(2) are parallel to

18 We can assure ourselves that a cutting plane passes through the vertex of a cone by making the vertex (point V in 28.4) _____
one corner (or one   _____ (your words).
point) of the plane

19 Similarly, we can be assured that a cutting plane is parallel to the axis of a cylinder by constructing one side of the plane _____
parallel to   _____ (your words).
the axis

20 Another criterion for solving this intersection problem is that one or more lines of the cutting planes must pass through the bases of the solids. These are the lines which define the element or elements cut by each plane.
In Figure 28.4, the plane bases of both the cone and cylinder appear
EV; TS   as _____ in the front view and as _____ views in the top view.

21 Four cutting planes were used to solve the problem. They are VAB, VAC, VAD, and VAE. In all four cases one line of each plane lies in the bases
AB, AC, AD,   of both solids. These are lines _____, _____, _____, and _____.
(and) AE

22 The front view (EV of bases) shows that these lines lie in the bases. The top view shows where the four cutting planes _____ the
intersect (cut)   circular bases.

23 Point A was found in both views by constructing line VA parallel to line
ST; cylinder   _____, the axis of the _____.

24 For line VA to be truly parallel to axis ST, it must show as being parallel in *both* views. Point A_F was found first. It is the intersection of a line through point V_F, constructed parallel to axis $S_F T_F$, with a horizontal line through the bases of the solids. An identical construction was then made through point V_T and point A_T was located by direct _____
projection   from point A_F.

25 Thus four cutting planes satisfy the three criteria for establishing elements on the two intersecting solids:
(1) They contain the _____ of the cone.
(2) They are _____ to the axis of the cylinder.
(3) Each plane has one line that lies in the _____ of both
(1) vertex;   solids.
(2) parallel;
(3) base

INTERSECTION OF CURVED SURFACES 421

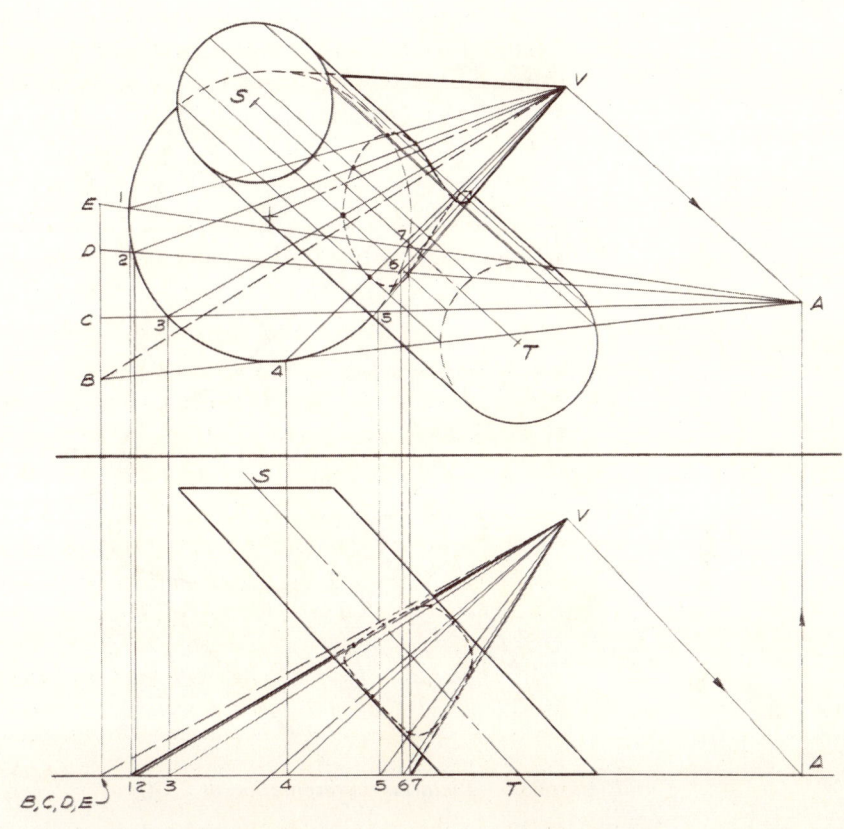

Figure 28.4 An oblique cone intersecting an oblique cylinder.

26 Cutting planes VAB and VAE are limiting planes and VAC and VAD are general planes.
    Plane VAB is tangent to the circular base of the cone. Any cutting plane passed in front of VAB _____ intersect the cone and _____ intersect the cylinder. (*Choose one: would/would not.*)

would not; would

27 Similarly, plane VAE controls the rear limit of the intersection in that line AE is _____ to the base of the _____.

tangent; cylinder

28 The intersections of lines AB, AC, AD, and AE with the two circular bases define the ends of elements cut by the four planes. Each plane gives the following number of elements:

	Cylinder	Cone
Plane VAB	_____	_____
Plane VAC	_____	_____
Plane VAD	_____	_____
Plane VAE	_____	_____

	Cyl.	Cone
(VAB):	2	1
(VAC):	2	2
(VAD):	2	2
(VAE):	1	2

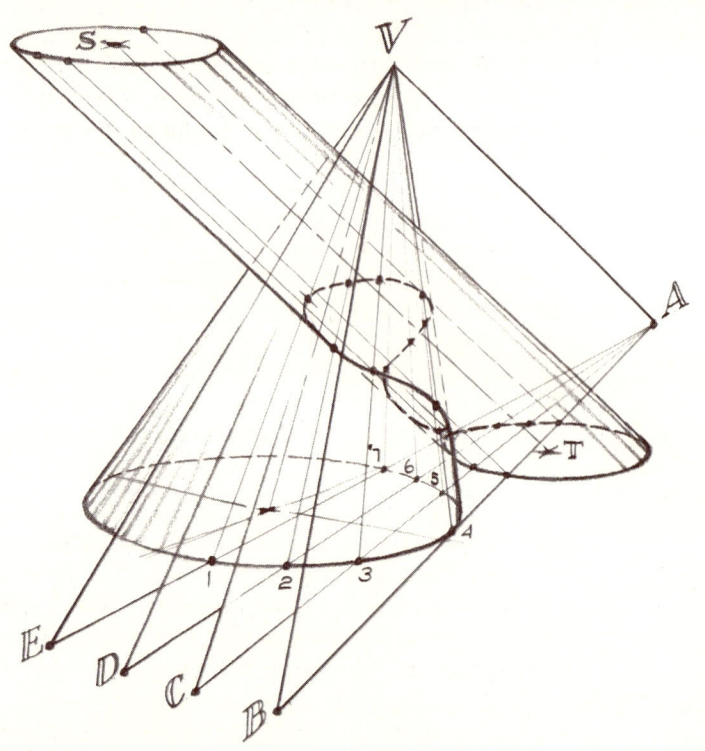

Figure 28.5 Pictorial representation of 28.4.

29 By matching sets of elements created by a single cutting plane, we see that the following number of points on the line of intersection are created by each cutting plane:

    Plane VAB _____
    Plane VAC _____
    Plane VAD _____
    Plane VAE _____

(VAB): 2
(VAC): 4
(VAD): 4
(VAE): 2

30 Joining these 12 points in correct sequence with a(n) _____ _____ gives the line of intersection between the cylinder and the cone.

smooth curve

Note that the third sides of the four cutting planes (BV, CV, DV, and EV) were never used. Only line BV was drawn as a dashed line. These lines are not needed in the solution since we know that two intersecting lines create a plane and we were only interested in line VA and lines AB, AC, AD, and AE. The general cutting planes VAC and VAD can be placed anywhere between the limiting planes VAB and VAE.

Figure 28.5 shows that the four cutting planes represent a fan-shaped array of planes that encompass the full intersection.

INTERSECTION OF CURVED SURFACES 423

CONCEPT OF THE CUTTING SPHERE

We have been concerned only with cutting planes. Figure 28.6 shows the concept of a sphere cutting a surface of revolution. Examples are shown for the most common surfaces of revolution, the cone and the cylinder.

axis	**31**	In both cases, a sphere is drawn whose center lies on the _____ of the surface of revolution.
circle	**32**	The intersection of the sphere with both the cone and the cylinder is a plane _____.
EV (edge views); TS (true-shape, true circles)	**33**	Consider the orthographic views of Figure 28.6. The plane circles appear as _____ in F and as _____ views in T.
perpendicular	**34**	This means that the plane circles cut by the spheres are _____ to the axes of the solids.
2; different	**35**	In the case of the cone, _____ plane circles are produced by the intersection of the solids. Are their diameters of equal or different size? _____.

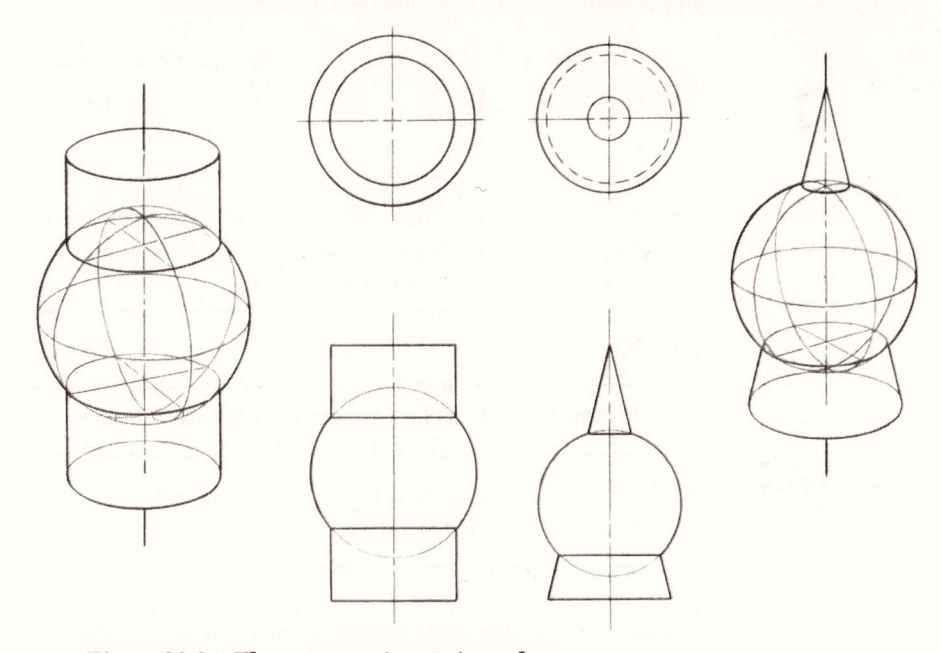

Figure 28.6   The concept of a cutting sphere.

We can use this concept of cutting spheres to find the line of intersection between two surfaces of revolution *as long as the axes of the two solids intersect.*

424 DESCRIPTIVE GEOMETRY/UNIT 5

SURFACES OF REVOLUTION (AXES INTERSECTING)

Consider Figure 28.7A and Figure 28.8. They show an ice-cream-cone-shaped object (vertical axis PQ) and a cone (inclined axis RS). Axes PQ and RS intersect at point X. In 28.7A the objects are positioned so that both axes appear as TL in the front orthographic view.

EV (edge views)

36 The two objects of Figure 28.7A are surfaces of revolution with axes intersecting. They were positioned so that both axes show as TL in F and therefore plane circles cut by cutting spheres will show as _____ in F.

---

perpendicular

37 If the centers of the cutting spheres are placed at the point of intersection, X, of the two axes, the plane circles cut on the two objects will be _____ to the axes of their respective objects.

In the orthographic view of Figure 28.7A five cutting spheres and their intersections with the two objects are shown. Three of the five spheres are numbered 1 through 3. Their intersections with the composite shape (ice cream cone) are labeled 1a through 3a. Similarly, the intersections of the spheres with the inclined cone are labeled 1b through 3b. In Figure 28.8 only sphere No. 2 and its intersections are shown.

---

circles

38 Consider cutting sphere No. 2 first. Its intersections with the two objects are two plane _____ which appear as edge views in F.

B_F

39 These two plane circles (2a and 2b) both lie on sphere 2. Therefore they must intersect each other. The point where the two edge views intersect in the front view is labeled _____.

---

2 (two)

40 How many points of intersection are represented by point B_F, the apparent intersection between the two plane circles, 2a and 2b? _____ (Consult Figure 28.8 and Figure 28.7B)

---

1 (one)

41 Consider now cutting sphere No. 1. How many points does it contribute to the intersection? _____.

Sphere No. 1 is a limiting sphere while No. 2 is a general cutting sphere. A further analysis of the figures will show that sphere No. 3 contributes three points. One point is the lower limit of the intersection of the inclined cone and the conical portion of the composite shape. The other two points are near the transition between the spherical (ice cream) part of the object and the conical part.

INTERSECTION OF CURVED SURFACES

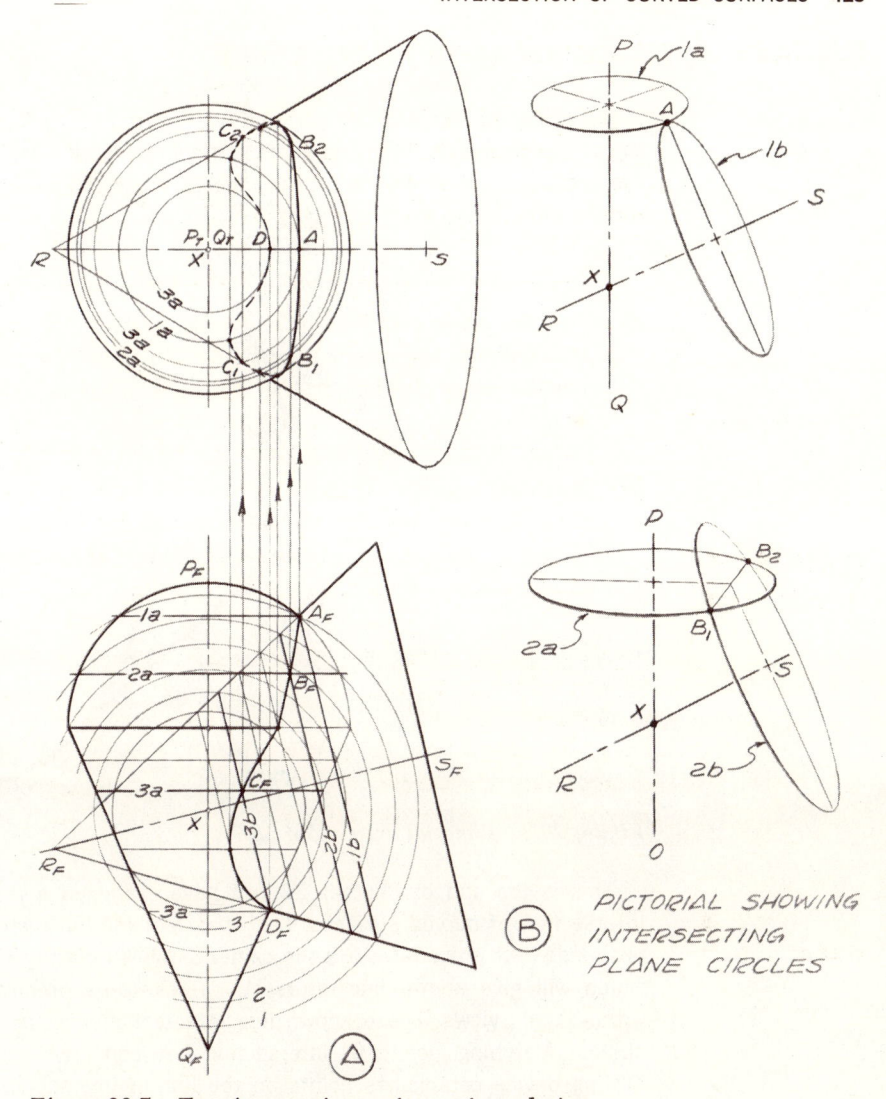

Figure 28.7 Two intersecting surfaces of revolution.

Axis PQ appears as a point in the top view. Therefore, all circles (1a, 2a, and 3a) cut on the composite shape will appear as true circles since these intersections with the four cutting spheres are perpendicular to PQ. Finding the top view of the line of intersection involves constructing circles 1a, 2a, and 3a in the top view and projecting the points A, B, C, and D from the front view.

■ **42** Sketch 71 is an exercise in finding points on the line of intersection between two solids. Using sketching techniques, find at least two points in the top and front views of the three problems given. Finding at least two points suggests the establishment of one cutting plane or cutting sphere. Check your solution.

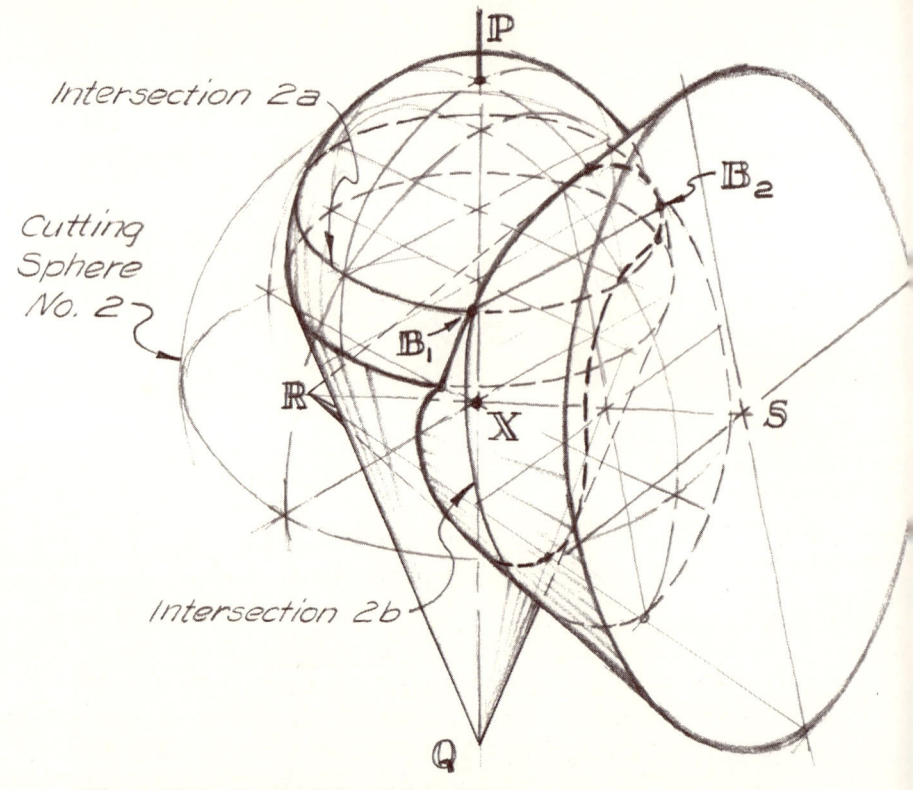

Figure 28.8 Pictorial drawing of 28.7A.

This is the end of Chapter 28. We have examined a few examples of the intersection of curved surfaces. Each problem will have variations from these examples, but procedures are the same. Establish cutting planes (or spheres) which will give known line intersections (elements or circles) on the given orthographic views. These known line intersections will be elements or plane circles. Matching points of intersection between lines created by a single cutting device establishes points on the line of intersection between the two solid figures.

SUMMARY OF DESCRIPTIVE GEOMETRY

This is also the end of our discussion of descriptive geometry—the precise solution of space problems by orthographic means. We have introduced operations on points, lines, and surfaces. Though problems may have infinite variations, the fundamental procedures of descriptive geometry are standard for any problem:

(1) *Visualize* the known parameters and *draw* them in at least two orthographic views.

(2) *Reason* out a succession of constructions or auxiliary views to give desired information.

(3) *Project and measure* according to standard orthographic procedures of projection and measurement to achieve a final solution.

UNIT 6 *graphic mathematics*

The last five chapters in this volume present the subject of graphic mathematics—the display and analysis of numerical information by graphic means.

Chapter 29 shows the accepted standards for plotting tabular numerical information into *graphs* for analysis of the relationships between two or more variable quantities. It then shows how *charts* are prepared to display numerical data as an effective means for communicating this data to a wide variety of audiences.

Chapter 30 presents a discussion of *scales* and shows how to construct a scale on any straight or curved line according to any mathematical function of a single variable quantity. The description is extended to the construction of *conversion scales*—a single line graduated to solve a mathematical equation of two variables.

Chapter 31 develops the use of scales further by showing how more than one scale can be arranged to solve equations involving three or more variables. This is the subject of *nomography*. A nomograph is a graphic computer; it is most useful when repeated calculations of multivariable equations are required.

Chapter 32 shows how to find by graphic means the equation that defines a set of observed experimental data. This is the subject of *empirical equations*.

Chapter 33 introduces *graphic calculus*—the presentation of techniques for constructing an integral curve and a differential curve of any given curve even though the equation of the given curve is unknown.

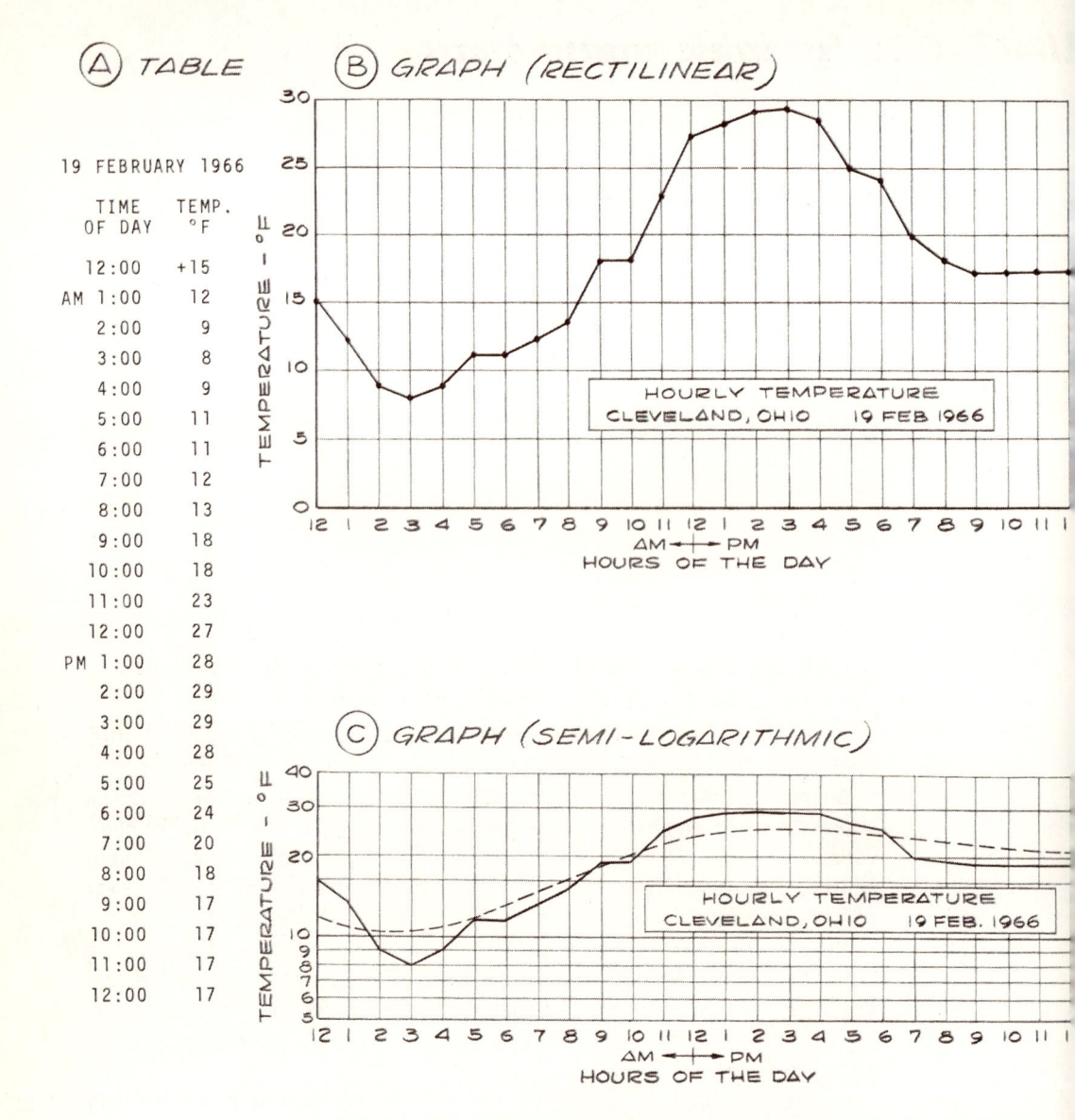

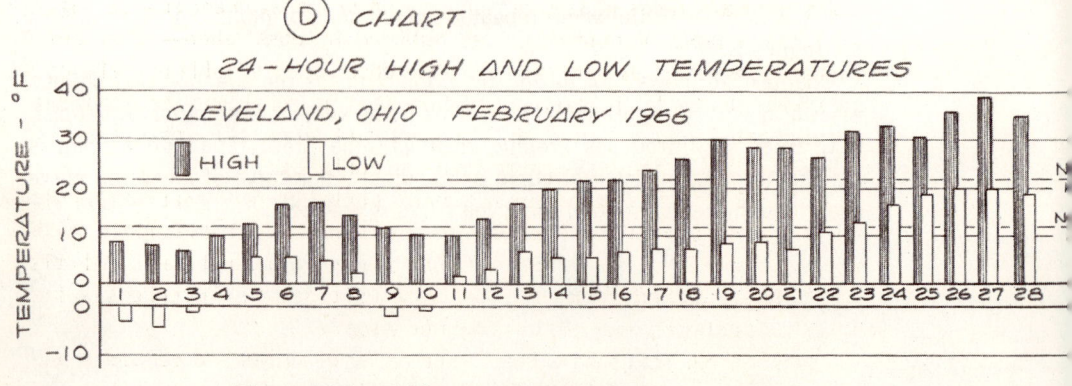

Figure 29.1 Tables, graphs, and charts.

chapter 29 graphs and charts

INTRODUCTION

A *graph* is a diagram that shows the relationship between two variable quantities. Often the term "graph" is used synonymously with the term "chart." In this text, a *chart* is defined as a visual representation of the elements of a relationship or a system. These elements may consist of numerical information, but may also take other forms.

A mathematic *graph* is usually drawn as a curve in a frame of reference formed by two of the axes of coordinate geometry—the horizontal x axis and the vertical y axis. A *chart* may be presented in this form, but can take other forms depending upon the type of information being displayed and the nature of the viewing audience.

Figure 29.1 shows the most important devices used by engineers and scientists to analyze and present numerical information. These are:

(A) Tabular data
(B) A rectilinear graph
(C) A semi-logarithmic graph
(D) A chart

The two variables are time and temperature. They could be any two variables from any fields of science, engineering or management science. Figure 29.1A is a *table* of numerical data obtained by observation—either direct human observation or automatic machine observation. A table is useful as a reference source for specific information such as, what was the exact temperature at 2:00 A.M.? However, the *pattern* of temperature change for the 24-hour period is hard to visualize from the tabular data.

At 29.1B, the data have been accurately plotted on rectilinear coordinates. A meteorologist would be interested in this *graph* since it gives him a visual picture showing *how* the temperature has changed in magnitude in 24 hours. The data are plotted from a standard base (zero degrees Fahrenheit) so that magnitudes at each hour can be compared visually.

The observed points have been noted as small circles and connected with straight lines. A curved line is not drawn through the points because no

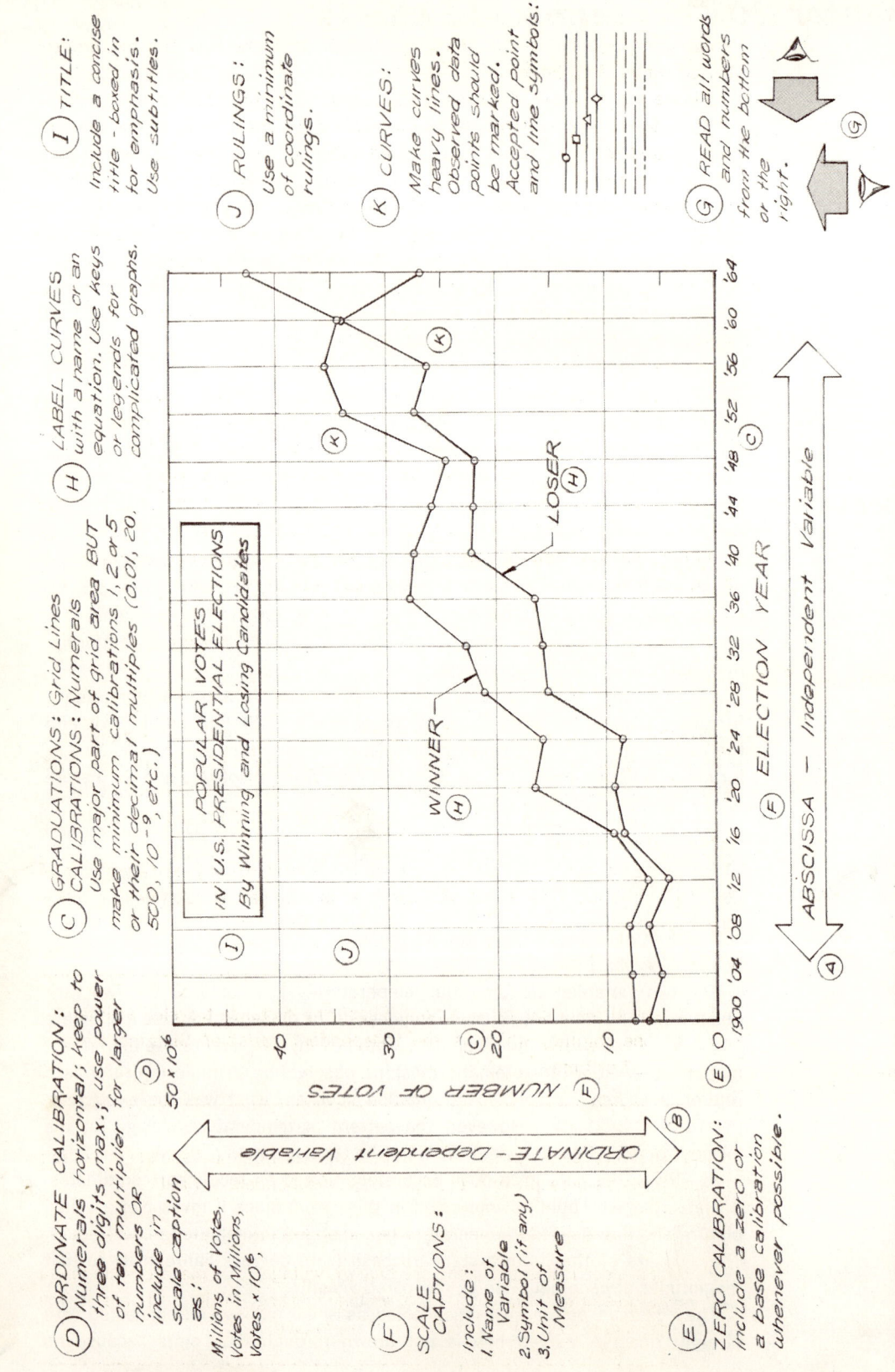

Figure 29.2 Accepted standards of graph construction.

GRAPHS AND CHARTS 431

information is available in the original data to show how the temperature varied within each hour.

Figure 29.1C shows the same data plotted as a graph having a logarithmic vertical scale. The same curve would result if the logarithms of the temperature data had been taken and plotted against hours. Because of this log scale, it is called a *trend curve* or *rate curve*. It shows the rate at which the temperature changes rather than the magnitudes. The smooth curve (dashed line) indicates the trend of the temperature change and might be used, with trends from previous days, to predict future temperature-change patterns.

The two graphs of 29.1B and C are drawn according to standard form as derived and accepted by the technical profession. The presentation of these standards is the main topic of this chapter.

Figure 29.1D is a *chart* of the high and low temperatures for a given station during the month of February. The table of 29.1A provided only two pieces of information for this chart, the high and low for 19 February 1966.

Temperatures have been drawn as vertical bars to assist the nontechnical person in visualizing temperature as a magnitude above or below a standard and familiar base (zero degrees). This is the type of chart that might appear in a daily newspaper to indicate temperature ranges day by day. Average monthly high and low temperatures for the recording station are shown to point out how each day's temperature agrees with the average.

GRAPHING STANDARDS

Figure 29.2 is a graphic summary of the accepted standards for graph plotting. A careful examination of this figure will acquaint you with the most important standards. You will see a number of circled letters pointing out important features of a graph. Some elaborations on these eleven points are given in the following text:

(A) The horizontal (x) axis is the *abscissa* and usually carries the *independent variable*. This is the variable that is changed or recorded at will in an experiment or observation.

---

1   If you drive a car for one hour measuring distance traveled at intervals of one minute, which is the independent variable? Distance traveled ____? or time ____?

time

(B) The vertical (y) axis is the *ordinate* and carries the *dependent variable* which varies as a result of a change in the independent variable.

---

2   In an electrical experiment, we change voltage on a piece of equipment and measure current at each voltage setting. Which variable is the dependent variable? Current ____? or voltage ____?

current

---

432 GRAPHIC MATHEMATICS/UNIT 6

3 Heat is applied to a cylindrical steel bar. We record a table of data showing expansion of the bar in inches and temperature in degrees Fahrenheit. In one hour, 20 readings of each variable are recorded. We wish to plot a graph showing expansion vs. temperature.
   (1) Expansion is the _____ variable.
   (2) Temperature is the _____ variable.

(1) dependent;
(2) independent

(C) Before plotting a curve, the range of the two variables should be noted to take advantage of the maximum portion of the grid area. However, care must be taken to calibrate the main grid lines with easily decipherable numbers. The numbers 1, 2, and 5 plus their multiples and decimal multiples (0.01, 0.2, 10, 100, 200, 50, 5000, etc.) are accepted standards. Most other numbers (7, 11, 23, etc.) make accurate reading between grid lines virtually impossible.

Figure 29.3 is a graduated scale calibrated in four different ways. Make a mark on each scale for the value 15.20. You should find that locating 15.20 is easiest on the 2 and 5 scales, but fairly difficult on the 7 and 11 scales.

(D) Use a maximum of three digits on scale calibrations. If the numbers are larger than three digits, use a power-of-ten multiplier, preferably with the maximum calibration. It is acceptable to include the multiplier with the scale caption.

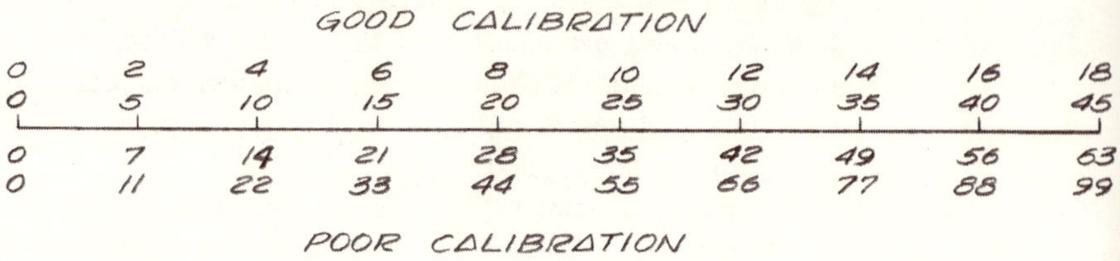

```
                    GOOD  CALIBRATION
  0    2    4    6    8    10   12   14   16   18
  0    5    10   15   20   25   30   35   40   45

  0    7    14   21   28   35   42   49   56   63
  0    11   22   33   44   55   66   77   88   99
                    POOR  CALIBRATION
```

Figure 29.3 *Calibrate the scales of a graph with easily decipherable numbers.*

4 If five variables range as shown below, how would you label the highest point of each according to standard D above?
   Variable A: 0 to 100   _____
   Variable B: 0 to 10,000   _____
   Variable C: 0 to 40,000,000   _____
   Variable D: 0 to 0.00025   _____
   Variable E: 0 to 1/10   _____

(A): 100
(B): 1×10^4
    (100×10^2)
(C): 40×10^6
(D): 2.5×10^{-4}
    (25×10^{-5})
(E): 0.10

(E) The zero calibration should be included on both the ordinate and the abscissa wherever possible. If visual comparisons of the ordinate data are to be made, the zero calibration must *always* be included.

---

5   Figure 29.4 shows two graphs that compare values of X and Y. We are interested in the magnitude of Y for values of X such as X_1 and X_2 above a zero base line. Which graph shows this best, (A) _____ or (B) _____ ?

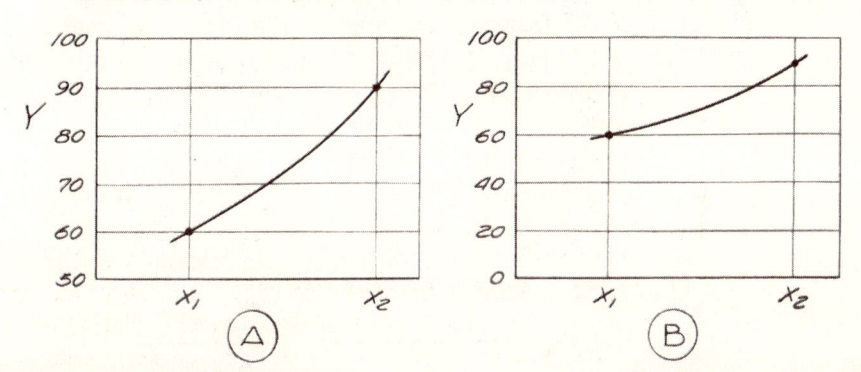

Figure 29.4   Inclusion of the zero (or base) calibration ensures correct interpretation of ordinate magnitudes.

---

6   The value of Y at X_1 is $Y_1 =$ _____. The value of Y at X_2 is $Y_2 =$ _____.

---

7   Thus the ratio of $Y_1/Y_2 = {}^{60}/_{90}$ or $2/3$. Comparing Figure 29.4A and B, what are the *apparent visual ratios* of Y_1/Y_2:
   (A) $Y_1/Y_2 =$ _____.
   (B) $Y_1/Y_2 =$ _____.

(F) Scale captions should include:
   (1) Name of the variable
   (2) Mathematic symbol (if any) in parentheses
   (3) Unit of measure

   EXAMPLES: POWER OUTPUT (E), MICROWATTS
   VELOCITY (V) IN MPH

Use only accepted standard abbreviations.

(G) All numerals, titles, scale captions, and notes must be read from either the bottom or the right-hand side of the graph. In most instances, the ordinate caption is the only item read from the right. Ordinate calibrations (numbers) should read from the bottom.

(H) Label all curves with a name or an equation. Avoid keys and legends except on the most complicated graphs.

(I) Include a *brief* title on each graph. Use subtitles to clarify or modify the title. It is acceptable to put titles on the grid area (enclosed in a rectangle).

(J) Vary the line weight for the various elements of the graph as follows:
- Heavy—the curve or curves
- Medium—the ordinate and abscissa
- Light—grid lines

Keep grid lines to a minimum. If grid lines are not absolutely essential to the reading of the graph, short ticks (1/8-in.) on the ordinate and abscissa are acceptable.

(K) Observed points must be marked. (See inset on Figure 29.2.) A small circle (1/16-in. diameter or less) is standard. If more than one curve is being plotted on a single graph, small triangles, squares, and diamonds can be used. Different curves can be drawn with different lines (solid, dashed, dot-dash, etc.). Calculated points should not be shown.

A graph is a highly efficient device for communicating complex numerical and statistical information. Nothing should be done in the drawing of a graph to hinder easy reading. All words and numerals should be neatly lettered in guide lines using standard engineering block capital letters.

---

abscissa;
independent

8 The x or horizontal coordinate on a graph is called the _____ and usually carries the _____ variable.

---

ordinate;
dependent

9 The y or vertical coordinate is called the _____ and usually carries the _____ variable.

GRAPH PAPER

Figure 29.5A shows the most common grid patterns that are available on commercial graph paper. They are 4, 5, 8, and 10 lines per inch and 5 and 10 lines per centimeter. The selection of a grid depends upon the degree of precision required for plotting and reading and the best scales to cover the ranges of the two variables. If graph paper is not available, a grid can be easily laid out and drawn using a T square and triangle. Preparing your own grid gives you the advantage of being able to adjust the coordinates to the data and to construct a graph of any desired size. Often the grid lines are not needed on the finished graph. Drawing a graph on tracing paper laid over a prepared grid permits you to use the grid underlay in plotting the points but the grid lines need not be traced. The graph of Figure 29.2 was drawn this way. On the finished graph, only the important grid lines were retained.

Other common grids are shown in 29.5B. One is the polar grid for plotting equations in polar coordinates. These are equations of the form $f_1(\theta) = f_2(r)$ where θ is the polar angle about a point (the pole), and r is the radius vector emanating from the pole. The other is the logarithmic grid (log-log and semi-log). (More about logarithmic scales in Chapters 30 and 32.)

Printed grid paper is available in three colors, blue, green, and red. Blue grid lines will not reproduce by photographic means. Green lines will and red lines have the best reproducibility but are tiring to the eyes when viewed for a prolonged period.

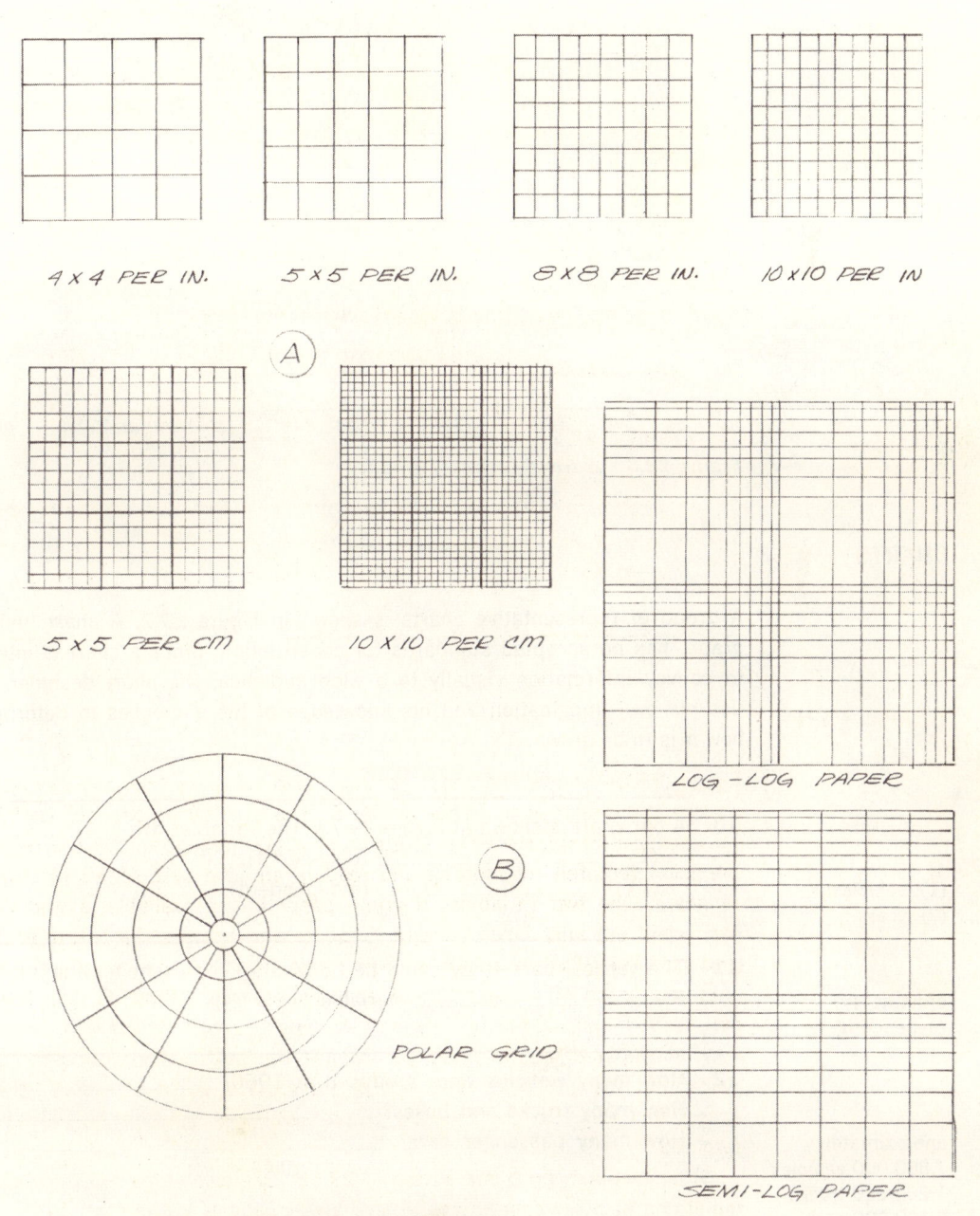

Figure 29.5   Types of grid papers.

TREND GRAPHS

In Figure 29.6 the popular vote data of Figure 29.2 is plotted on semi-log paper in two curves, the Republican candidate and the Democratic candidate. This is in contrast to 29.2 which showed the magnitude of popular votes for the winner and the loser regardless of political party. A smooth curve has been drawn to represent the trend toward increasing voter activity in both parties. This trend has been projected into the 1968 and 1972 elections.

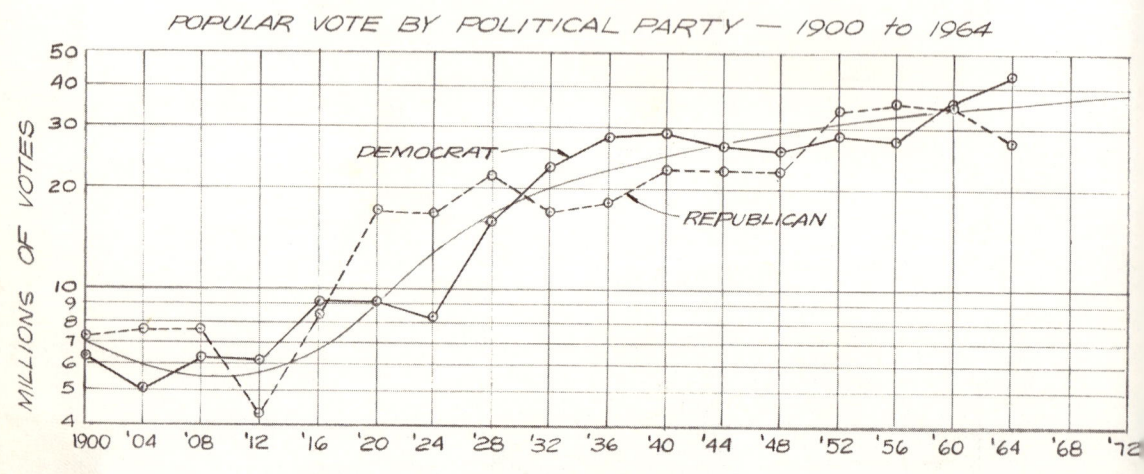

Figure 29.6   A trend (or rate) graph.

CHARTS

A group of representative charts is shown in Figure 29.7. A chart, unlike a graph, has no accepted standards for construction. Since a chart is intended to convey information visually to a wide audience, the chart designer must use his own imagination and his knowledge of his audiences in determining how it is to be drawn.

10   A *bar chart* is shown at Figure 29.7A. The variables are:
(1) _____ (dependent).
(2) _____ (independent).

(1) vehicles;
(2) years

11   The vertical bars (they could be horizontal) represent the magnitude of the _____ variable.

ordinate (y)
(dependent)

12   How many vehicles were produced in 1960? _____
How many trucks and buses? _____
How many passenger cars? _____

approximately:
7,800,000 vehicles
1,200,000 trucks
6,600,000 cars

GRAPHS AND CHARTS   437

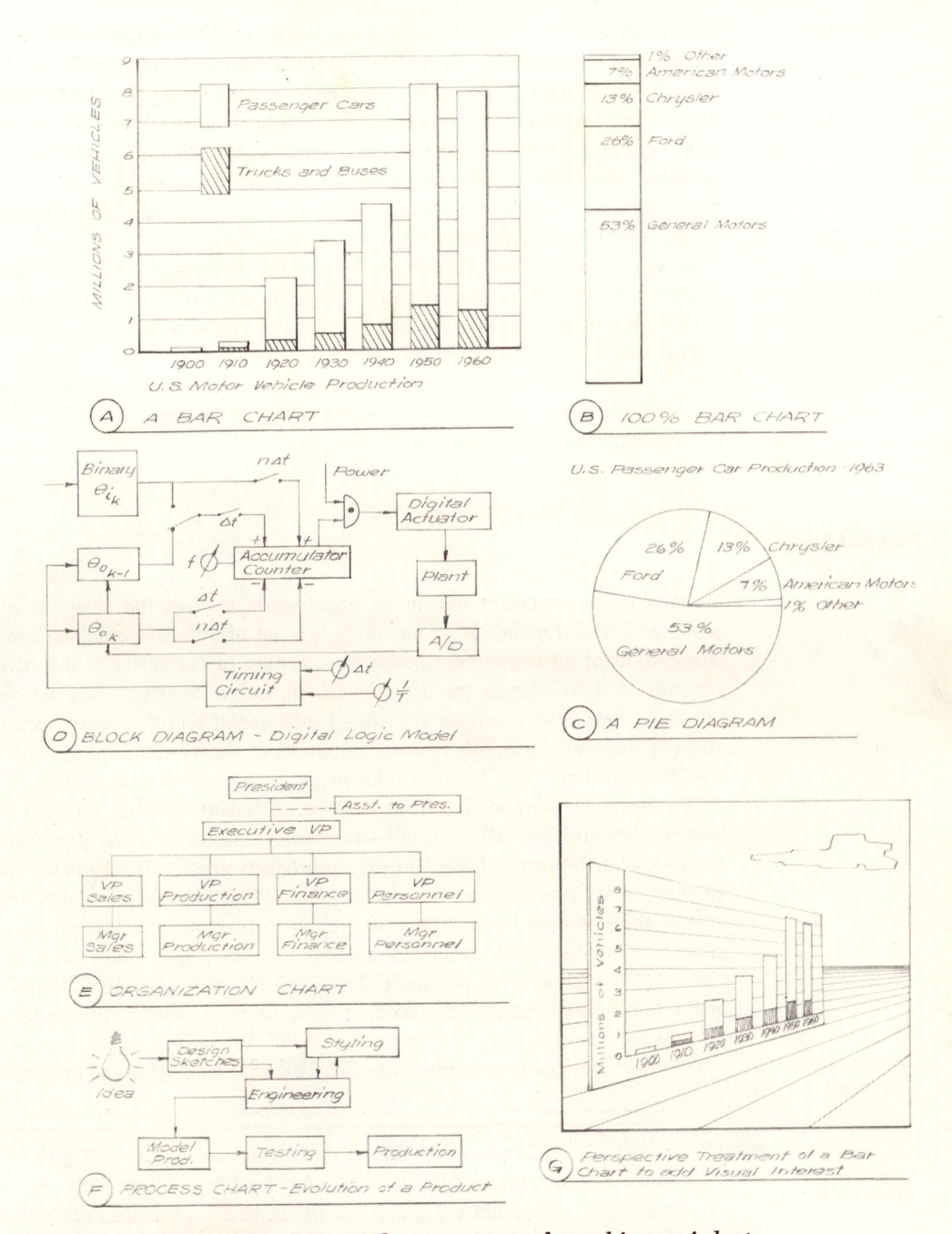

Figure 29.7 Some of the more commonly used forms of charts.

A *100% bar chart* is at 29.7B. This type of chart compares the magnitudes of the parts to the total (100%).

Another type of 100% chart, the *pie diagram*, is shown at 29.7C.

The *block diagram* is one of the most useful charts in science and engineering. Whether used to show the logic model of a digital controller as at 29.7D, the organization of a company (29.7E), or the steps in a process (29.7F), these easy-to-draw charts prove useful for organizing and displaying complex systems or ideas. *Flow charts* are a form of block diagram and derive their name from the fact that they are used to depict the flow of a real process —usually a chemical process.

It was stated earlier that the designer's imagination was the only serious limitation to the drawing of a chart. Any type of chart can be treated in an artistic or glamorous way depending upon the audience for whom the information is intended. At Figure 29.7G, for example, the bar chart of 29.7A has been put into perspective form to increase its visual interest—to cause more people to look at it. Most charts can be improved by the use of color—color for coding or just for decoration.

SUMMARY

A graph is an important tool in technical work. It gives the scientist or the engineer a visual model of an equation or a set of numbers obtained through observation of an experiment. A graph can help to (1) verify or disprove an experiment, (2) indicate trends upon which future decisions may be made, (3) show how theory can be translated into useful and achievable practice, and (4) uncover discontinuities not apparent in theory, plus many other details that are important to successful research and development work.

The technical worker is concerned principally with graphs. He uses them first to communicate with himself—to display results in a manner that will assist him in his work. Later, he may reproduce graphs in technical reports so that others may obtain a view of the project. If his work has significance on the social scene, he may be called upon to prepare charts describing his project to nontechnical audiences so that they may gain some idea of the technical implications of the work. These charts can take many forms and might appear as displays, projection slides, parts of motion picture film, or video tape.

Whether chart or graph, the purpose is the *communication of information*.

■ 13   Sketch 72 shows a grid. Taking data from the graph of Figure 29.2, make a graph of the *total popular vote* for the elections between 1944 and 1964. Complete the graph according to the standards presented by Figure 29.2, including scale labels, title, etc. Check your solution.

chapter 30 scales

We do not proceed through a day without reading a *scale*. Everytime we read the time, check temperature, adjust a radio, or control the speed of an auto we make reference to a graphic scale that converts a phenomenon into useful numbers. It is important to know how to create a scale that displays a variable quantity according to a desired mathematical function of that variable and over a given range of the variable.

SCALE NOMENCLATURE

Figure 30.1 shows a simple scale and the nomenclature used in designing scales. The line is the *stem*; it can be straight or curved. The marks (ticks) on the stem are the *graduations*. The fineness of the graduation depends upon the precision with which we wish to read the scale.

Each graduation represents a number called the *calibration*. The calibration relates the graduation to a standard or familiar base. When the mercury column stands at 72 on a thermometer, experience relates for us the physical situation to a comfort situation through the calibration, 72.

The left end of the stem is the *origin*. The calibration at the origin may or may not be zero.

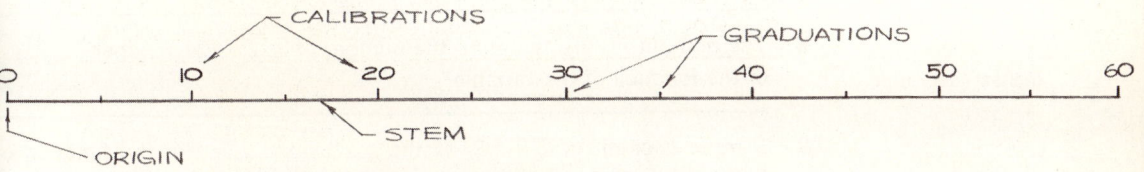

Figure 30.1 Scale nomenclature.

440 GRAPHIC MATHEMATICS/UNIT 6

graduations

1 The marks on the stem of a scale are called _____.

2 The number we assign to each graduation on a scale is called the _____.

calibration

½

3 The scale of Figure 30.1 is graduated every _____ inch.

calibrated

4 Every second graduation is _____.

5 The calibrations are 10, 20, ..., 60. If we wished to have unit calibrations (1, 2, 3, ..., 60) we would have to _____ the scale every _____ inch.

graduate; ¹⁄₁₀

THE SCALE EQUATION

Figure 30.2 gives the basis for drawing any scale. The procedure is to determine a distance, d, for every desired graduation of the scale. This distance is the measure, in inches, from the graduation, $x = 0$, to the desired graduation

NOTE: The graduation $x = 0$ may or may not be the origin (left end of the stem).

d is found by calculation, using the scale equation, $d = mf(x)$, where m = the scale modulus and $f(x)$ is the function of the variable, x, and represents the mathematical condition of the variable (x^2, $\sqrt{x}$, $x/2$, $\log x$, etc.).

Scale Modulus

The concept of scale modulus is important in this and the succeeding chapters. Note that, by definition, it is the *increment of scale length* for a unit change of the *function of the variable*. Its units are *inches per unit*. The student often gets these units reversed and considers modulus to be the number of units per inch.

6 The scale modulus (m_x) can be found (see Figure 30.2) by solving for m_x in the equation $R_f m_x = L$. The modulus is the increment of scale _____ for a unit change in the _____ of the variable, x.

length; function

7 If $m_x = \frac{1}{4}$, then we would measure _____ inch when $f(x)$, the function of x, changes from 1 to 2, 10 to 11, 100 to 101, etc.

¼ (inch)

8 The units of m_x are therefore the number of _____ per _____ of the function of the variable.

inches (per) unit

9 $R_f m_x = L$ or $m_x = L/R_f$. R_f is the _____ of the function of the variable, $f(x)$.

range

SCALES 441

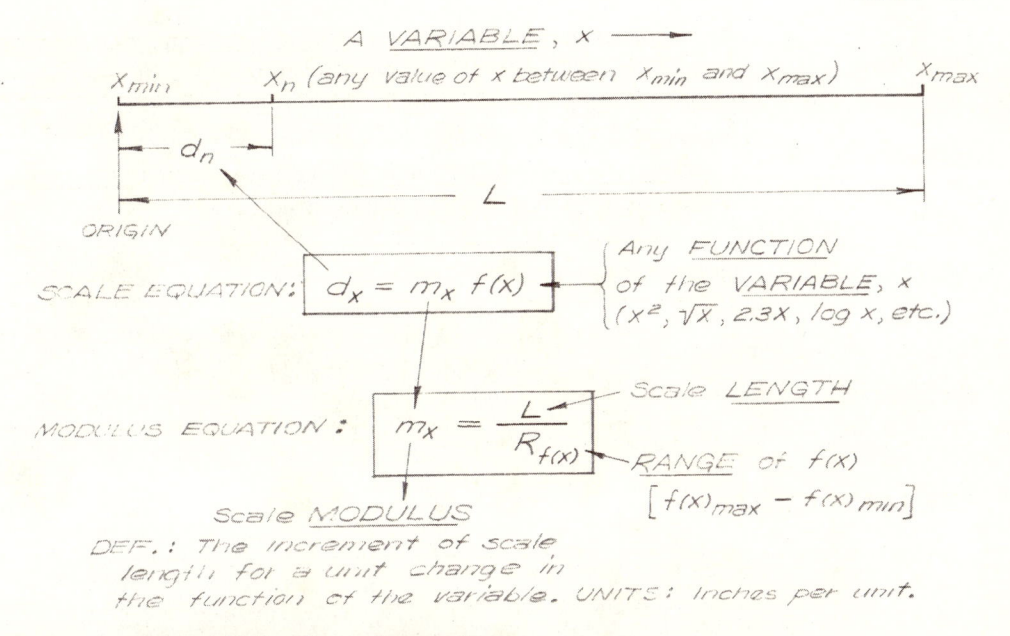

Figure 30.2 The scale equation.

$R_x = 10-0 = 10$

10 The range of a variable or a function of a variable is the maximum value minus the minimum value. If x ranges from 0 to 10, then $R_x = x_{max} - x_{min} = $ _____.

$R_f = 100-0 = 100$

11 If f(x) ranges from 0 to 100, then $R_f = $ _____.

$R_f = 10^2-0^2 = 100$

12 Let x range from 0 to 10 and let $f(x) = x^2$. Then $R_f = f(x)_{max} - f(x)_{min}$ = _____ − _____ = _____.

$m_x = 5/(100-0)$
$\quad = 5/100 = 1/20$
$\quad = 0.05$

13 Thus if x ranges from 0 to 10 and $f(x) = x^2$ and we let the scale length $L = 5$ in., then $m_x = L/R_f = $ _____ = _____.

$d_x = 0.05x^2$

14 If $m_x = 0.05$ in./unit and $f(x) = x^2$, then the scale equation is $d_x = $ _____.

---

origin; variable

15 We are interested in graduating the scale for values of x, which ranges from 0 to 10. By substituting any value of x into $d_x = 0.05x^2$ we find d_x, the distance measured from the _____ to any value of the _____, x.

■ **16** Sketch 73A gives a 5-in. stem and a table of values of x from 0 to 10. Calculate the corresponding values of d_x according to $d_x = 0.05x^2$, then graduate and calibrate the scale. Check your solution.

442 GRAPHIC MATHEMATICS/UNIT 6

The scale of Sketch 73A is called a nonlinear scale. The graduations are not evenly spaced. It is an important scale in technical work in that many variables change according to their square, $f(x) = x^2$.

Figure 30.3 is a simple example of a linear scale, $f(x) = x$. The scale and the calculations are shown. Intuition would have enabled us to graduate and calibrate this scale as well as the equations did.

LET $f(x) = x$ AND LET x RANGE FROM 0 TO 6.
SCALE LENGTH, $L = 6$ IN.

$$m_x = \frac{L}{R_f} = \frac{L}{f_{max}(x) - f_{min}(x)} = \frac{6}{6-0} = 1 \text{ IN./UNIT}$$

SCALE EQUATION: $d_x = m_x f(x) = 1 \cdot x = x = d_x$

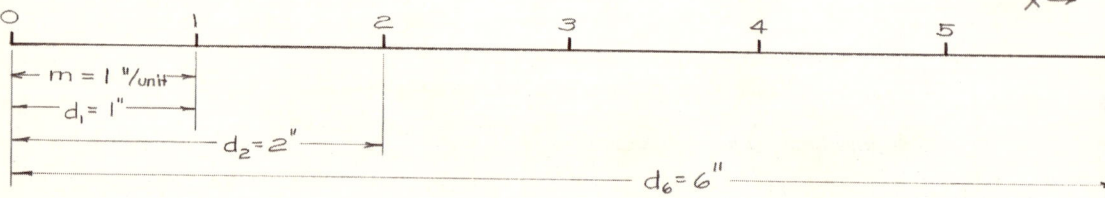

Figure 30.3 A linear scale.

■ 17 Another example of a linear scale is given in Sketch 73B. The calculations are complete. A table is shown giving unit values of x over the given range of the variable, x.

The table has two blank rows, $d_x = x/2$ and $d_x = x/2 + 2$. Proceed as follows:

(1) Complete the table for $d_x = x/2$ and graduate and calibrate the 6-in. scale.

(2) If you cannot see how to graduate the scale from this data, then complete the table for $d_x = x/2 + 2$ and proceed with the graduation. No solution is given. If your computations are correct, the correctness of your solution should be evident.

The example of Sketch 73B illustrates the fact that the origin (left end of the stem) of a scale need not have a calibration of zero. The origin is actually −4. Using the first equation, $d_x = x/2$, we graduated the scale from a point 2 in. to the left of the zero calibration, $d_x = 0$ when $x = 0$.

Using the second equation, $d_x = x/2 + 2$, we found that $d_x = 0$ at the origin or, at the minimum value of x, $x = -4$. This latter used the *special* form of the scale equation which states:

$$d_x = m_x [f(x) - f(x)_{min}]$$

Using the scale equation in this form assures us that every value of d_x is measured from the origin, $d_x = 0$.

Also note that in this example two moduli are shown, the *actual modulus* and the *effective modulus*. m_x was found to equal ¼ in. per unit and by definition, this is the scale length for a unit change in $f(x)$. However, we find that by substituting m_x into the scale equation, we get $d_x = ¼(2x) = ½x$. The value ½ is called the *effective modulus* and is actually used to graduate the scale. An examination of the finished scale will show that ½ in. is the scale length for a unit change in the *variable*, x.

The *effective modulus* is used to graduate the scale. The *actual modulus* will be used in the next chapter to position the scales of a nomograph.

Logarithmic Scales

■ 18   Sketch 74 presents two examples of logarithmic scales. The log scale is the most common nonlinear scale encountered in technical work.

In 74A, x varies from 1 to 10. This represents one logarithmic cycle. In 74B, x varies from 10 to 10,000 or three log cycles.

A stem ($L = 6$ in.) is provided for both 74A and B. Complete the calculations, fill in the tables, graduate and calibrate the two scales. Check your solutions.

19   An examination of the solutions of Sketch 74A and B shows that the modulus of a log scale is equal to the length of one logarithmic _____.

cycle

20   This is always true. Scale modulus is defined as the increment of scale length for a unit change in the _____ of a variable.

function

21   The function of a log scale is always the log of the variable, $\log x$. The function shows a unit change for every multiple-of-10 change in the variable x.

Thus, if x changes from 10 to 100, 0.01 to 0.1, or 2.75 to 27.5, then the change in $\log x$ is _____.

1 (one)

The scale equation, $d = L \log x$ (Sketch 74A) is useful whenever a log scale is needed and not available. Knowing the value of the logarithms of the numbers from 1 through 10, you can draw a scale of any length, L. The values

      $\log 1 = 0$ $\log 6 = 0.778$
      $\log 2 = 0.301$ $\log 7 = 0.845$
      $\log 3 = 0.477$ $\log 8 = 0.903$
      $\log 4 = 0.602$ $\log 9 = 0.954$
      $\log 5 = 0.699$ $\log 10 = 1.00$

should be memorized much as one memorizes the decimal equivalents of fractions of an inch, the value of π, or other useful constants.

GRAPHIC GRADUATION

Certain scales can be graduated much more easily by graphic means than through the use of the scale equation. All linear scales fall in this category. Figure 30.4A shows the construction necessary to graduate a power scale. Any convenient measuring scale can be used. An accurate log scale can be found on a slide rule, in textbooks, and many other places. A logarithmic modulus chart is given in 30.4B. Transferred to the edge of a strip of paper it can be used to graduate a scale as shown in the figure. The graduations within each of the three cycles of Sketch 74B are best plotted graphically.

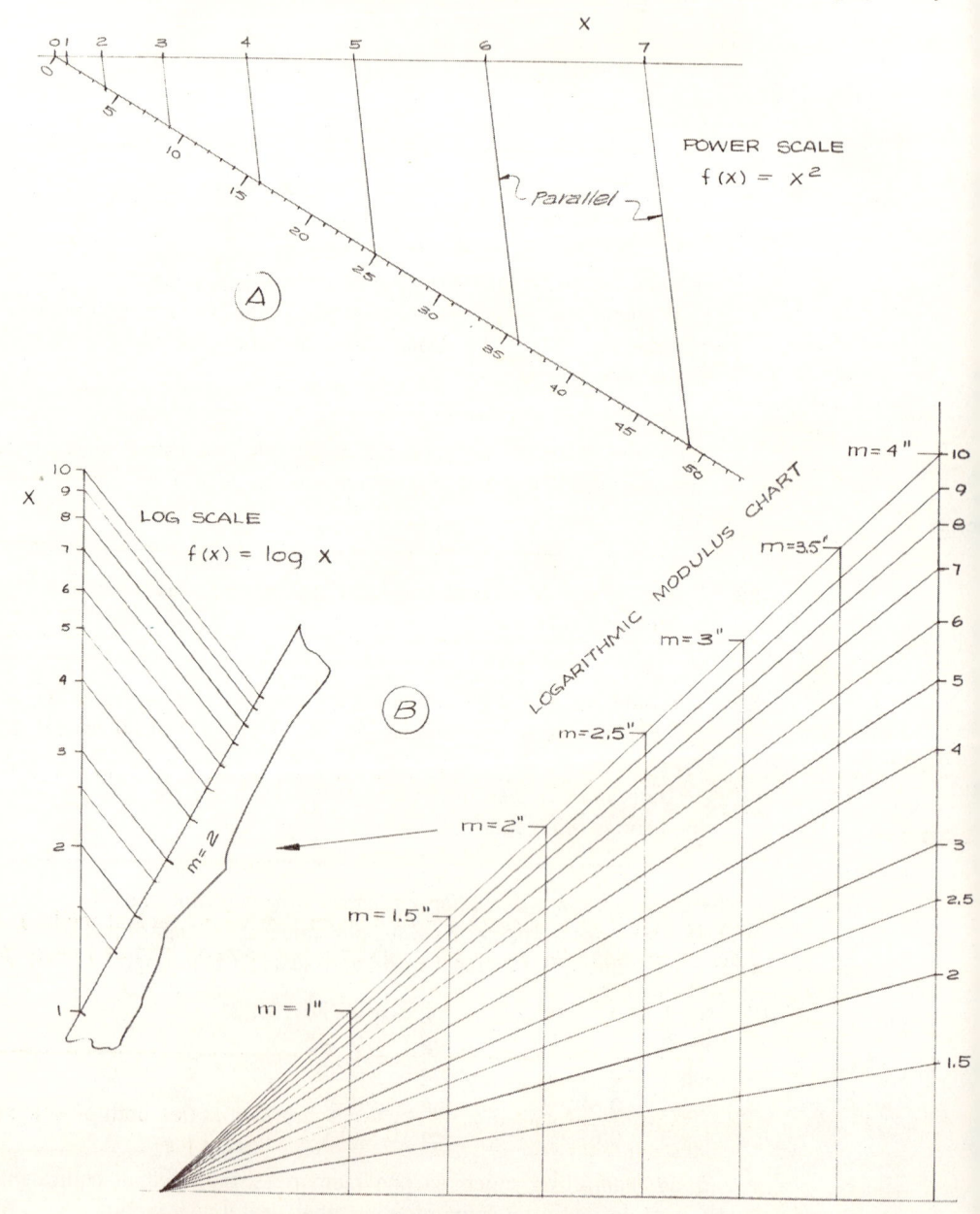

Figure 30.4 Graphic graduation of scales.

CONVERSION SCALES

If we introduce a second variable, we can graduate a single stem with calibration of the two variables and obtain a *conversion scale*. Figure 30.5 shows such a scale which converts inches to centimeters or vice versa. The equation, $2.54I = C$, is of the general form, $f_1(x) = f_2(y)$ where f_1 and f_2 are particular functions of the variables x and y respectively. The range of I (0 to 10 in.) was given along with the scale length, L. Letting $f(I) = I$ and $f(C) = C/2.54$, we may proceed to graduate and calibrate both sides of the scale so that the result is a graphic means to convert from inches to centimeters.

EQUATIONS OF THE FORM, $f_1(x) = f_2(y)$

PROBLEM: Design a double scale to convert inches to centimeters over a range of inches from 0 to 10 on a stem 6 in. long.

EQUATION: $2.54 I = C$ or $I = \dfrac{C}{2.54}$ $C; 0 \rightarrow 25.4$ (calc.)

$f(I) = I$ $\boxed{m_I = \dfrac{6}{10} = 0.6 = m_C}$ $\therefore d_I = 0.6 I$

NOTE: $m_I = m_C$. A necessary condition for Conversion Scales since $f(I) = f(C)$ and d_I must equal d_C.

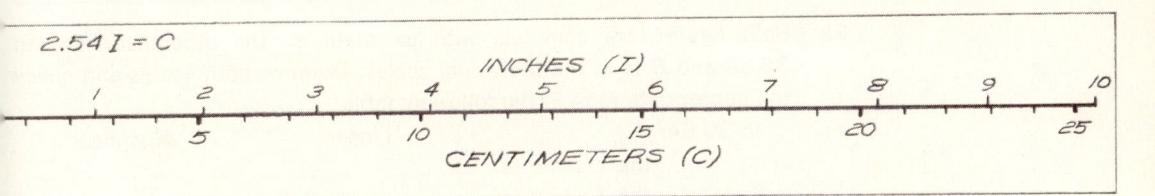

$f(C) = \dfrac{C}{2.54}$ $\therefore d_C = 0.6 \dfrac{C}{2.54} = 0.24 C$ [$0.24 = m_C$ (effective)]

Figure 30.5 A conversion scale which changes inches to centimeters and vice versa.

Note that it is a necessary condition that $m_I = m_C$ (actual moduli). Complete calculations for m_C would prove this, but knowing it to be true simplifies the calculations.

---

22   Figure 30.6A and B presents two conversion scales both of which solve the equation $A = \pi r^2$, the expression for finding the _____ of a circle of a given _____.

area; radius

446 GRAPHIC MATHEMATICS/UNIT 6

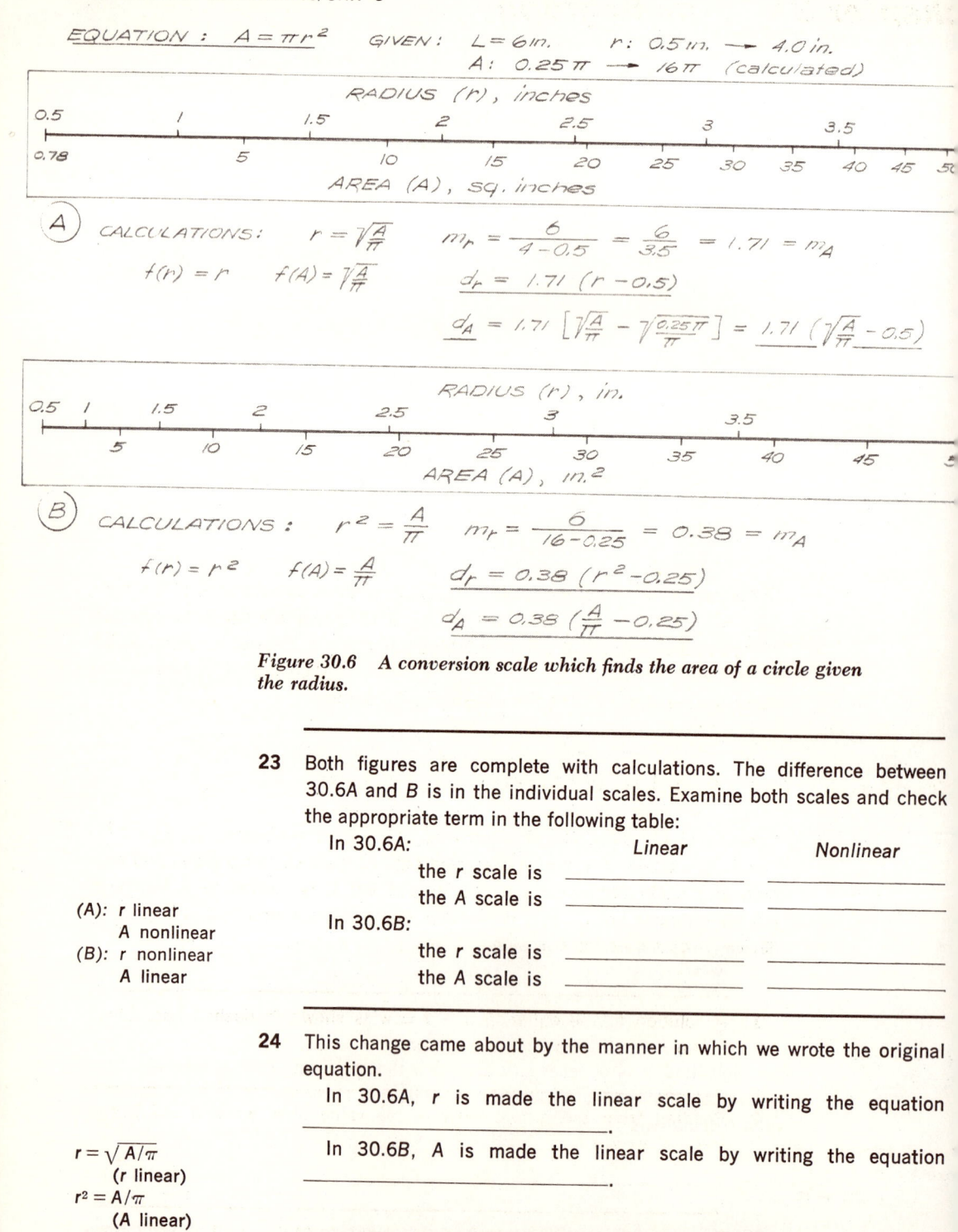

EQUATION: $A = \pi r^2$ GIVEN: $L = 6$ in. $r: 0.5$ in. → 4.0 in.
         $A: 0.25\pi$ → 16π (calculated)

RADIUS (r), inches
0.5   1   1.5   2   2.5   3   3.5
0.78   5   10   15   20   25   30   35   40   45   50
AREA (A), sq. inches

(A) CALCULATIONS: $r = \sqrt{\frac{A}{\pi}}$ $m_r = \frac{6}{4 - 0.5} = \frac{6}{3.5} = 1.71 = m_A$

$f(r) = r$ $f(A) = \sqrt{\frac{A}{\pi}}$ $d_r = 1.71 \, (r - 0.5)$

$d_A = 1.71 \left[\sqrt{\frac{A}{\pi}} - \sqrt{\frac{0.25\pi}{\pi}} \right] = 1.71 \left(\sqrt{\frac{A}{\pi}} - 0.5 \right)$

RADIUS (r), in.
0.5   1   1.5   2   2.5   3   3.5
5   10   15   20   25   30   35   40   45
AREA (A), in.²

(B) CALCULATIONS: $r^2 = \frac{A}{\pi}$ $m_r = \frac{6}{16 - 0.25} = 0.38 = m_A$

$f(r) = r^2$ $f(A) = \frac{A}{\pi}$ $d_r = 0.38 \, (r^2 - 0.25)$

$d_A = 0.38 \left(\frac{A}{\pi} - 0.25 \right)$

Figure 30.6 A conversion scale which finds the area of a circle given the radius.

---

23 Both figures are complete with calculations. The difference between 30.6A and B is in the individual scales. Examine both scales and check the appropriate term in the following table:

	Linear	Nonlinear
In 30.6A:		
the r scale is	___	___
the A scale is	___	___
In 30.6B:		
the r scale is	___	___
the A scale is	___	___

(A): r linear
    A nonlinear
(B): r nonlinear
    A linear

---

24 This change came about by the manner in which we wrote the original equation.

In 30.6A, r is made the linear scale by writing the equation _____.

In 30.6B, A is made the linear scale by writing the equation _____.

$r = \sqrt{A/\pi}$
(r linear)
$r^2 = A/\pi$
(A linear)

This is the end of Chapter 30. We will be concerned with these fundamental considerations about scales in the next three chapters.

chapter 31   *nomography*

In the last chapter we ended with a graphic solution to equations of the form $f_1(x) = f_2(y)$, that is, a particular function of x is equal to a particular function of y. Suppose we add a third variable and obtain a general equation which we will write in the form $f_1(u) + f_2(v) = f_3(w)$. This chapter will show how to construct *nomographs* to solve equations of this form.

CONCURRENCY GRAPHS AND NOMOGRAPHS

Consider Figure 31.1. The simplest form of the general equation $[f_1(u) + f_2(v) = f_3(w)]$, namely, $u + v = w$ is plotted both as a concurrency graph and as a nomograph. On rectilinear coordinates (31.1A) it is plotted as a family of curves (straight lines). This is called a concurrency graph and is a graphic device which solves the equation.

1. A solution to the equation $u + v = w$ is shown in dashed line. Given that $u = $ _____ and $v = $ _____, then $w = $ _____.

$u = 2; v = 3;$
$w = 5$

2. Reading from the graph, what is the value of w when $u = 2.7$ and $v = -1.4$? $w = $ _____.

I don't know exactly, or
$w = 1.3$ (calculated)

3. To get an accurate answer to the problem $(2.7 + (-1.4) = ?)$ from the graph, we would first have to _____
_____ (your words).

draw a curve for
$v = -1.4$

447

448 GRAPHIC MATHEMATICS/UNIT 6

A *nomograph* of the equation $u + v = w$ is shown at 31.1B. It consists of three parallel scales arranged so that any straight line drawn across all three will be a solution of $u + v = w$. A nomograph is more accurate and easier to use than a concurrency graph since we are not limited to specific values of any of the three variables. The accuracy of the nomograph for *any* values of u, v, and w is dependent only upon the precision and detail used in graduating the three scales.

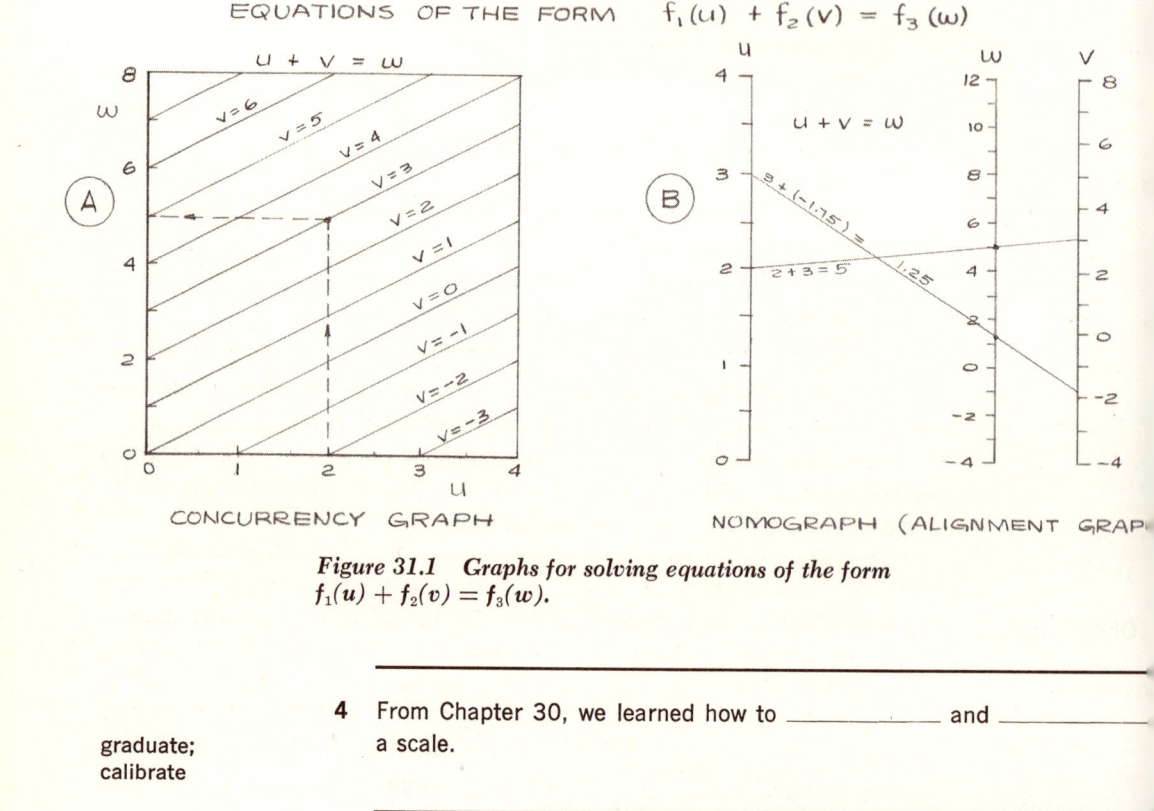

EQUATIONS OF THE FORM $f_1(u) + f_2(v) = f_3(w)$

A — CONCURRENCY GRAPH

B — NOMOGRAPH (ALIGNMENT GRAPH)

Figure 31.1 *Graphs for solving equations of the form $f_1(u) + f_2(v) = f_3(w)$.*

---

graduate;
calibrate

4 From Chapter 30, we learned how to _____ and _____ a scale.

---

(1) function,
(2) range;
(3) length

5 Let's review the construction of a scale. To graduate and calibrate any scale we must first be given:
   (1) The _____ of the variable.
   (2) The _____ of the variable.
   (3) The scale _____.

$d_x = m_x f(x)$
(x is the variable)

6 Given the function and range of a variable and a desired scale length, we can write a scale equation. In its most general form, the scale equation is _____. (Call the variable x)

NOMOGRAPHY   449

origin (left end of stem)	7	The distance *d* is the measure in inches along the scale to any desired graduation. It is measured from the graduation at x = 0. x = 0 may or may not be at the _____ of the scale.
$d_x = m_x[f(x) - f_{\min}(x)]$	8	If we wish *d* to be measured from the origin, we write the scale equation in the special form: $d_x = m_x$ _____.
$(d =) 0$	9	Using $d_x = m_x[f(x) - f_{\min}(x)]$ assures us that $d =$ _____ at the minimum value of the variable, x.
function	10	The term "*m*" is the scale modulus. Modulus is defined as "the increment of scale length per unit change in the _____ of the variable."
inches (per) unit	11	The units of modulus are _____ per _____.
$m = L/R_f$	12	The equation for calculating *m* is $m =$ ____/____.
calculating $f(x)_{\max} - f(x)_{\min}$ using the given max and min values of x	13	R_f is the range of the function of the variable. Our given information includes the range of the variable. We can find R_f by _____ _____ (your words).
graduate	14	We have talked about two different moduli, the actual modulus and the effective modulus. In nomography, the actual modulus is used to *position* the scales. The effective modulus is used to _____ each scale.
½ (effective); ¼ (actual)	15	Suppose we wish to graduate a 5-in. scale, $f(x) = 2x$ and the range of x is from 0 to 10. $R_f = 20 - 0 = 20$ $m = L/R_f = 5/20 = ¼$ $\therefore d = ¼(2x) = ½x$ The effective modulus is _____. The actual modulus is _____.

With this brief review of scale construction, let's see how we can construct a nomograph. Given the equation $u + v = w$, the scale lengths L, and the ranges of u and v, only two elements needed for the construction of a nomograph are missing:

(1) Information relating to the *placement* of the scales
(2) The modulus of the w scale

Figure 31.2 shows how we can obtain this missing data.

BASIS FOR CONSTRUCTING A NOMOGRAPH

A formula for placing the scales is derived in Figure 31.2A. Given three parallel scales with an arbitrary spacing between the u and v scales of $a + b$ we show, by similar triangles, that

$$\frac{m_u}{m_v} = \frac{a}{b}$$

Knowing m_u and m_v and combining this with the fact that $a + b = K$ (a constant), we can find a and b and thus place the w scale between the u scale and the v scale.

The formula for finding the value of the modulus of the w scale is derived in Figure 31.2B. Again using similar triangles, we find that $m_w = \dfrac{m_u m_v}{m_u + m_v}$. Using this to write the scale equation, $d_w = m_w f(w)$, we may complete the nomograph.

(A) **SPACING THE SCALES**

a and b are given (arbitrary)

By Similar Triangles: $\boxed{\dfrac{m_u}{m_v} = \dfrac{a}{b}}$

NOTE: m_u, m_v, and m_w are ACTUAL moduli.

(B) **CALCULATING m_w**

By Similar Triangles:

$$\frac{m_w}{m_u} = \frac{b}{a+b}, \quad \text{but } a = \frac{b m_u}{m_v}$$

$$\therefore \frac{m_w}{m_u} = \frac{b}{\frac{b m_u}{m_v} + b} = \frac{1}{\frac{m_u}{m_v} + 1}$$

$$\therefore \boxed{m_w = \frac{m_u m_v}{m_u + m_v}}$$

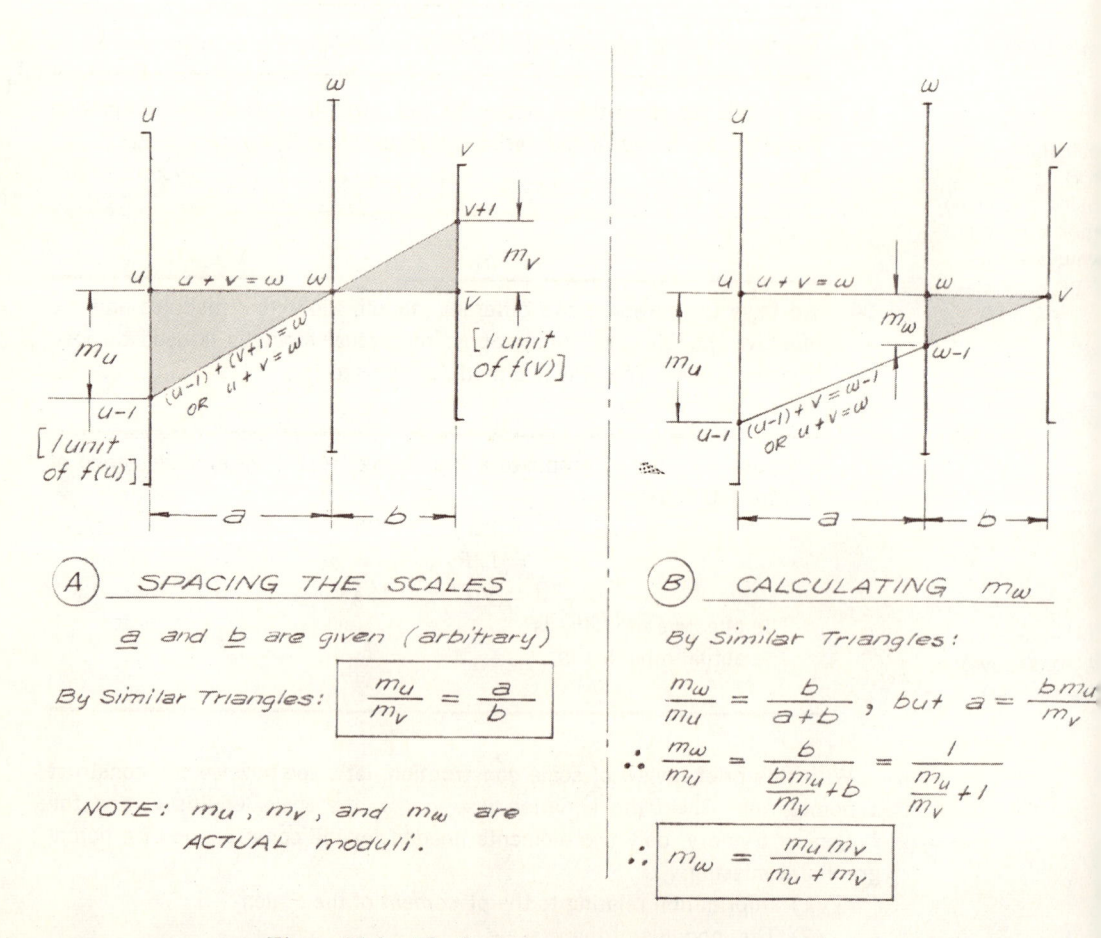

Figure 31.2 The basis for constructing a nomograph.

NOMOGRAPHY 451

Nomograph for $u + v = w$

A simple example is shown at Figure 31.3. The three scale equations were found and the w scale was placed by finding that $a/b = 6/5$. Note the manner in which the w scale was placed.

11 (eleven)

| 16 | We leave 6/5 as a fraction. A scale was placed across the distance $a + b$ (arbitrary) dividing it into _____ units. |

6; a

| 17 | From the left-hand u scale, we counted off _____ units to correspond to the distance _____. |

5; b

| 18 | This leaves _____ units as the distance _____. |

This is another example of the usefulness of the graphic method for proportioning distances.

$a = 2.5 - 1.14 =$
1.36 in.

| 19 | The distances a and b can be calculated and then measured directly in inches. We originally set $a + b = 2.5$ in. (an arbitrary and convenient distance). Knowing that $a/b = 6/5$, we also know that $a = 6b/5$ and therefore $6b/5 + b = 2.5$ or $11b = 12.5$. $b = 1.14$ in. and $a =$ _____ inches. |

$L = 3$ in.
$a + b = 2.5$ in.
$u; 0 \rightarrow 10$
$v; 0 \rightarrow 12$

$m_u = \dfrac{L}{R_f} = \dfrac{3}{10} = 0.3$

$f(u) = u \quad \therefore \quad d_u = 0.3u$

$m_v = \dfrac{L}{R_f} = \dfrac{3}{12} = 0.25$

$f(v) = v \quad \therefore \quad d_v = 0.25v$

$\dfrac{m_u}{m_v} = \dfrac{a}{b} = \dfrac{3/10}{3/12} = \dfrac{6}{5}$

$m_w = \dfrac{m_u m_v}{m_u + m_v} = \dfrac{0.3 \times 0.25}{0.55}$

$m_w = 0.136$

$f(w) = w \quad \therefore \quad d_w = 0.136w$

CALCULATIONS

EQUATION: $u + v = w$

Graphic proportioning method used to position the w scale.

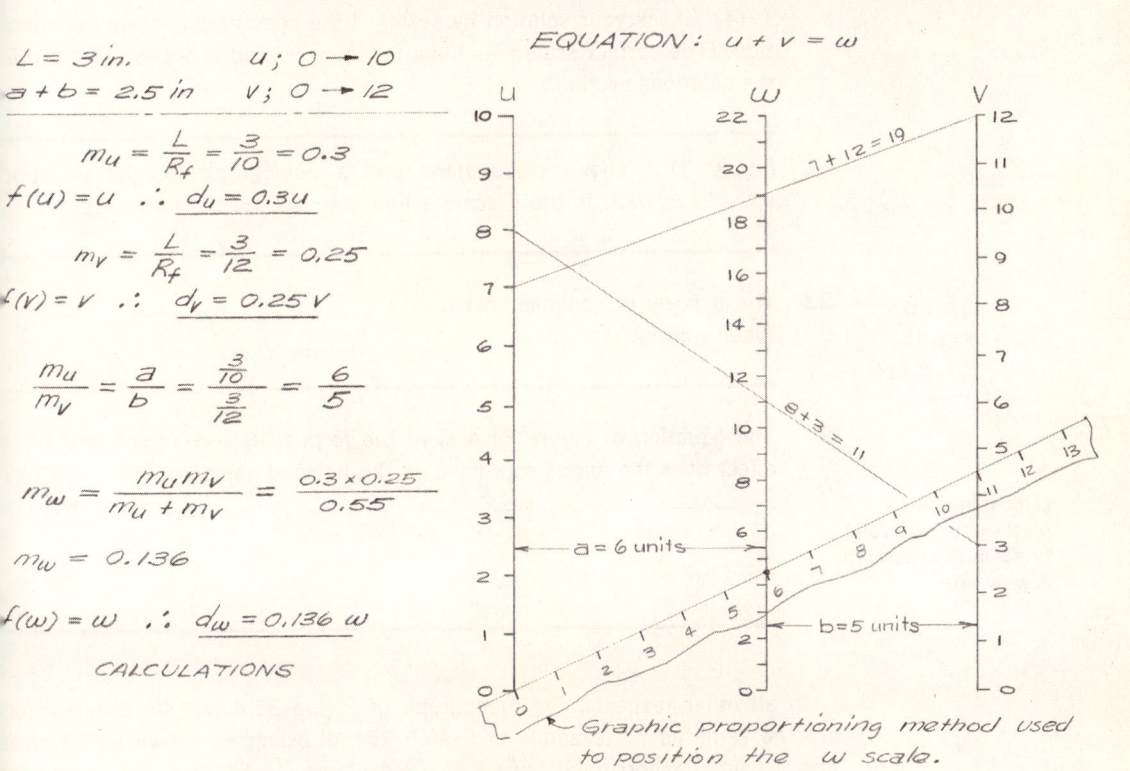

Figure 31.3 A nomograph for the equation $u + v = w$.

452 GRAPHIC MATHEMATICS/UNIT 6

The complete calculations are given in the example of Figure 31.3. The nomograph is finished when the three scales have been positioned, graduated and calibrated. The proof of the construction is that it must solve the equation. Note that the two lines drawn across the figure show that $7 + 12 = 19$ and $8 + 3 = 11$, both of which satisfy the original equation, $u + v = w$. If the nomograph does not solve the equation, a mistake was made in the calculations or in the construction.

---

20. Sketch 75 shows an equation of the form $u + v = w$. It is: $u/3 + v = 2w$.

    $f(u) = $ _____
    $f(v) = $ _____
    $f(w) = $ _____

$f(u) = u/3$
$f(v) = v$
$f(w) = 2w$

■ 21. In Sketch 75 you are given enough information to construct a nomograph for the equation $u/3 + v = 2w$. Four-inch u and v scales are drawn 3 in. apart ($a + b = 3$). Proceed as follows:
   (1) Complete the calculations shown.
   (2) Position the w scale.
   (3) Graduate and calibrate the three scales.
   (4) Check your solution by seeing if the nomograph solves the equation. (The correct scale equations for the u, v, and w scales are given in the solutions section.)

---

22. Figure 31.4 shows calculations and a nomograph for the equation $u^2 - 2v = w/2$. Is the u scale a linear or nonlinear scale? _____

nonlinear

---

23. The u scale is nonlinear because _____ (your words).

$f(u) = u^2$

---

24. The equation of Figure 31.4 is of the form $f_1(u) - f_2(v) = f_3(w)$. What effect does the minus sign have on the finished nomograph? _____
_____ (your words).

(The minus sign
reverses the v scale)
(v scale is calibrated
downward)

In all other respects, the nomograph of Figure 31.4 and its construction are the same as the example of Sketch 75. All examples shown so far have had a common horizontal base line. (See Figure 31.4.) This is achieved by writing the scale equation in the form $d_x = m_x[f(x) - f(x)_{\min}]$.

NOMOGRAPHY

$= 3$ in.
$+ b = 2.5$ in.

$u; 1 \to 5$
$v; -5 \to 5$

Calculations:

$m_u = \dfrac{3}{25-1} = \dfrac{1}{8}$

$f(u) = u^2 \quad \therefore \quad d_u = \dfrac{1}{8}(u^2 - 1)$

$m_v = \dfrac{3}{10-(-10)} = \dfrac{3}{20}$

$f(v) = 2v \quad \therefore \quad d_v = \dfrac{3}{20}(2v + 10)$

$d_v = \dfrac{3}{10} v + 1.5$

$\dfrac{a}{b} = \dfrac{m_u}{m_v} = \dfrac{\frac{1}{8}}{\frac{3}{20}} = \dfrac{5}{6}$

$m_w = \dfrac{\frac{1}{8} \cdot \frac{3}{20}}{\frac{1}{8} + \frac{3}{20}} = 0.068$

$f(w) = \dfrac{w}{2} \quad w; -18 \to 70 \text{ (calc.)}$

$\therefore d_w = 0.034 w + 0.61$

Calibrate Downward

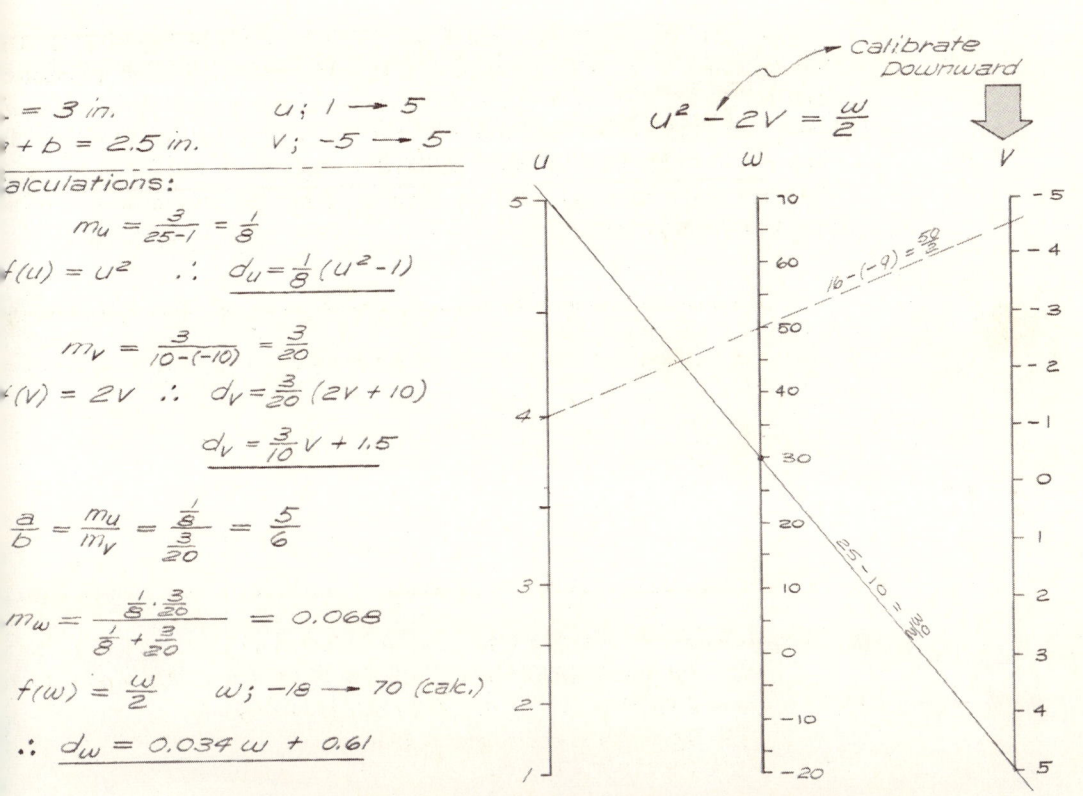

$u^2 - 2v = \dfrac{w}{2}$

Figure 31.4 A nomograph for the equation $u^2 - 2v = w/2$.

Nomograph for $uv = w$

Let us now consider an equation of the form $f_1(u) \times f_2(v) = f_3(w)$. In its simplest form $uv = w$. A concurrency graph for this equation is shown at Figure 31.5A. Again we have a family of curves that can be used to solve the equation as long as we have enough curves representing the desired values of v.

25 Figure 31.5B is a nomograph of the equation $uv = w$. Note that all three scales are _____ scales.

og
(logarithm)

26 The logarithm of a simple product such as u times v is equal to the sum of the logarithms of u and v. Thus, if we take the logarithm of both sides of the equation $uv = w$, we obtain the equation, _____
_____.

og u + log v =
og w

27 This is an equation of the familiar form, $f_1(u) + f_2(v) = f_3(w)$ where:
f(u) = _____
f(v) = _____
f(w) = _____

f(u) = log u
f(v) = log v
f(w) = log w

454 GRAPHIC MATHEMATICS/UNIT 6

28 We may now proceed as in previous examples, remembering that all three scales will be log scales.

From Chapter 30 we learned that the modulus of a log scale is always equal to the length of one logarithmic _____.

cycle

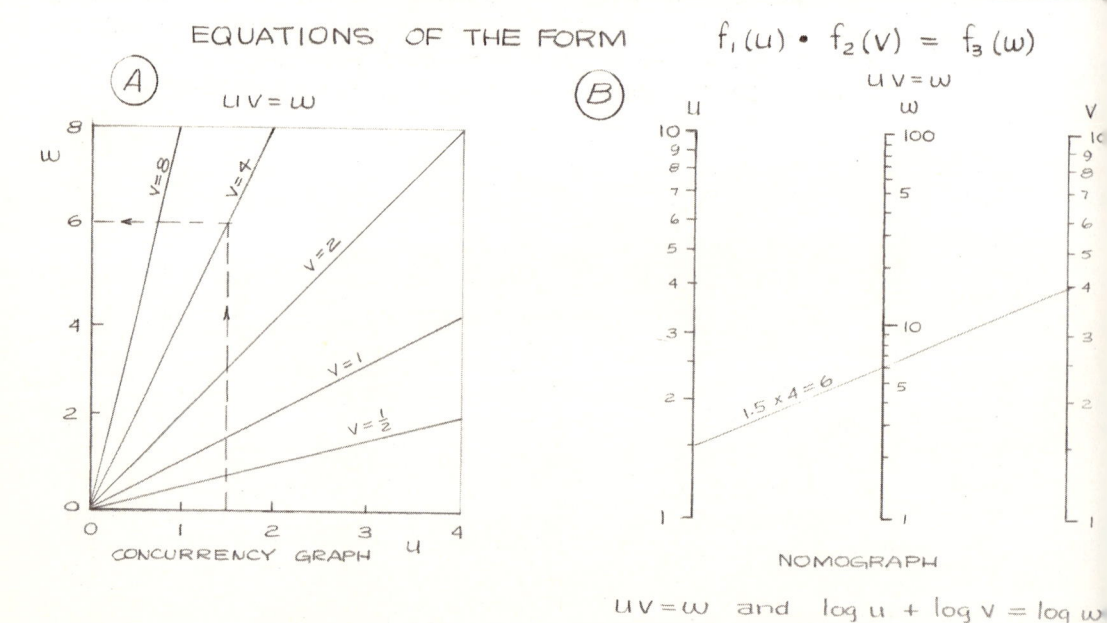

Figure 31.5 Graphs for solving equations of the form $f_1(u) \times f_2(v) = f_3(w)$.

29 How many logarithmic cycles are represented in the following ranges of numbers:

(a) 0 to 10? _____
(b) 0.001 to 10? _____
(c) 10 to 10,000,000? _____
(d) 5.28 to 5280? _____

(a) 1
(b) 4
(c) 6
(d) 3

30 Since the modulus of a log scale equals the length of one cycle, we can graduate the scale by cycles. The nine divisions within one cycle are most easily proportioned by _____ means.

graphic

Figure 31.6 presents a nomograph to find the volume of a cylinder over given ranges of the radius r and the height h. The equation is $V = \pi r^2 h$. Note that the scales are all neatly labeled with the graphing techniques presented in Chapter 29. The name of the variable, the symbol, and the unit are given for each scale.

Taking the logarithm of each side of the equation $V = \pi r^2 h$ gives $\log v = \log \pi + 2 \log r + \log h$. The constant term, $\log \pi$, is transposed to the left side of the equation and included as part of the function of V, or $f(V) = \log V - \log \pi$. Note that the scale equation for V is $d_v = \frac{3}{4} \log V - 0.373$

Reference to the nomograph will show that ¾ in. is the effective modulus of the v scale (cycle length) and that 0.373 in. is the amount that the calibration $v = 1$ is dropped below the horizontal base line.

Since a log scale is first graduated in cycles using decimal multiples or divisions of 10, it is important to locate the calibration $v = 1$ or $v = 10$. $v = 10$ was found after the r and h scales were fully graduated by solving for h when $r = 1$ and $v = 10$. This gave a value, $h = 3.18$ which could be plotted on the h scale. A straight line drawn across the v scale to $r = 1$ located $v = 10$. Both methods for starting the v scale are equally effective.

$$V = \pi r^2 h$$

$$\log V = \log \pi + 2\log r + \log h$$

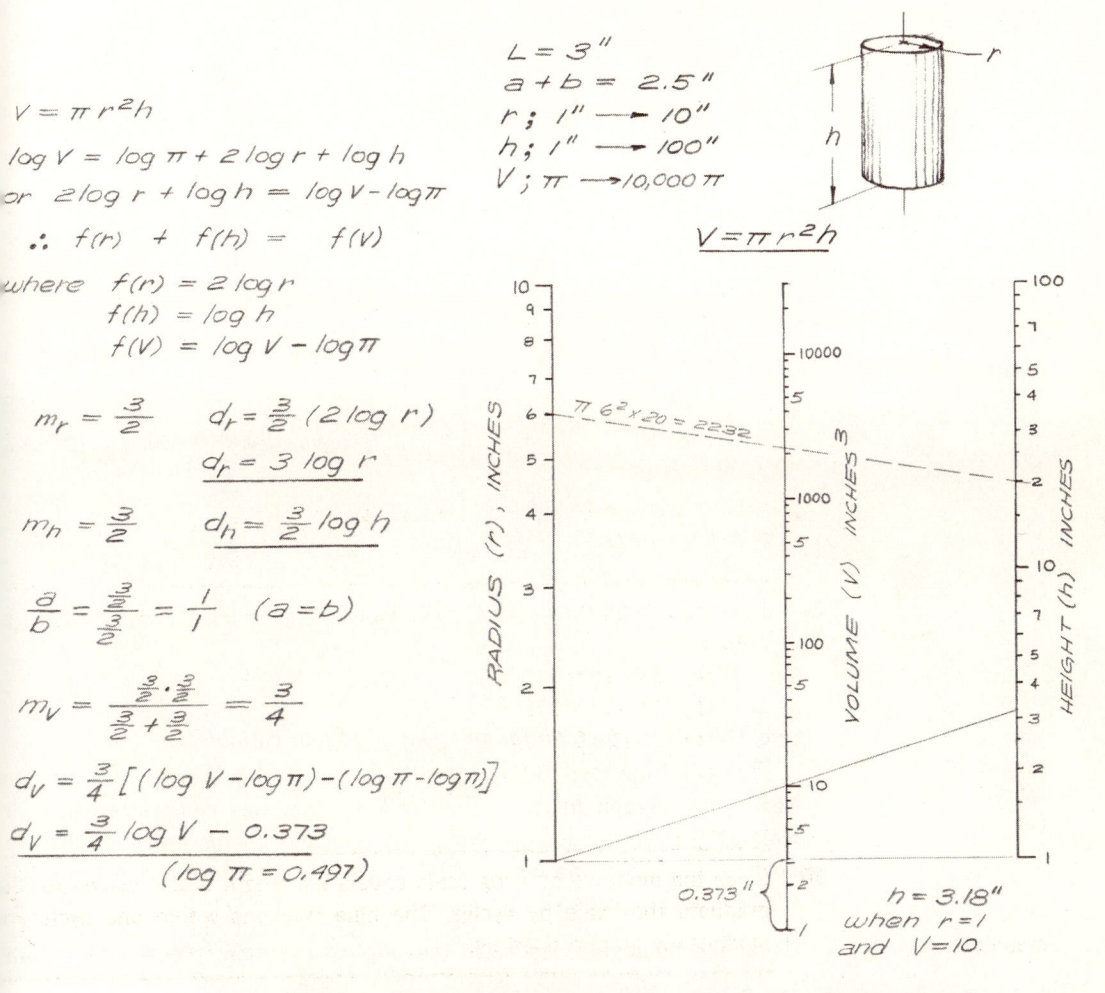

or $2\log r + \log h = \log V - \log \pi$

$$\therefore f(r) + f(h) = f(V)$$

where $f(r) = 2\log r$
$f(h) = \log h$
$f(V) = \log V - \log \pi$

$$m_r = \frac{3}{2} \qquad d_r = \frac{3}{2}(2\log r)$$

$$d_r = 3\log r$$

$$m_h = \frac{3}{2} \qquad d_h = \frac{3}{2}\log h$$

$$\frac{a}{b} = \frac{3/2}{3/2} = \frac{1}{1} \quad (a = b)$$

$$m_V = \frac{\frac{3}{2} \cdot \frac{3}{2}}{\frac{3}{2} + \frac{3}{2}} = \frac{3}{4}$$

$$d_V = \frac{3}{4}[(\log V - \log \pi) - (\log \pi - \log \pi)]$$

$$d_V = \frac{3}{4}\log V - 0.373$$

$(\log \pi = 0.497)$

$L = 3''$
$a + b = 2.5''$
$r; 1'' \to 10''$
$h; 1'' \to 100''$
$V; \pi \to 10{,}000\pi$

$V = \pi r^2 h$

$\pi\, 6^2 \times 20 = 2232$

$0.373''$

$h = 3.18''$ when $r = 1$ and $V = 10$.

Figure 31.6 A nomograph for finding the volume of a cylinder given the radius and the height ($V = \pi r^2 h$).

■ **31** Sketch 76 asks for a nomograph for the electrical power equation $P = I^2 R$. All necessary information is given. Complete the calculations and construct the nomograph graduating log cycles only (0.1, 1, and 10 in the case of I). Use the ¼-in. grid for making measurements. Check your solution by seeing if the nomograph solves the equation. Answers are given for a/b, m_P, and the three scale equations.

Four or More Variables

The techniques developed so far can be used to construct nomographs for equations of more than three variables by combining elements of the equation. A nomograph for the simple equation $t + u + v = w$ is shown in Figure 31.7. First the equation is broken into two parts, $u + v = A$ and $t + A = w$. The nomograph for $u + v = A$ was constructed with the A scale as the center scale.

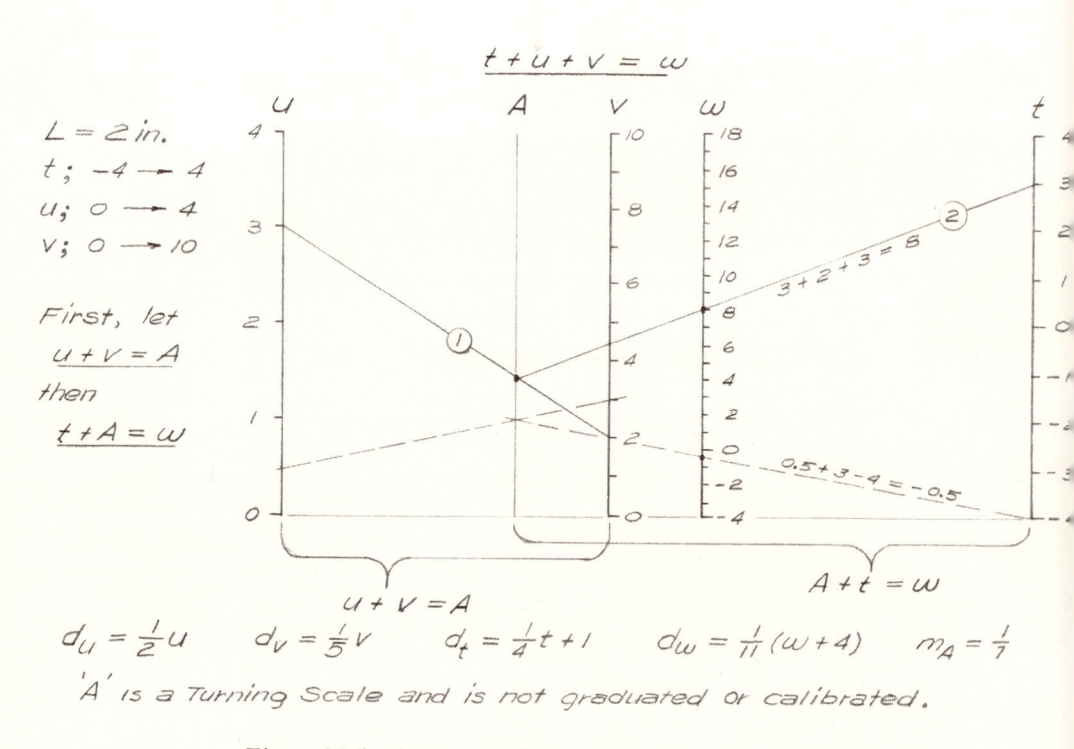

Figure 31.7 A nomograph for an equation of four variables.

Next a nomograph for the equation $A + t = w$ was constructed with the A scale as the left-hand scale. Note that even though the modulus, m_A, was calculated, the scale was not graduated. We are not interested in the value of A. The A scale is called a turning scale.

Using the nomograph to solve the original equation, proceed as follows:

(1) Select values of $u = 3$ and $v = 2$ and draw line 1 across the A scale.

(2) Select a value of $t = 3$ and draw line 2 across the w scale to a connection with the intersection on A found previously.

(3) Read the answer at the intersection of the t scale and line 2. $w = 3 + 2 + 3 = 8$.

If we had more than four variables, we could carry this procedure on and construct a nomograph to solve the equation.

Other Forms for Nomographs

A nomograph can take many forms. Figure 31.8 shows four of these and the general form of the equations that they solve. A detailed description of the techniques for constructing these nomographs can be found in a text on nomography.

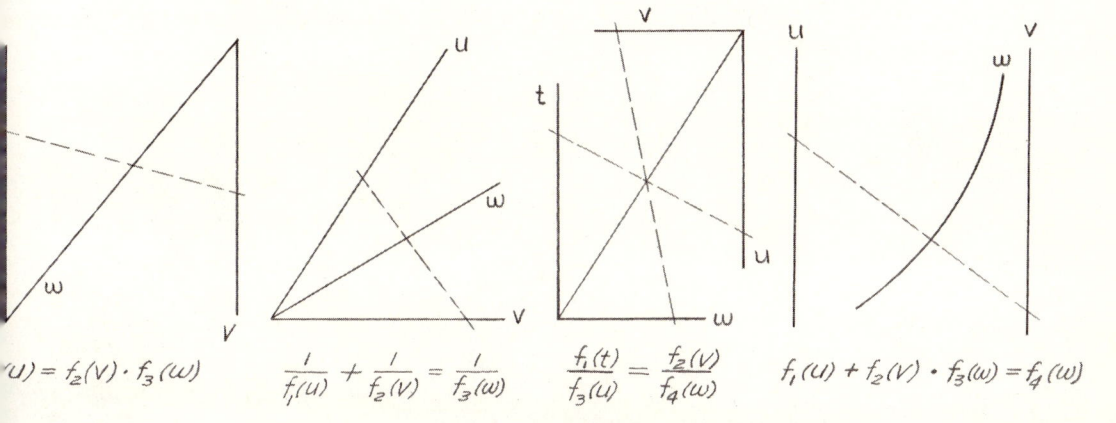

$f_1(u) = f_2(v) \cdot f_3(w)$ $\qquad$ $\dfrac{1}{f_1(u)} + \dfrac{1}{f_2(v)} = \dfrac{1}{f_3(w)}$ $\qquad$ $\dfrac{f_1(t)}{f_3(u)} = \dfrac{f_2(v)}{f_4(w)}$ $\qquad$ $f_1(u) + f_2(v) \cdot f_3(w) = f_4(w)$

Figure 31.8 Other forms of nomographs.

chapter 32 empirical equations

Empirical data is data gained from observation or experience. The experimental researcher relies heavily on observed data and experimental results. If the observed data is numerical, he can plot the data for any two variables as a graph to determine if the relationship between the two sets of numbers exhibits a definite graphic pattern.

Most physical and chemical phenomena follow mathematical laws and therefore can be plotted on a graph in a definite pattern if the experiment was performed accurately. This chapter shows how we can write an equation for certain kinds of empirical data through a graphic analysis of the data.

EQUATION OF A STRAIGHT LINE: $y = mx + b$

The most common general-form equation is $y = mx + b$, the equation of straight line, where m is the *slope* of the line ($\Delta y/\Delta x$) and b is its intersection with the y axis ($y = b$ when $x = 0$).

straight line (This is called the *best curve* through the observed points.)	1	Figure 32.1A shows a set of numerical x and y data plotted on rectlinear coordinates. The curve drawn is a(n) _____ _____
$y = mx + b$	2	We know that the general form of the equation of a straight line is _____.
slope; $x = 0$	3	In $y = mx + b$, m is the _____ of the line and b is the value of y when _____.
$b = 1.4$	4	The value of b in Figure 32.1 is _____.

458

$y = mx + b$

m = slope of the line $\left(\frac{\Delta y}{\Delta x}\right)$
b = y-axis intercept
    $y = b$ when $x = 0$

FINDING THE SLOPE (m)

Graphic Method

1. Measure Δy and Δx in inches
2. Change to 'units' by dividing each by its modulus. Thus,

$$m = \frac{\frac{\Delta y}{m_y}}{\frac{\Delta x}{m_x}} = \frac{\Delta y}{\Delta x} \cdot \frac{m_x}{m_y}$$

Selected-Point Method

$m = \frac{y_2 - y_1}{x_2 - x_1}$

For (A), $m = \frac{8.3 - 2}{3.5 - 0.31} = 1.975$

Graphic

(A) $m = \frac{1\frac{9}{16}}{1\frac{19}{32}} \cdot \frac{\frac{1}{2}}{\frac{1}{4}} = 1.96$

(B) $m = \frac{\frac{13}{16}}{2\frac{9}{16}} \cdot \frac{\frac{3}{4}}{\frac{1}{8}} = 1.90$

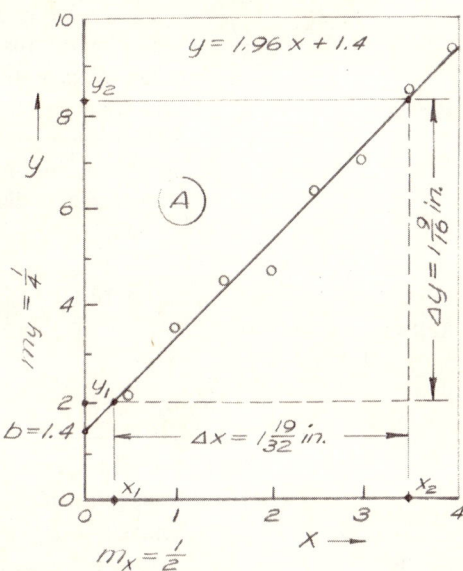

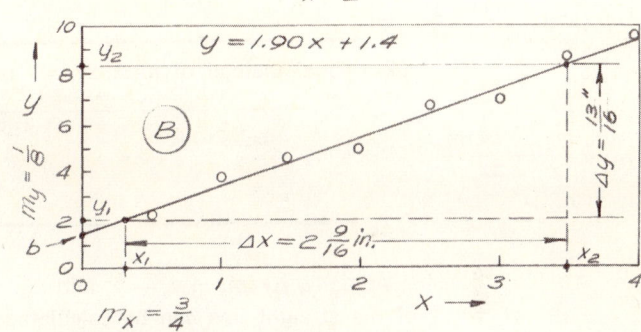

Figure 32.1 $y = mx + b$, the equation of a straight line.

Δy (and) Δx

5 The slope, m, of a straight line can be found by selecting two points that lie on the line, points x_1y_1 and x_2y_2 and constructing a right triangle (slope triangle) whose sides are _____ and _____.

Δy/Δx

6 The actual numerical slope can be calculated from $m = $ _____.

is not

7 However, consider Figure 32.1B. It shows the same data plotted on a different set of rectilinear coordinates and a best curve drawn through the points. The visual slope, Δy/Δx, _____ the same as that of Figure 32.1A. (Choose one: is/is not.)

yes

8 Should the arithmetic slope (true numerical slope) be the same for the curves of Figures 32.1A and 32.1B? Yes _____? No _____?

460 GRAPHIC MATHEMATICS/UNIT 6

9 Yes, the value of m is a constant for this data since there is only one equation of the form $y = mx + b$ which will satisfy the data. What then, causes the change in the visual slope? In both graphs, Δy and Δx are measured in _____ .

inches

10 However, the moduli of the x and y scales of 32.1A are different from the corresponding moduli of 32.1B.

In 32.1A, $m_x =$ _____ in./unit.
$m_y =$ _____ in./unit.
In 32.1B, $m_x =$ _____ in./unit.
$m_y =$ _____ in./unit.

32.1A: $m_x = \frac{1}{2}$;
$m_y = \frac{1}{4}$;
32.1B: $m_x = \frac{3}{4}$;
$m_y = \frac{1}{8}$

11 In each case, if we divide the measured values of Δy and Δx by their respective moduli, we convert distances measured in inches to distances measured in _____ .

units

$\left(\dfrac{\text{in.}}{\text{in./unit}} = \text{units} \right)$

12 Thus, instead of $m = \Delta y / \Delta x$, the expression for slope becomes $m =$ _____ .

$m = \dfrac{\frac{\Delta y}{m_y}}{\frac{\Delta x}{m_x}} = \dfrac{\Delta y}{\Delta x} \cdot \dfrac{m_x}{m_y}$

Actually, $\Delta y / m_y = y_2 - y_1$ and $\Delta x / m_x = x_2 - x_1$. The true arithmetic slope of the straight line can be calculated from $m = (y_2 - y_1)/(x_2 - x_1)$. This is called the method of *selected points* and, in Figure 32.1, it is the easiest method. However, the general graphic form

$$m = \dfrac{\Delta y}{\Delta x} \cdot \dfrac{m_x}{m_y}$$

should be remembered and will be found useful in other examples. In Figure 32.1A, $m = 1.96$ (graphic method) and $m = 1.975$ (selected-points method). In 32.1B, $m = 1.90$ by the graphic method. This is a reasonable agreement considering the errors that can enter through construction, measuring, and reading. The best equation for the straight line is $y = 1.4 + 1.98x$.

BEST CURVE AND SCALE MODULI

From the example of Figure 32.1 it is evident that two operations are critical in finding the equation of empirical data by graphic means:

(1) Determining the best straight line through a set of experimentally derived points.
(2) Reading the scale modulus from the scale itself.

EMPIRICAL EQUATIONS 461

Figure 32.2 shows a simple way of determining the best line. Bracket the complete set of points by two light lines representing maximum and minimum positions of the desired line. The best line will be about midway between these two. There should be as many points on one side of the line as on the other, or, more accurately, the sum of the perpendicular distances from the points above the line should equal the sum of the perpendiculars from the points below the line. It is not necessary to measure these distances and add them. A visual inspection of the number of points on either side of a trial line and their respective distances from the line will provide results consistent with the accuracy of any graphic construction.

There are more sophisticated systems for finding the best curve through a set of data, but certainly for a first approximation the simple visual method is adequate.

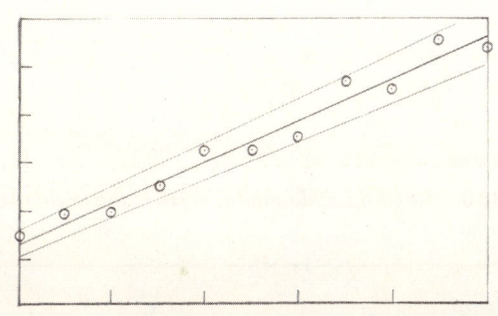

Figure 32.2 Finding the best straight line through a set of observed points.

---

effective

13 The scale moduli that we are concerned with are _____ moduli. (*Choose one: actual/effective.*)

---

variable

14 The effective modulus of a scale is defined as the increment of scale length for a unit change in the _____.

---

in./unit

15 The units of the scale modulus are _____.

---

(1) 0.01;
(2) 10;
(3) ¼

16 Given a completely graduated and calibrated scale, the easiest way to determine its modulus is first to determine the number of units per inch of scale. Modulus is the *reciprocal* of this number.
   What are the moduli for the following? A scale has:
   (1) 100 units per inch; $m = $ _____ in./unit.
   (2) 0.1 units per inch; $m = $ _____ in./unit.
   (3) 4 units per inch; $m = $ _____ in./unit.

462 GRAPHIC MATHEMATICS/UNIT 6

17 Figure 32.3 gives four calibrated scales. Write the modulus for each
  (1) m = _____ in./unit.
  (2) m = _____ in./unit.
  (3) m = _____ in./unit.
  (4) m = _____ in./unit.

(1) ⅓ = 0.33
(2) ⅒ = 0.1
(3) ¹⁄₁₀,₀₀₀ = 0.0001
(4) 2

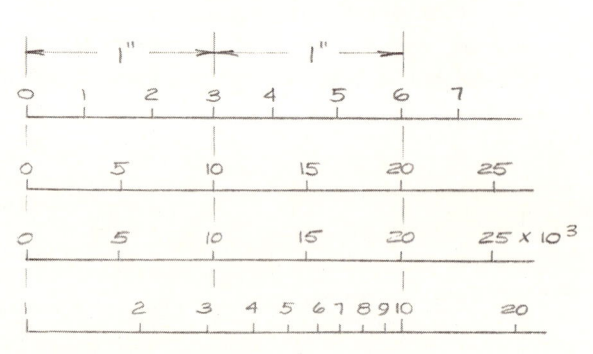

Figure 32.3 *Reading scale modulus from the scale itself.*

POWER EQUATION: $y = bx^m$

Figure 32.1 illustrated the case in which the empirical data are plotted as straight line on rectilinear coordinates. Consider the equation $y = 2x^2$ in Figure 32.4A. On rectilinear coordinates it is plotted as a parabola with its vertex at the origin. It is of the general form $y = bx^m$ and is called the *power equation*. Many phenomena in nature follow the form of the power equation so it is reasonable to assume that a set of empirical data could fall into this category.

18 Write the logarithm of both sides of the equation $y = bx^m$. _____

log y =
 log b + m log x

19 The equation $\log y = \log b + m \log x$ is of the form $y = mx + b$ we consider that $f(x) =$ _____ and $f(y) =$ _____

$f(x) = \log x$;
$f(y) = \log y$

20 If we took the logarithms of all the x data and y data used to obtain the parabola of Figure 32.4A and plotted them on rectilinear coordinates, we would obtain _____ (your words).

a straight line

EMPIRICAL EQUATIONS 463

Plots as a Straight Line on Log-Log Paper

$$y = bx^m$$
$$\log y = m \log x + \log b$$
$$Y = mX + B$$

$$\text{Slope}, m = \frac{\Delta y}{\Delta x} \cdot \frac{m_x}{m_y} = \frac{2\frac{13}{16}''}{1\frac{13}{32}''} \cdot \frac{2}{2} = 2$$

$$m_x = m_y = 2 \text{ in./unit (Log Cycle)}$$

when $x = 1$, $\log x = 0$.
thus $\log y = \log b$ or $y = b = 2$

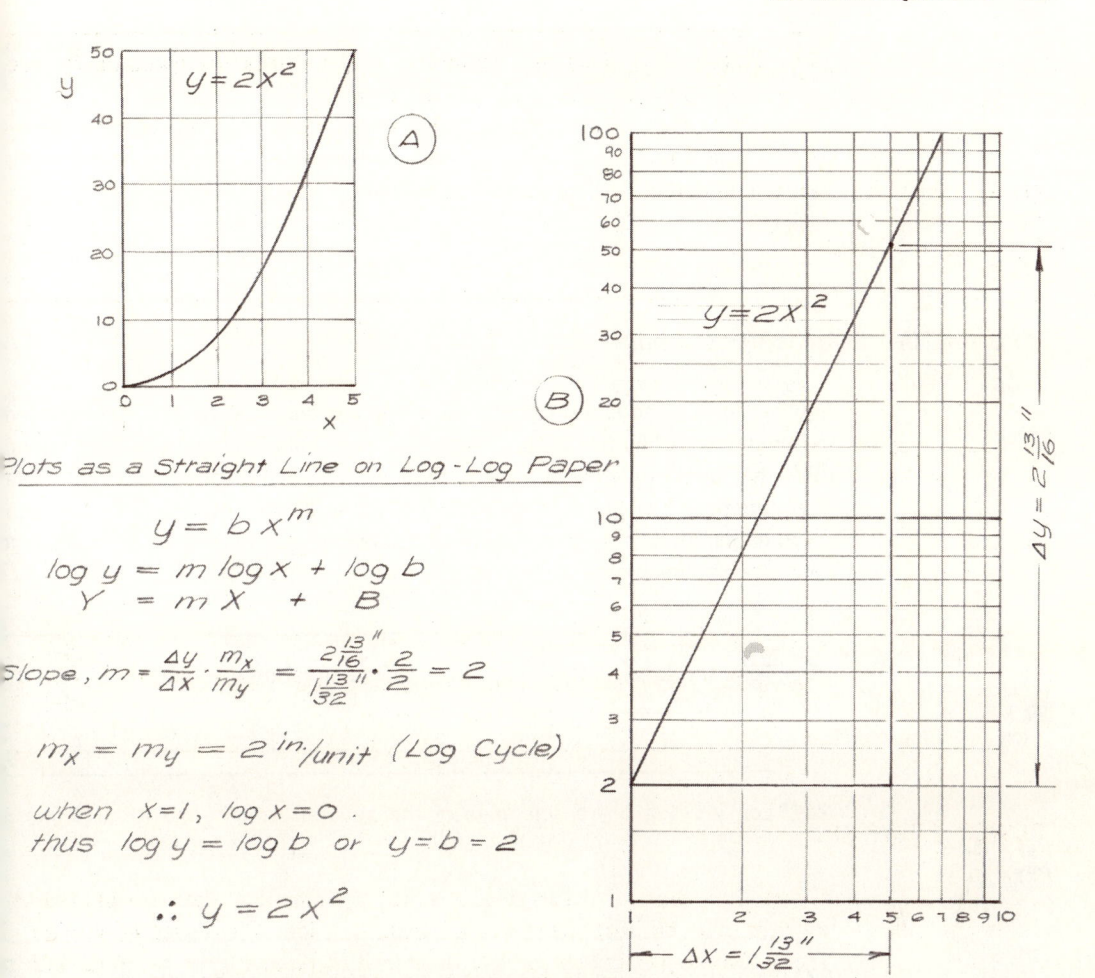

$$\therefore y = 2x^2$$

Figure 32.4 $y = bx^m$, the power equation.

log-log
(logarithmic)

21 Figure 32.4B shows an easier way to obtain this straight line. The data has been plotted on _____ paper.

$m_x = m_y$

22 The complete graphic solution to obtain the equation $y = 2x^2$ is shown. Δy and Δx were measured in inches. Conversion to units is not necessary since $m_x =$ _____.

log $y = \log b$ (or)
$y = b$

23 In $\log y = \log b + m \log x$, when $x = 1$, $\log y =$ _____ or $y =$ _____.

$(x =) 1$

24 On a log-log graph of a straight line, the value of b is found at the intersection of the line and the ordinate, $x =$ _____.

464 GRAPHIC MATHEMATICS/UNIT 6

The graphic method is easier than the method of selected points in this example. The selected-point method would yield

$$m = \frac{\log y_2 - \log y_1}{\log x_2 - \log x_1}$$

necessitating the finding of the logarithm of the coordinates of the points selected.

EXPONENTIAL EQUATION: $y = bm^x$

Another form of equation that appears often in technical work is the *exponential equation*. It has the general form $y = bm^x$. This equation applies to quantities that decay or grow at a rate proportional to the amount already present. The equation $y = 2(2)^x$ is shown, in Figure 32.5A, plotted on rectilinear coordinates. It is one of a family of curves having a hyperbolic shape and approaching the x axis asymptotically, crossing the y axis at $y = 2$.

---

log y =
 log b + x log m

25 Take the logarithm of both sides of the equation $y = bm^x$. _____

$f(y) = \log y$
$f(x) = Kx$
(K = const. = log m)

26 Log $y = \log b + x \log m$ is of the general straight-line form if $f(y) = $ _____ and $f(x) = $ _____

---

took the log of
the y data

27 The equation would be plotted as a straight line on rectilinear coordinates if we _____ (your words)

However, we can again make use of special paper (semi-log in this case) to plot the x and y data directly and still obtain a straight line as shown in Figure 32.5B.

The moduli of the x and y scales are not equal, so to find the slope m by graphic means, we must use the expression

$$m = \frac{\Delta y}{\Delta x} \cdot \frac{m_x}{m_y}$$

The resulting computation yields a value for log m. The value of m is found by taking the antilog of log m.

A more general form of the exponential equation is $y = b(e)^{mx}$ in which the constant m is equal to the instantaneous rate of growth expressed as percentage. Taking the log of this equation, we get: $\log y = \log b + mx \log e$

This again is of the form $y = mx + b$ where $f(y) = \log y$ and $f(x) = mx$. Plotting the data on semi-log paper will yield a straight line. log e = 0.434

EMPIRICAL EQUATIONS 465

[Graph A: rectilinear plot showing curve $y = 2(2)^x$ with y-axis 0–70, x-axis 0–5]

[Graph B: semi-log plot showing straight line $y = 2(2)^x$, with $\Delta y = 2\frac{1}{4}''$ and $\Delta x = 2\frac{1}{2}''$]

Plots as a Straight Line on Semi-Log Paper

$$y = bm^x$$
$$\log y = x \log m + \log b$$
$$Y = Mx + B$$

SLOPE: $M = \log m$

$$\log m = \frac{\Delta y}{\Delta x} \cdot \frac{m_x}{m_y} = \frac{2\frac{1}{4}}{2\frac{1}{2}} \cdot \frac{\frac{1}{2}}{1\frac{1}{2}} = 0.30$$

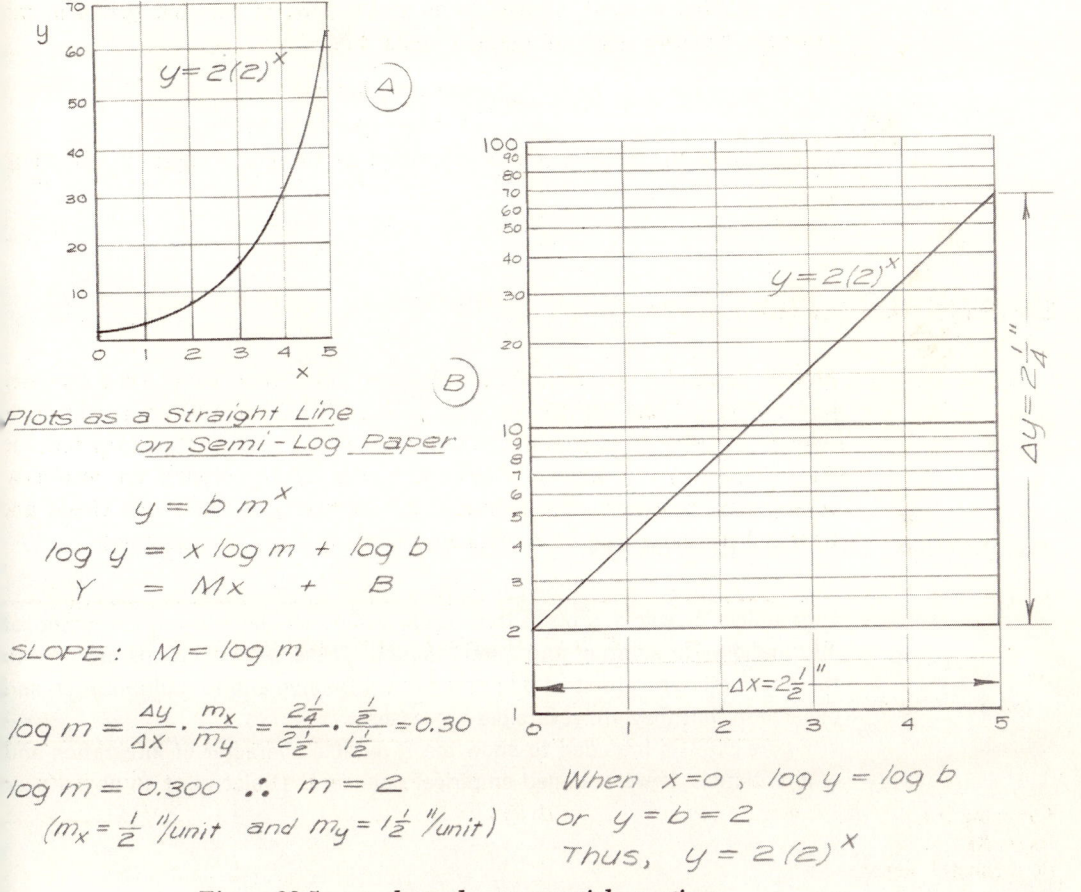

$\log m = 0.300$ ∴ $m = 2$
($m_x = \frac{1}{2}''$/unit and $m_y = 1\frac{1}{2}''$/unit)

When $x = 0$, $\log y = \log b$
or $y = b = 2$
Thus, $y = 2(2)^x$

Figure 32.5 $y = bm^x$, the exponential equation.

■ 28   Sketch 77 is an exercise in finding the equation of empirical data that can be plotted as a straight line on rectilinear paper or as a rectified straight line on log-log or semi-log paper.

Find the equations for (A), (B), and (C). Note in (B) that it was necessary to perform construction to extend the range of the ordinate in order to find the intercept with the y axis. You will need a scale to measure Δy and Δx.

The correct equations will be found in the solutions section.

chapter 33   graphic calculus

INTEGRATION AND DIFFERENTIATION

The subject of calculus presents two operations for the analysis of numerical information—integration and differentiation. These two operations are akin to the arithmetic operations of addition and subtraction and to multiplication and division in that they are reversible processes.

Figure 33.1 is intended to show the reversible character of integration and differentiation. If we obtained empirical data on the velocity of an object over

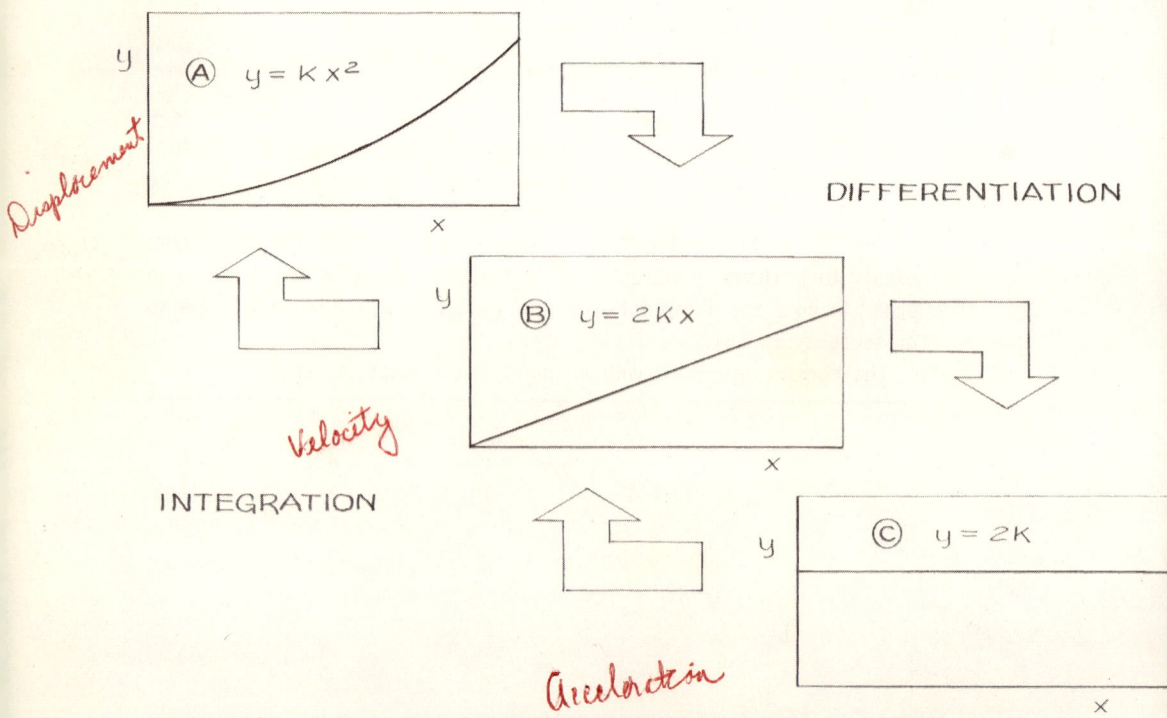

Figure 33.1 Integration and differentiation are reversible processes.

a time period, we could plot this as a velocity curve (B). Through graphic calculus, we can obtain the displacement (distance) curve (A) by integrating (B) graphically and also obtain the acceleration curve (C) by differentiating (B). All of this can be done without knowing any of the equations of motion.

Graphic calculus is an important tool for the engineer or scientist. It gives him a means of making quick approximations of the integral or differential of any empirical curve. The results are not precise. Computer techniques will give answers to a high degree of precision. However, a quick graphic solution is often advisable to determine whether the problem is worth computer time.

This chapter shows how you can integrate or differentiate a known curve by graphic means.

GRAPHIC INTEGRATION

Figure 33.2 shows the basic considerations for graphic integration by the pole-and-ray method. Integration is a summing process. It involves finding the *area* under a curve and expressing this area in appropriate units. An integral curve is thus a continuous measure of the area under a given curve.

As an example, look at the closed curve superimposed on a ¼-in. grid (Figure 33.2A). The area inside the curve can be found by counting the full

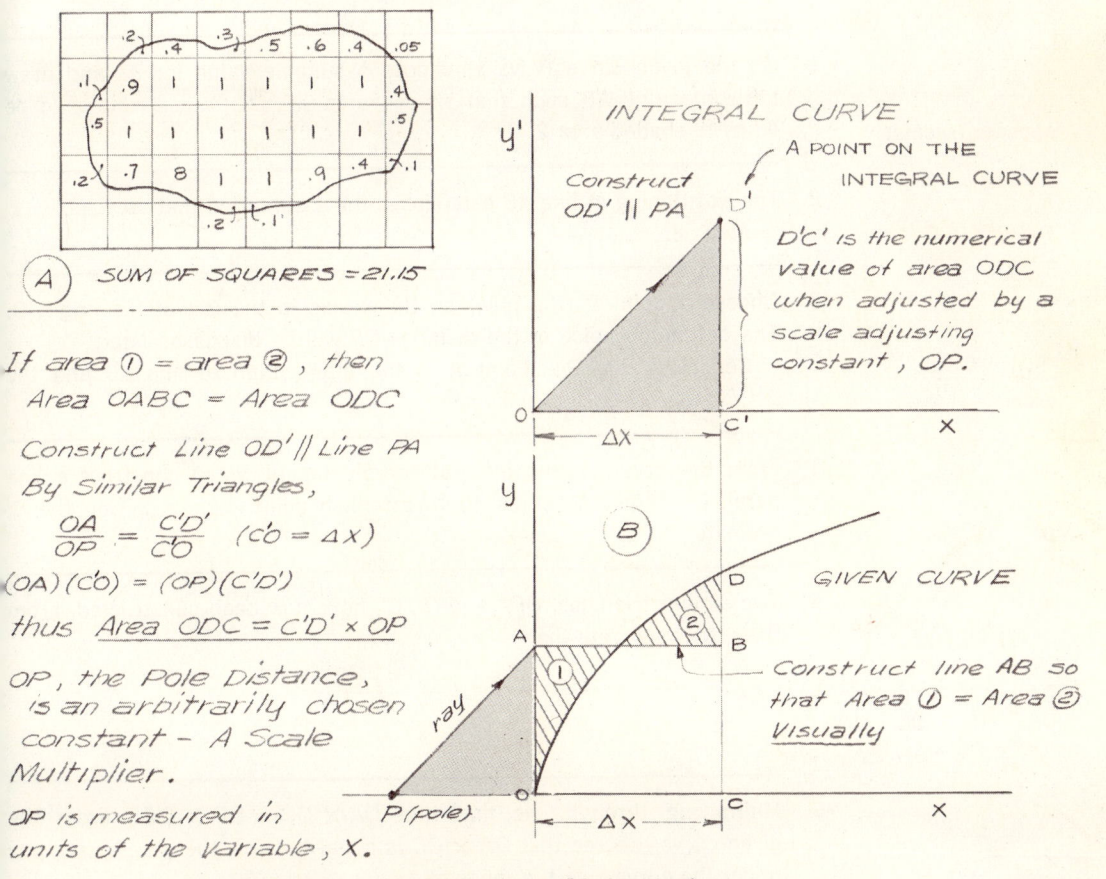

Figure 33.2 The basis for graphic integration.

468 GRAPHIC MATHEMATICS/UNIT 6

squares and estimating the part squares. The sum of all the full and part squares yields an area of 21.15 squares. If this were the outline of a small lake drawn to a scale of 1" = 400 ft, then each square represents 10,000 sq ft and the total area of the lake is 211,500 sq ft. This is an easy but crude method of computing the area. It is significant to note that had the squares been made much smaller, the accuracy of the computation would increase (and so would the job of counting the squares).

Graphic integration is based upon this idea. We make an assumption about the area under a given curve over a range, Δx, of the x-axis variable. We do this with the understanding that as Δx decreases in size, our assumption is more valid.

Pole-and-Ray Method

Study the given curve in Figure 33.2B. We make an assumption that rectangle OABC has the same area as ODC. We then prove that area OABC can be represented by the ordinate C'D' on the integral curve multiplied by an arbitrarily chosen constant, OP. This is the pole-and-ray method of graphic integration. The procedure for graphic integration is given in the following frames:

---

appears (seems)

1 On the given curve (y vs x) choose a suitable value for Δx and draw horizontal line AB such that shaded area 1 _____ to be equal to shaded area 2.

OABC = ODC

2 The positioning of line AB establishes the assumption that area _____ = area _____.

---

y-axis

3 Choose a point P (the pole) on the x axis to the left of the origin O. The optimum choice of the distance OP will be discussed later.
Construct line PA. Point A is the intersection of line AB and the _____.

D'

4 From the origin of the integral coordinates (y' vs x) construct a line parallel to line PA across Δx to establish point _____ on the C' ordinate.

$\dfrac{OA}{OP} = \dfrac{C'D'}{OC'} = \dfrac{C'D'}{OC}$
(OC' = OC = Δx)

5 Two similar triangles, OPA and OD'C', have now been constructed. From these we may write that
$\dfrac{OA}{OP} =$ _____.

---

area

6 Multiplying through, we find that OA × OC = OP × C'D'. From the given curve, we see that OA × OC is equal to ODC, the _____ under the curve over the range of Δx.

GRAPHIC CALCULUS 469

7. Thus area $OCD = C'D'$ multiplied by the pole distance _____ .

    OP

8. The pole distance *OP* was arbitrarily selected. It is a scale multiplier and is used to establish the calibration of the *y'* axis.

    . The significant point of the foregoing procedure is that the ordinate $C'D'$ represents a *numerical* value of the _____ (your words) over the abscissa range, Δx.

    area under the curve

The graphic integration is continued by choosing a second value of Δx adjacent to ordinate *CD* and repeating the above procedure.

This is the basic process. Figure 33.3 shows a complete solution and shows how the scale multiplier *OP* is used to establish the calibration of the *y'* scale.

Selecting a Pole Distance

In Figure 33.3A a rough approximation is made to the integration of the given curve. This is useful in choosing an optimum pole distance, *OP*. A full scale value (10 units) was selected for Δx. Line *AB* was drawn so that the sum of areas 1 and 2 *appears* to be equal to area 3. The area of rectangle *OABC* is $3.8 \times 10 = 38$. This rough guess at the total area under the given curve represents the approximate full-scale ordinate of the integral curve.

Line *OD'* was constructed so that distance *CD'* represents a *reasonable height* for the integral curve. Next a line parallel to *OD'* was constructed from point *A* to an intersection with the *x* axis extended. Point *P* was now plotted by choosing a point near this intersection that gave a *pole distance OP equal to an integral number of units of the x axis* (6 in this instance).

9. The distance *OP* was transferred to the given curve at Figure 33.3B. The length of *OP* is 6 units of _____ (your words).

    the *x*-axis graduation

10. The procedure for graphic integration was followed for five equal values of Δx. The value of Δx used is $\Delta x = $ _____ .

    $\Delta x = 2$

11. The accuracy of the integral curve can be increased by _____ the value of Δx.

    decreasing

12. Ray No. 1 establishes an ordinate at $x = 2$ on the integral curve that is numerically equal to area 1. Ray No. 2 establishes an ordinate at $x = 4$ that is numerically equal to _____ _____ (your words).

    area 1 + area 2

13. The total construction at 33.3B yields five points through which a smooth curve can be drawn. Any point on this integral curve represents the numerical value of the _____ _____ under the given curve to the left of the point.

    total area

470 GRAPHIC MATHEMATICS/UNIT 6

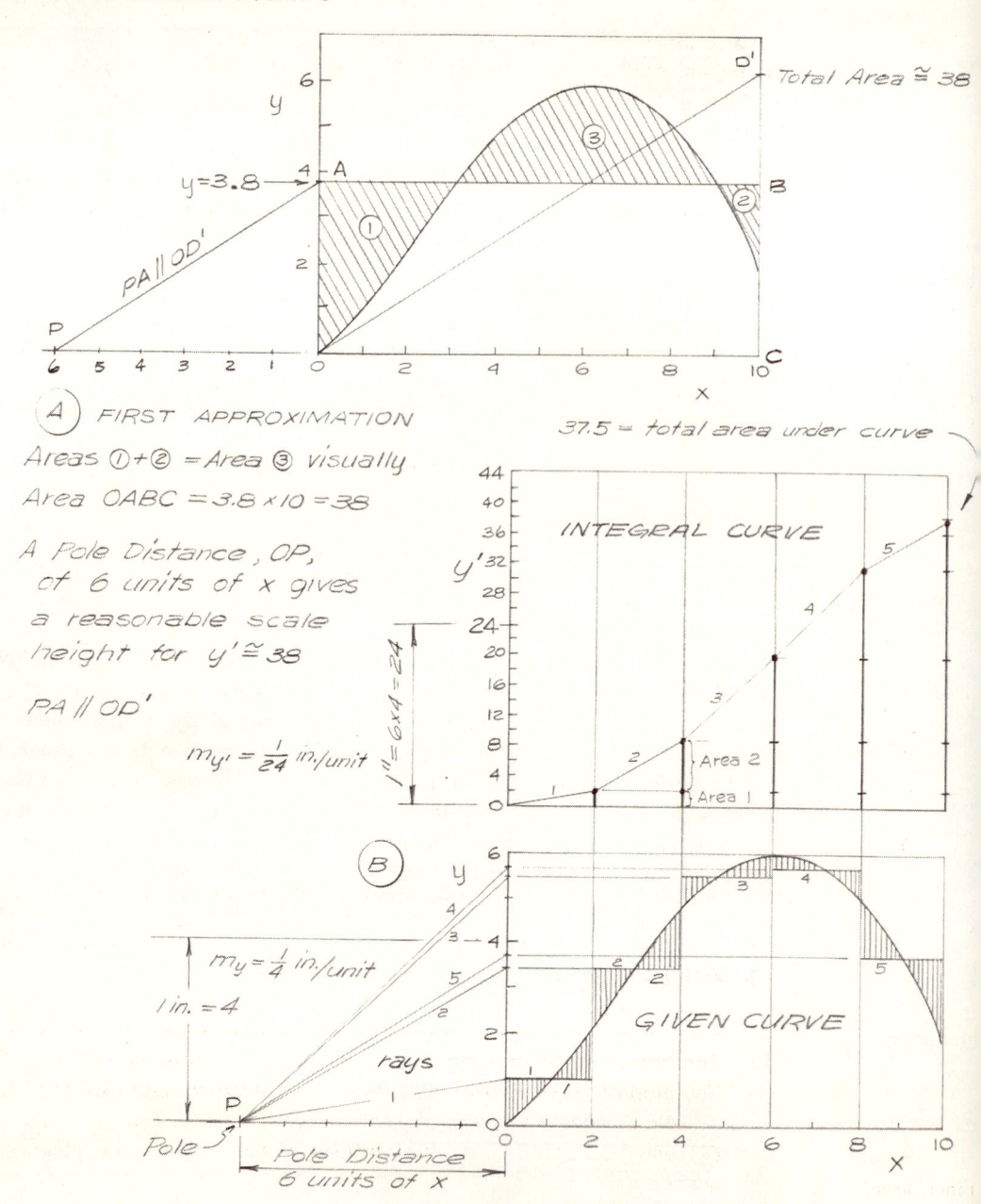

Figure 33.3 The pole-and-ray method of graphic integration.

It is important to realize that the integral curve (Figure 33.3B) is meaningless if the ordinate scale of the integral curve is not calibrated correctly. The distance OP controls this calibration. In 33.3B, 1 inch of the given ordinate (y) is 4 units. Using OP = 6 as a scale multiplier, we get 6 × 4 = 24 and, therefore, *1 inch of the integral ordinate (y') is 24 units*. Establishing this point allows us to calibrate the full y' scale and the problem is complete. Note that the total area came to 37.5 which shows that our rough guess at 33.3A of 38 units was reasonably accurate.

GRAPHIC CALCULUS 471

An examination of the y and y' scales of Figure 33.3B will also show that the modulus of the y scale divided by the pole distance (in x units) equals the modulus of the y' scale. Thus, $m_y/6 = m_{y'}$, or in the specific example, $m_{y'} = \frac{1}{24}$ in. per unit.

---

14 Figure 33.4 shows the effect of varying the pole distance, OP. As OP increases, the ordinate value of the integral curve _____.
(*Choose one:* increases/decreases.)

decreases

---

15 The distance OP should be selected to give a reasonable height to the integral curve. OP is always measured in _____ (your words) never in inches or any other unit of measure.

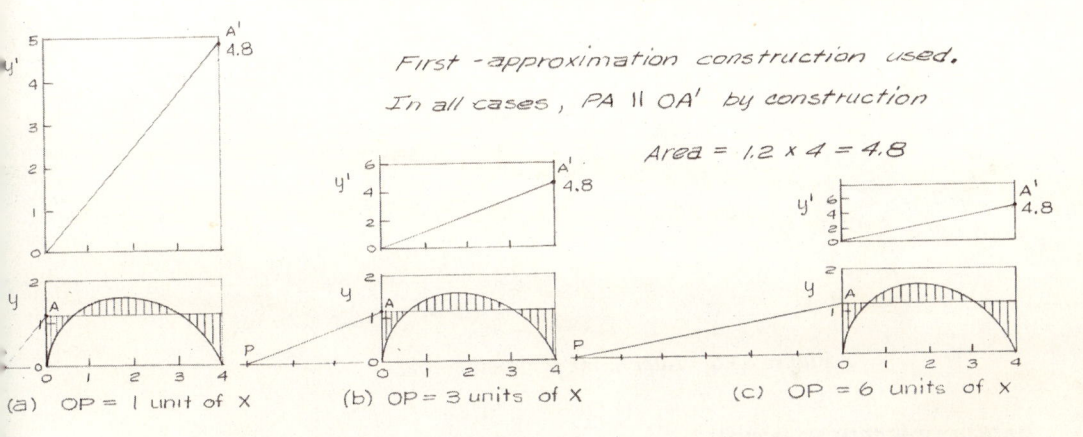

units of x

Figure 33.4 *The effect of the pole distance (OP) on the scale of the integral curve.*

Integrating a Closed Curve

16 The closed curve of Figure 33.2A is shown enlarged in Figure 33.5. The graphic integration of the area within the closed figure is complete. Note that the x axis was passed through the shape. The _____ (A) and _____ (B) portions of the outline were integrated separately.

upper, lower

---

17 The two curves were then _____ graphically to produce the total integral curve, (A) + (B).

added (summed)

---

18 The pole distance selected was 4 units of x. The modulus of the y axis, $m_y = \frac{3}{8}$ in. per unit. The modulus of the area axis $m_A = $ _____ inches per unit.

$\frac{3}{8} \div 4 = \frac{3}{32}$ in./unit

---

19 The total area within the closed curve is 21.6 square units. This is in reasonable agreement with the total of 21.15 square units achieved by the rough-approximation (_____-squares) method of 33.2A.

counting-(squares)

472 GRAPHIC MATHEMATICS/UNIT 6

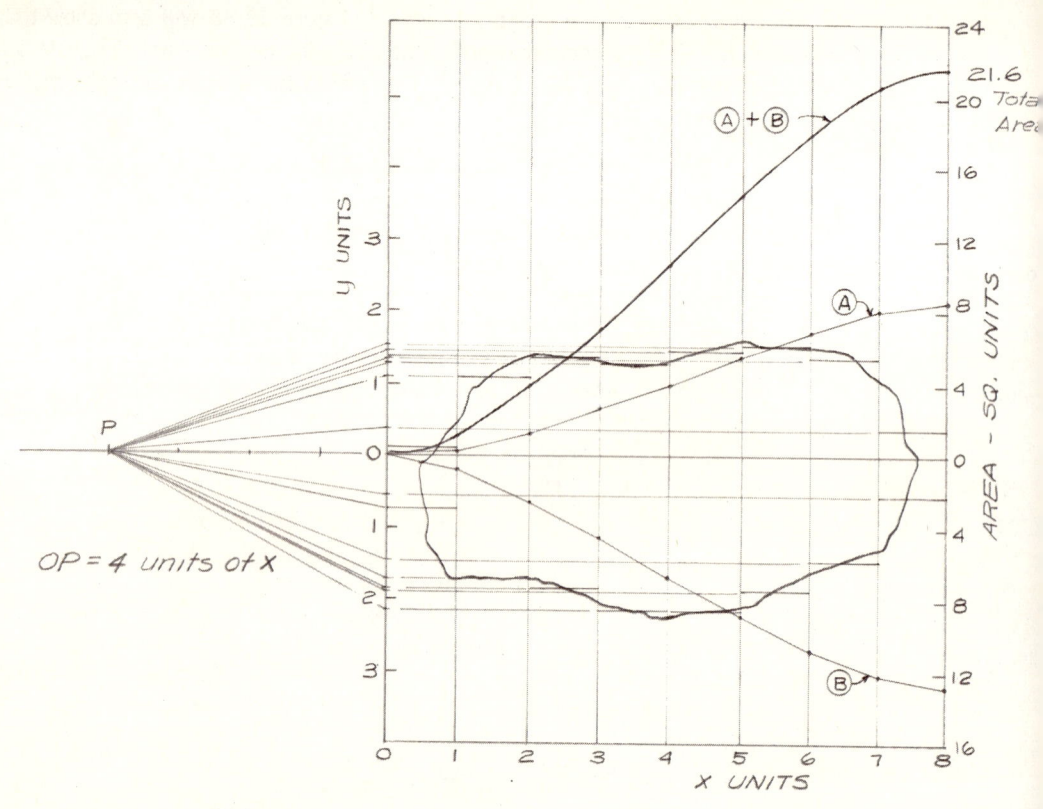

Figure 33.5   Integrating a closed curve.

GRAPHIC DIFFERENTIATION

At the beginning of this chapter we said that integration and differentiation are reversible processes. Although the concepts of the two processes seem completely different, you will see from the graphic analysis of differentiation that it is truly the reverse of the integration process.

The differential curve (often called the derivative curve) is a *curve of slopes*. Any point on a differential curve represents the numerical equivalent of the slope of the given curve at that particular value of the abscissa. A simple example is shown at Figure 33.6A. A curve is divided into three segments AB, BC, and CD. If we make an assumption that the tangent to the curve is parallel to the chord at the midpoint of an increment, Δx, then we find that the slopes of these tangents are 2, 1, and 0.5 respectively. Since the differential curve is a curve of slopes, we can plot these three points at appropriate places on the abscissa and draw the differential curve.

Pole-and-Ray Method

Figure 33.6B shows the basis for graphic differentiation by the pole-and-ray method. Two coordinate systems are set up, x vs y and x vs y'. A given curve is drawn at x vs y.

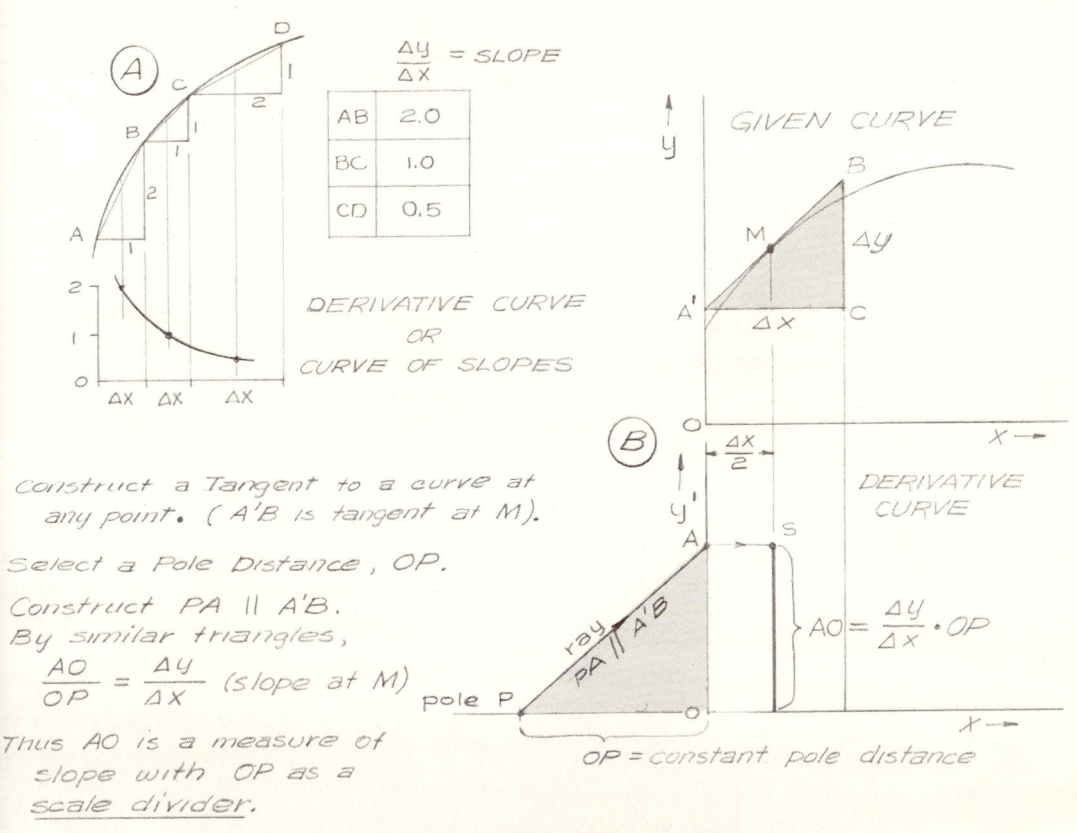

Figure 33.6 *The basis for graphic differentiation.*

	20	A line A'B is constructed tangent to the given curve at point _____.
M		
	21	A slope triangle (A'BC) is constructed with this tangent as the hypotenuse. The slope of the tangent is _____.
$\Delta y/\Delta x$ (BC/A'C)		
	22	Next we construct line PA parallel to line A'B (the tangent) from a preselected pole position, P, on the x axis (extended) of the derivative curve. Point A is the intersection of this line with the _____.
y' axis		
	23	We have constructed two similar triangles, A'BC and PAO. Thus we can say that $\dfrac{\Delta y}{\Delta x} =$ _____.
$\dfrac{\Delta y}{\Delta x} = \dfrac{AO}{OP}$		
	24	Since $\Delta y/\Delta x$ is the slope of the given curve at point M, distance AO is a numerical measure of the _____ of the tangent A'B when adjusted by the *scale divider*, OP.
slope		
	25	The pole distance OP will aid us in _____ the y' scale so that plotted ordinates are truly measures of the slope of the given curve at specific points.
calibrating		

474 GRAPHIC MATHEMATICS/UNIT 6

multiplier;
divider

26 In graphic integration, pole distance OP is a scale _____
In graphic differentiation, OP is a scale _____.

To complete the construction of 33.6B, the distance AO is transferred to an ordinate dropped from the point of tangency, M, by projecting point A horizontally to point S. Point S is one point on the derivative curve.

Drawing Tangents

It is virtually impossible to construct a true tangent to a given curve at a given point. Even the sharpest pencil point draws a line of finite thickness and a true tangent could be found only if the lines had zero thickness. There are, however, some methods that approximate the true tangent.

An interesting method is shown at 33.7A. A small mirror is held perpendicular to the paper as shown. The mirror is rotated about point A until the

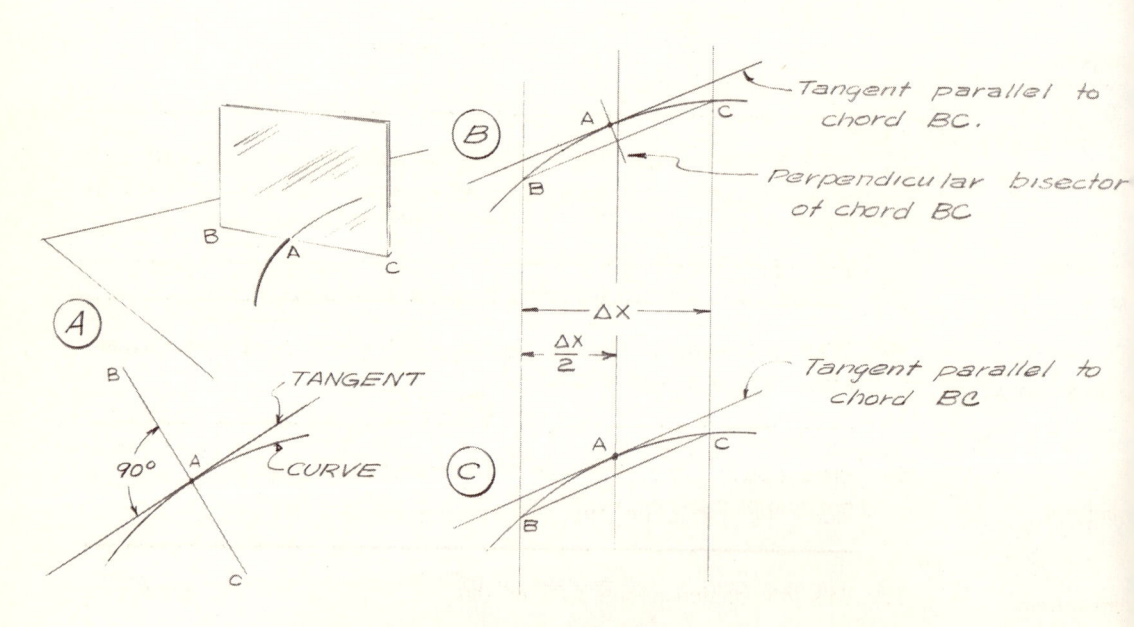

Figure 33.7   Methods for drawing a tangent to a curved line.

image of the curve in the mirror appears to be a smooth extension of the part of the curve in front of the mirror. When this position is found, line BC (the base of the mirror) is "perpendicular" to the curve at point A. If line BC is sketched using the mirror as a guide, the tangent can be drawn through point A perpendicular to BC.

The method shown at 33.7B is based upon the assumption that the tangent of a segment of a curve is parallel to the chord at a point A which is the intersection of the curve and the perpendicular bisector of the chord.

GRAPHIC CALCULUS 475

A still simpler method (Figure 33.7C) assumes that the tangent is parallel to the chord at the *midpoint* of the increment Δx. In the next few examples, we will use this last method.

The assumptions made in these last two methods increase in validity as Δx decreases in size.

Graphic Differentiation of a Given Curve

Figure 33.8 shows an example of a complete graphic differentiation of a given curve. The procedure is as follows:

---

27 **STEP 1:** Select values of Δx over the full range of x. When the given curve is sharply changing, it is wise to keep Δx small. As the curve flattens out, Δx can become larger.

STEP 2: Construct ordinates through the midpoints of each Δx segment. Draw the chord for each segment of the given curve bounded by a Δx value.

tangent;
midpoint

This chord is an approximation to the _____ to the curve at the _____ of each Δx segment.

---

28 **STEP 3:** Select a pole distance, OP. The effect of the distance OP on a derivative curve is opposite to its effect on an integral curve. Thus, as the length of OP increases, the height of the derivative curve _____ (*Choose one:* increases/decreases.)

increases

---

29 **STEP 4:** Construct a line parallel to chord 1 from point P to an intersection (point A) with the _____.

y' axis

---

30 Project point A horizontally to an intersection with the _____ ordinate of the first Δx segment.

midpoint

---

31 This intersection is one point on the _____ curve.

derivative

---

32 Repeat the above steps for all the rest of the Δx segments. Join the points obtained with a(n) _____.

smooth curve

---

33 Calibrate the y' scale using the pole distance OP as a scale divider. $OP = 2$ units of x.

On the y scale of Figure 33.8, 1 in. = 4 units. Therefore, on the y' scale, 1 in. = _____.

(1 in. =) 4/2 =
2 units

476 GRAPHIC MATHEMATICS/UNIT 6

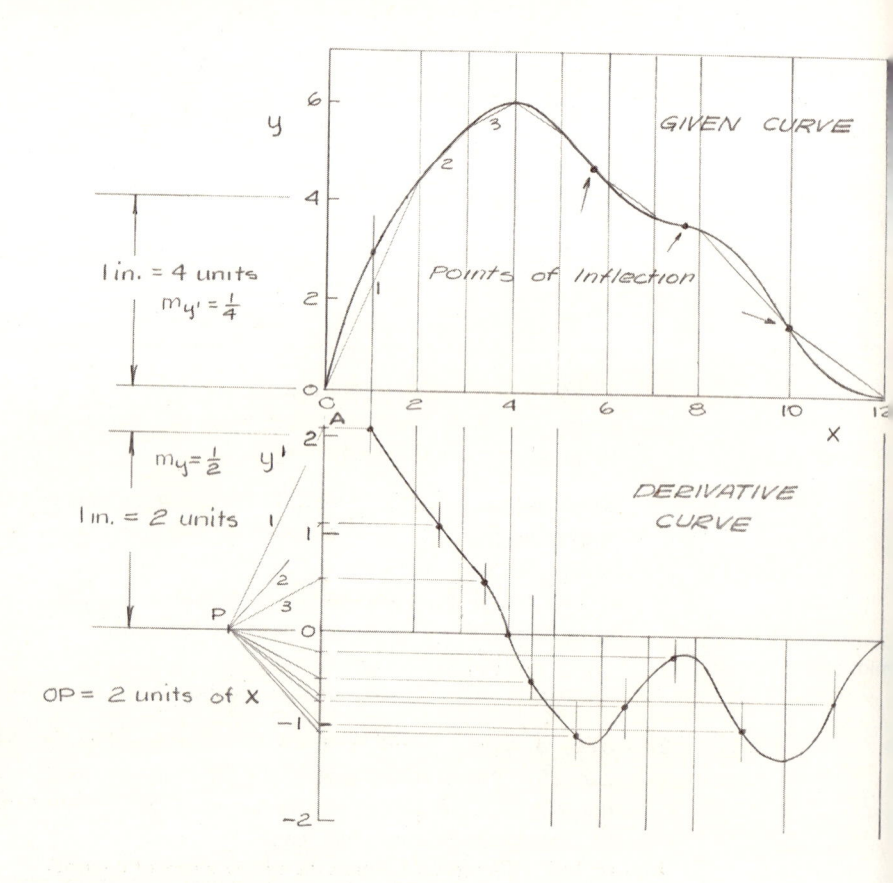

Figure 33.8 The pole-and-ray method of graphic differentiation.

An examination of 33.8 shows that the derivative curve is quite sensitive to slight changes in the given curve. Every point of inflection (change from an increasing slope to a decreasing slope or vice versa) results in a peak on the derivative curve. Figure 33.9 shows some of these effects.

Slope of a Curve

A number of curves are shown in Figure 33.9A along with a qualitative example of the meaning of the slope of the original curve. Understanding these effects is important to the effective use of graphic differentiation.

At Figure 33.9B a given curve is differentiated qualitatively just by reasoning out what happens to the slope of the curve at all points. Note the two points at which an instantaneous change occurs.

GRAPHIC CALCULUS 477

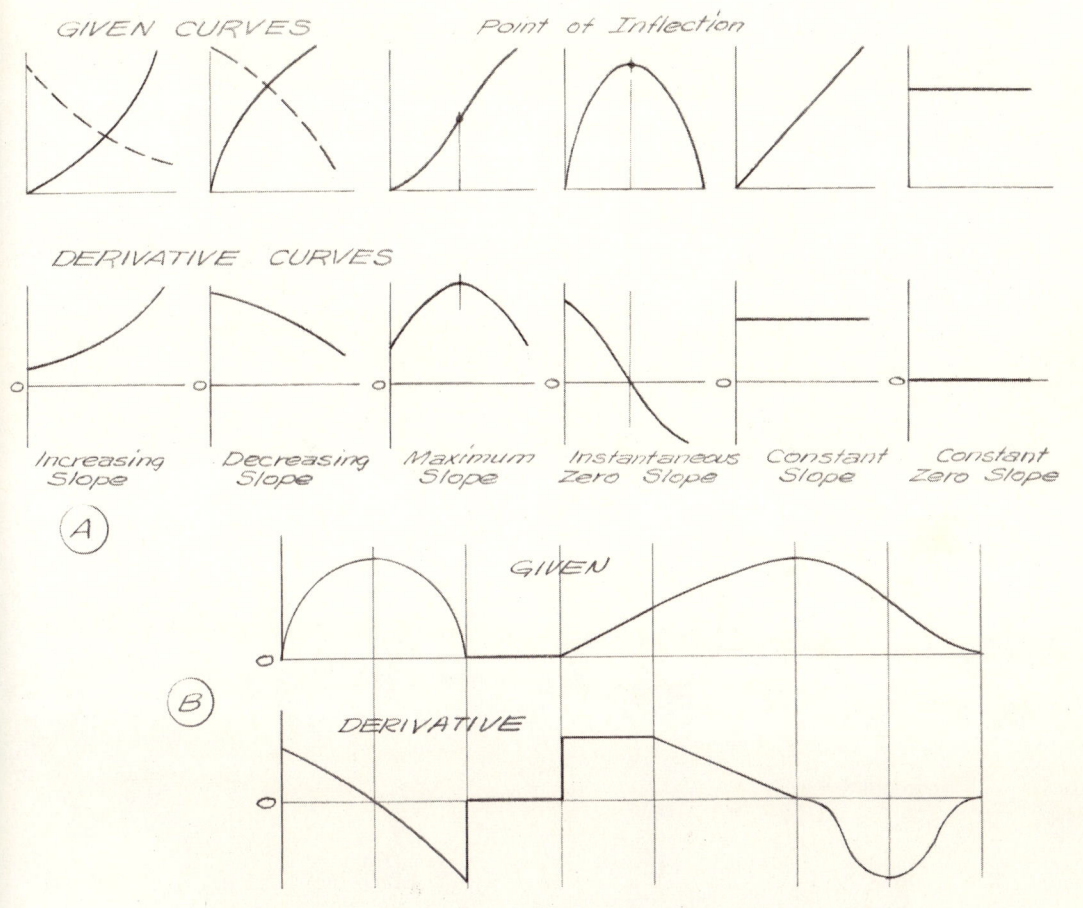

Figure 33.9 The graphic meaning of the slope of a curve.

■ **34** A time-velocity curve is given in Sketch 78. This represents the recorded velocity of an automobile during an 80-second test. The integral curve will show the distance traveled at any time in the 80-second test. It is the time-distance curve. The derivative curve will show the acceleration pattern over the time of the test, or the time-acceleration curve.

PROBLEM: Find the time-distance and time-acceleration curves by graphic techniques.

Both of these can be found by freehand techniques with a fair degree of accuracy. The hardest part of freehand construction is transferring parallel lines. This can be aided by setting a straightedge parallel to the given line and then sliding it carefully to the new position maintaining the original setting.

The peaks, constant slopes, and zero slopes of the derivative curve can be found intuitively. Use graphic construction to establish the magnitudes of the known elements of the curve.

A complete solution is given.

GRAPHICS AND THE COMPUTER

This is the end of *Programmed Graphics*. But for you, it should be the beginning of your use of graphics as a language of communication. No matter what your role or interests are in the technical professions, you should continue to develop proficiency in the use of graphics to communicate technical information. Proficiency comes with practice.

You should recognize now that the central problem of technical drawing is knowing where to place points on a piece of paper. We started with a dot-to-dot drawing and now end with an emphasis on the same idea—prompted by consideration of the role of the computer in the future of graphic communications.

The first picture was drawn by a computer about ten years ago. The last five years has seen tremendous activity in the development of means for displaying computer-generated drawings. The future of computer graphics holds a promise of relieving man from all of the tedium of precise graphic construction while giving him complete flexibility in the use of graphics for design, analysis, synthesis, and communication.

The aerospace industries have been the principal developers and users of computer graphics. Their range of usage includes the complete analysis and design of a new airframe right down to an accurate scale drawing suitable for manufacturing purposes, as well as an animated movie of the view the pilot will see as this projected new plane lands and takes off from any specified airport or from the deck of a carrier. Using computers to translate the geometry of the new airplane into a series of successive motion picture frames, we can produce such movies long before the airplane prototype is flying.

A computer is a machine which accepts input data, performs computations on it according to machine programs or programs specified by human users,

480 GRAPHICS AND THE COMPUTER

and then submits output data in a desired form. In the case of drawings as computer output, the input data is coordinate information, xyz for three dimensional objects and xy for two-dimensional objects. The output information can be a table of coordinate numbers from a high-speed printer or a teletypewriter, drawings in pencil or ink from an x-y plotter, or electronically produced lines on the face of a cathode-ray tube (CRT).

The figure below shows an early attempt by the author to generate a drawing on a computer. It is a reproduction of the objects of Figure 28.3, two intersecting cylinders and four cutting planes, taken from a different point of view. It consists of

(A) A reproduction of a portion of the printed computer output. (A total of 64 points was used to define the objects.)

(B) The x and y output values plotted on a two-dimensional coordinate system.

(C) Sixty-four points connected in the correct sequence to yield a perspective view of the intersecting cylinders.

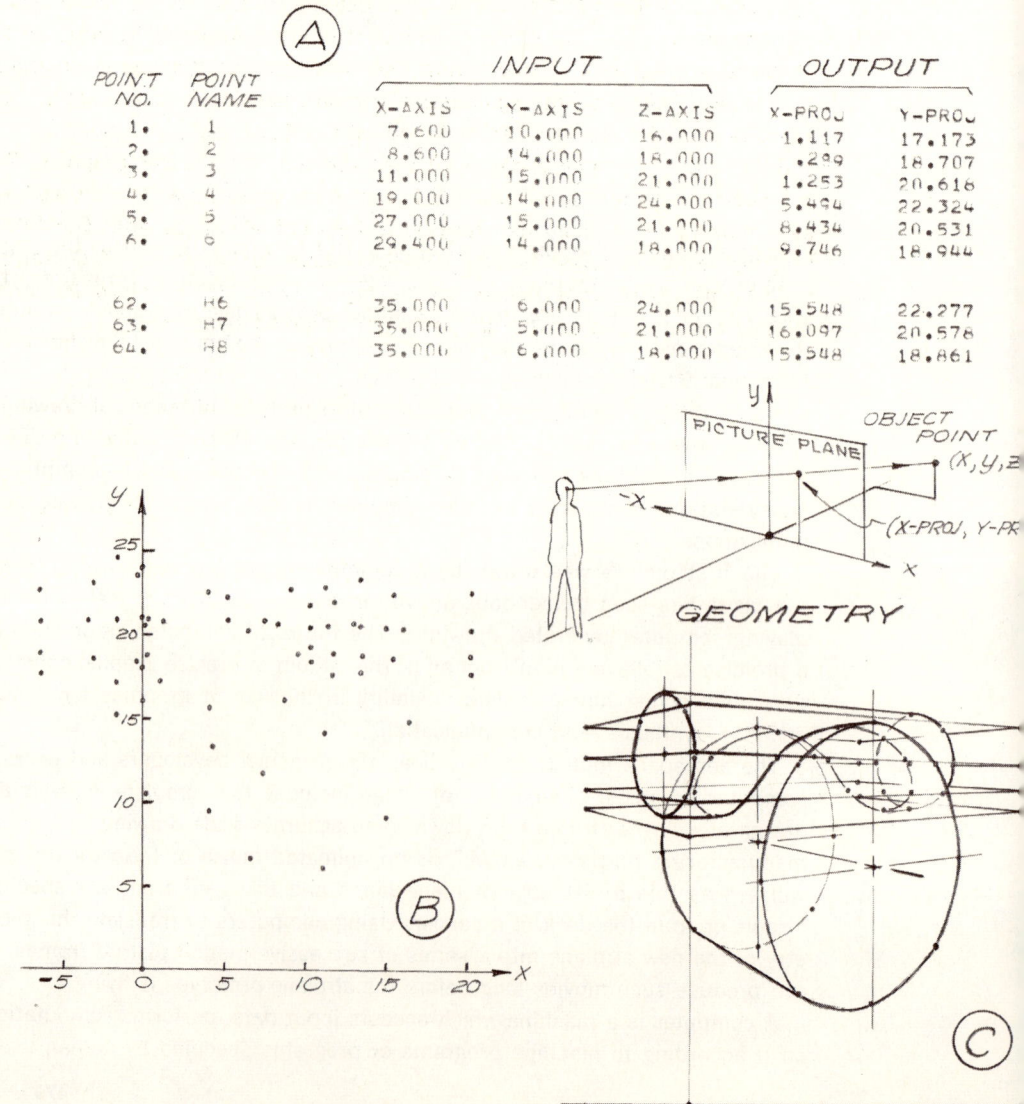

Ⓐ

POINT NO.	POINT NAME	INPUT X-AXIS	Y-AXIS	Z-AXIS	OUTPUT X-PROJ	Y-PROJ
1.	1	7.600	10.000	16.000	1.117	17.173
2.	2	8.600	14.000	18.000	.299	18.707
3.	3	11.000	15.000	21.000	1.253	20.618
4.	4	19.000	14.000	24.000	5.494	22.324
5.	5	27.000	15.000	21.000	8.434	20.531
6.	6	29.400	14.000	18.000	9.746	18.944
62.	H6	35.000	6.000	24.000	15.548	22.277
63.	H7	35.000	5.000	21.000	16.097	20.578
64.	H8	35.000	6.000	18.000	15.548	18.861

Ⓑ

Ⓒ

GEOMETRY

A computer program was written asking the computer to accept input data and then to compute, according to the geometry of projected perspective, the x and y values of the intersection of the observer's line of sight to each point with a picture plane. If we wished to draw a different perspective view of the objects, we would have only to change the observer's distance from the picture plane and the height of the horizon line in the computer program and run it again to obtain a new set of values for x and y. Coupling the computer output directly to an x-y plotter or a CRT would produce a drawing rather than a table of coordinate values.

Displaying a drawing on a computer terminal requires and generates large amounts of data. Unless these data are structured carefully, the computations for even an average drawing job can swamp the data storage capacity of the largest computer. Bringing together graphics and the computer will call for new ways to look at technical drawing and its geometry. *Programmed Graphics* has provided a fundamental view of the geometry of technical drawing as a computer must be made to see it. The computer output shown above should be familiar to you—the same technique has been stressed in most of the chapters of this book. The program that produced the drawing is universal. If it is resubmitted to the computer along with new input data, another perspective view of any object can be produced at will.

This is but a simple example of the potential of computer graphics. The computer is only a tool. However, it is one of the most powerful tools we have ever known and we must be prepared to take full advantage of it. Our future thinking must be tuned to new concepts about the structure of a drawing and the geometric data needed to display it. Computer graphics will bring graphic communications into usages never dreamed of before.

PART THREE *sketch solutions*

The following pages contain solutions for all sketch exercises which call for a solution. You are to evaluate your own work on the sketches for which printed solutions are not provided. In most instances, these solution sketches are typical in that there are many possible solutions to any drawing problem. Variations between your solution and the given one will be minor if you are following the text instructions carefully and are correctly applying the principle under study. You must judge whether your solution is equivalent.

The solution sketches are arranged numerically. Sketch exercise numbers (circled) are positioned on the outside edges of the pages for convenience in locating solutions. Each solution has a reference to chapter and frame number.

SCALE

All solutions are presented at a scale smaller than your solution, being three-quarters (75 per cent) of original size. Part of learning to draw is gaining an ability to recognize the difference between proportion and scale (see Chapter 6). Though of reduced size, all solution sketches are the same proportion as your original. When inch measurements are important to a solution, an inch scale is given.

TECHNIQUE

Most of the sketches have been done by freehand techniques. You are urged to use the same techniques in developing your solutions (see Chapter 4). Where increased precision was required, a straightedge, a compass, and a circle template were used. Heavy lines and lettering were done with an HB pencil, thin lines with an H pencil.

SKETCH SOLUTIONS **485**

No solution given. The accuracy of your drawing should be obvious. Ref: Chapter 1 / Frame 1   (1)

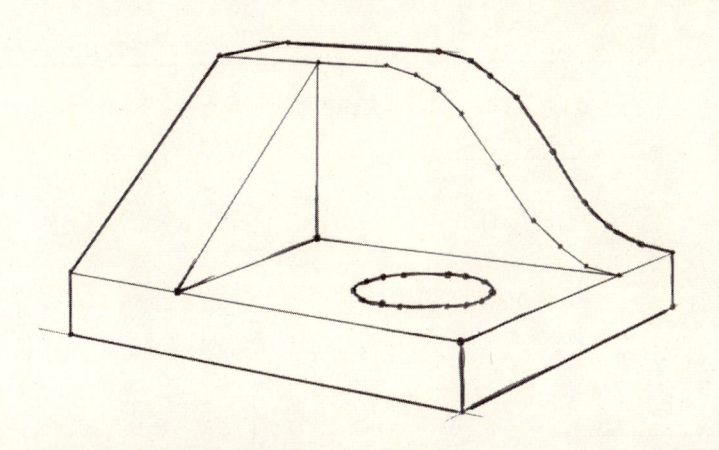

Ref: Ch. 1 / Fr. 13   (2)

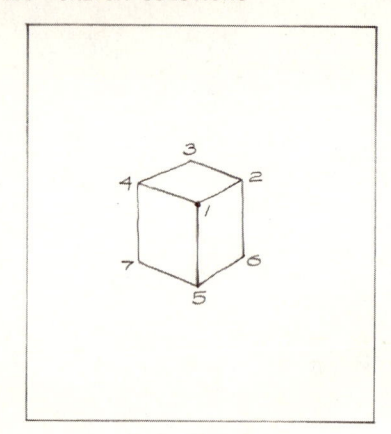

486 SKETCH SOLUTIONS

③ PICTORIAL SKETCH OF A CUBE — Ch. 1/36

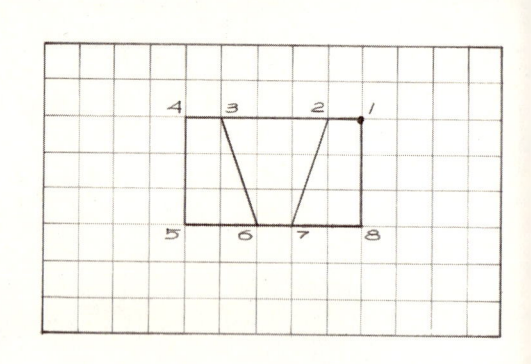

④ ORTHOGRAPHIC VIEW OF THE BLOCK — Ch. 1/55

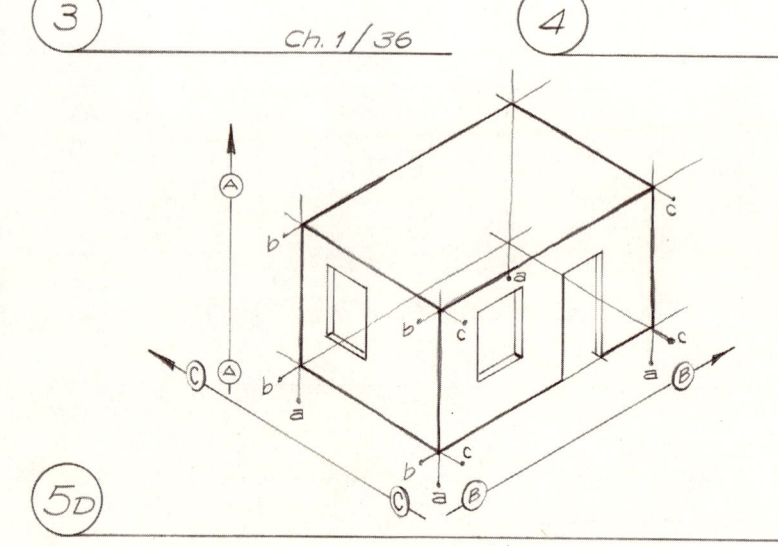

⑤D Ch. 2/24

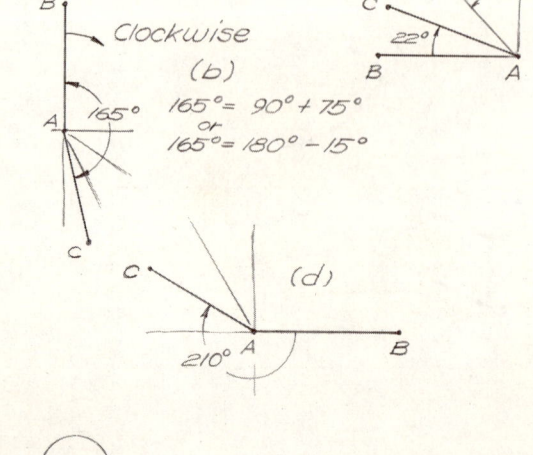

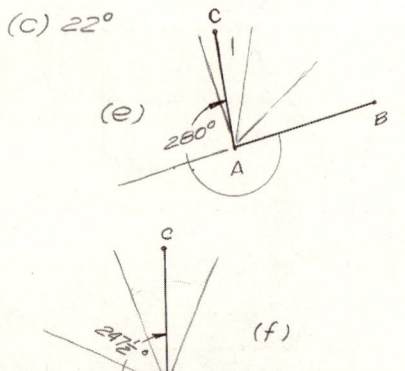

⑥D Ch. 2/37

(b) Clockwise 165° = 90° + 75° or 165° = 180° − 15°

(c) 22°

(d) 210°

(e) 280°

(f) 247½°

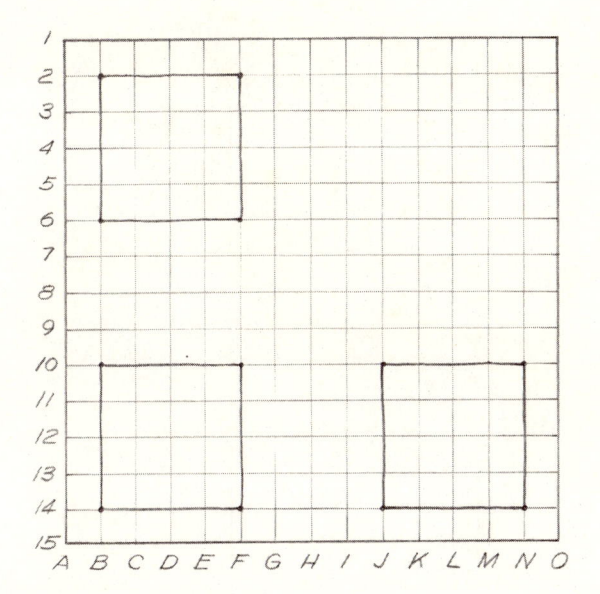

SKETCH SOLUTIONS 487

Ch. 3/1 ⑫

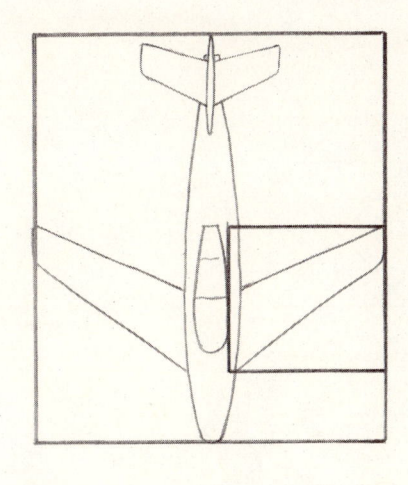

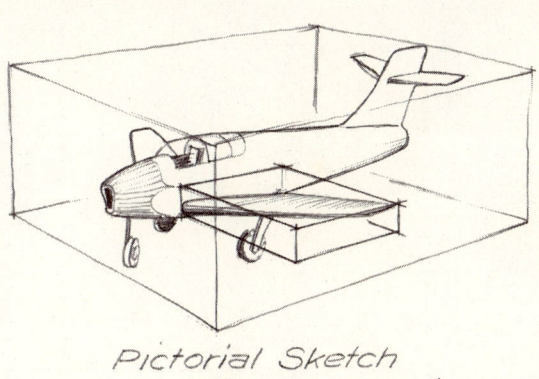

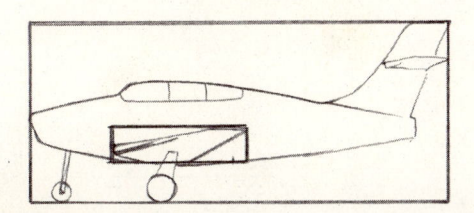

Pictorial Sketch
(Perspective)

Ch. 3/9 ⑬

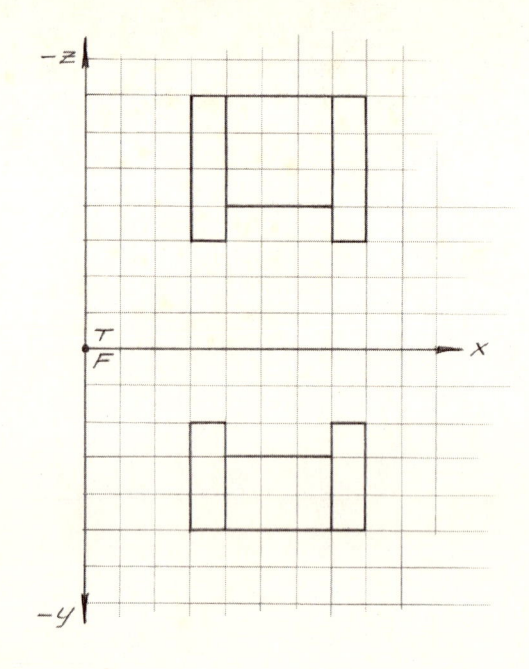

488 SKETCH SOLUTIONS

(14) Ch. 3/42

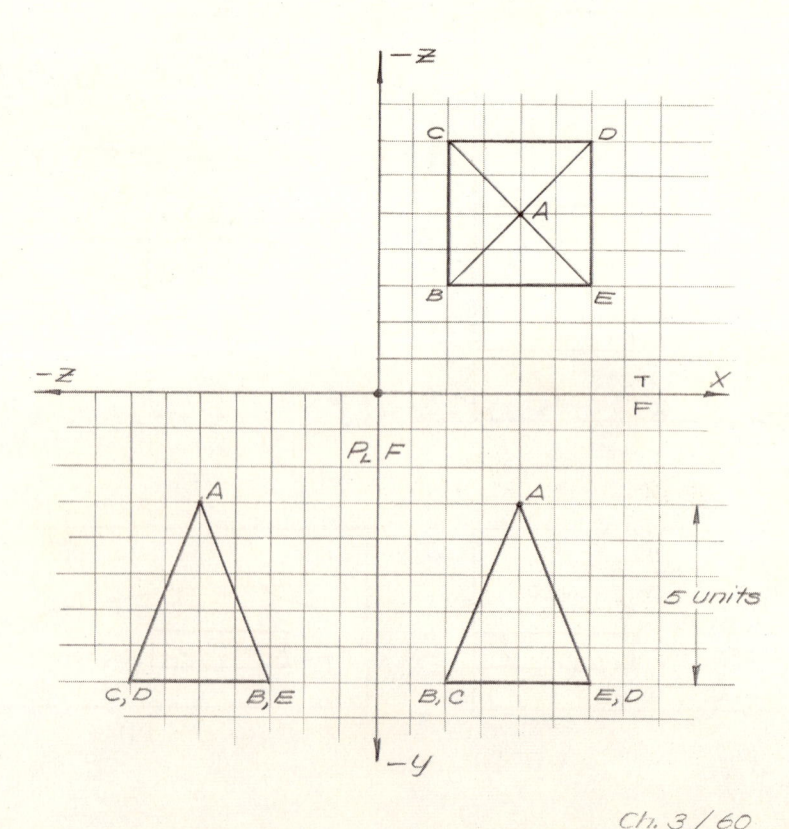

(15) Ch. 3/60

SKETCH SOLUTIONS **489**

REAR

TOP

$H=2$

P_L *FRONT* P_R

$W=5$ $D=3$

One Typical Solution

BOTTOM

Ch. 3/72

(16)

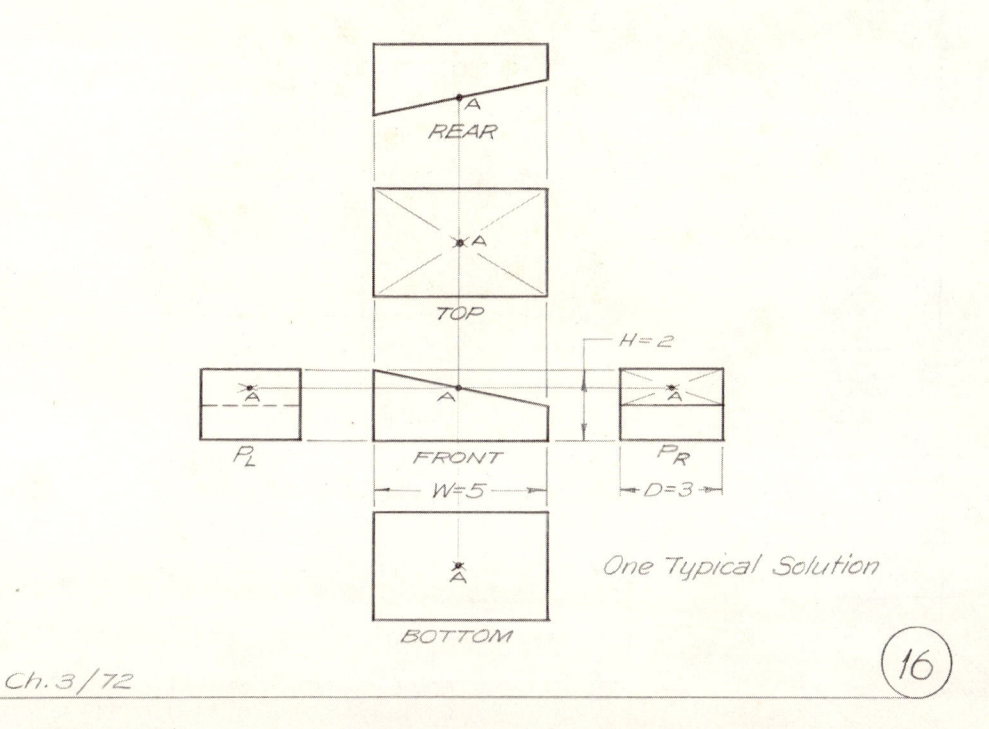

Correct Line Precedences and Hidden-Line Junctures were used in this Solution.

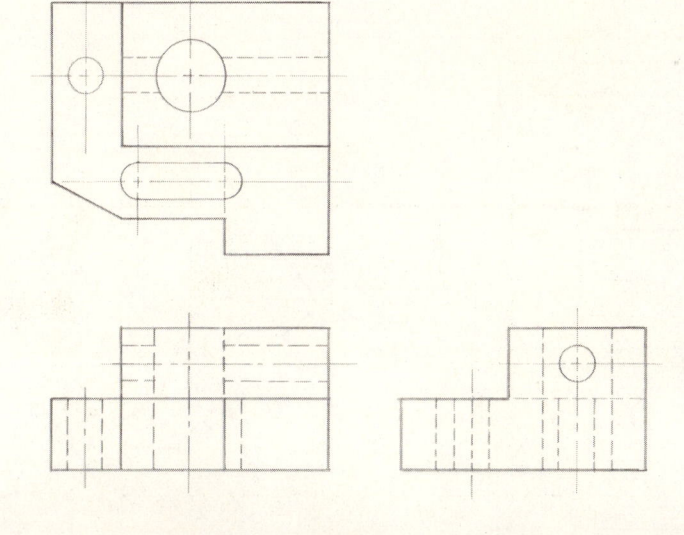

Ch. 4/24

(17)

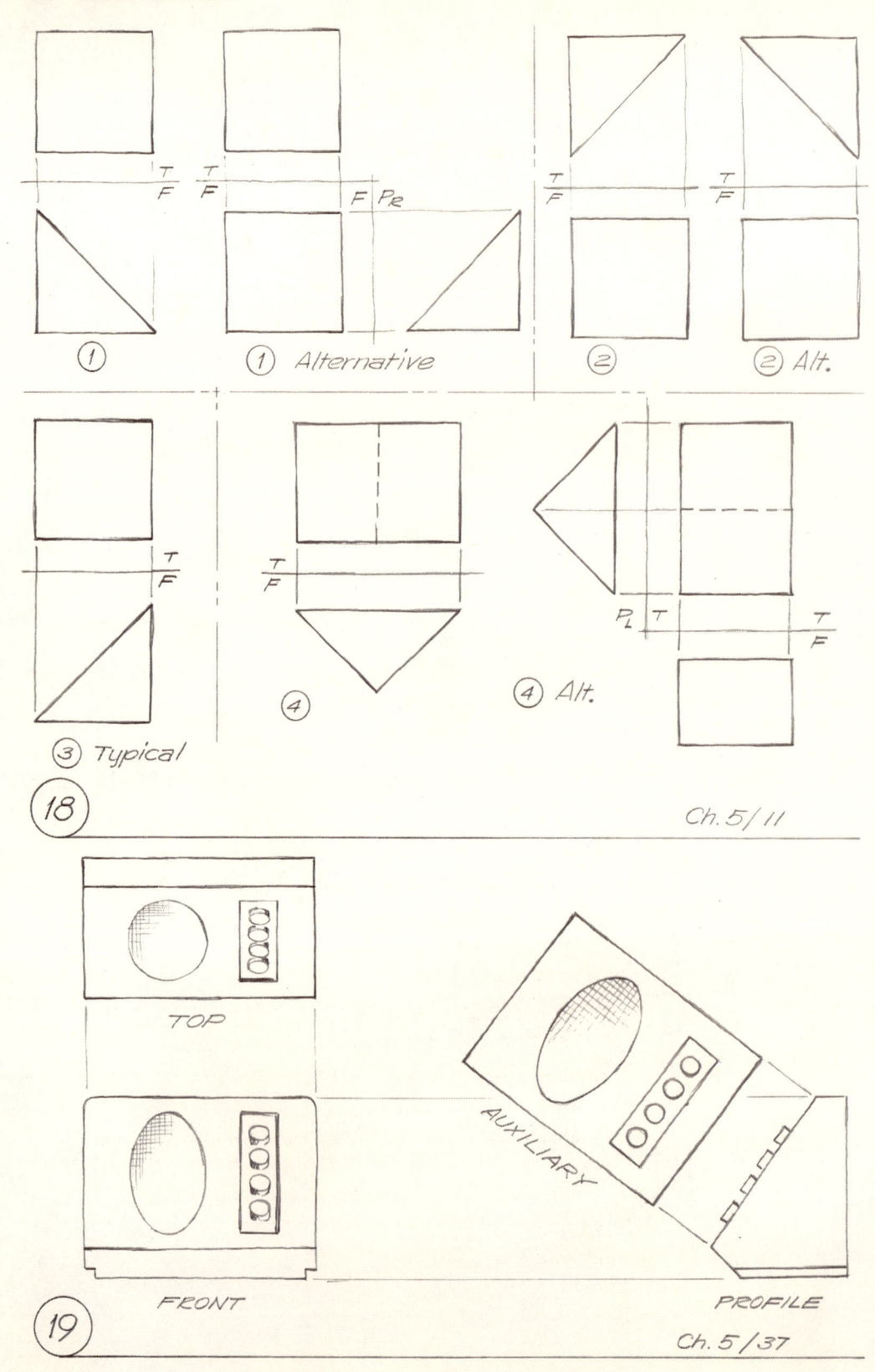

490 SKETCH SOLUTIONS

(1)

(1) Alternative

(2)

(2) Alt.

(3) Typical

(4)

(4) Alt.

(18) Ch. 5/11

TOP

FRONT

AUXILIARY

PROFILE

(19) Ch. 5/37

SKETCH SOLUTIONS **491**

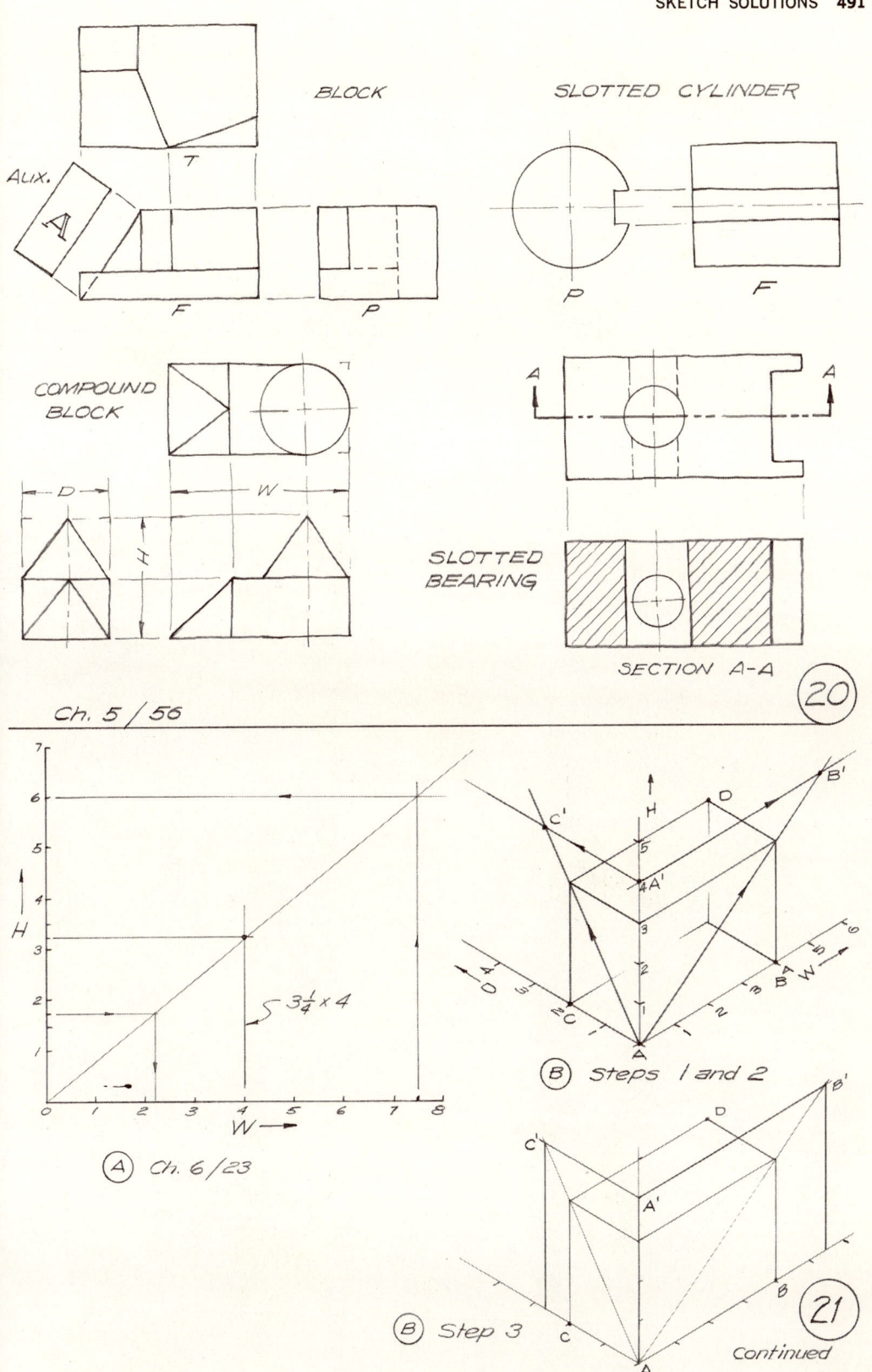

Ch. 5 / 56 ⟨20⟩

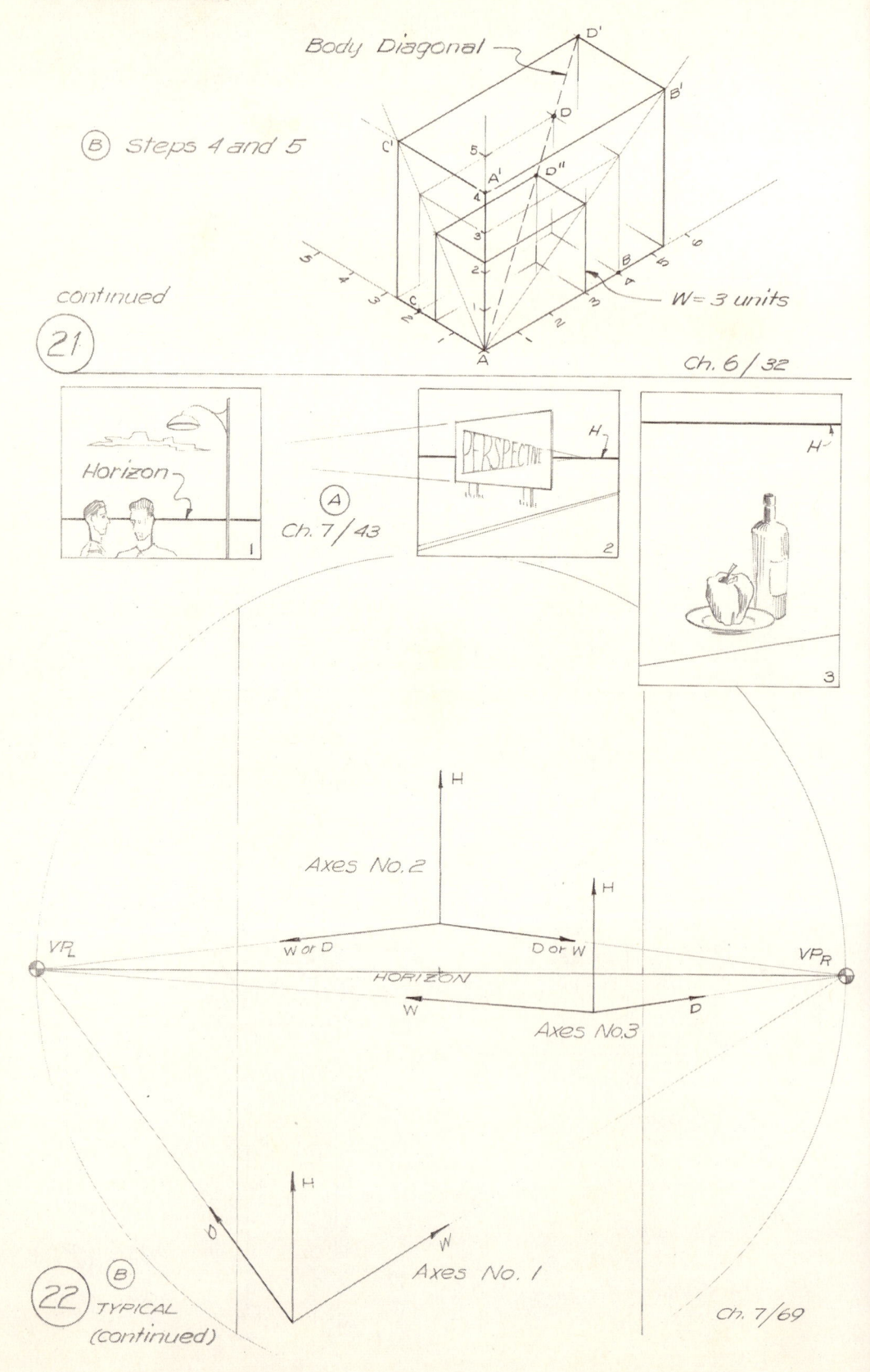

492 SKETCH SOLUTIONS

Ⓑ *Steps 4 and 5*

Body Diagonal

continued

㉑

W = 3 units

Ch. 6/32

Ⓐ Ch. 7/43

Horizon

Axes No. 2

Axes No. 3

VP_L W or D   D or W VP_R

HORIZON

W   D

Axes No. 1

㉒ Ⓑ *TYPICAL (continued)*

Ch. 7/69

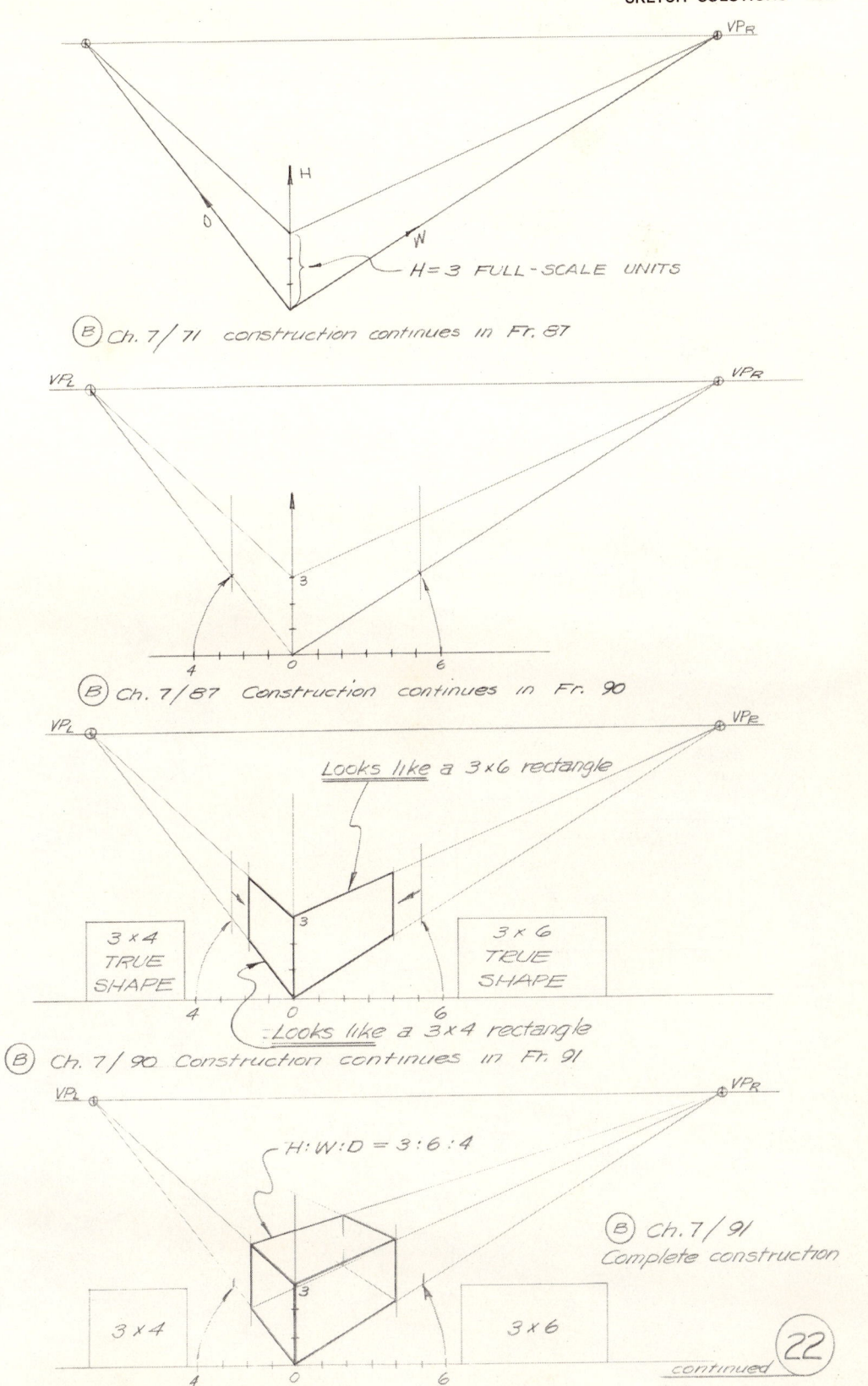

B Ch. 7/71 construction continues in Fr. 87

H = 3 FULL-SCALE UNITS

B Ch. 7/87 Construction continues in Fr. 90

Looks like a 3×6 rectangle

3×4 TRUE SHAPE

3×6 TRUE SHAPE

Looks like a 3×4 rectangle

B Ch. 7/90 Construction continues in Fr. 91

H:W:D = 3:6:4

3×4

3×6

B Ch. 7/91 Complete construction

continued 22

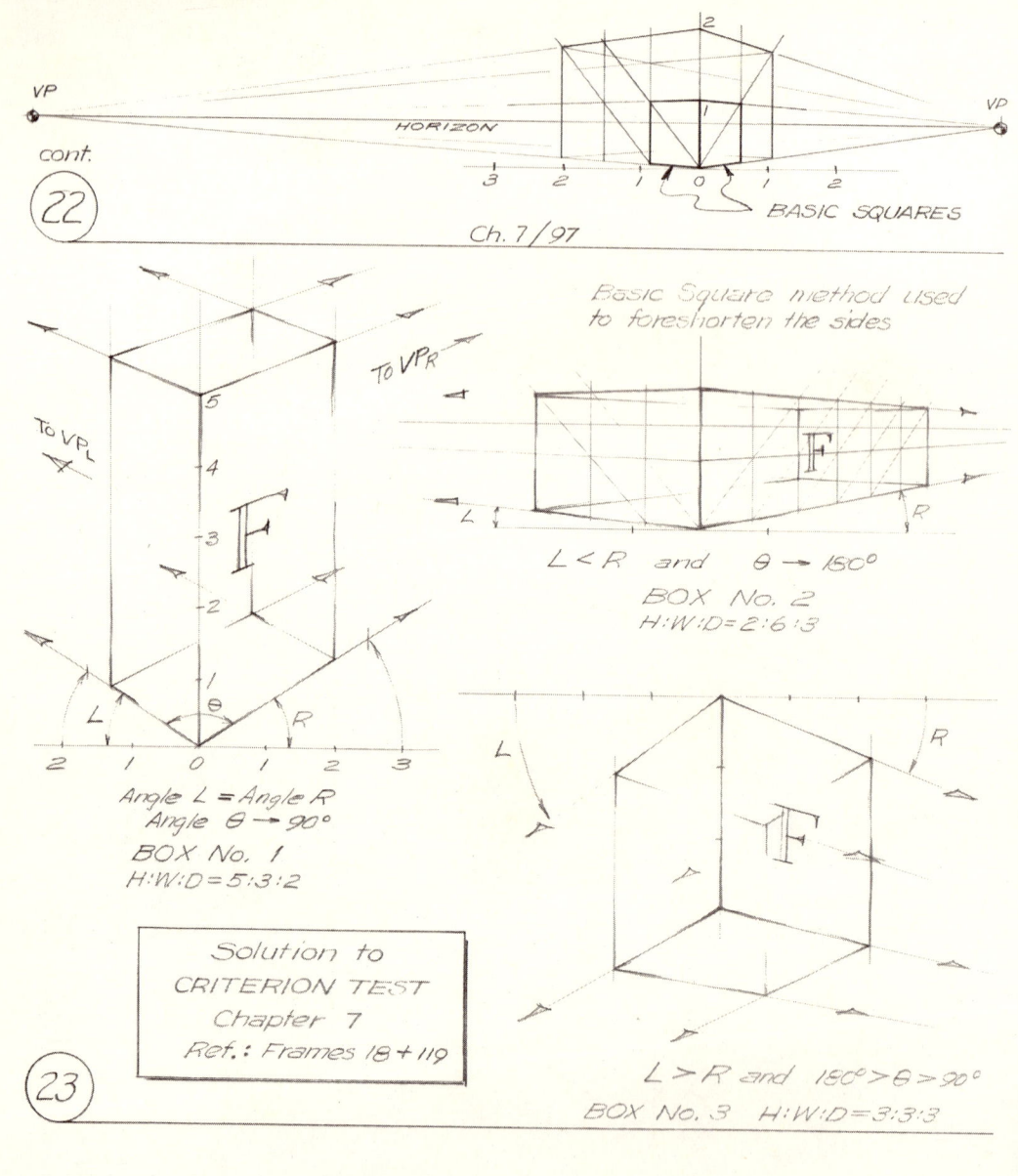

494 SKETCH SOLUTIONS

(22) cont.

Ch. 7/97

BASIC SQUARES

Basic Square method used to foreshorten the sides

$L < R$ and $\theta \rightarrow 150°$
BOX No. 2
$H:W:D = 2:6:3$

Angle L = Angle R
Angle $\theta \rightarrow 90°$
BOX No. 1
$H:W:D = 5:3:2$

$L > R$ and $180° > \theta > 90°$
BOX No. 3 $H:W:D = 3:3:3$

(23) Solution to CRITERION TEST Chapter 7
Ref.: Frames 18 + 119

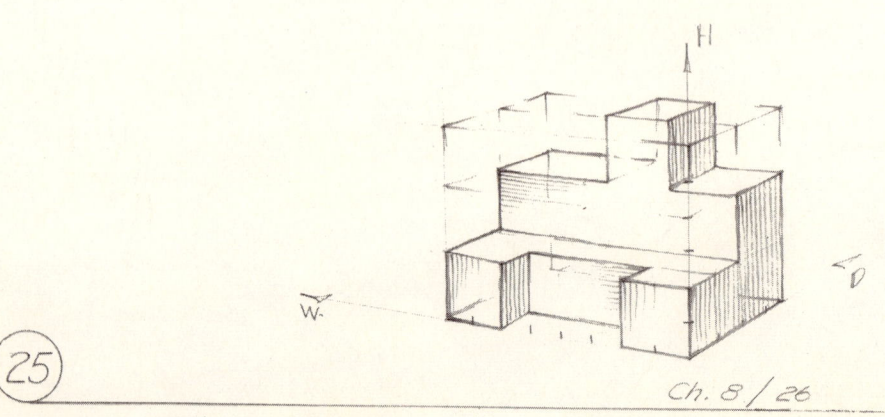

(25)

Ch. 8/26

SKETCH SOLUTIONS **495**

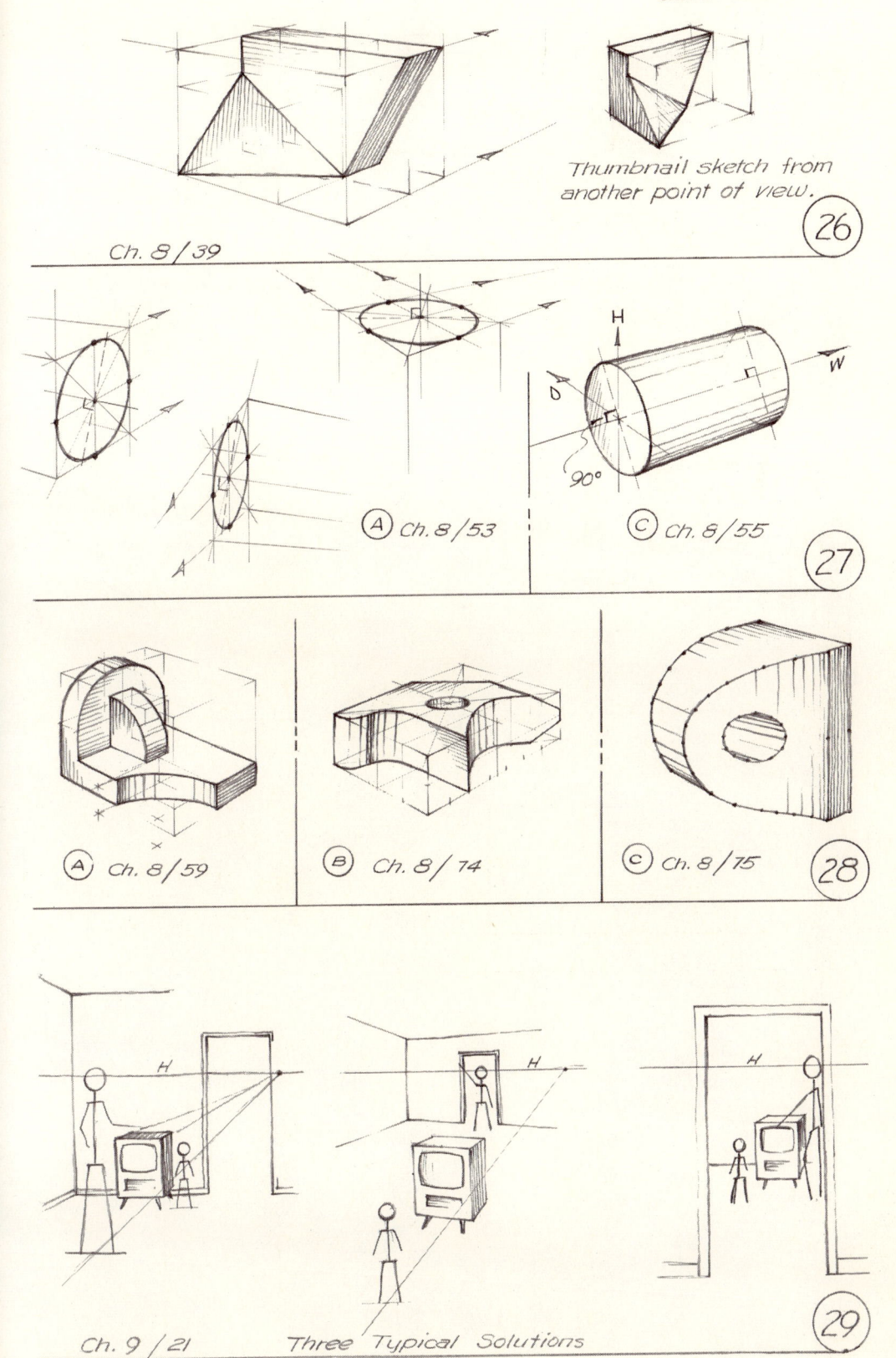

Ch. 8/39

Thumbnail sketch from another point of view.

(26)

(A) Ch. 8/53 (C) Ch. 8/55

(27)

(A) Ch. 8/59 (B) Ch. 8/74 (C) Ch. 8/75

(28)

Ch. 9/21 Three Typical Solutions

(29)

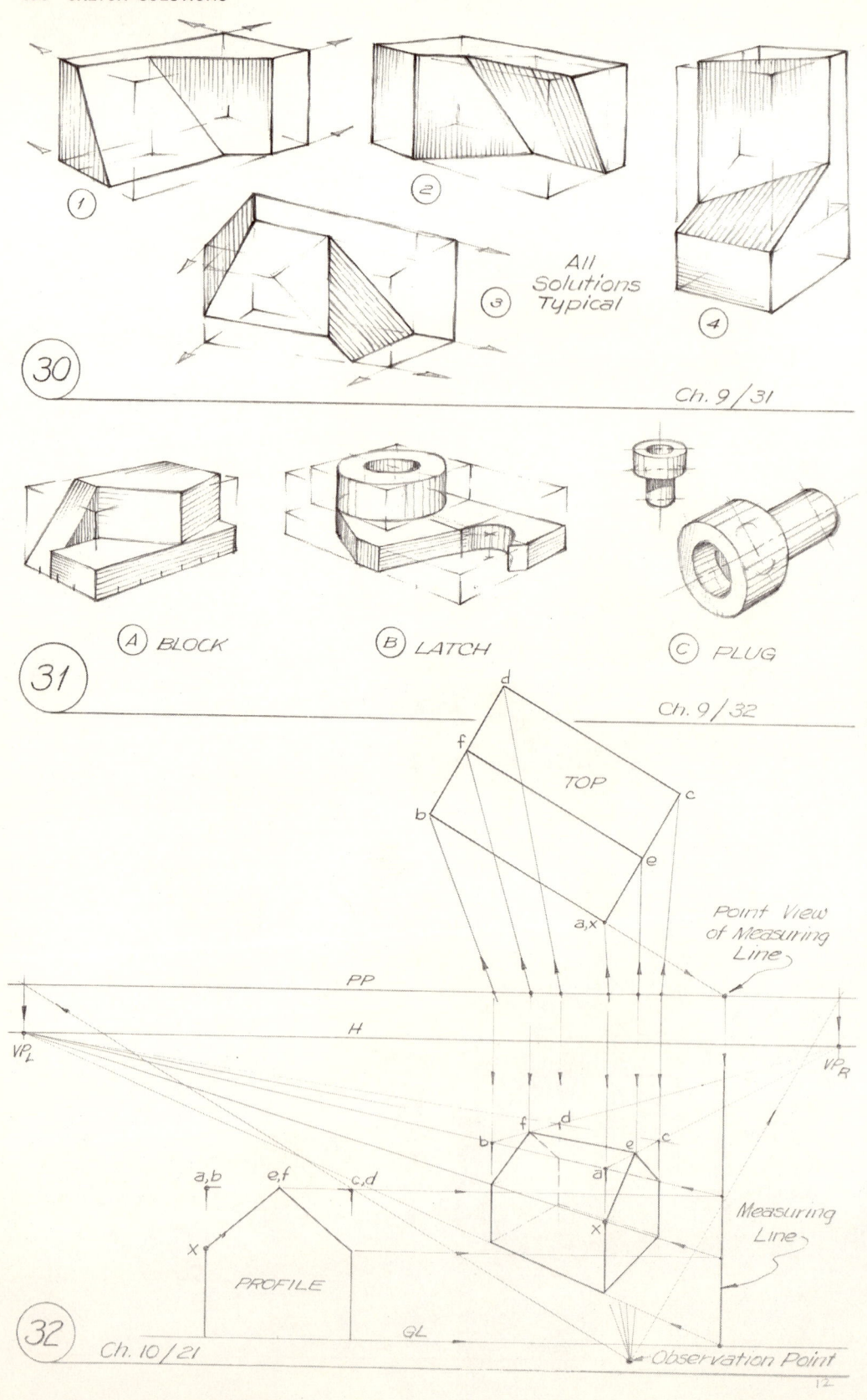

496 SKETCH SOLUTIONS

30

① ② ③ ④ All Solutions Typical

Ch. 9/31

31

Ⓐ BLOCK Ⓑ LATCH Ⓒ PLUG

Ch. 9/32

32

Ch. 10/21

TOP — d, f, b, c, e, a,x

Point View of Measuring Line

PP
H
VP_L
VP_R

PROFILE

GL

Measuring Line

Observation Point

SKETCH SOLUTIONS **497**

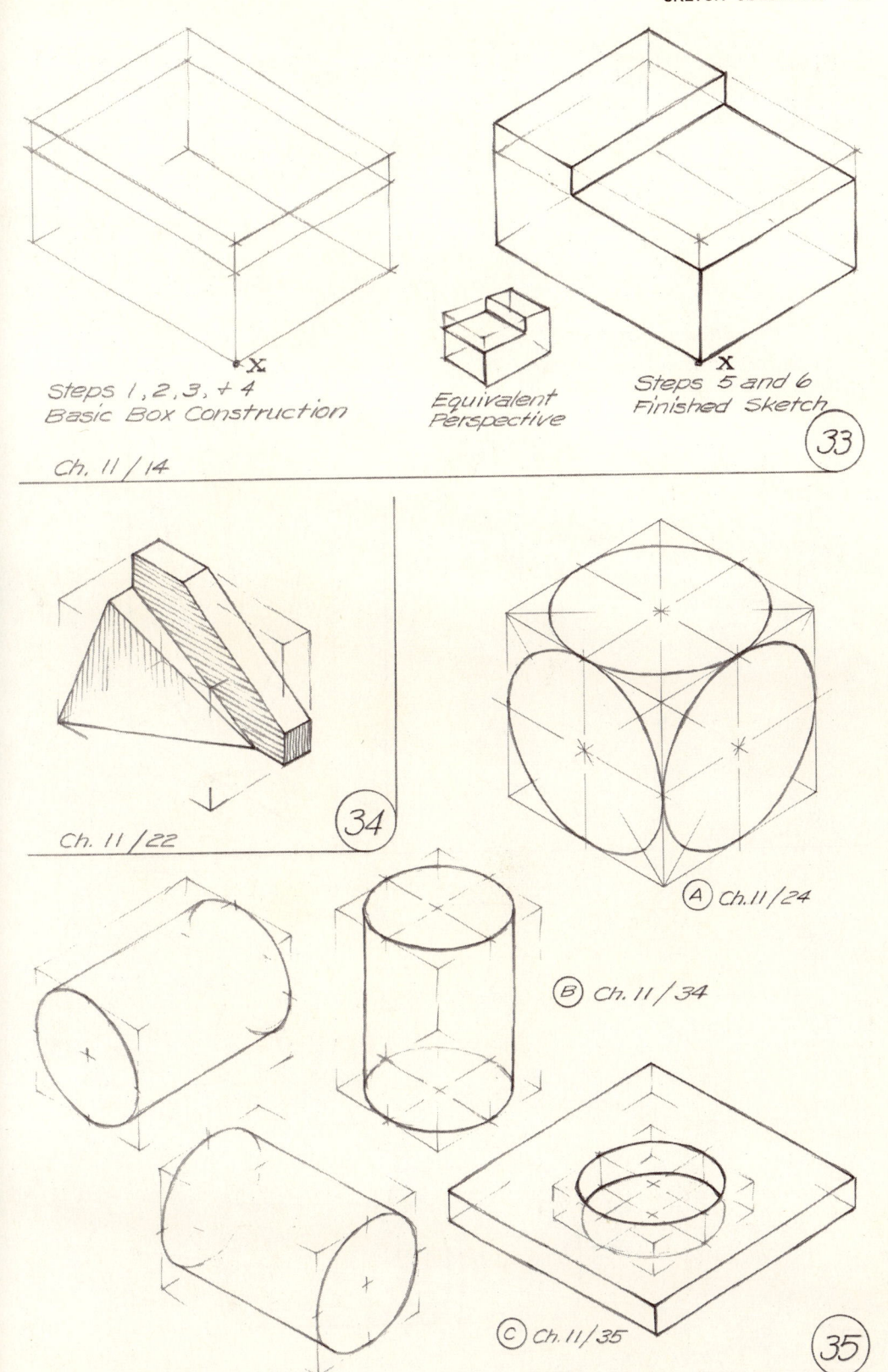

Steps 1, 2, 3, + 4
Basic Box Construction

Equivalent Perspective

Steps 5 and 6
Finished Sketch

(33)

Ch. 11/14

Ch. 11/22 (34)

Ⓐ Ch. 11/24

Ⓑ Ch. 11/34

Ⓒ Ch. 11/35 (35)

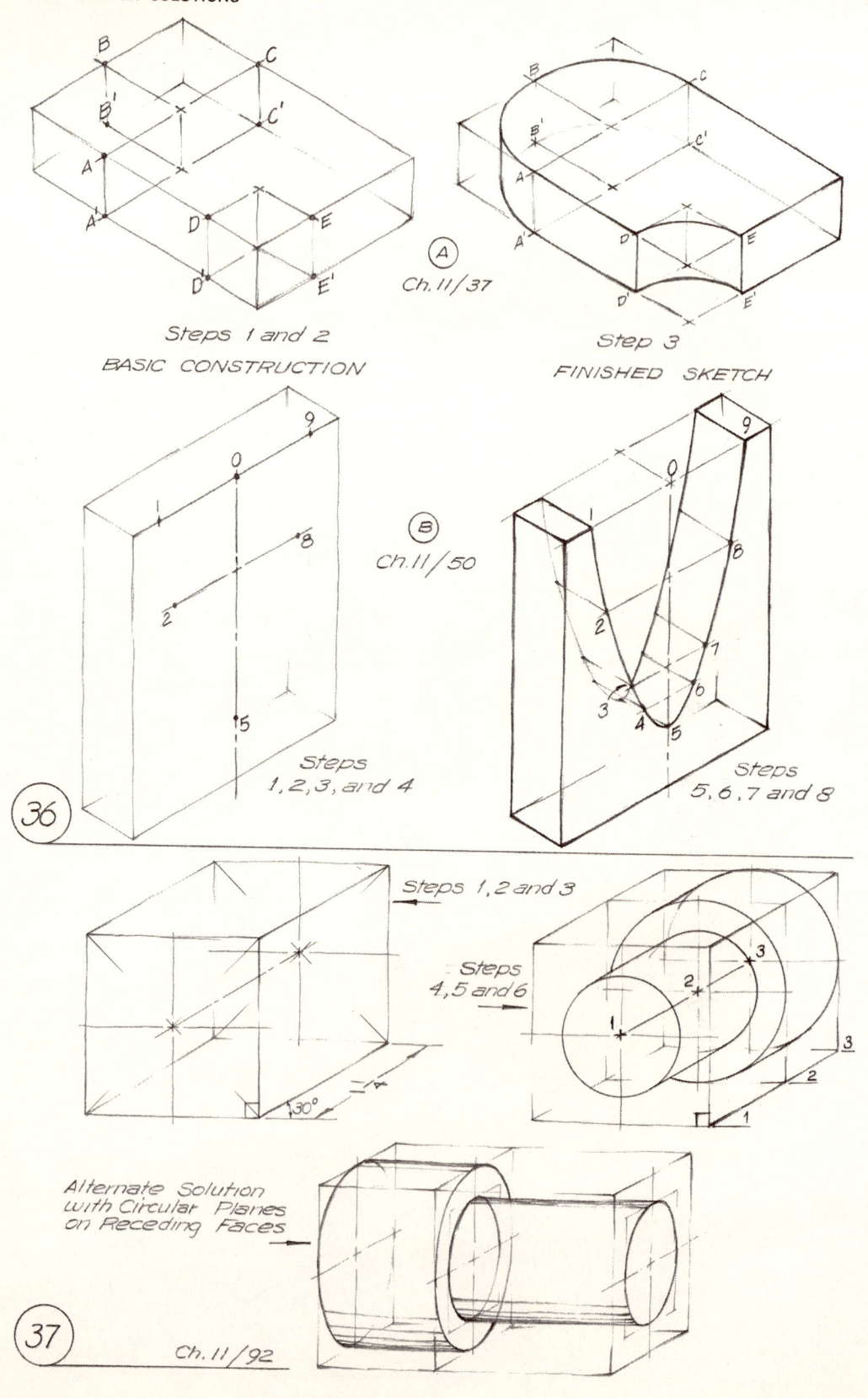

498 SKETCH SOLUTIONS

Ⓐ Ch. 11/37

Steps 1 and 2
BASIC CONSTRUCTION

Step 3
FINISHED SKETCH

Ⓑ Ch. 11/50

Steps 1, 2, 3, and 4

Steps 5, 6, 7 and 8

㊱

Steps 1, 2 and 3

Steps 4, 5 and 6

Alternate Solution with Circular Planes on Receding Faces

㊲ Ch. 11/92

SKETCH SOLUTIONS 499

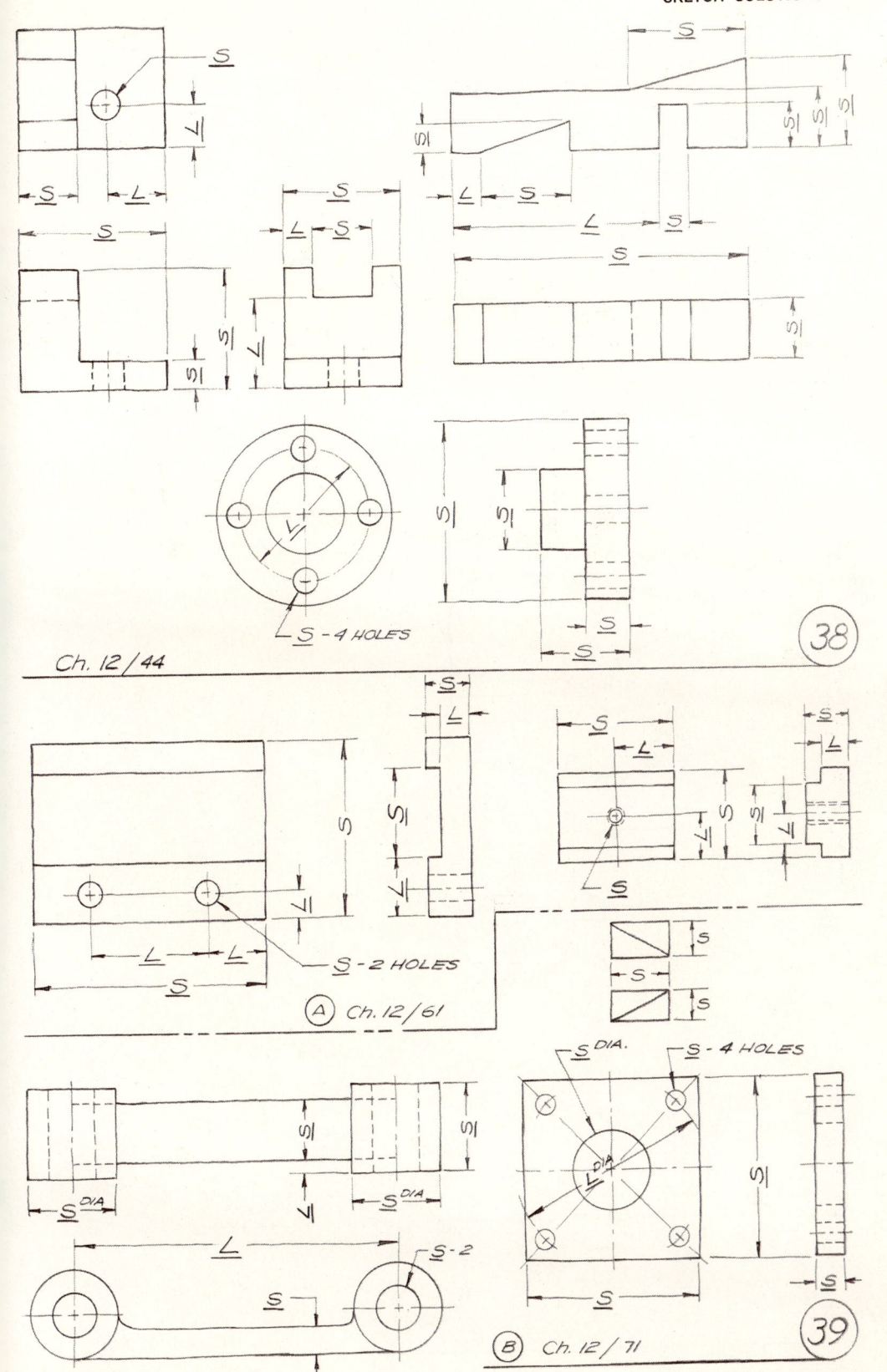

Ch. 12/44 (38)

(A) Ch. 12/61

(B) Ch. 12/71 (39)

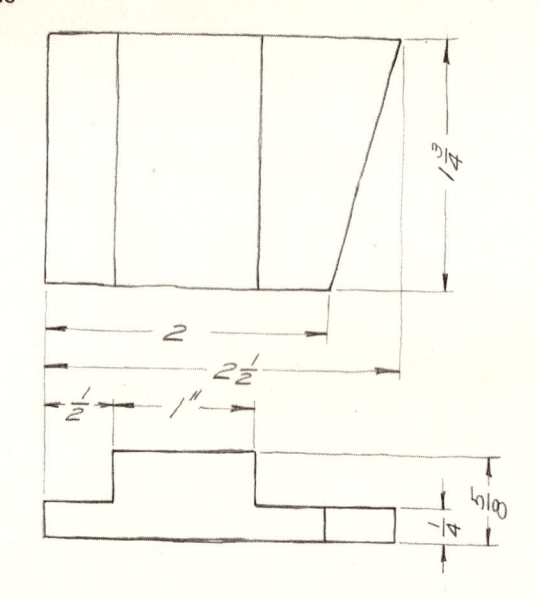

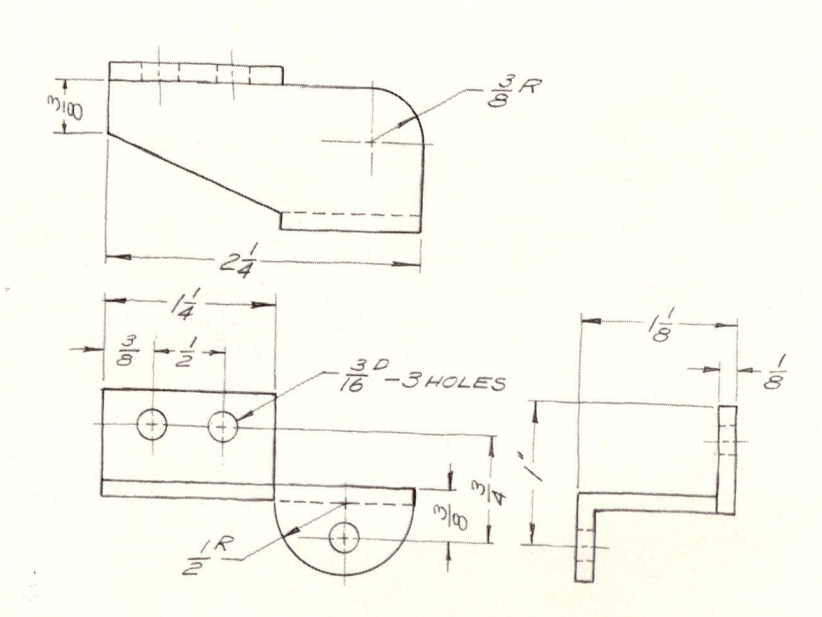

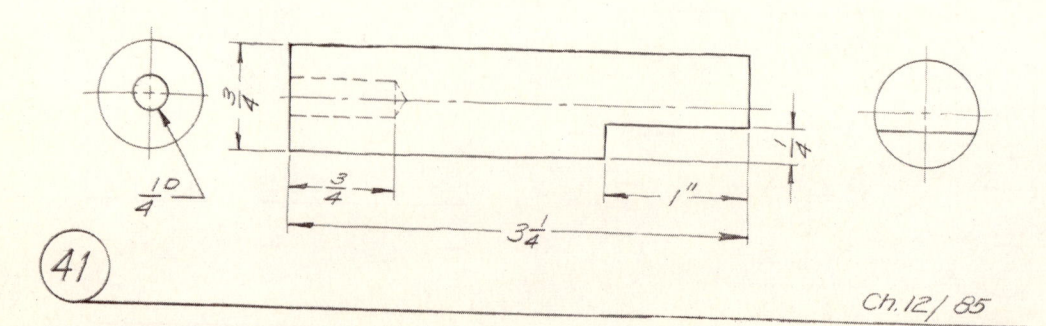

500 SKETCH SOLUTIONS

(41)

Ch. 12/85

SKETCH SOLUTIONS **501**

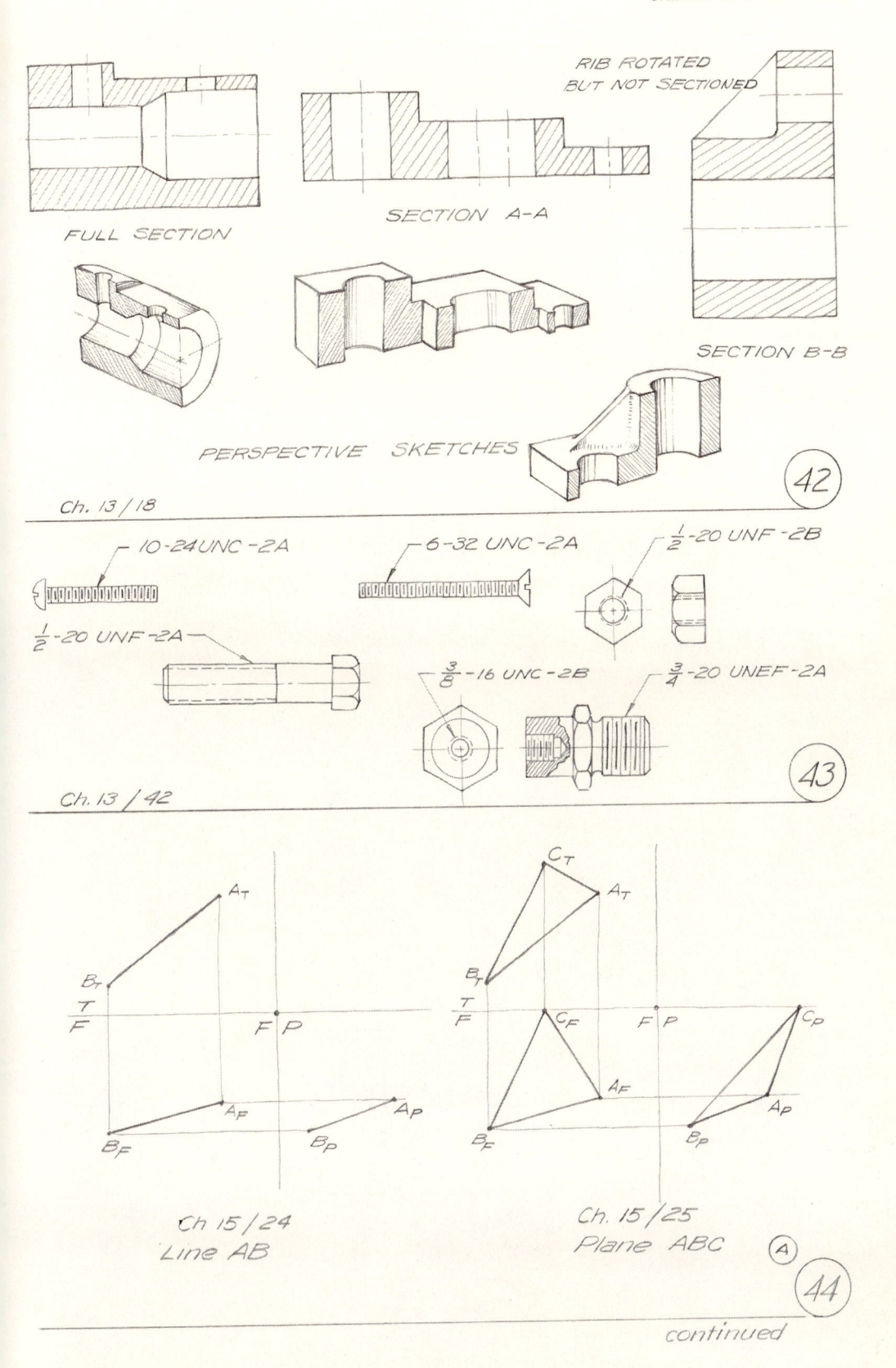

FULL SECTION

SECTION A-A

RIB ROTATED BUT NOT SECTIONED

SECTION B-B

PERSPECTIVE SKETCHES

Ch. 13/18 (42)

— 10-24 UNC-2A
— 6-32 UNC-2A
— $\frac{1}{2}$-20 UNF-2B
$\frac{1}{2}$-20 UNF-2A
— $\frac{3}{8}$-16 UNC-2B
— $\frac{3}{4}$-20 UNEF-2A

Ch. 13/42 (43)

Ch. 15/24
Line AB

Ch. 15/25
Plane ABC

(A) (44)

continued

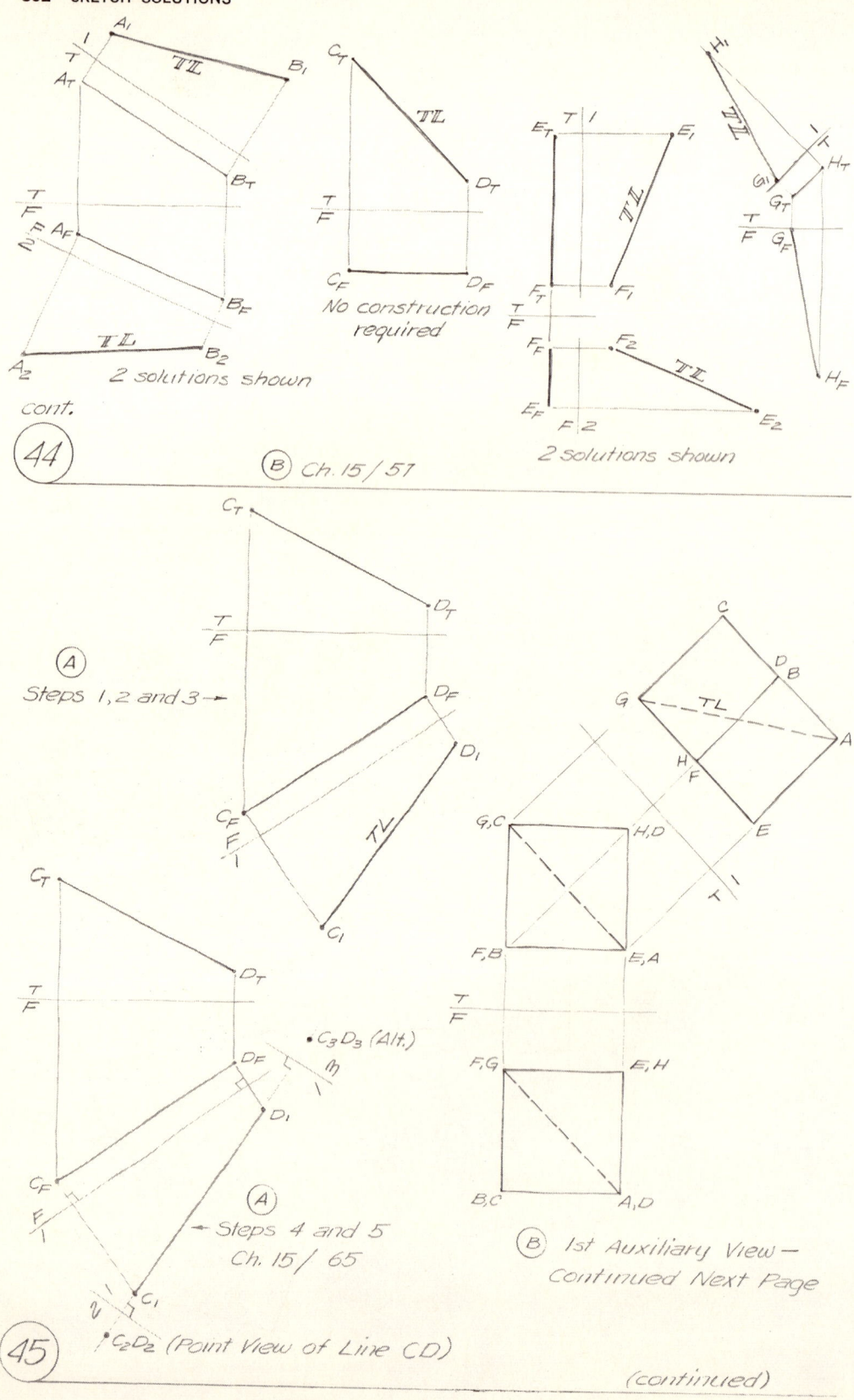

502 SKETCH SOLUTIONS

(44) cont.

(B) Ch. 15/57 — No construction required

2 solutions shown

2 solutions shown

(45)

(A) Steps 1, 2 and 3 →

(A) ← Steps 4 and 5
Ch. 15/65

$C_2 D_2$ (Point View of Line CD)

$C_3 D_3$ (Alt.)

(B) 1st Auxiliary View —
Continued Next Page

(continued)

SKETCH SOLUTIONS **503**

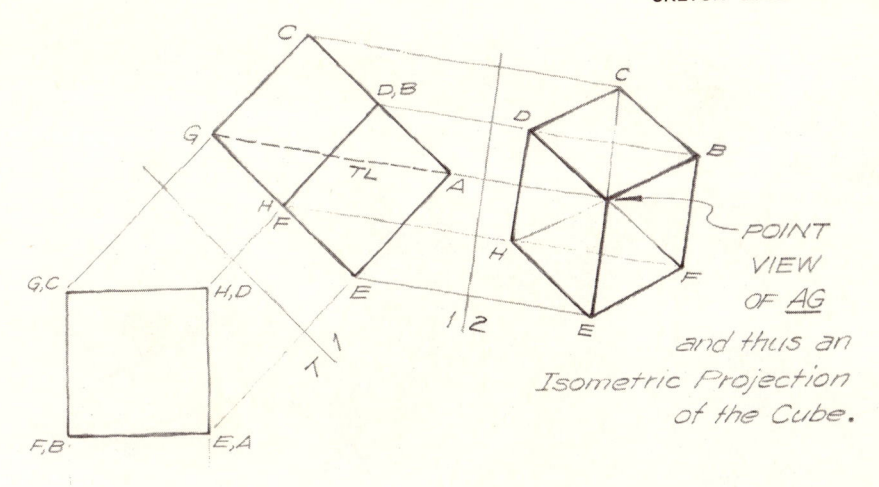

POINT VIEW OF <u>AG</u> and thus an Isometric Projection of the Cube.

FRONT View omitted. Not used in construction of Auxiliary View No. 2. See preceding page.

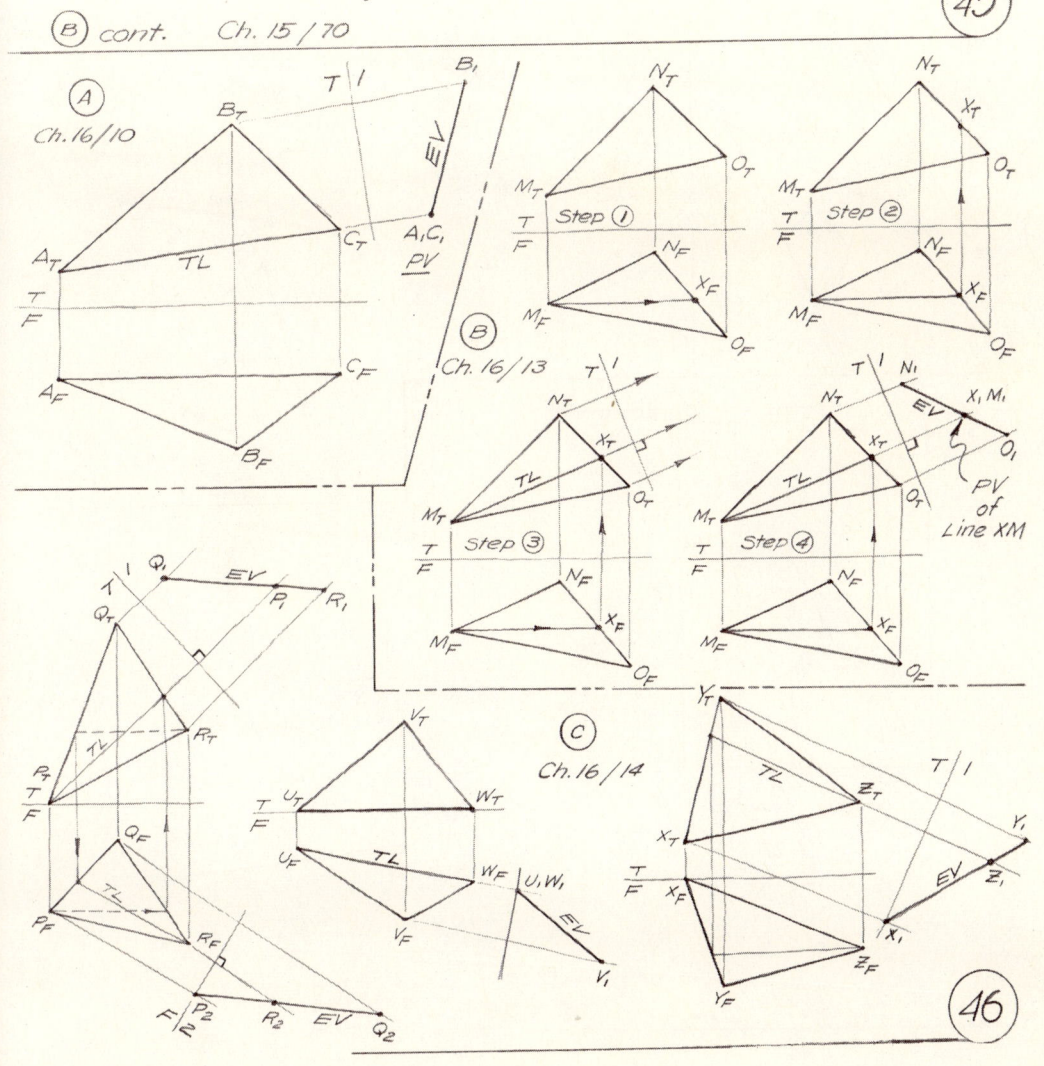

(45)

(B) cont. Ch. 15 / 70

(A) Ch. 16/10

(B) Ch. 16/13

Step ①
Step ②
Step ③
Step ④   PV of Line XM

(C) Ch. 16/14

(46)

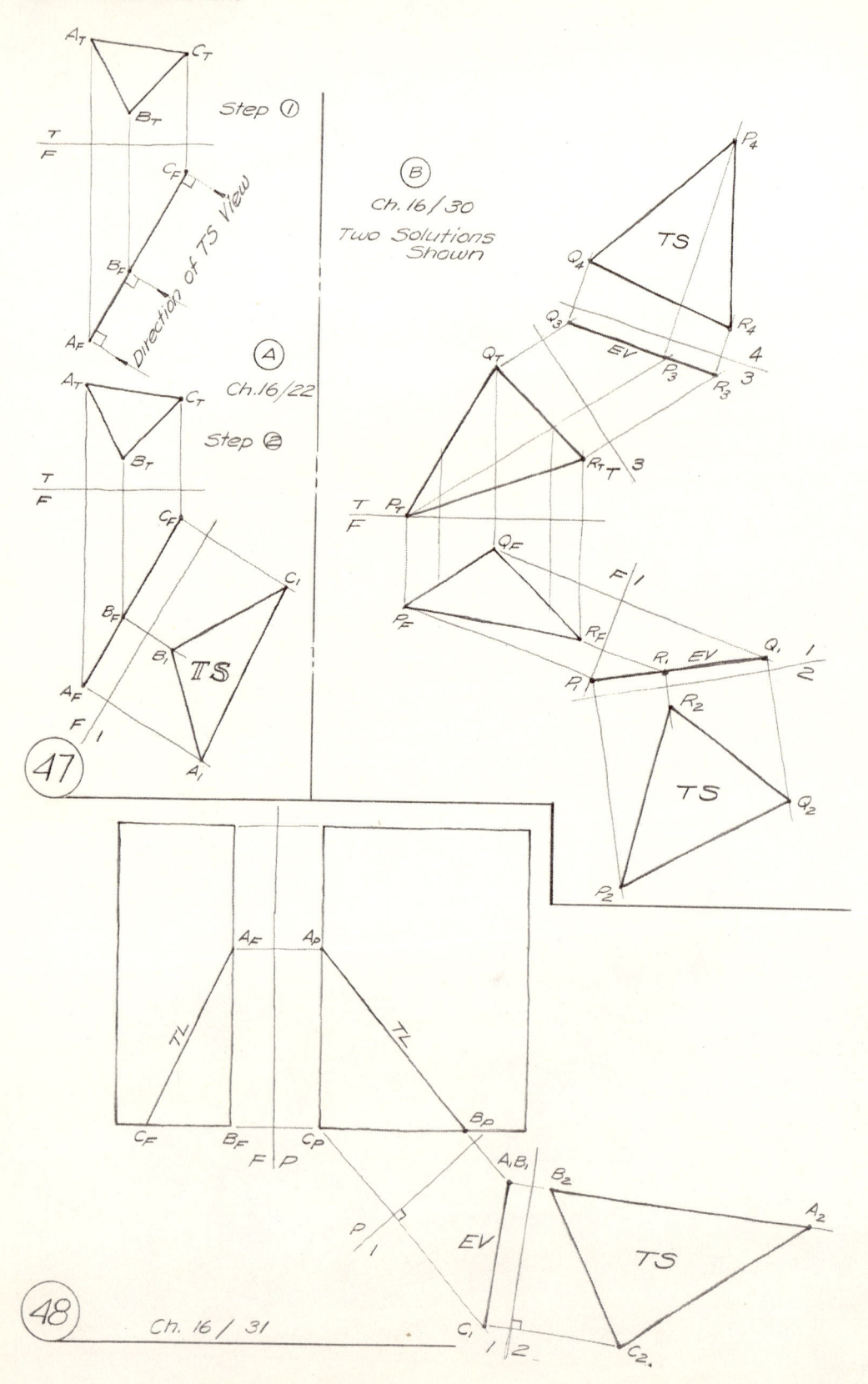

504 SKETCH SOLUTIONS

SKETCH SOLUTIONS 505

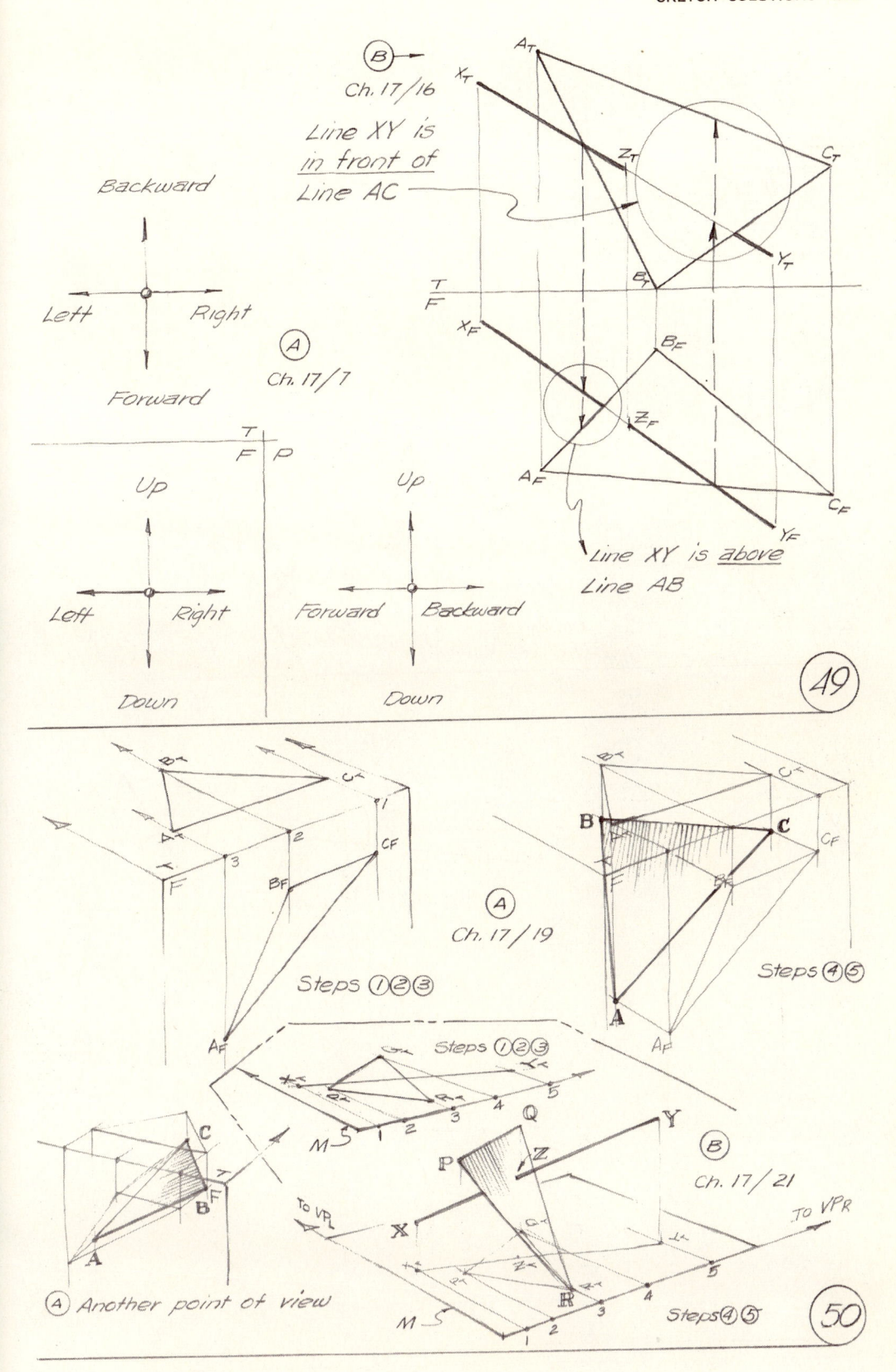

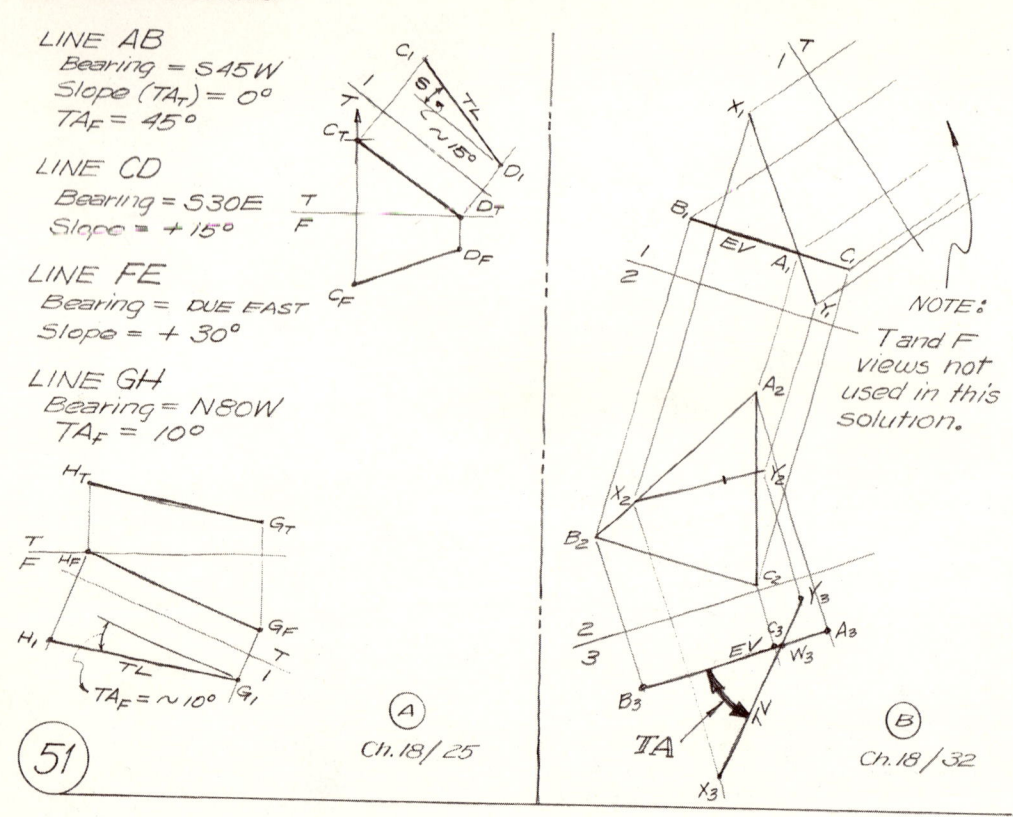

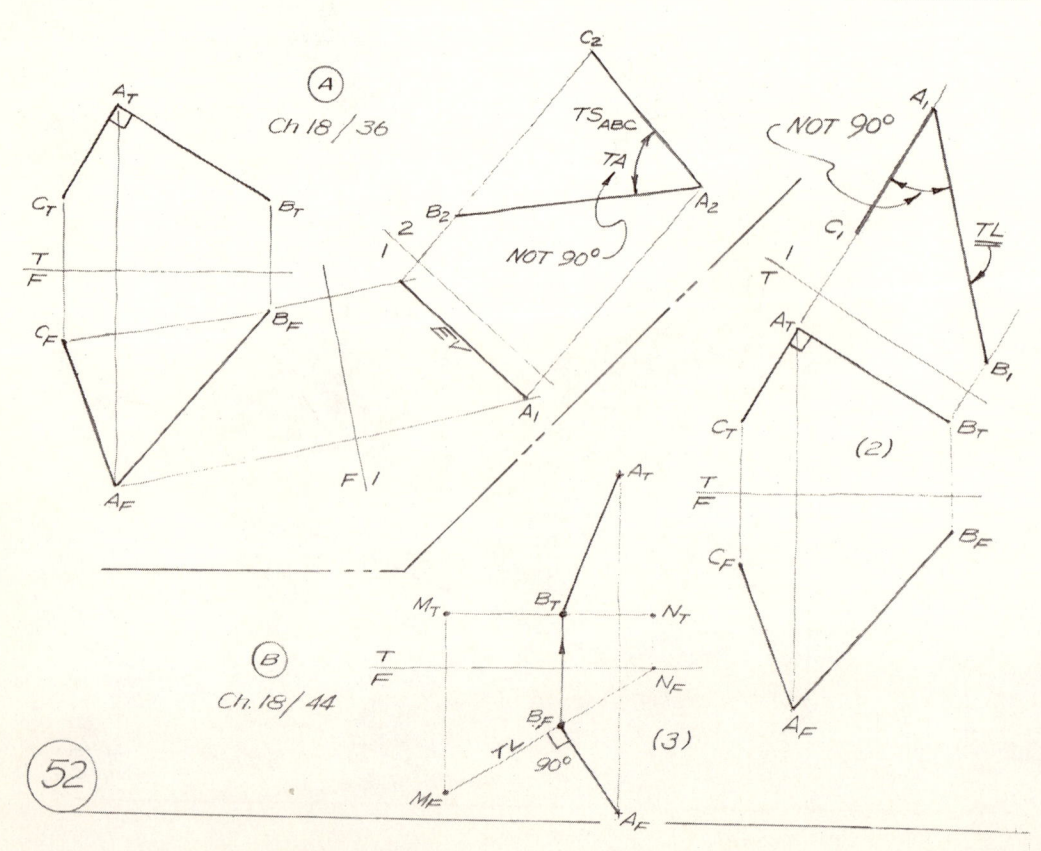

506 SKETCH SOLUTIONS

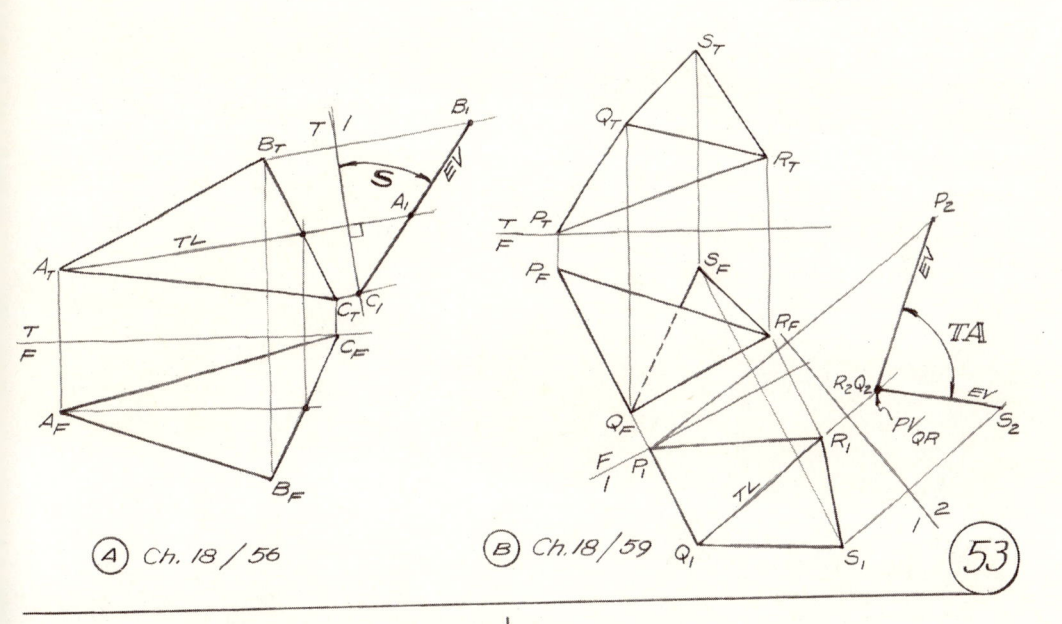

SKETCH SOLUTIONS **507**

(A) Ch. 18/56

(B) Ch. 18/59

(53)

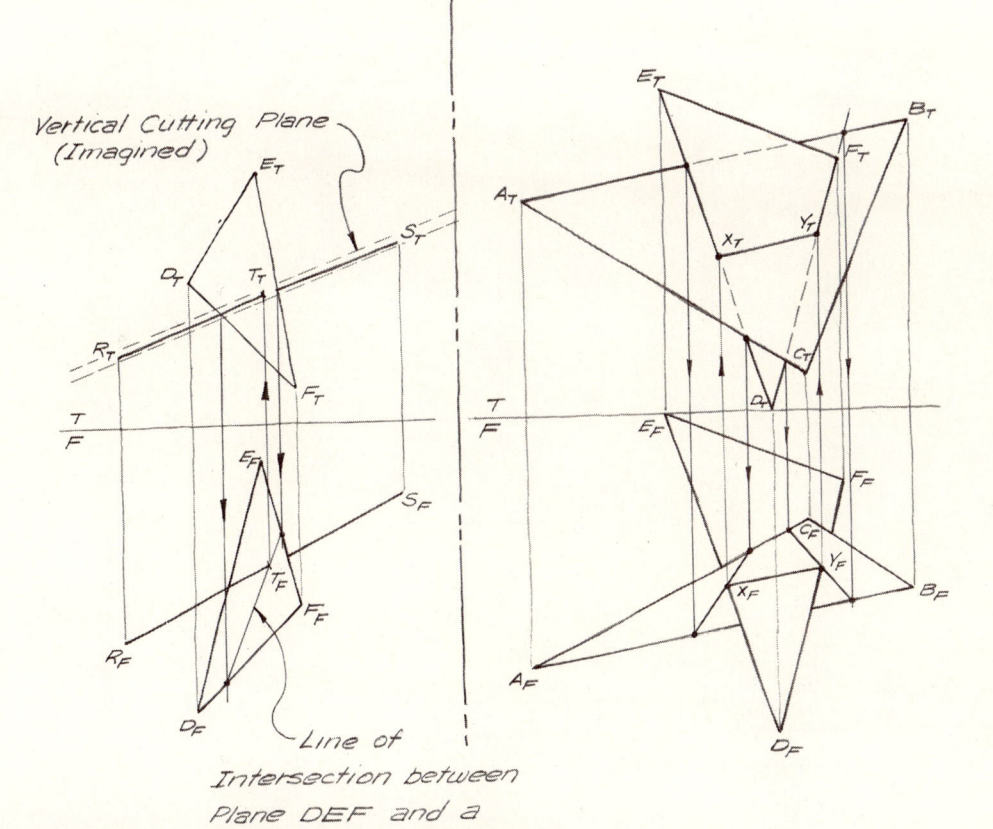

Vertical Cutting Plane (Imagined)

Line of Intersection between Plane DEF and a Vertical Cutting Plane which contains Line RS

(A) Ch. 18/77

(B) Ch. 18/78

(54)

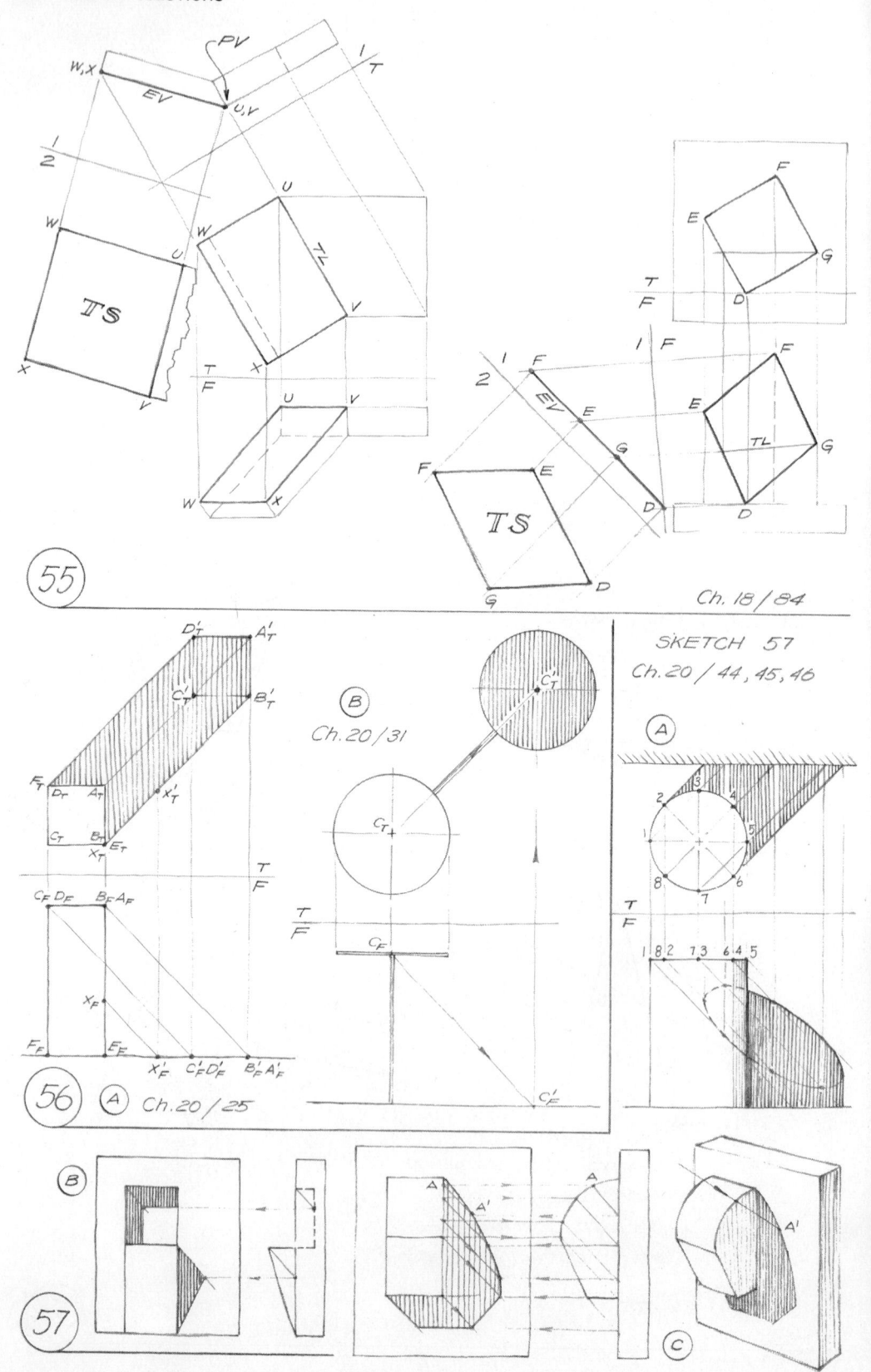

508 SKETCH SOLUTIONS

(55)

Ch. 18/84

(56) (A) Ch. 20/25

(B) Ch. 20/31

SKETCH 57
Ch. 20/44, 45, 46

(A)

(57) (B) (C)

SKETCH SOLUTIONS 509

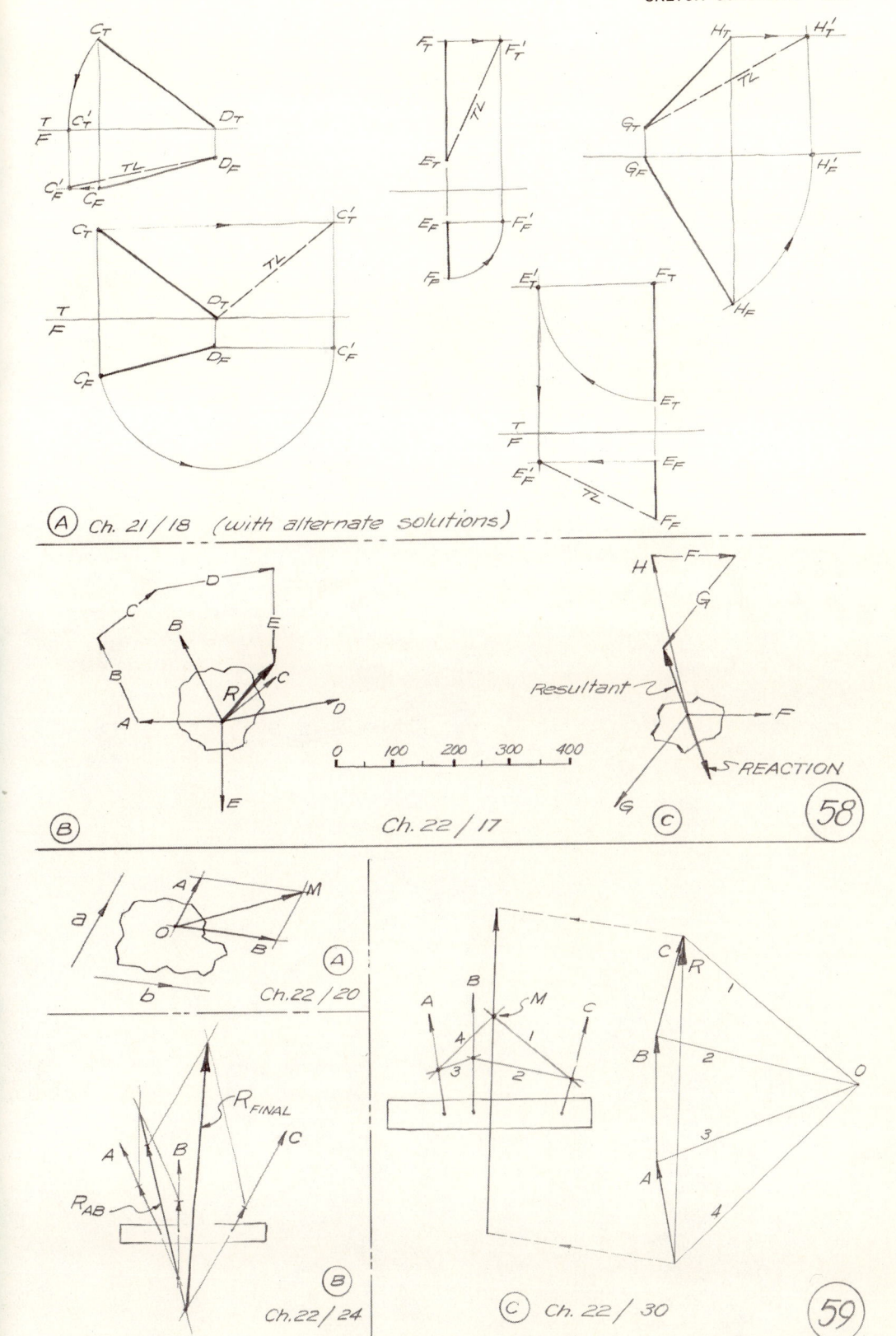

(A) Ch. 21/18 (with alternate solutions)

(B) Ch. 22/17

(C)

58

(A) Ch. 22/20

(B) Ch. 22/24

(C) Ch. 22/30

59

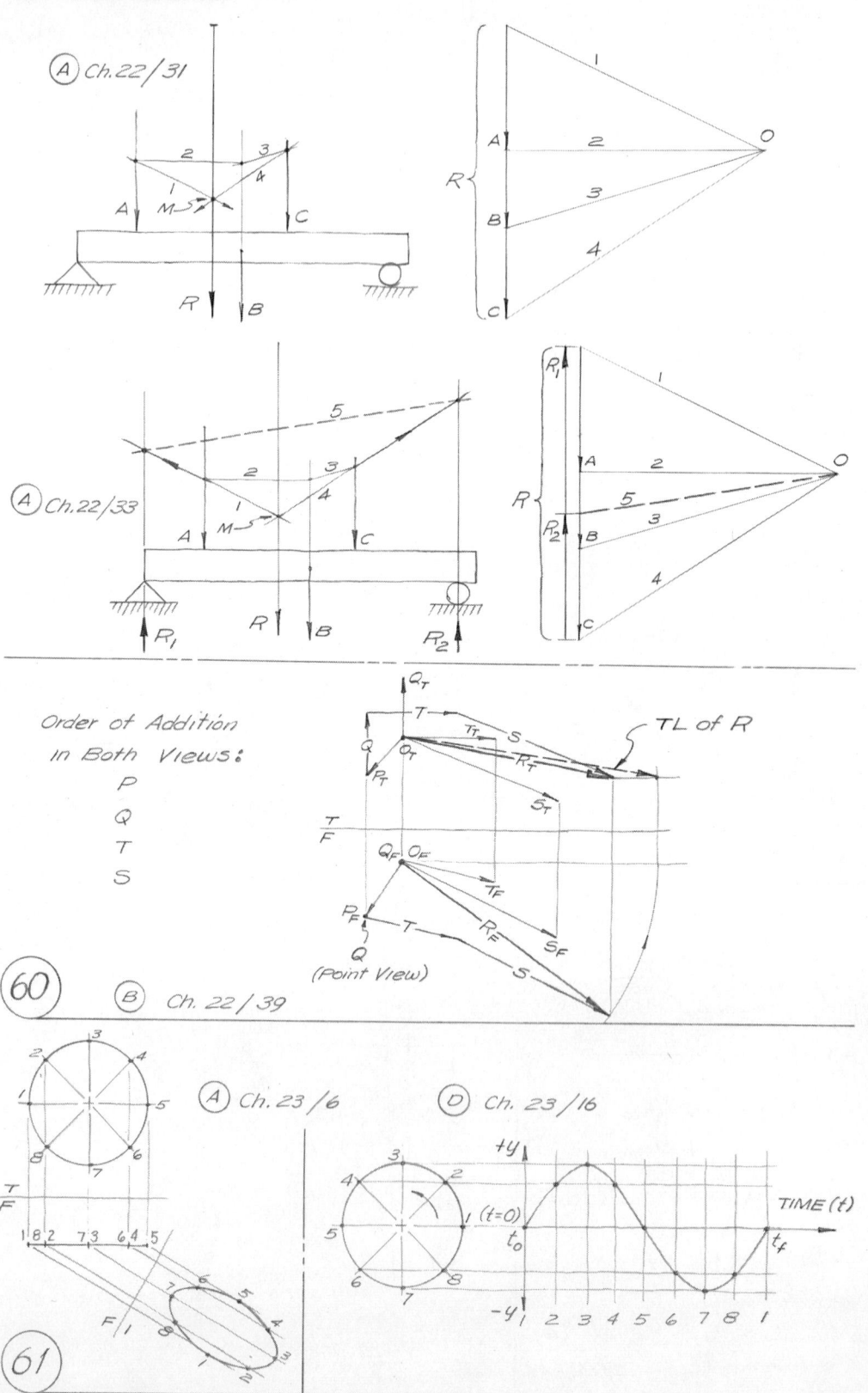

510 SKETCH SOLUTIONS

(A) Ch. 22/31

(A) Ch. 22/33

Order of Addition
in Both Views:
P
Q
T
S

(60) (B) Ch. 22/39

(Point View)

TL of R

(A) Ch. 23/6

(D) Ch. 23/16

(61)

TIME (t)

SKETCH SOLUTIONS 511

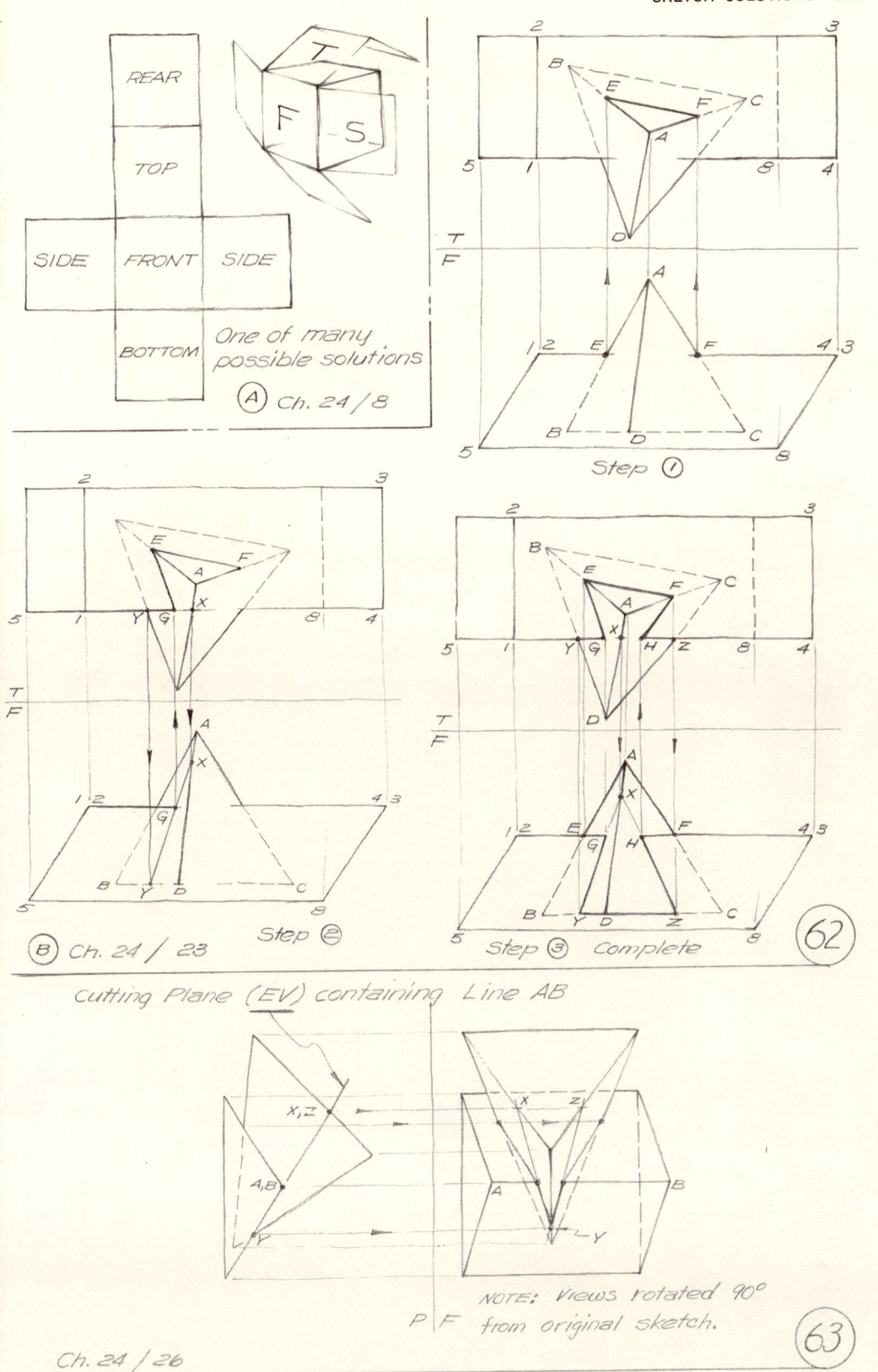

One of many possible solutions
(A) Ch. 24/8

Step ①

(B) Ch. 24/23   Step ②

Step ③ Complete

62

Cutting Plane (EV) containing Line AB

NOTE: Views rotated 90° from original sketch.

Ch. 24/26

63

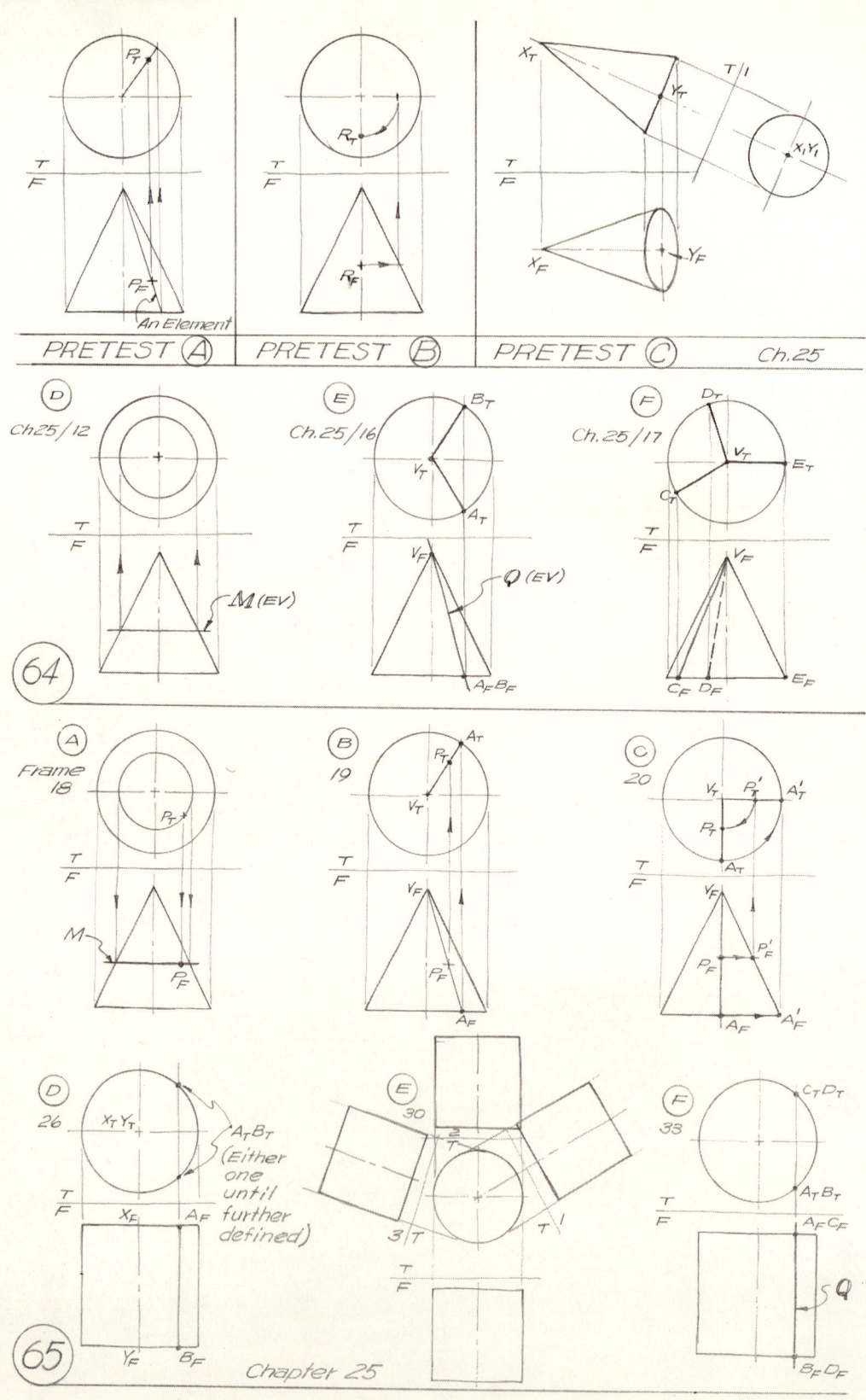

512 SKETCH SOLUTIONS

SKETCH SOLUTIONS 513

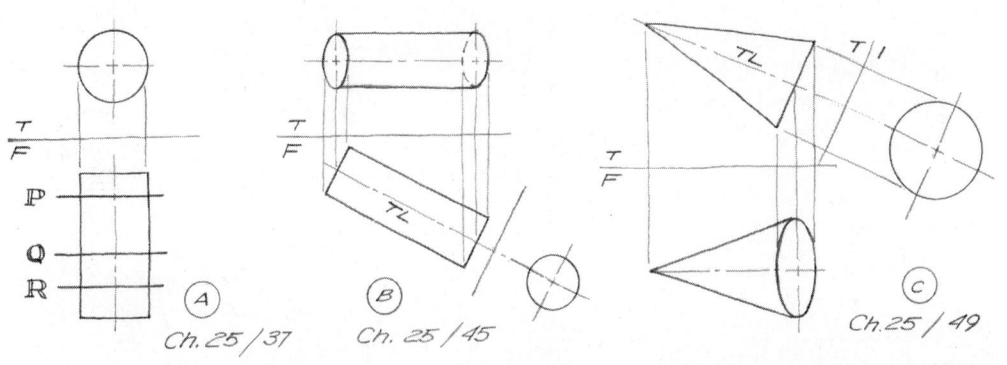

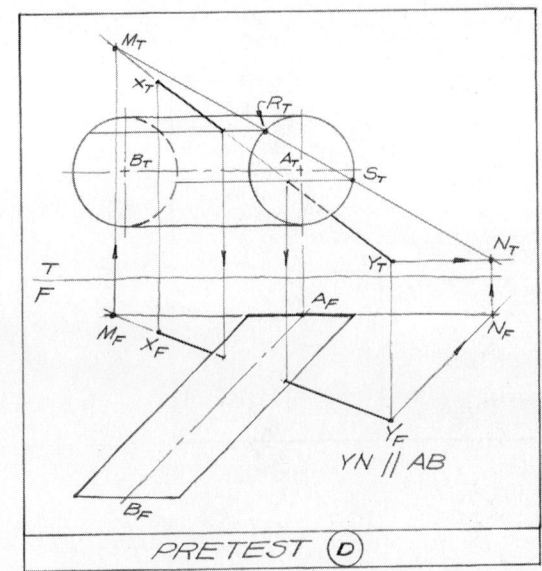

(A) Ch. 25/37
(B) Ch. 25/45
(C) Ch. 25/49

PRETEST (D)

YN ∥ AB

Also: Solution to Steps 3 and 4 of Frame 57. Steps 1 and 2 below.

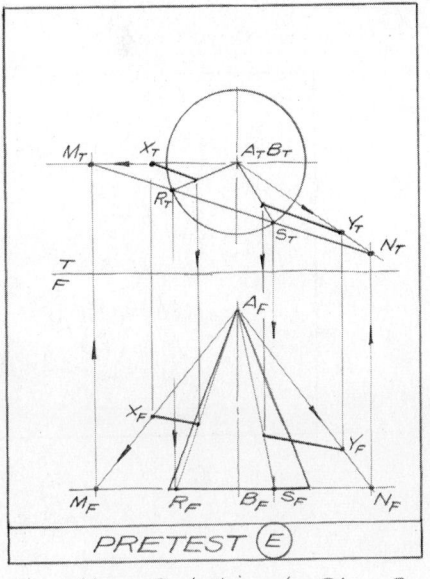

PRETEST (E)

Also: Solution to Step 2 of Frame 60. See next page for Step 1 sol.

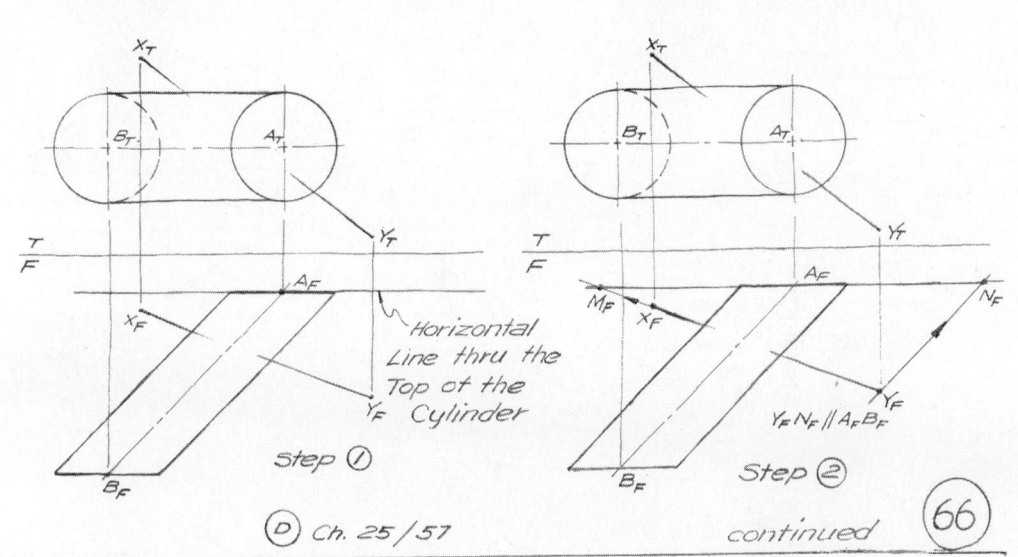

Step ① — Horizontal Line thru the Top of the Cylinder

Step ② — $Y_F N_F \parallel A_F B_F$

(D) Ch. 25/57   continued

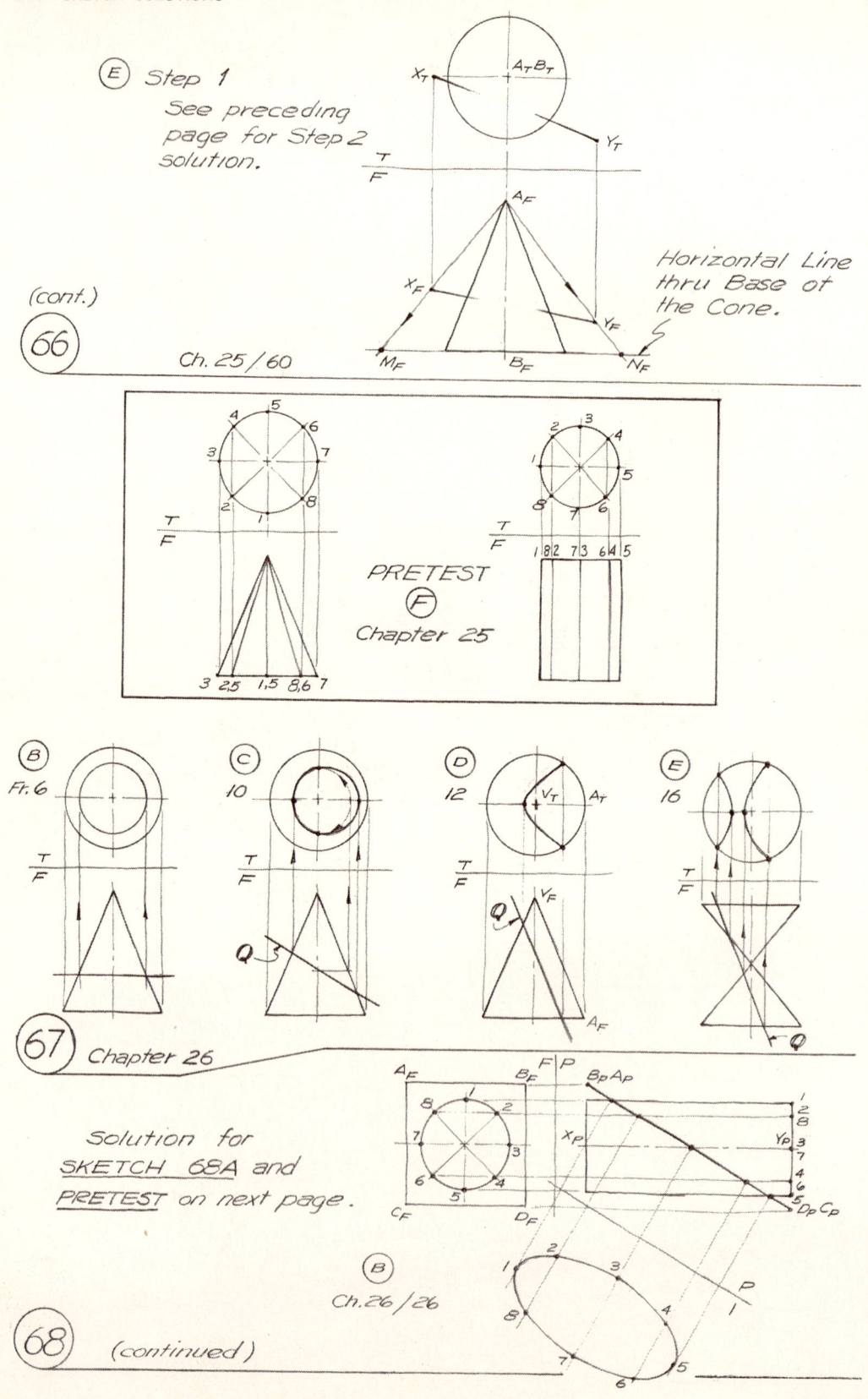

514 SKETCH SOLUTIONS

(66) (cont.) Ⓔ Step 1 See preceding page for Step 2 solution. Ch. 25/60

Horizontal Line thru Base of the Cone.

PRETEST Ⓕ Chapter 25

(67) Chapter 26

Ⓑ Fr. 6
Ⓒ 10
Ⓓ 12
Ⓔ 16

(68) (continued)

Solution for SKETCH 68A and PRETEST on next page.

Ⓑ Ch. 26/26

SKETCH SOLUTIONS 515

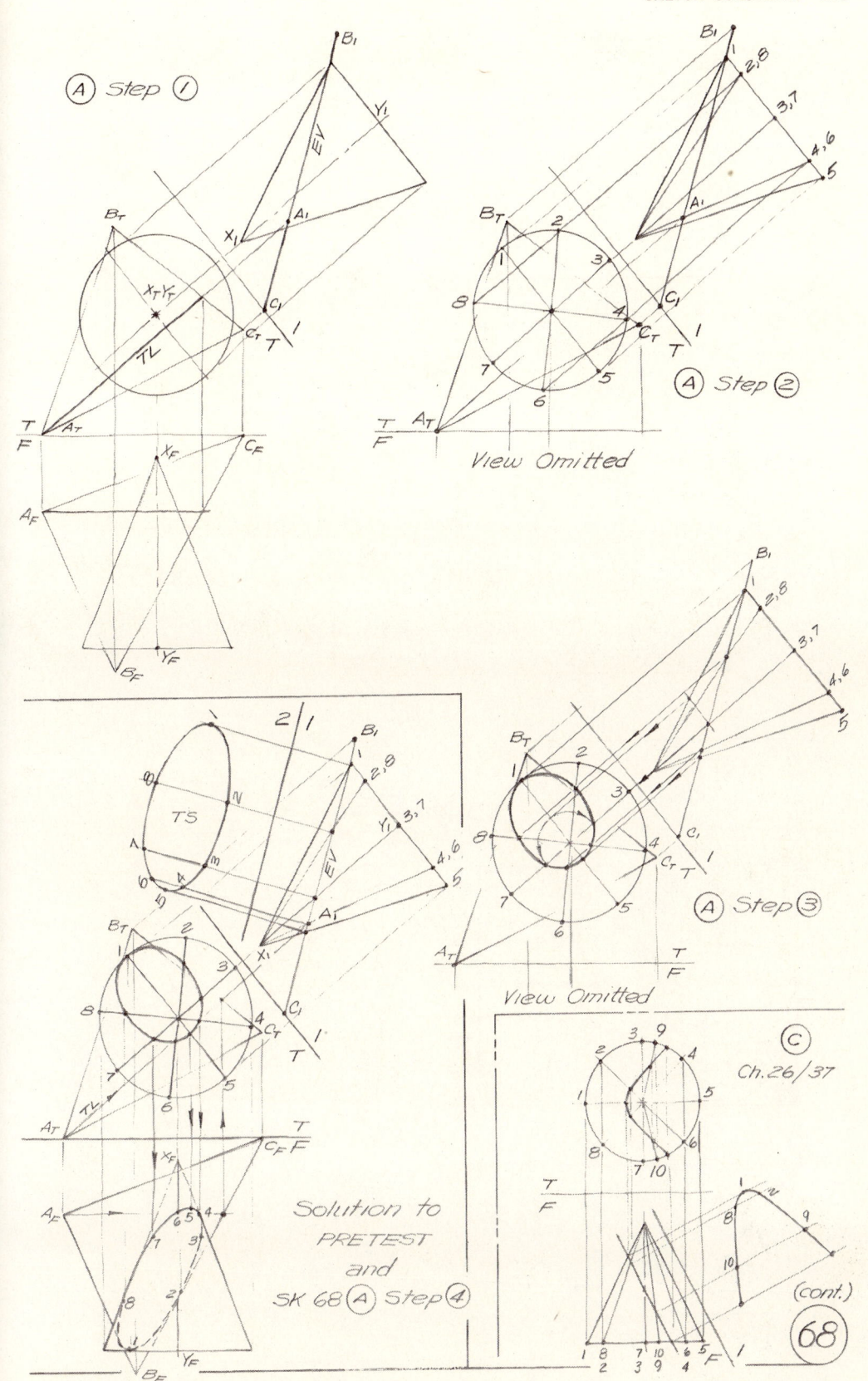

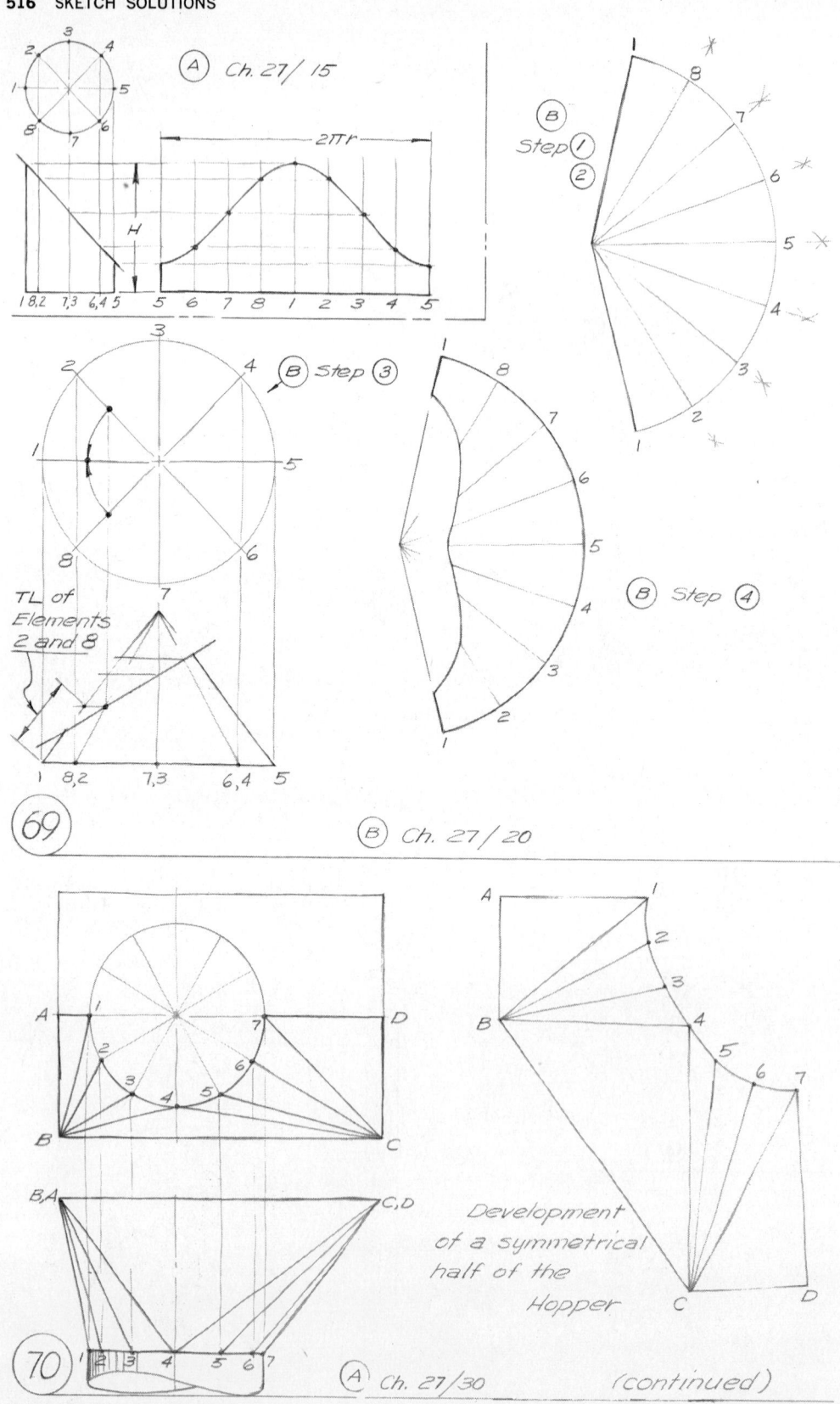

516 SKETCH SOLUTIONS

(69) (A) Ch. 27/15

(B) Step 1, 2
(B) Step 3
(B) Step 4

TL of Elements 2 and 8

(B) Ch. 27/20

(70) (A) Ch. 27/30

Development of a symmetrical half of the Hopper

(continued)

SKETCH SOLUTIONS 517

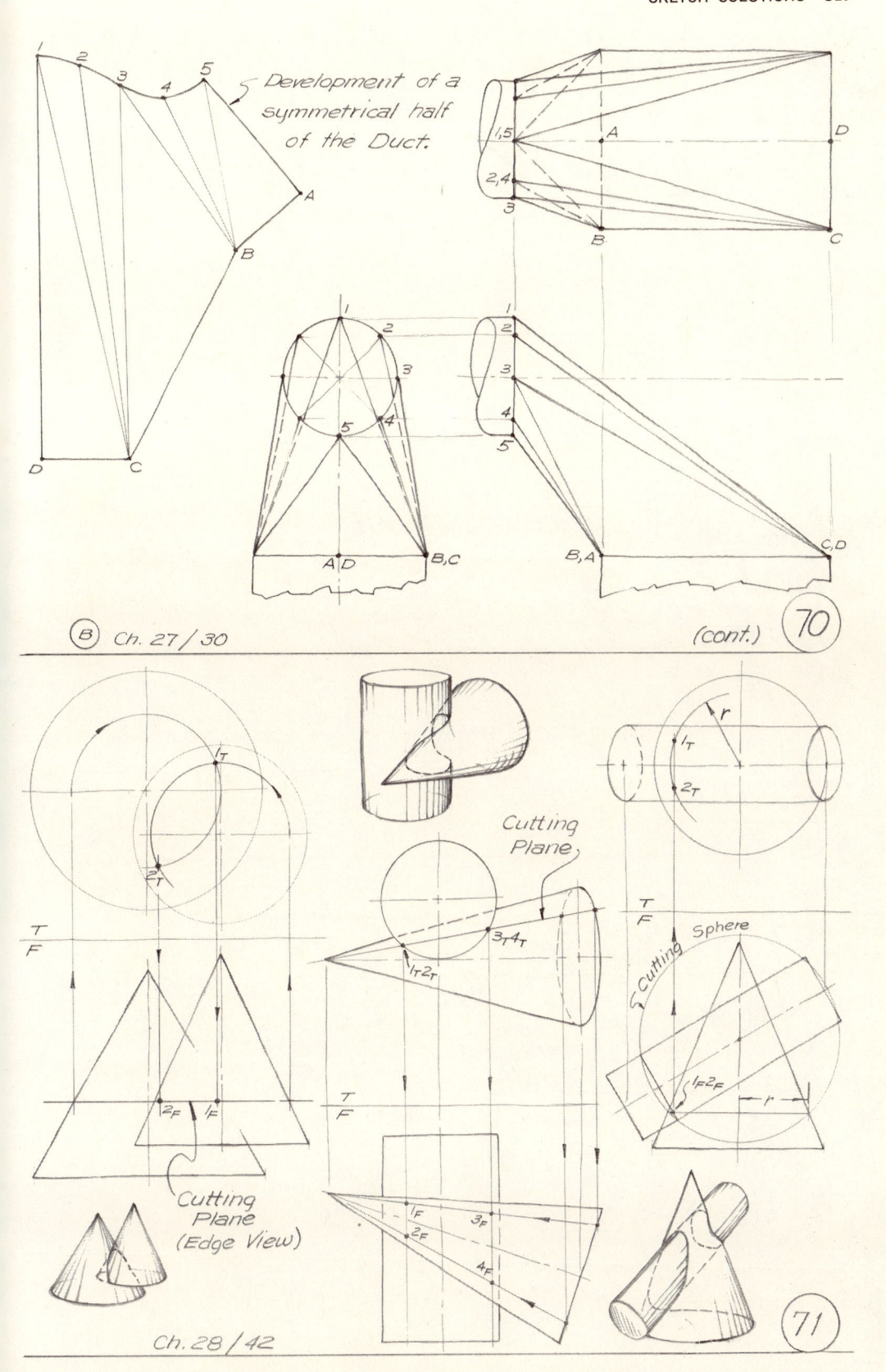

Development of a symmetrical half of the Duct.

B Ch. 27/30 (cont.) 70

Cutting Plane
Cutting Plane (Edge View)
Cutting Sphere

Ch. 28/42 71

518 SKETCH SOLUTIONS

(72)

TOTAL POPULAR VOTE IN U.S. PRESIDENTIAL ELECTIONS 1944 to 1964

Ch. 29/13

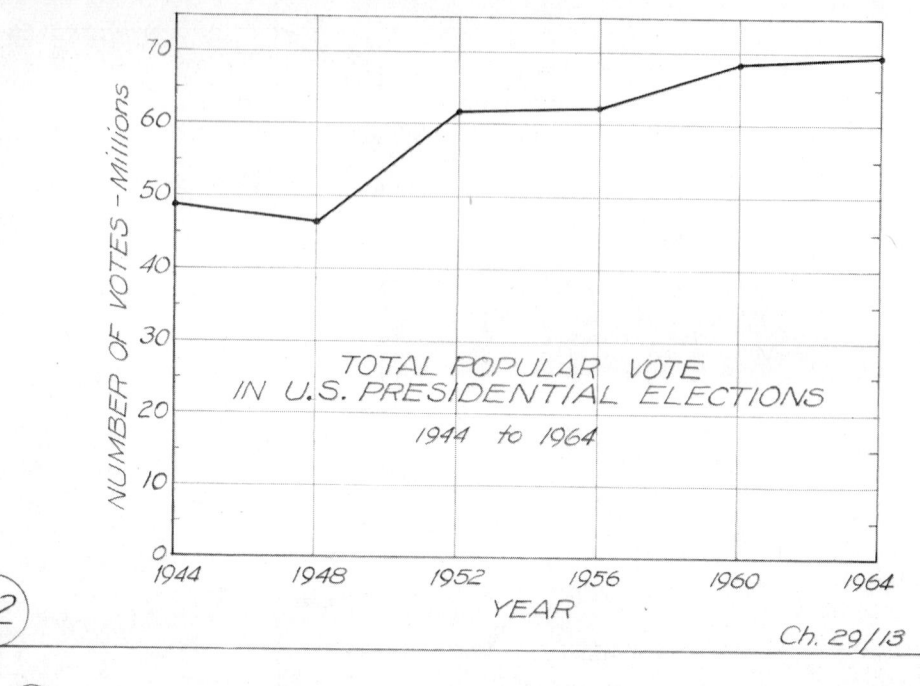

(A) Ch. 30/16

Typical calculation: When $X = 4$, $d_x = 0.05 \cdot 16 = 0.80$ in.

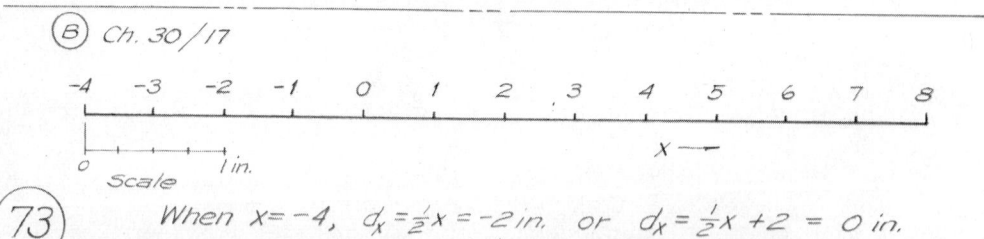

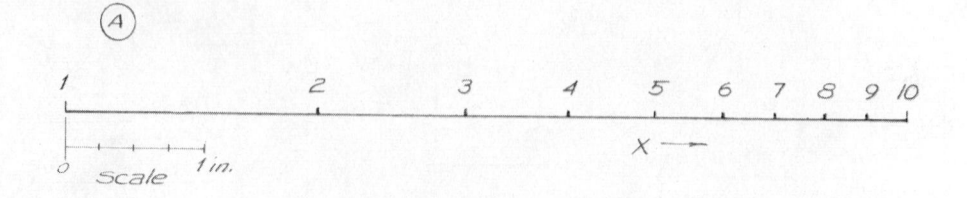

(B) Ch. 30/17

(73) When $x = -4$, $d_x = \frac{1}{2}x = -2$ in. or $d_x = \frac{1}{2}x + 2 = 0$ in.

(A)

$d_x = 6 \log X$, $m_x = 6$ in. per unit

when X = 1   2   3   4   5   6   7   8   9   10
d_x =   0 1.81 2.86 3.61 4.19 4.67 5.07 5.42 5.72 6 in.

(74)

(continued)

SKETCH SOLUTIONS 519

Ⓑ

[scale diagram: 10, 100, 1000, 10,000 with x → arrow; graphic graduation within each cycle shown on diagonal]

$m_x = \dfrac{6}{8-2} = 1$ $d_x = 1[2 \log x - 2] = 2(\log x - 1)$

when $x = 10 \quad 100 \quad 1000 \quad 10,000$ (cont.)

$d_x = 0 \quad 2 \quad 4 \quad 6$ in.   ⑦④

Ch. 30 / 18

$\boxed{\dfrac{u}{3} + v = 2w}$ $m_u = 2$ $d_u = \dfrac{2u}{3}$

$\dfrac{a}{b} = \dfrac{2}{\frac{1}{3}} = \dfrac{6}{1}$ $m_v = \dfrac{1}{3}$ $d_v = \dfrac{1}{3}(v-6) = \dfrac{v}{3} - 2$

$m_w = \dfrac{2}{7}$ $d_w = \dfrac{2}{7}(2w-6) = \dfrac{4}{7}(w-3)$

⑦⑤

Ch. 31 / 21

$\boxed{P = I^2 R}$ $m_I = 1$ $d_I = 1[2 \log I + 2] = 2(\log I + 1)$

$\dfrac{a}{b} = \dfrac{1}{\frac{4}{3}} = \dfrac{3}{4}$ $m_R = \dfrac{4}{3}$ $d_R = \dfrac{4}{3}(\log R - 2)$

$m_P = \dfrac{4}{7}$ $d_P = \dfrac{4}{7} \log P$

⑦⑥

Ch. 31 / 31

Ⓐ $y = 1.5 x + 8$ Ⓑ $y = 50 x^{2.5}$ Ⓒ $y = 2.1(-2)^x$

⑦⑦

Ch. 32 / 28

520 SKETCH SOLUTIONS

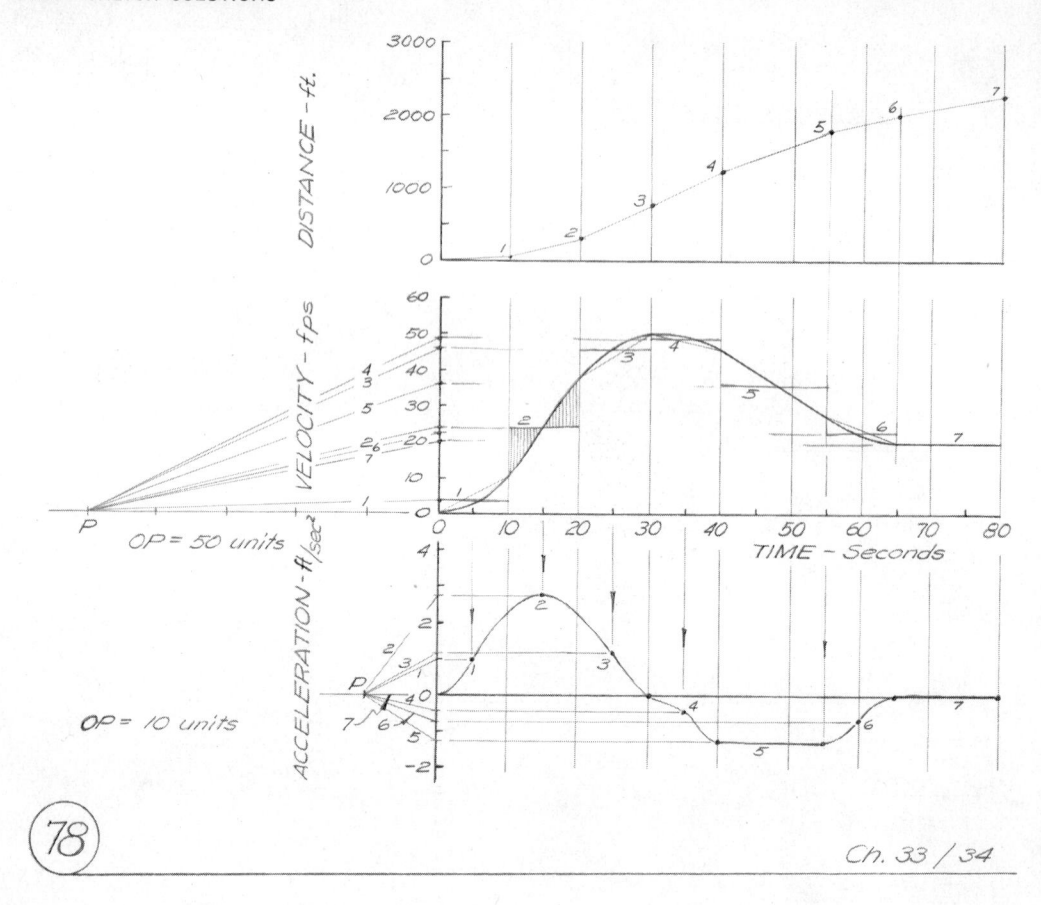

(78)   Ch. 33 / 34